Concepts and Challenges in

The Biophysics of Hearing

PREVIOUS WORKSHOPS AND PUBLICATIONS IN THIS SERIES

#1: July 13 – 15, 1983 (Delft University of Technology, The Netherlands)
Mechanics of Hearing. E. de Boer and M. Viergever (eds.), Delft University Press, Den Hague.

#2: Aug 13 – 16, 1985 (Boston University, Massachusetts, USA)
Peripheral Auditory Mechanisms. J.B. Allen, J. Hall, A. Hubbard, S.T. Neely and A. Tubis (eds.), Springer-Verlag, Berlin.

#3: July 3 – 8, 1988 (Keele University, United Kingdom)
Cochlear Mechanisms: Structure, Function and Models. J.P. Wilson and D.T. Kemp (eds.), Plenum Press, New York.

#4: June 25 – 29, 1990 (University of Wisconsin-Madison, USA)
The Mechanics and Biophysics of Hearing. P. Dallos, C.D. Geisler, J. Matthews, M.A. Ruggero and C. Steele (eds.), Springer-Verlag, Berlin.

#5: June 28 - July 3, 1993 (Paterswolde, The Netherlands)
Biophysics of Hair Cell Sensory Systems. H. Duifhuis, J.W. Horst, P. van Dijk and S. van Netten (eds.), World Scientific, Singapore.

#6: June 24 – 28, 1996 (University of California - Berkeley, USA)
Diversity in Auditory Mechanics. E. Lewis, G. Long, R. Lyon, P. Narins, C. Steele and E. Hecht-Poinar (eds.), World Scientific, Singapore.

#7: July 25-30, 1999 (Miyagi - Zao Hotel, Sendai, Japan).
Recent Developments in Auditory Mechanics. H. Wada, T. Takasaka, K. Ikeda, Y. Susuki and T. Koike (eds.), World Scientific, Singapore.

#8: July 27 - Aug 1, 2002 (Titisee, Germany).
The Biophysics of the Cochlea: Molecules to Models. A.W.Gummer (ed)., World Scientific, Singapore.

#9: July 23 – 28, 2005 (Portland, Oregon, USA)
Auditory Mechanisms: Processes and Models. A.L. Nuttall, T. Ren, P. Gillespie, K. Grosh, E. de Boer (eds.), World Scientific, Singapore.

Concepts and Challenges in The Biophysics of Hearing

Proceedings of the 10th International Workshop on the Mechanics of Hearing

Keele University, Staffordshire, UK 27 – 31 July 2008

Editors

Nigel P Cooper

Keele University, UK

David T Kemp

University College London, UK

World Scientific

NEW JERSEY • LONDON • SINGAPORE • BEIJING • SHANGHAI • HONG KONG • TAIPEI • CHENNAI

Published by

World Scientific Publishing Co. Pte. Ltd.
5 Toh Tuck Link, Singapore 596224
USA office: 27 Warren Street, Suite 401-402, Hackensack, NJ 07601
UK office: 57 Shelton Street, Covent Garden, London WC2H 9HE

British Library Cataloguing-in-Publication Data
A catalogue record for this book is available from the British Library.

CONCEPTS AND CHALLENGES IN THE BIOPHYSICS OF HEARING (With CD-ROM)
Proceedings of the 10th International Workshop on the Mechanics of Hearing

ISBN-13 978-981-283-377-8
ISBN-10 981-283-377-3

Printed in Singapore by B & JO Enterprise

CONTENTS

SECTION III
NEW MEASUREMENT TECHNIQUES

SECTION IV
MICROMECHANICS: BM, TM AND SUB-TECTORIAL SPACE

SECTION V
MODELLING THE COCHLEAR AMPLIFIER AND THE COCHLEA'S DYNAMICS

SECTION VI
HAIR CELLS AND ELECTRO-MECHANICAL TRANSDUCTION

SECTION VII
HAIR BUNDLES AND MECHANO-ELECTRICAL TRANSDUCTION

SECTION VIII
DISCUSSION

PREFACE

Hearing research is a vast and fascinating subject. This book focuses on the input mechanism of the auditory system, the cochlea, which receives and manipulates sound energy in order to translate the information it contains into nerve signals to be processed and interpreted by the brain. The cochlea is an exceptionally complex, sensitive and delicate mechanism which has attracted and challenged biologists, physicists, engineers and mathematicians for well over 100 years. It is also the site of the majority of sensory hearing loss. The incidence of cochlear dysfunction exceeds one in a thousand at birth and rises exponentially with more than half of people over 60 years of age suffering from significant levels of hearing loss which we are at present unable to prevent. A deeper understanding of the cochlea is essential if we are to reduce the incidence of deafness and its negative impacts on the individual, the family and society.

This book provides an account of present day attempts to understand how cochlear mechanisms function. It contains the proceedings of an international expert meeting, 'The Mechanics of Hearing Workshop', which took place at Keele University, in the UK, at the end of July, 2008. The book includes both theoretical and experimental research papers on cochlear mechanisms at the molecular, cellular and systems levels.

The workshop was the tenth in a triennial series which began in Delft, The Netherlands, in 1983. That first meeting was a direct response to a rapid succession of new cochlear observations, including sharp mechanical frequency tuning, mechanical nonlinearity, sound re-emission, and a vibration amplifying capability in the cochlea, all of which appeared to be incompatible with contemporary concepts of cochlear function. Guiding this first meeting was the conviction that the dramatic and disparate experimental findings of the previous 5 years could be best understood by developing new mathematical models to explain and unify the experimental evidence. This approach owes much to the foresight of Egbert de Boer, who not only organised the first meeting, but has contributed consistently to all subsequent meetings and to this volume.

The books arising from this series of workshops (see page ii) document both the tremendous progress that has been made and the enormous challenges which still face researchers of cochlear function. The 2008 workshop came exactly 20 years after the last such workshop to be held in Keele. Then as now the focus was on the structure, function and modelling of cochlear mechanisms. The 1988 Keele meeting was notable for the attendance of two giants in this field, Sir James Lighthill and Thomas Gold, both no longer with us. The 2008 meeting honoured Pat Wilson, the organiser of the 1988 workshop and a pioneer in the re-evaluation of cochlear function at Keele in the 1970s and 80s.

Subsequent workshops widened the scope of contributions to include the biophysics of hair cell sensory systems, the diversity in auditory mechanics, and the molecular biology of sensory mechanisms and processes. The 2008 workshop adopted a similar broad approach. Topics covered in this book include:- the transmission of stimulation into, within and out of the inner ear; the role of sensory cell motilities in amplifying this stimulation; mechanisms involved in the modulation and control of cochlear sensitivity; the micro-mechanics of the basilar and tectorial membranes and the organ of Corti; and mechano-electrical and electromechanical transduction by the system's sensory cells and organelles. The great diversity of topics represented is an apt reflection of the complexity of the cochlea. It is impossible to say that any particular area is more important than others, but nevertheless we invited three participants to present keynote lectures. Jont Allen explored the role of the cochlea in human speech recognition. The intrinsic analytical properties of the cochlea must certainly have had a profound effect on which vocalic features could and could not be incorporated into a reliable communication code, and Allen's analysis leads to new challenges for cochlear researchers. Under the title of "Firing up the amplifier", Joe Santos-Sacchi's presentation gave deep insights into the molecular motors that drive electromotility in cochlear hair cells, focusing on the effects of temperature. Finally, Pascal Martin revealed a unity in the various incarnations of active hair-bundle motility that are exhibited by vertebrate hair cells.

It is a measure of the collective achievement of the participants of these meetings that most students of auditory science are now routinely taught about the cochlear amplifier and the role of hair cell motility and nonlinearity in the propagation of stimulation along the cochlea, and also of the reverse process which gives rise to otoacoustic emissions. These are concepts which were speculative, if not heretical, at the commencement of this series of workshops. It is a measure of the maturity and progress within the field of cochlear biophysics that some of these concepts are now themselves being challenged by new discoveries and models. The present workshop's Discussion session, which was led by Chris Shera and David Mountain, exposed these controversies to open debate, and this debate is fully transcribed in the final section of this book. Not only was the nature of and evidence for the 'cochlear amplifier' called into question, but even von Békésy's travelling wave was no longer considered wholly adequate to explain all propagation of energy inside the cochlea. The role of modelling was also constructively challenged. Experimentalists challenged modellers to identify more clearly the testable predictions arising from their theories, and modellers challenged experimentalists to specify useful goals for theoreticians to address. This section is essential reading for anyone wanting to see where our understanding of the cochlea is heading, and which cherished concepts may have to be relinquished.

The questioning of commonly held concepts about cochlear mechanisms at this workshop is a refreshing and positive outcome, and will perhaps herald a new generation of active cochlear models. The new and greatly refined technologies for observing the cochlea and its components described in this book are revealing information beyond the wildest dreams of the participants of the first workshop. This process can only take us closer to a complete understanding of the cochlea.

David Kemp & Nigel Cooper
University College London & Keele University October, 2008

ACKNOWLEDGMENTS

Plenary Lecturers

J.B. Allen, J. Santos-Sacchi, P. Martin

Organizing committee

N.P. Cooper, D.T. Kemp, J. Norton

International advisers and assistants

D.C. Mountain, C.A. Shera, C. Bergevin, J. Baumgart, A.W.Gummer.

Local assistants

R. Knapper, R. Wüst, M.J. Palmer, M.G. Evans, D.N. Furness

Session chairs and co-chairs

A.L. Nuttall, W. Dong, M.A. Ruggero, D.T. Kemp, E.S. Olson, F. Böhnke, H. Duifhuis, C. Bergevin, J.J. Guinan, M.G. Evans, J.F. Ashmore, R. Hallworth, W.E. Brownell, C. Köppl, A.W. Gummer, J.T. Fulton, I.J. Russell, A. Fridberger, J.T. Albert, N.P. Cooper, P. van Dijk, B. Nadrowski, S.J. Elliott, A.J. Aranyosi, C.R. Steele, S. Puria, C.A. Shera, D.C. Mountain.

Sponsors

The Oticon Foundation, Hustru Ida Emilies Fond,
Kongebakken 9, DK-2765 Smørum, Denmark

Otodynamics Ltd., 30-38 Beaconsfield Road, Hatfield,
Herts, AL10 8BB, United Kingdom

Starkey Laboratories Ltd., William F. Austin House,
Bramhall Technology Park, Pepper Road, Hazel Grove,
Stockport, SK7 5BX, United Kingdom

MED-EL, Worldwide Headquarters, Fürstenweg 77a,
A-6020 Innsbruck, Austria

Polytec GmbH, Polytec-Platz 1-7,
D-76337 Waldbronn, Germany

Tucker-Davies Technologies, 11930 Research Circle,
Alachua, FL 32615, USA

10th International Workshop on the Mechanics of Hearing. Keele University, UK, July 27-31, 2008.

PHOTO LEGEND

1 Shun Kumano
2 Pavel Mistrik
3 Stanley Huang
4 James Fulton
5 Tianying Ren
6 Christopher Bergevin
7 Reinhart Frosch
8 Michael Slama
9 Manuela Nowotny
10 Dáibhid Ó Maoiléidigh
11 Shanthini Mahendrasingham
12 Kai Dierkes
13 Jonathan Ashmore
14 Allyn Hubbard
15 Egbert de Boer
16 Fred Nuttall
17 Mario Ruggero
18 Jean-Pierre Hardelin
19 Joe Santos-Sacchi
20 Torsten Marquardt
21 Pascal Martin
22 Pim van Dijk
23 Jont Allen
24 Charles Steele
25 Richard Chadwick
26 Anders Fridberger
27 Bill Brownell
28 Bastian Epp
29 Eric LePage
30 Tony Gummer
31 David Mountain
32 Roland Gärtner
33 Chris Shera
34 Martin Homer
35 Rick Hallworth
36 Christian Gerstenberger
37 Franz-Erich Wolter
38 Johannes Baumgart
39 Christine Köppl
40 Mario Fleischer
41 Wiktor Jedrzejczak
42 Jeffrey Cheng
43 Martin Braun
44 Lisa Olson
45 Diek Duifhuis
46 John Rosowski
47 David Kemp
48 Alexandra Stefanovic
49 Andrew Bell
50 Rick Schoffelen
51 Yi-Wen Liu
52 Kuni Iwasa
53 Yong Zhang
54 Koji Iida
55 Emery Ku
56 Elisa Caberlotto
57 Ombeline de La Rochefoucauld
58 Diana Turcanu
59 Robert Szalai
60 Glenis Long
61 Marcia Mellado-Lagarde
62 Adria LeBoeuf
63 Heidi Nakajima
64 Wim Decraemer
65 Renata Sisto
66 Sunil Puria
67 Sarah Verhulst
68 Jörg Albert
69 KJB Lee
70 Jian Zuo
71 Wei Dong
72 Dave Furness
73 Carole Hackney
74 Mary-Ann-Cheatham
75 Jackie How
76 Lin Bian
77 Nigel Cooper
78 Janet Norton
79 Frank Böhnke
80 Ron Knapper
81 Peter Steyger
82 Karl Grosh
83 Bas Meenderink
84 Björn Nadrowski
85 Shaum Bhagat
86 Claus-Peter Richter
87 Seth Newburg
88 Julien Meaud
89 Steve Neely
90 James Harte
91 Alex Hellmuth
92 Steve Elliott
93 Cem Dogan
94 Ben Warren
95 Jeff Lichtenhan
96 Tony Gardner-Medwin
97 Ben Lineton
98 AJ Aranyosi
99 Marcel van der Heijden
100 Alan Champneys
101 John Guinan
102 Peter Lampacher
103 Denny Freeman
104 Alex Spector

Not in photograph

- Hashir Aazh
- Cornelius Abel
- Mike Evans
- Jonathan Fisher
- Nuria Gavara
- Roozbeh Ghaffari
- Rachel Gueta
- Andriy Koslov
- Manfred Kössl
- Andrei Lukashkin
- Michio Murakoshi
- Itay Rousso
- Ian Russell
- Masashi Unoki
- Andre van Schaik
- Hiroshi Wada
- Pat Wilson
- Anna Wittekindt

WORKSHOP DELEGATES

Arranged alphabetically by surname (photo id# in brackets)

Hashir Aazh (-) *hashir.aazh@nhs.net*
Audiology Department, Ealing Hospital, Uxbridge Road, Southall, London, UB1 3HW, United Kingdom

Cornelius Abel (-) *abel@bio.uni-frankfurt.de*
Institute for cell Biology and Neuroscience, Siesmayerstr 70A, 60323 Frankfurt, Germany

Jörg Albert (**68**) *joerg.albert@ucl.ac.uk*
UCL Ear Institute, University College London, 332 Gray's Inn Rd., London, WC1X 8EE, United Kingdom

Jont Allen (**23**) *jontalle@uiuc.edu*
ECE Department, University of Illinois, 2061 Beckman Institute, Urbana IL 61853, USA

AJ Aranyosi (**98**) *aja@mit.edu*
77 Massachusetts Ave., Room 36-893, Cambridge MA 02139, USA

Jonathan Ashmore (**13**) *j.ashmore@ucl.ac.uk*
Department of Physiology and UCL Ear Institute, Gower Street, London WC1E 6BT, United Kingdom

Johannes Baumgart (**38**) *johannes.baumgart@tu-dresden.de*
Institute for Aerospace Engineering, Faculty of Mechanical Engineering, Technische Universität Dresden, 01062 Dresden, Germany

Andrew Bell (**49**) *andrew.bell@anu.edu.au*
Centre for Visual Sciences, Research School of Biological Sciences, The Australian National University, Canberra, ACT 0200, Australia

Christopher Bergevin (**6**) *cbergevin@math.arizona.edu*
University of Arizona, Department of Mathematics, Office #321, 617 N. Santa Rita Ave., P.O. Box 210089, Tucson, AZ 85721-0089, USA

Shaum Bhagat (**85**) *sbhagat@memphis.edu*
School of Audiology and Speech-Language Pathology, The University of Memphis, 807 Jefferson Avenue, Memphis, Tennessee 38105, USA

Lin Bian (**76**) *Lin.Bian@asu.edu*
Dept. of Speech and Hearing Science, Arizona State University, Tempe, AZ 85287-0102, USA

Frank Böhnke (**79**) *frank.boehnke@lrz.tum.de*
HNO Klinik, Klinikum r.d. Isar, Ismaningerstr. 32, D-81675 München, Germany

Martin Braun (**43**) *nombraun@telia.com*
Neuroscience of Music, Gansbyn 14, S-67195 Klassbol, Sweden

William Brownell (**27**) *brownell@bcm.tmc.edu*
Otolaryngology, Baylor College of Medicine, One Baylor Plaza, Houston, TX 77030, USA

Elisa Caberlotto (**56**) *elisa.caberlotto@gmail.com*
Unite de Genetique et Physiology de l'audition - INSERM UMRS 587, Institut Pasteur, 25 rue du Dr. Roux, 75724 PARIS Cedex 15, France

Richard Chadwick (**25**) *chadwick@helix.nih.gov*
NIDCD, NIH, Auditory Mechanics Section, 10 Center Drive MSC 1417, Bldg 10 Rm 5D49, Bethesda , MD 20892, USA

Alan Champneys (**100**) *a.r.champneys@bristol.ac.uk*
Department of Engineering Mathematics, University of Bristol, Bristol BS8 1TR, United Kingdom

Mary Ann Cheatham (**74**) *m-cheatham@northwestern.edu*
2-240 Frances Searle Bldg., 2240 Campus Drive, Northwestern University, Evanston IL 60208-3550, USA

Jeffrey Cheng (**42**) *Tao_Cheng@meei.harvard.edu*
Eaton-Peabody Lab, Massachusetts Eye and Ear Infirmary, 243 Charles Street, Boston, MA 02114, USA

Nigel Cooper (**77**) *n.p.cooper@keele.ac.uk*
School of Life Sciences, Keele University, Keele, Staffordshire, ST5 5BG, United Kingdom

Egbert de Boer (**15**) *e.d.boer@hccnet.nl*
Oldambtstraat 7, 1079 PT, Amsterdam, The Netherlands

Ombeline de La Rochefoucauld (**57**) *or2107@columbia.edu*
Columbia University, P&S 11-452, 630 West 168th Street, New York, NY 10032, USA

Wim Decraemer (**64**) *wim.decraemer@ua.ac.be*
University of Antwerp, CGB, Groenenborgerlaan 171, B-2020 Antwerpen, Belgium

Kai Dierkes (**12**) *kai@mpipks-dresden.mpg.de*
Max Planck Institute for the Physics of Complex Systems, Nöthnitzer Straße 38, 01187 Dresden, Germany

Cem Dogan (**93**) *cdcd@gmx.net*
Institut für Mensch-Maschine-Kommunikation, FG Graphische Datenverarbeitung, Leibniz Universität Hannover, Welfengarten 1, D-30167 Hannover, Germany

Wei Dong (**71**) *wd2015@columbia.edu*
P & S 11-452, 630 West 168th Street, Columbia University, New York, NY 10032, USA

Diek Duifhuis (**45**) *H.Duifhuis@rug.nl*
Zonnebloemweg 21, 9765 HW Paterswolde, The Netherlands

Steve Elliott (**92**) *sje@isvr.soton.ac.uk*
Institute of Sound and Vibration Research, University of Southampton, Southampton, SO17 1BJ, United Kingdom

Bastian Epp (**28**) *bastian.epp@uni-oldenburg.de*
Carl von Ossietzky Universität Oldenburg, Institut für Physik, AG AKU/NEURO, Carl-von-Ossietzky-Str. 9-11, 26111 Oldenburg, Germany

Mike Evans (**-**) *m.g.evans@keele.ac.uk*
School of Life Sciences, Keele University, Keele, Staffordshire, ST5 5BG, United Kingdom

Jonathan Fisher (**-**) *jfisher@rockefeller.edu*
Howard Hughes Medical Institute, Laboratory of Sensory Neuroscience, The Rockefeller University, Box 314, 1230 York Avenue, New York, NY 10065, USA

Mario Fleischer (**40**) *fleischer@ifkm.mw.tu-dresden.de*
Technische Universität Dresden, Institut für Festkörpermechanik, 01062 Dresden, Germany

Denny Freeman (**103**) *freeman@mit.edu*
77 Massachusetts Ave. RM 36-889, Cambridge, MA. 02139, USA

Anders Fridberger (**26**) *anders.fridberger@ki.se*
Karolinska Institutet, Center for Hearing and Communication Research, M1 Karolinska University Hospital, SE-171 76 Stockholm, Sweden

Reinhart Frosch (**7**) *reinifrosch@bluewin.ch*
Sommerhaldenstrasse 5B, CH-5200 Brugg, Switzerland

James Fulton (**4**) *Jtfulton@cox.net*
c/o Hearing Concepts, 1106 Sandpiper Dr., Corona Del Mar, CA 92625, USA

Dave Furness (**72**) *d.n.furness@keele.ac.uk*
School of Life Sciences, Keele University, Keele, Staffordshire, ST5 5BG, United Kingdom

Tony Gardner-Medwin (**96**) *ucgbarg@ucl.ac.uk*
Dept.Physiology, UCL, London, WC1E 6BT, United Kingdom

Roland Gärtner (**32**) *gaertner@ifkm.mw.tu-dresden.de*
Technische Universität Dresden, Institut für Festkörpermechanik, 01062 Dresden, Germany

Nuria Gavara (**-**) *gavaran@nidcd.nih.gov*
Section on Auditory Mechanics, NIDCD/NIH, Building 10, Room 5D/49, 10 Center Drive, MSC 1417, Bethesda, MD 20892, USA

Christian Gerstenberger (**36**) *cgerstenberger@gdv.uni-hannover.de*
Institut für Mensch-Maschine-Kommunikation, FG Graphische Datenverarbeitung, Leibniz Universität Hannover, Welfengarten 1, D-30167 Hannover, Germany

Roozbeh Ghaffari (**-**) *rooz@mit.edu*
77 Massachusetts Ave. RM 36-889, Cambridge, MA. 02139, USA

Karl Grosh (**82**) *grosh@umich.edu*
Department of Mechanical Engineering, University of Michigan, Ann Arbor, MI 48109-2125, USA

Rachel Gueta (**-**) *rachel.fachima@weizmann.ac.il*
Weizmann Institute of Science, Structural Biology Dept., 76100 Israel

John Guinan (**101**) *jjg@epl.meei.harvard.edu*
Eaton Peabody Lab, Mass. Eye & Ear Infirmary, Harvard Medical School, 243 Charles St., Boston MA, 02114-3002, USA

Anthony W. Gummer (**30**) *anthony.gummer@uni-tuebingen.de*
HNO-Klinik, University of Tübingen, Elfriede-Aulhorn-Str. 5, 72076 Tübingen, Germany

Carole Hackney (**73**) *cmh89@cam.ac.uk*
Department of Physiology, Development and Neuroscience, University of Cambridge, Downing Street, Cambridge, CB2 3EG, United Kingdom

Richard Hallworth (**35**) *hallw@creighton.edu*
Department of Biomedical Sciences, Creighton University School of Medicine, Omaha, NE 68178, USA

Jean-Pierre Hardelin (**18**) *hardelin@pasteur.fr*
Unite de Genetique et Physiology de l'audition - INSERM UMRS 587, Institut Pasteur, 25 rue du Dr. Roux, 75724 PARIS Cedex 15, France

James Harte (**90**) *jha@elektro.dtu.dk*
Centre for Applied Hearing Research, Acoustic Technology, DTU Elektro, Ørsted Plads, Building 352, Technical University of Denmark, DK-2800 Kgs. Lyngby, Denmark

Alexander Hellmuth (**91**) *alexander.hellmuth@medel.com*
Vibrant MED-EL, Fuerstenweg 77, 6020 Innsbruck, Austria

Martin Homer (**34**) *martin.homer@bristol.ac.uk*
Department of Engineering Mathematics, University of Bristol, Bristol BS8 1TR, United Kingdom

Jackie How (**75**) *jh@isvr.soton.ac.uk*
Signal Processing & Control Group, Institute of Sound and Vibration, University of Southampton, Highfield, Southampton, SO17 1BJ, United Kingdom

Stanley Huang (**3**) *sh2365@columbia.edu*
P & S 11-452, 630 West 168th Street, Columbia University, New York NY 10032, USA

Allyn Hubbard (**14**) *aeh@bu.edu*
Boston University, ECE Department, 8 St Mary's Street, Boston MA 02215, USA

Koji Iida (**54**) *iida@wadalab.mech.tohoku.ac.jp*
Department of Bioengineering and Robotics, Tohoku University, 6-6-01 Aoba-yama, Sendai 980-8579, Japan

Kuni Iwasa (**52**) *iwasa@nih.gov*
Biophysics Section, NIDCD, NIH, Rockville, Maryland 20850, USA

Wiktor Jedrzejczak (**41**) *w.jedrzejczak@ifps.org.pl*
Institute of Physiology and Pathology of Hearing, ul. Zgrupowania AK 'Kampinos' 1, 01-943 Warszawa, Poland

David Kemp (**47**) *d.kemp@ucl.ac.uk*
UCL Centre for Auditory Research/ Ear Institute, 332 Grays Inn Road, London WC1X 8EE, United Kingdom

Christine Köppl (**39**) *ckoeppl@physiol.usyd.edu.au*
Department of Physiology, Anderson Stuart Building (F13), University of Sydney, NSW 2006, Australia

Andriy Koslov (**-**) *akozlov@rockefeller.edu*
Howard Hughes Medical Institute, Laboratory of Sensory Neuroscience, The Rockefeller University, Box 314, 1230 York Avenue, New York, NY 10065, USA

Manfred Kössl (**-**) *koessl@bio.uni-frankfurt.de*
Institute for cell Biologiy and Neuroscience, Siesmayerstr 70A, 60323 Frankfurt, Germany

Emery Ku (**55**) *ek@isvr.soton.ac.uk*
Institute of Sound and Vibration Research, University of Southampton, Highfield, Southampton, SO17 1BJ, United Kingdom

Shun Kumano (**1**) *kumano@wadalab.mech.tohoku.ac.jp*
Department of Bioengineering and Robotics, Tohoku University, 6-6-01 Aoba-yama, Sendai 980-8579, Japan

Peter Lampacher (**102**) *peter.lampacher@medel.com*
Vibrant MED-EL, Fuerstenweg 77, 6020 Innsbruck, Austria

Adria LeBoeuf (**62**) *aleboeuf@rockefeller.edu*
Laboratory of Sensory Neuroscience, The Rockefeller University, 1230 York Ave, Box 314, New York, NY 10021, USA

Kong-Ju-Bock Lee (**69**) *kjblee@ewha.ac.kr*
Department of Physics, Ewha Womans University, Seoul, 120-750, Korea

Eric LePage (**29**) *ericlepage@oaericle.com.au*
PO Box 2564, Mount Claremont, W.A. 6010, Australia

Jeffery Lichtenhan (**95**) *jlichtenhan@gmail.com*
Eaton Peabody Lab, Mass. Eye & Ear Infirmary, Harvard Medical School, 243 Charles St., Boston MA, 02114-3002, USA

Ben Lineton (**97**) *bl@isvr.soton.ac.uk*
Institute of Sound and Vibration Research, University of Southampton, Southampton, SO17 1BJ, United Kingdom

Yi-Wen Liu (**51**) *liuy@boystown.org*
Boys Town National Research Hospital, 555 North 30th Street, Omaha, NE 68131, USA

Glenis Long (**60**) *glong@gc.cuny.edu*
Speech-Language-Hearing Program, Graduate Center of the City University of New York, 365 Fifth Ave, New York, NY 10016, USA

Andrei Lukashkin (**-**) *a.lukashkin@sussex.ac.uk*
School of Life Sciences, University of Sussex, Falmer, Brighton, BN1 9QG, United Kingdom

Shanthini Mahendrasingam (**11**) *s.mahendrasingam@cns.keele.ac.uk*
Central Electron Microscope Unit, Institute of Science and Technology in Medicine, School of Life Sciences, Keele University, Keele, Staffordshire ST5 5BG, United Kingdom

Torsten Marquardt (**20**) *t.marquardt@ucl.ac.uk*
UCL Ear Institute, University College London, 332 Gray's Inn Rd., London WC1X8EE, United Kingdom

Pascal Martin (**21**) *pascal.martin@curie.fr*
Laboratoire Physico-Chimie Curie, Institut Curie recherché, 26, rue d'Ulm, 75005 Paris, France

Julien Meaud (**88**) *jmeaud@umich.edu*
Mechanical Engineering Dept., University of Michigan, 2202 GG Brown Bldg., 2350 Hayward Ave., Ann Arbor, MI 48109-2125, USA

Bas Meenderink (**83**) *s.meenderink@erasmusmc.nl*
Dept. of Neuroscience, Erasmus MC, Dr. Molewaterplein 50, Rotterdam, 3015 GE, The Netherlands

Marcia Mellado Lagarde (**61**) *M.M.Mellado-Lagarde@sussex.ac.uk*
School of Life Sciences, University of Sussex, Falmer, Brighton, BN1 9QG, United Kingdom

Pavel Mistrik (**2**) *p.mistrik@ucl.ac.uk*
UCL Ear Institute, Gray's Inn Road, London WC1X8EE, United Kingdom

David Mountain (**31**) *dcm@bu.edu*
Boston University, Department of Biomedical Engineering, 44 Cummington St., Boston, MA 02215, USA

Michio Murakoshi (**-**) *michio@wadalab.mech.tohoku.ac.jp*
Department of Bioengineering and Robotics, Tohoku University, 6-6-01 Aoba-yama, Sendai 980-8579, Japan

Björn Nadrowski (**84**) *bjoern.nadrowski@uni-koeln.de*
Zoologisches Institut, Weyertal 119, 50931 Köln, Germany

Heidi Nakajima (**63**) *heidi_nakajima@meei.harvard.edu*
Eaton Peabody Lab, Mass. Eye & Ear Infirmary, Harvard Medical School, 243 Charles St., Boston MA, 02114-3002, USA

Stephen Neely (**89**) *neely@boystown.org*
Boys Town National Research Hospital, 555 North 30th Street, Omaha, NE 68131, USA

Seth Newburg (**87**) *setho@bu.edu*
Boston University, Department of Biomedical Engineering, 44 Cummington St., Boston, MA 02215, USA

Manuela Nowotny (**9**) *nowotny@bio.uni-frankfurt.de*
Institut fur Zellbiologie und Neurowissenschaft, AK Neurobiologie und Biosensorik, Siesmayerstrasse 70 A, 60323 Frankfurt/Main, Germany

Alfred Nuttall (**16**) *nuttall@ohsu.edu*
OHSU, Oregon Hearing Research Center, 3181 SW Sam Jackson Park Rd., Portland, Oregon 97239, USA

Dáibhid Ó Maoiléidigh (**10**) *dmelody@mpipks-dresden.mpg.de*
Max-Planck-Institut fur Physik komplexer Systeme, Room 1B17, Noethnitzer Str. 38, 01187 Dresden, Germany

Elizabeth Olson (**44**) *eao2004@columbia.edu*
P & S 11-452, 630 West 168th Street, Columbia University, New York NY 10032, USA

Sunil Puria (**66**) *puria@stanford.edu*
Stanford University, 496 Lomita Mall, Durand Building, Rm 283, Stanford, CA 94305, USA

Tianying Ren (**5**) *rent@ohsu.edu*
OHSU, Oregon Hearing Research Center, 3181 SW Sam Jackson Park Rd., Portland, Oregon 97239, USA

Claus-Peter Richter (**86**) *cri529@northwestern.edu*
303 E. Chicago Ave., Chicago, IL 60611, USA

John Rosowski (**46**) *John_Rosowski@meei.harvard.edu*
Eaton-Peabody Lab, Massachusetts Eye and Ear Infirmary, 243 Charles Street, Boston, MA 02114, USA

Itay Rousso (**-**) *itay.rousso@weizmann.ac.il*
Dept. of Structural Biology, Weizmann Institute of Science, Rehovot, 76100, Israel

Mario Ruggero (**17**) *mruggero@northwestern.edu*
Department of Commication Sciences and Disorders, Northwestern University, 2240 Campus Drive, Evanston, IL 60208, USA

Ian Russell (**-**) *I.J.Russell@sussex.ac.uk*
School of Life Sciences, University of Sussex, Falmer, Brighton, BN1 9QG, United Kingdom

Joe Santos-Sacchi (**19**) *joseph.santos-sacchi@yale.edu*
Dept. of Surgery (Otolaryngology), Yale University School of Medicine, BML 244, 333 Cedar St., New Haven, CT 06510, USA

Rick Schoffelen (**50**) *r.l.m.schoffelen@med.umcg.nl*
Dept of Otorhinolaryngology, University Medical Center Groningen, P.O. Box 30001, 9700RB Groningen, The Netherlands

Christopher Shera (**33**) *shera@epl.meei.harvard.edu*
Eaton-Peabody Laboratory, 243 Charles Street, Boston, MA 02114-3096, USA

Renato Sisto (**65**) *renata.sisto@ispesl.it*
ISPESL - Dipartimento Igiene del Lavoro, Via di Fontana Candida, 1, 00040 Monte Porzio Catone (RM), Italy

Michael Slama (**8**) *mslama@mit.edu*
Eaton-Peabody Laboratory, 243 Charles Street, Boston, MA 02114-3096, USA

Alex Spector (**104**) *aspector@jhu.edu*
720 Rutland Ave., Traylor 411, Department of Biomedical Engineering, Johns Hopkins University, Baltimore, Maryland 21205, USA

Charles Steele (**24**) *chasst@stanford.edu*
Mechanics and Computation Group, 496 Lomita Mall, Durand Building, Room 219, Stanford, CA 94305, USA

Aleksandra Stefanovic (**48**) *astefanovi@rockefeller.edu*
430E 63 St 3M, New York, NY 10021, USA

Peter Steyger (**81**) *steygerp@ohsu.edu*
OHSU, Oregon Hearing Research Center, 3181 SW Sam Jackson Park Rd., Portland, Oregon 97239, USA

Robert Szalai (**59**) *r.szalai@bristol.ac.uk*
Department of Engineering Mathematics, University of Bristol, Bristol BS8 1TR, United Kingdom

Diana Turcanu (**58**) *diana.turcanu@uni-tuebingen.de*
HNO-Klinik, University of Tübingen, Elfriede-Aulhorn-Str. 5, 72076 Tübingen, Germany

Masashi Unoki (**-**) *unoki@jaist.ac.jp*
School of information science, Japan Advanced Institute of Science and Technology, 1-1 Asahidai, Nomi, Ishikawa 923-1211, Japan

Marcel van der Heijden (**99**) *m.vanderheyden@erasmusmc.nl*
Department of Neuroscience, room Ee 1285, Erasmus MC, P.O.Box 2040, 3000 CA Rotterdam, Netherlands

Pim van Dijk (**22**) *p.van.dijk@med.umcg.nl*
Dept. of Otorhinolaryngology, University Medical Center Groningen, P.O. Box 30001, 9700 RB Groningen, The Netherlands

Andre van Schaik (**-**) *andre@ee.usyd.edu.au*
School of Electrical and Information Engineering, The University of Sydney, Sydney, NSW 2006, Australia

Sarah Verhulst (**67**) *sv@oersted.dtu.dk*
Centre for Applied Hearing Research (CAHR), Technical University of Denmark, Ørsteds Plads, bldg. 352, DK-2800 Kgs. Lyngby, Denmark

Hiroshi Wada (**-**) *wada@cc.mech.tohoku.ac.jp*
Department of Bioengineering and Robotics, Tohoku University, 6-6-01 Aoba-yama, Sendai 980-8579, Japan

Ben Warren (**94**) *bw21@sussex.ac.uk*
School of Life Sciences, University of Sussex, Falmer, Brighton, BN1 9QG, United Kingdom

Pat Wilson (**-**)
1, Thistleberry Villas, Newcastle-under-Lyme, Staffordshire, ST5 2HP, United Kingdom

Anna Wittekindt (**-**) *Anna.Wittekindt@bio.uni-frankfurt.de*
Institute for Cell Biology and Neuroscience, Siesmayerstrasse 70, 60323 Frankfurt, Germany

Franz-Erich Wolter (**37**) *few@gdv.uni-hannover.de*
Institut für Mensch-Maschine-Kommunikation, FG Graphische Datenverarbeitung, Leibniz Universität Hannover, Welfengarten 1, D-30167 Hannover, Germany

Yong Zhang (**53**) *xyzhang@phya.yonsei.ac.kr*
Institute of Physics and Applied Physics, Yonsei University, Seoul 120-749, Korea

Jian Zuo (**70**) *jian.zuo@stjude.org*
Dept. of Developmental Neurobiology, St. Jude Children's Research Hospital, Memphis, TN 38105, USA

SECTION I

SOUND TRANSMISSION TO AND FROM THE INNER EAR, AND WAVE PROPAGATION WITHIN IT

C.A. Shera and E. de Boer

TIME AVERAGE HOLOGRAPHY STUDY OF HUMAN TYMPANIC MEMBRANE WITH ALTERED MIDDLE EAR OSSICULAR CHAIN

JEFFREY T. CHENG, MICHAEL E. RAVICZ, JOHN J. ROSOWSKI

EPL, Massachusetts Eye & Ear Infirmary, Harvard Medical School
243 Charles St., Boston, MA, 02114,USA

NESIM HULLI, MARIA S. HERNANDEZ-MONTES, COSME FURLONG

Department of Mechanical Engineering, Worcester Polytechnic Institute
100 Institute Road, Worcester, MA, 01609,USA

Computer-assisted time average holographic interferometry was used to study the vibration of the human tympanic membrane (TM) in cadaveric temporal bones before and after alterations of the ossicular chain. Simultaneous laser Doppler vibrometer measurements of stapes velocity were performed to estimate the conductive hearing loss caused by ossicular alterations. The quantified TM motion described from holographic images was correlated with stapes velocity to define relations between TM motion and stapes velocity in various ossicular disorders. The results suggest that motions of the TM are relatively uncoupled from stapes motion at frequencies above 1000 Hz.

1 Introduction

The tympanic membrane (TM) is the first middle-ear structure involved in the transduction of air-borne sound to sound pressure in the cochlear fluid. Alterations of the middle ear by trauma or disease (such as TM perforation and ossicular disorders) reduce the efficiency of sound coupling between the TM and inner ear and result in various degrees of conductive hearing loss [1, 2]. While the clinical diagnosis of conductive hearing loss can be made from audiometry, it is not always easy to differentiate various causes of the conductive hearing loss especially in patients with intact TMs. Nakajima et al. [2] used temporal bone studies to quantify the relationship between umbo and stapes velocity in three cases of artificial ossicular chain fixation (malleus, incus and stapes fixations); the results indicated that the relative reduction of stapes and umbo velocity depended on the locations of the ossicular fixation. However, those results are limited by the measurement of a single point on the TM surface (umbo) and how such fixations affect the vibration of the entire TM is not known. In this study, we used time average holography interferometry (TAHI) [3] to investigate the effects of ossicular manipulations on the motions of the TM and the stapes over a range of 0.1 to 25 kHz. The results improve our understanding of the function of the TM and may help in the diagnosis of various ossicular disorders and plans for treatment.

2 Method

2.1 *Time Average Holographic Interferometry (TAHI)*

Computer-assisted TAHI uses a digital camera synched with an optical phase shifter to capture two-dimensional interference images while stepping the optical phase of one of the interfering beams in cyclic steps of 90° [4]. Digital manipulation of four consecutive phase-stepped interference patterns yields one time average hologram for every 4 camera frames, where frame rates of 40 fps allow the calculation of a new hologram every 0.1 s. This rapid computation rate reduces the effects of very low-frequency motion noise in the system and allows observation of TM motion at video rates.

The TAHI image describes the magnitude of motion of a surface via a gray-scaled surface map in which the intensity of the image at any point $I(p)$ is proportional to the square of the 0th order Bessel function J_0 of the displacement amplitude $z(p)$ [5]:

$$I(p) \propto J_0^2(4\pi\, z(p)/\lambda) \qquad (1)$$

where λ is the wavelength of the laser, and the wave number is $k=2\pi/\lambda$.

The mapping between the image intensity and the displacement magnitude couples the smallest displacement of the surface with the highest intensity, or the brightest area of the image. As the displacement magnitude at a local area increases, the image intensity at that location goes through a series of relative maxima and minima, which appear as a series of lighter areas bordered by dark fringes. Each fringe defines an iso-displacement contour whose magnitude corresponds to a zero of Eq. (1).

2.2 *Temporal Bone Preparation*

A fresh human temporal bone without history of otologic disease was prepared. The bony external auditory canal (EAC) was drilled away, down to the level of the tympanic ring in order to expose 80 to 90% of the TM surface. All soft tissues near the lateral surface of the membrane in the canal were carefully removed while keeping the epidermal layer of the TM untouched. The stapes was accessed by widely opening the mastoid and removing the segment of the facial nerve between the petrous bone and the posterior rim of the TM. The roof of the middle-ear air space was opened to access the head of the malleus and body of the incus. Mucosal folds covering the inside of the middle ear cavity and the ossicles were gently stripped away to improve later ossicular fixations. A hollow metal needle tube was glued to the edge of the TM annulus for placement of a probe microphone. The TM and middle ear cavity were kept moist by frequent spraying of saline and regular immersion in saline.

2.3 *Holographic Interferometry Measurement*

The temporal bone was positioned at the opening of the sound coupler (Fig. 1) within the interferometer head such that the tympanic ring was perpendicular to the object beam of the laser, with easy access to the stapes via the facial recess and the head of the malleus

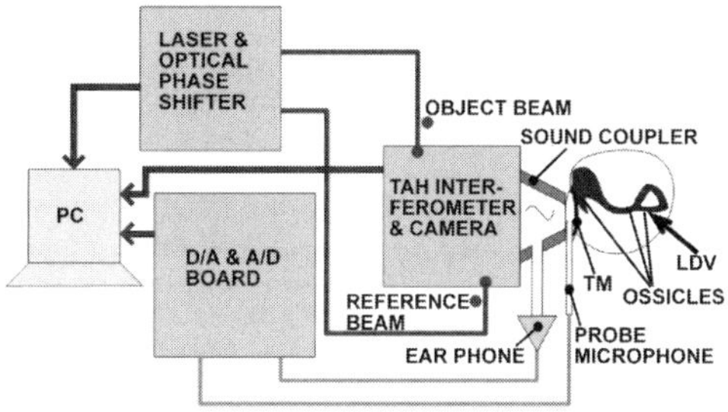

Figure 1. Experimental Setup.

and incus. An earphone piped to the coupler provided the stimulus to the TM, and a probe microphone positioned at the edge of the TM monitored the stimulus level.

Tones of frequencies of 0.2 to 25 kHz and levels between 75 and 130 dB SPL were used as stimuli. The sound pressures varied with frequency to cover the range between minimal (the first signs of TM motion) and near maximal (large motions of the entire TM surface) TM excitation (usually a range of 20 dB in 2 dB steps). To increase the light reflected from the near transparent TM, the TM surface was painted with a solution of TiO_2 powder. This painting has a small effect on middle-ear motion.

2.4 Laser Doppler Vibrometry (LDV) Measurement of stapes motion

The laser beam from the LDV was focused on the posterior crus of the stapes via the open facial recess with an angle of 45 to 60 degree relative to the plane of the stapes footplate. Simultaneous measurements of the stapes velocity and ear canal sound pressure were measured in response to a series of stepped tones (0.2 to 25 kHz).

2.5 Ossicular Chain Fixation

The ossicular fixations mimic conductive pathologies produced by middle ear disease [2]. We first fixed the malleus head to the epitympanic wall with dental cement. After TAHI and LDV measurements, the cement was removed to return the ossicular chain to the baseline condition and TAHI and LDV measurements were repeated. Finally the stapes was fixed and both TAHI and LDV measurements were repeated.

2.6 Analysis of TAHI Images for TM Sensitivity to Sound

The average intensity of TAHI images was used to quantify TM motion. The outline of the TM was identified first on TAHI images. The intensity was then averaged over the TM surface and normalized by the average intensity of the TM without stimulation. According to Eq. (1), when the displacement increases, the image intensity decreases due to an increase in gray level on the TM, such that the normalized intensity approaches 100% (the no stimulus value) at the lowest stimulus level. We select the 80% level of normalized image intensity as a threshold and compare the sound levels needed to achieve this threshold at different frequencies and various ossicular states.

3 Results and Discussion

Figure 2 illustrates TAHI images produced at six frequencies (increasing from the top to bottom) and four different bone conditions: column (A) the control case, (B) malleus head fixation, (C) malleus head freed and (D) stapes fixation. The manubrium is outlined in the bottom panel of each column. At frequencies of 1 kHz and lower, there are a few areas of maximal displacement defined by several concentric dark fringes on the TM surface. These "simple" patterns are consistent with previous results from Tonndorf and Khanna [3, 6]. At frequencies between 1 and 8 kHz, the patterns become "complex" with multiple inter-digitating local maxima. When the frequency is higher than 10 kHz, we see "ordered" patterns with many displacement maxima arranged like "pearls" on "strings". The strings are arranged roughly in concentric circles around the manubrium, and the pearls are arranged radially on the strings. The distance between the strings and the size of the pearls decreases as frequency increases. Such "ordered" patterns are suggestive of regular two-dimensional standing waves, a phenomenon that has not been observed

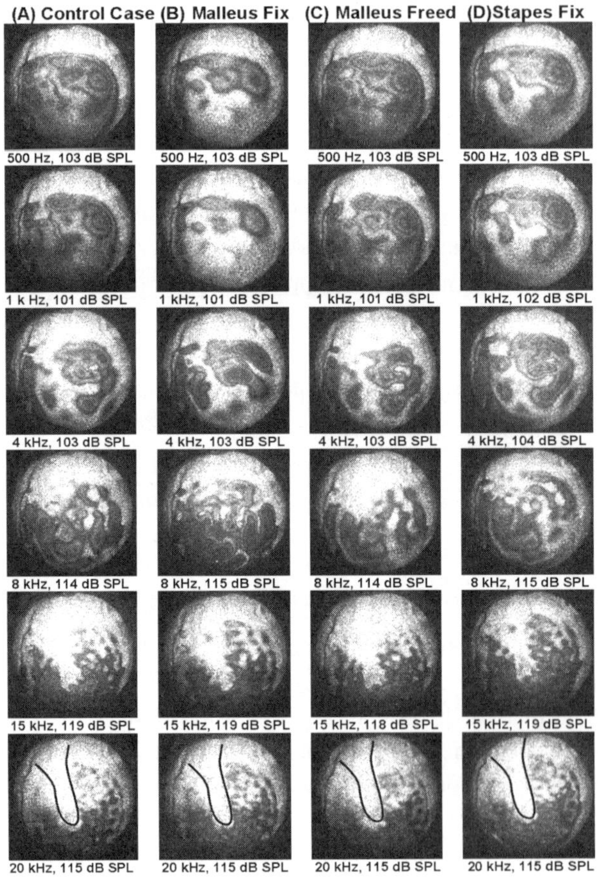

Figure 2. TAHI Images of Human TM before and after Ossicular Fixations.

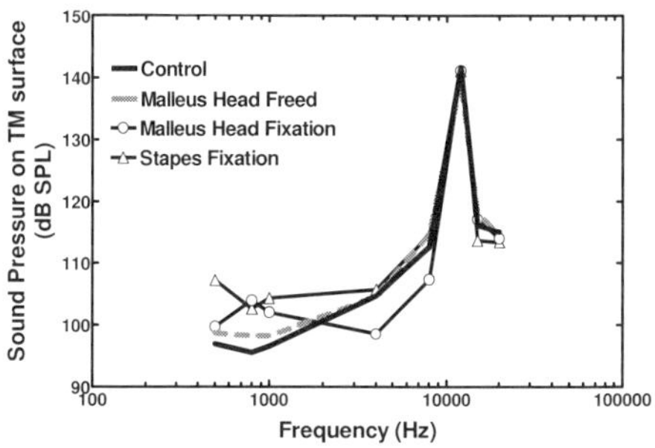

Figure 3. TM Sensitivity to Sound.

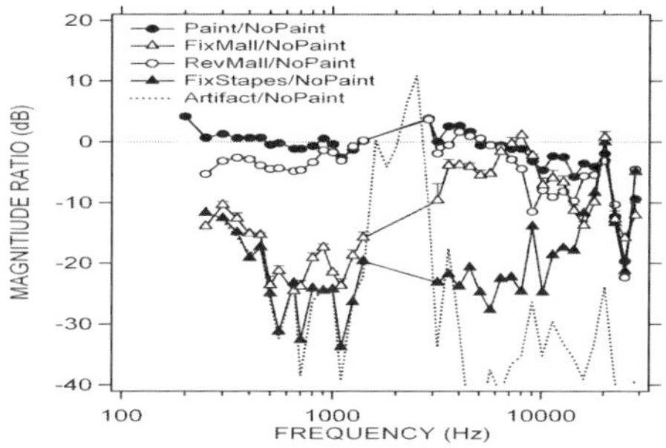

Figure 4. Stapes Velocity from LDV.

before. Figure 2 also illustrates the effects of ossicular manipulations on TM vibration. Column A (the control case) differs significantly from column B (malleus fixed) at f < 4 kHz but not at higher frequencies. Much of these differences are explainable by attenuations of the entire TM motion. Column C (malleus freed) is similar to Column A. Column D (stapes fixed) shows patterns intermediate between A and B at f < 4 kHz and small differences at higher frequencies.

Figure 3 shows computed 80% image intensity thresholds. In the control case (the thick dark line without symbols), the threshold sound pressures are lower at low frequencies, consistent with a higher sensitivity of TM motion at low frequencies [7]. Above 1 kHz, the thresholds generally increase, peak around 12 kHz, and then decrease till 20 kHz. This maximum at 12 kHz needs further investigation. After malleus head fixation (the solid line with open circles), the thresholds are increased at low frequencies compared to the control case, which is consistent with published data describing a decrease in umbo movement after malleus-head fixation [2]. However, the increased sensitivity of TM motion between 3 and 8 kHz after malleus fixation is not consistent with published umbo motion data [2]. After the malleus is freed (the gray dashed line), the thresholds return to near the control case. After stapes fixation (the solid line with triangles), the thresholds at f < 4 kHz are increased, consistent with a decrease in TM motion [2], while above 4 kHz, the thresholds are equal to the control case. The stapes fixation produced larger changes in TM motion than malleus fixation, which is contrary to previous umbo motion measurements [2].

Changes in stapes velocity produced by the same ossicular manipulations and the effect of painting the TM are illustrated in Fig. 4. The illustrated velocities are all relative to the velocity measured in the NoPaint condition. Painting the TM (filled circles) produced little change in stapes motion (differences of +2 to -5 dB at f < 20 kHz). Malleus fixation (open triangles) produced 20 dB or larger decreases in stapes motion at f < 1 kHz and little change at f > 4 kHz. Malleus freed (the open circles) returns the stapes velocity to near the initial case. Stapes fixation (the filled triangles) produced a reduction in stapes motion of at least 20 dB except at the highest and lowest frequencies. These changes are similar to those observed by Nakajima et al. [2], and are significantly larger in magnitude than the differences in 80% thresholds in Fig. 3. Also illustrated in Figure 4 is the relative velocity of the petrous bone. This estimate of motion artifact in our system places lower bounds to discriminate the differences of motion. The artifact is generally 20 to 35 dB smaller than the stapes motion, except in a band between 1 and 3 kHz where the artifact dominates the measured motions. We excluded data in this band from our analysis. The cause of this artifact peak may be related to motion between the LDV laser mounted on an operating microscope and the bone coupled to the TAHI system.

4 Conclusion

The vibrations of the human TM stimulated by pure tones over 0.2 to 25 kHz were studied through TAHI and grouped into three patterns: simple, complex and ordered. The TM motions at various conditions were compared with stapes velocity. The results show the ossicular manipulations had effects on stapes motion that were visible at higher frequencies than the effects observed on TM motion. This is particularly true in stapes fixation, which reduced stapes velocity by at least 20 dB at frequencies less than 10 kHz, but had little effect on TM motion at frequencies of 4 kHz or higher.

Acknowledgments

This work was supported by NIDCD, WPI, MEEI and the Mittal Fund. We thank Dr. Hideko H. Nakajima for preparing the temporal bone and discussions.

References

1. Voss SE, Rosowski JJ, Merchant SN, Peake WT, 2001. Middle-ear function with tympanic perforations. I. Measurements and mechanisms. J. Acoust. Soc. Am. 110, 1432-1444.
2. Nakajima HH, Ravicz ME, Merchant SN, Peake WT, Rosowski JJ, 2005. Experimental ossicular fixations and the middle ear's response to sound: evidence for a flexible ossicular chain. Hear. Res. 204, 60-77.
3. Rosowski JJ, Furlong C, Ravicz ME, Rodgers MT, 2007. Real-time opto-electronic holographic measurements of sound-induced tympanic membrane displacements. In: Proceedings of the 4th Int. Symp. of Middle Ear Mechanics. World Scientific, pp. 295-305.
4. Furlong C, Pryputniewicz RS, 1998. Hybrid computational and experimental approach for the study and optimization of mechanical components. Opt. Eng. 37(5), 1448-55.
5. Kreis T, 2005. Handbook of Holographic Interferometry: Optical and Digital Methods. WILEY-VCH GmbH & Co. KGaA.
6. Tonndorf J, Khanna SM, 1972. Tympanic membrane vibrations in human cadaver ears studied by time averaged holography. J. Acous. Soc. Am. 52(4), 1221-33.
7. Rosowski JJ, 1996. Models of external and middle-ear function. The Springer Handbook of Auditory Research, Volume 6: Auditory Computation. Edited by Hawkins HL, McMullen TA, Popper AN and Fay RR. New York, Springer-Verlag, pp15-61.

MEASUREMENTS OF MIDDLE EAR PRESSURE GAIN AND COCHLEAR INPUT IMPEDANCE IN THE CHINCHILLA

MICHAEL C. C. SLAMA[1,2], MICHAEL E. RAVICZ[2], HIDEKO H. NAKAJIMA[2,3], WEI DONG[4], JOHN J. ROSOWSKI[1,2,3]

[1]*Speech and Hearing Bioscience and Technology Program, Harvard-MIT Division of Health, Sciences and Technology, 77 Massachusetts Avenue Cambridge, MA 02139, USA*

[2]*Eaton-Peabody Laboratory, Massachusetts Eye and Ear Infirmary, 243 Charles St. Boston, MA 02114, USA*

[3]*Department of Otology and Laryngology, Harvard Medical School, 243 Charles St. Boston, MA 02114, USA*

[4]*Department of Otolaryngology, Head and Neck Surgery, Columbia University, 630 West 168th St. New York, NY 10032, USA*

Measurements of middle ear conducted sound pressure in the cochlear vestibule P_V have been performed in only a few species. Simultaneous measurements of sound-induced stapes velocity V_S are even more rare. We report simultaneous measurements of V_S and P_V in chinchillas. The V_S measurements are performed using single-beam laser-Doppler vibrometry; P_V is measured with fiber-optic pressure sensors like those described by Olson [1]. P_V and V_S have been measured in six animals, and the middle ear pressure gain (ratio of P_V to the sound pressure in the ear canal) and the cochlear input impedance (ratio of P_V to the product of V_S and area of the footplate) computed. Our measurements of middle ear pressure gain are similar to published data in the chinchilla at stimulus frequencies of 500 Hz to 3 kHz, but are different at other frequencies. Our measurements of cochlear input impedance differ somewhat from previous estimates in the chinchilla and show a resistive input impedance up to at least 10 kHz.

1 Introduction

Quantifying middle ear pressure gain (G_{ME}), stapes volume velocity (U_S) and cochlear input impedance (Z_C) in animals is crucial to our understanding of middle ear function:

- G_{ME} is the ratio between the sound pressure in the vestibule (input of the inner ear) P_V and the sound pressure in the ear canal near the tympanic membrane P_{TM}. This transfer function depends on frequency and quantifies the passive pressure amplification function of the middle ear.
- U_S is a measure of the time rate of change of the volume displaced by the stapes footplate.
- $Z_C = P_V/U_S$ is the inner ear load to the middle ear. The real part of this impedance is a primary determinant of the acoustic power delivered to the cochlea.

We measured P_V and U_S in the chinchilla in response to acoustic stimulation, using miniature fiber-optic pressure sensors and laser-Doppler vibrometry.

Measurements of P_V are constrained by the limited space available for pressure sensors and the fragility of middle ear structures. We chose fiber-optic pressure sensors because they have good sensitivity, good high frequency response, and because they are very small. Their small size (about 170 μm in diameter) is an important factor because:

- Disruption to the pressure field in the inner ear is negligible at frequencies in the chinchilla's hearing range, both because of the small size (170 μm is less than 1% of the wavelength in water at 30 kHz) and the relatively high impedance associated with such a small microphone: The volume displacement of the diaphragm produced by loud sounds (< 0.4 nL) is less than 0.1% of the fluid volume in the vestibule.
- Damage done to the middle ear and inner ear structures during insertion into the vestibule is minimal.

2 Methods

2.1 Animal Preparation

The animals were anesthetized with Nembutal and Ketamine. After a tracheotomy to facilitate respiration, an opening was made in the superior bulla. The tendon of the tensor tympani muscle and the facial nerve innervating the stapedius muscle were cut to prevent the effects of random contractions of these muscles during the experiment [2]. A second hole was made in the posterior bulla to view the stapes and round window. Part of the bony wall around the round window through which the facial nerve passes was removed in order to see the wall of the vestibule posterior to the stapes. In doing so, extreme care was taken to avoid pulling or damaging the stapedius tendon. A hole of approximate diameter 200 μm was made in the vestibule with a fine sharp pick for the fiber-optic pressure sensor.

The cartilaginous ear canal was cut and a brass tube was glued in the bony ear canal to allow repeatable couplings of the earphone delivering sound stimuli (broadband chirps and tones from 62 Hz to 30 kHz). The middle ear was open during the measurements.

2.2 Fiber-Optic Pressure Sensors

The fiber-optic pressure sensors were fabricated following the techniques of Olson [1]. They are composed of a glass capillary tube (167 μm outer diameter) with a gold-coated polymer diaphragm affixed to one end. A single optical fiber (100 μm o.d.) is inserted into the other end. The optical fiber is spliced to a "Y" coupling. A Light-Emitting Diode (LED) attached to one "Y" branch introduces incoherent light, and a photodiode attached to the other branch measures the light reflected from the diaphragm. Sound pressure flexes the diaphragm and modulates the reflected light.

Calibration of the sensors was performed in a vibrating water column, following the method of Schloss and Strasberg [3]. The sensor is immersed in a column of liquid, which is shaken vertically; the pressure at the diaphragm depends on both the depth of immersion and on the acceleration of the column.

The main issues with these sensors are their fragility and their stability, and we periodically recalibrated the sensor to determine any variation in its sensitivity. We report data only from sensors whose calibration was stable throughout the measurement session and whose sensitivity to temperature was small.

2.3 Laser Doppler Vibrometry

To measure stapes velocity V_S, we used a single-beam laser Doppler vibrometer (Polytec CLV 700) aimed at small (< 50 μm diameter) reflective plastic beads placed on the posterior crus and the footplate. Our surgical exposure of the stapes allowed nearly direct measurement of the piston-like component of stapes motion: the angle of the laser beam was about 30° relative to the piston direction. The volume velocity U_S was estimated by multiplying the measured V_S by the area of the chinchilla footplate (2 mm^2 [4]).

2.4 Pressure Near the Tympanic Membrane

The middle ear pressure gain G_{ME} is defined as the ratio between P_V and P_{TM}, the pressure near the tympanic membrane (TM). A reference microphone built into the sound coupler measured ear-canal sound pressure (P_{EC}) at the entrance of the brass coupling tube, about 5 mm from the TM. P_{TM} is different from P_{EC}, at high frequencies. To account for these differences, we measured the transfer function P_{TM}/P_{EC} in a dead ear and multiplied our measured P_{EC} with this function.

This correction affects our computations of G_{ME} and normalized stapes velocity (U_S/P_{TM}), but does not affect Z_C, whose computation does not involve P_{TM}.

3 Results

Thirteen animals were used in this study. Among these, 4 had their middle or inner ears damaged during surgery. In 2 other experiments the pressure sensor proved unstable. We therefore present G_{ME} results in 7 animals. We simultaneously measured U_S in 6 of these, so we present 6 Z_C sets of measurements. For each experiment, we restrict the results shown to the frequency range over which the measurements were above the noise floor, which we determined by testing the repeatability of both response magnitude and angle.

3.1 Middle Ear Pressure Gain

G_{ME} (Figure 1) was computed from simultaneous measurements of P_V and P_{EC}, and corrected to account for the differences between P_{TM} and P_{EC} as described above.

$|G_{ME}|$ was between 20 and 40 dB over almost the entire frequency range of measurement. The angle was near 0 between 0.3 and 3 kHz and

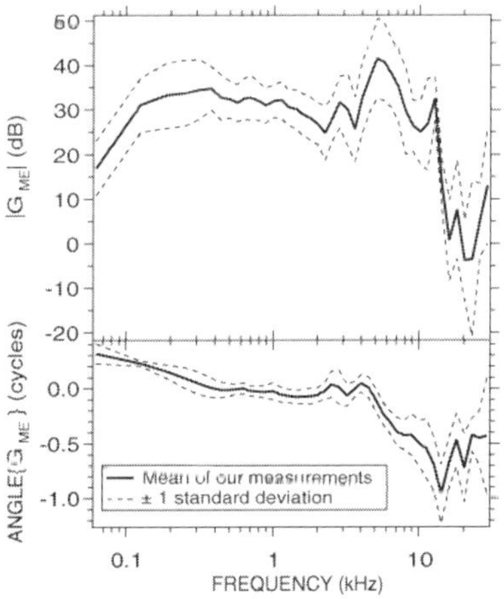

Figure 1. Middle Ear Pressure Gain.

accumulated with frequency above that, reaching -0.7 cycles by 10 kHz. The standard deviation (dashed line) was about 5 dB for $|G_{ME}|$ and <0.1 cycle for the angle.

$|G_{ME}|$ was comparable to the anatomical "transformer ratio" of 35 dB computed from the "area ratio" (of the TM to the area of the footplate) of 29 dB and the "lever ratio" (of malleus length to incus length) of 6 dB [4,5]. The angle differs significantly from 0 (the expected angle of an ideal transformer) at f<300 Hz and >3 kHz.

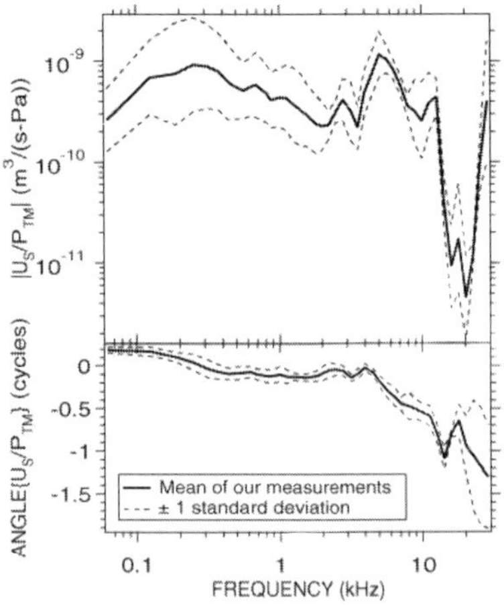

Figure 2. Normalized Stapes Velocity.

3.2 *Normalized Stapes Velocity*

U_S (Figure 2) was computed from the stapes velocity V_S and a mean stapes footplate area as described in the methods, and normalized by P_{TM}.

$|U_S|$ increased with frequency 60–300 Hz and the angle was near +0.25 cycles, consistent with a compliance. $|U_S|$ decreased slightly with frequency 0.3–2 kHz and the angle was between 0 and -0.25 cycles, consistent with a mass-resistance combination. There is a large notch in $|U_S|$ between 12 and 30 kHz, which will be discussed in the Discussion section.

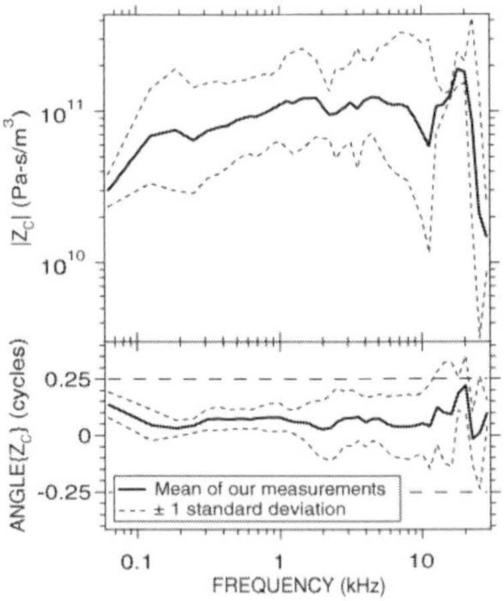

Figure 3. Cochlear Input Impedance.

3.3 *Cochlear Input Impedance*

Z_C (Figure 3) was computed from simultaneous measurements of P_V and U_S. The computation of Z_C does not use P_{TM}, therefore our results are not affected by the error in estimating P_{TM}.

$|Z_C|$ was about 10^{11} acoustic ohms, roughly constant with frequency up to 10 kHz, and the angle was near zero. This is consistent with a resistance. The angle had values between –0.25 and +0.25 cycles at all frequencies measured except where it was contaminated by noise. This is consistent with the input impedance of a passive system.

4 Discussion

4.1 *High Frequency Responses*

The high frequency responses we obtained for G_{ME} and normalized U_S are characterized by an increased variance in magnitude and/or angle relative to lower frequencies. For G_{ME}, the standard deviation was about 15 dB above 20 kHz for the magnitude, and as large as 0.5 cycles between 25 and 30 kHz for the angle. For normalized U_S, the variance of the magnitude above 20 kHz is similar to lower frequencies, but the standard deviation of the angle grows to 1.5 cycles.

These increased variances are not due to measurement noise, because only responses with good signal-to-noise ratios were kept. They can be explained by at least two factors:

- In earlier experiments, the earphone we were using did not provide a good signal-to-noise ratio at high frequency, so there are only 3 ears with good signal-to-noise ratio above 16 kHz. We need to repeat the measurements in more individuals to have better estimates of the mean response at high frequencies.
- For U_S, we are assuming piston-like motion of the stapes, and measuring velocity in only one direction. It is likely that this assumption is not valid at high frequencies. This could explain the notch we found in the magnitude between 12 and 30 kHz, as well as the sharp peak in $|Z_C|$ at these frequencies.

4.2 Effect of the Vestibular Hole

It was necessary to make a hole in the vestibule to introduce the pressure sensor and measure P_V. To assess the influence of the hole on U_S, we compared measurements of U_S before the hole was made and afterward with the pressure sensor in place. We found an increase in the magnitude of U_S/P_{TM} in the condition with the vestibular hole, which is consistent with the hole facilitating motion. Nonetheless, this increase was small and significant ($p<0.01$) only in a small region around 8 kHz.

To determine the influence of the hole on P_V, we tried in each experiment, with the pressure sensor in place, to plug the hole with dental impression material (Jeltrate), dental cement, or a sodium hyaluronate viscoelastic gel of high molecular weight (Healon GV 14 mg/mL). In most preparations, it was not possible to plug the hole effectively because of the limited space available with the sensor in place and because of the outward flow of perilymph pushing the material away. In one case, the Healon GV gel appeared to cover most of the hole, resulting in an increase in pressure, especially at frequencies below 1 kHz. After removing the gel, pressure went back to the lower level. Overall, the effects of the vestibular hole on P_V and U_S were small and limited in frequency.

4.3 Comparison with other Chinchilla Studies

Our G_{ME} results are very similar to a previous study by Décory [6] in chinchilla between 500 Hz and 3 kHz, for both the magnitude and the angle (Figure 4). Below 500 Hz, and between 3 kHz and 10 kHz, we found a larger $|G_{ME}|$.

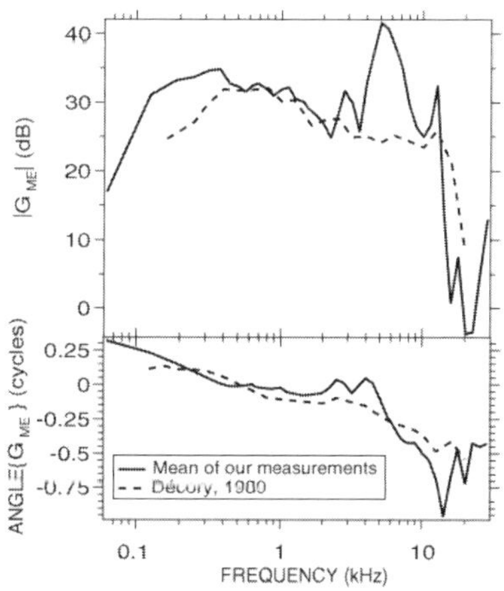

Figure 4. Comparison of our Middle Ear Pressure Gain with another chinchilla study.

Our measurements of stapes velocity compare very well with a study by Songer and Rosowski [7] for the magnitude as well as for the angle. Our results are also very similar to a study by Ruggero *et al.* [8] at frequencies below 12 kHz.

We compared our Z_C measurements with a model by Songer and Rosowski [7], as well as computations by Ruggero *et al.* [8], who used their own U_S measurements and Décory's P_V measurements in other animals (see Figure 4). The 3 data sets share many similarities (Figure 5): In particular, the impedances are mostly resistive and have an order of magnitude of about 10^{11} acoustic ohms. The differences in magnitude between our data and Ruggero *et al.*'s below 500 Hz and between 3 and 10 kHz are consistent with the larger G_{ME} we measured at these frequencies.

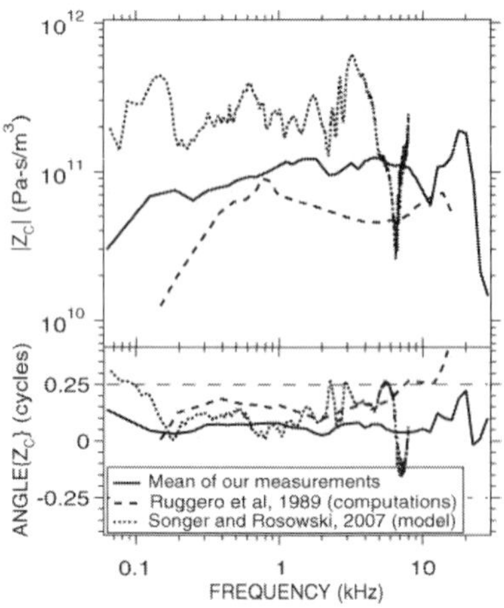

Figure 5. Comparison of our Cochlear Input Impedance with other chinchilla studies.

Acknowledgments

We thank Elizabeth Olson for her assistance in making the fiber-optic pressure sensors, Melissa Wood for her mastery of chinchilla surgery, the staff of the Eaton-Peabody Laboratory, and the staff of the Microsystems Technology Laboratories at MIT. Funded by NIDCD.

References

1. Olson, E.S., 1998. Observing middle and inner ear mechanics with novel intracochlear pressure sensors. J. Acoust. Soc. Am. 103 (6), 3445-3463.
2. Rosowski, J.J., Ravicz, M.E., Songer, J.E., 2006. Structures that contribute to middle-ear admittance in chinchilla. J. Comp. Physiol. A. 192, 1287-1311.
3. Schloss, F., Strasberg, M., 1962. Hydrophone calibration in a vibrating column of liquid. J. Acoust. Soc. Am. 34 (7), 958-960.
4. Vrettakos, P.A., Dear, S.P., Saunders, J.C., 1988. Middle ear structure in the chinchilla: A quantitative study. Am. J. Otolaryngol. 9, 58-67.
5. Fleischer, G., 1973. Studien am skelett des gehörorgans der säugetiere, einschliesslich des menschen. Säugetierkundl. Mitteilungen (München). 21, 131-239.
6. Décory, L., 1989. Origine des différences interspécifiques de susceptibilités au bruit. Thèse de Doctorat de l'Université de Bordeaux, France
7. Songer, J.E., Rosowski, J.J., 2007. Transmission matrix analysis of the chinchilla middle ear. J. Acoust. Soc. Am. 122 (2), 932-942.
8. Ruggero, M.A., Rich, N.C., Robles, L., Shivapuja B.G., 1990. Middle ear response in the chinchilla and its relationship to mechanics at the base of the cochlea. J. Acoust. Soc. Am. 87 (4), 1612-1629.

A MIDDLE-EAR REVERSE TRANSFER FUNCTION COMPUTED FROM VIBRATION MEASUREMENTS OF OTOACOUSTIC EMISSIONS ON THE EAR DRUM OF THE GUINEA PIG

ERNST DALHOFF, DIANA TURCANU, ANTHONY W. GUMMER

Section of Physiological Acoustics and Communication, Eberhard-Karls-University, Elfriede-Aulhorn-Str. 5, Tübingen, 72076, Germany

Using distortion products measured as vibration of the umbo and as sound pressure in the ear canal of guinea pigs, we calculated the corresponding reverse transfer function. We compare the measurements with a middle-ear model taken from the literature and adapted to the guinea pig. A reasonable fit could be achieved. We conclude that the reverse transfer function will be useful to aid fitting a middle-ear model to measured transfer functions of human subjects.

1 Introduction

Recently, using a custom-built laser Doppler vibrometer (LDV), we have shown that distortion product otoacoustic emissions (DPOAEs) can be measured as vibration of the human eardrum *in vivo* [1]. Here, we show measurements of the transfer function of the ear-canal pressure to umbo velocity in the guinea pig when the middle ear is driven in reverse by otoacoustic emissions. Measurements are compared with a middle-ear model taken from the literature and adapted to the giunea pig.

2 Methods

Normal and hearing-impaired guinea pigs were anaesthetized and their pinna together with the cartilaginous part of the ear canal removed. The tip of a probe microphone was glued within 2-4 mm of the tympanic membrane. Acoustic stimulation was delivered free-field. For the vibration measurements, we mounted our LDV [1] onto an operating microscope. For an averaging time of 40 s, we typically achieved noise levels of 0.2 pm for frequencies higher than 1.5 kHz. Using a multi-tone stimulus, the forward transfer function of umbo velocity relative to ear-canal sound pressure was obtained. Using two-tone stimulation with a frequency ratio of f2/f1=1.2, we computed from the DPOAEs at the combination tone frequencies 2f2-f1, 2f1-f2, 3f1-2f2, and f2-f1 the ratio of umbo velocity to ear-canal sound pressure. Taking the values from two-tone measurements recorded with different primary frequencies in the same animal, we assemble the reverse transfer function at frequencies ranging from 0.8-20 kHz. Here, forward and reverse transfer functions obey the same sign convention: velocity in the medial direction over condensation pressure.

We compared the measured forward and reverse transfer functions with a middle-ear model taken from the literature [2] for cat and adapted it to the guinea pig. This model is based on a transmission-line representation of the tympanic membrane. We started using values for the impedance of the cavities, area of tympanic membrane and stapes footplate area as given in [4] and [5], a lever ratio of 2 [6] and a resistive cochlear input impedance of 0.45 $M\Omega_{cgs}$ [7]. The characteristic impedance of the transmission line was chosen to be 700 Ω_{cgs}, approximately the characteristic impedance of a tube with 2.8-mm diameter, the diameter of the ear canal [8]. Then, we assembled the model stage-by-stage to match the characteristic impedances of the stages as outlined by Puria and Allen [2].

For the reverse impedance, we started with a tube of radius 1.3 mm and length 3 mm representing the ear canal without cartilaginous portion and accounting for the partial occlusion by the microphone probe tube. The tube is loaded by the radiation impedance.

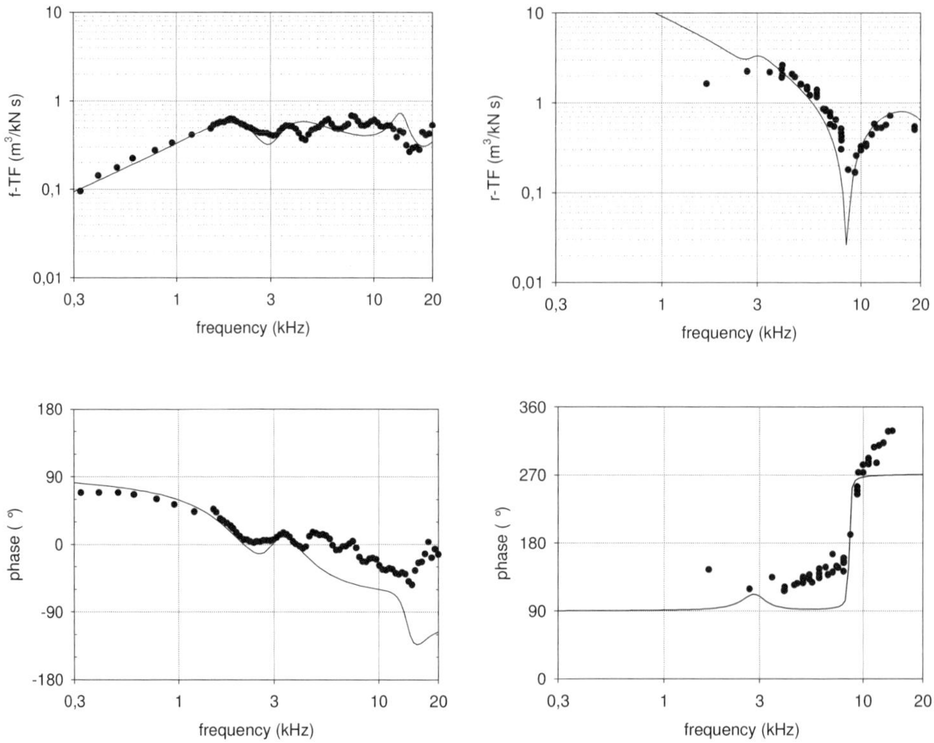

Fig. 1. Experimentally measured forward (f-TF) and reverse transfer function (r-TF) of one individual guinea pig (Gp66) along with a model fit (dots and continuous line, respectively). See text.

3 Results

The measured reverse transfer function shows two passbands, the first of which is centred at 3-4 kHz and the second at approximately 15 kHz; they are separated by a sharp notch (Q_{10dB}~2). There is good agreement between the model and the data above 2 kHz. The position of the notch at 8-9 kHz is mainly dependent on the delay of the tympanic membrane, which was 18 μs for Fig. 1, on the characteristic impedance of the transmission line and on the diameter of the ear canal.

Comparison of the forward transfer function with the model shows that, besides some limitations in the phase behaviour at frequencies above 4 kHz, a reasonable fit is obtained for frequencies up to 10 kHz. Most of the starting parameters mentioned above did not need to be changed; the main exceptions were for the tympanic-membrane area (A_{TM}=0.20 cm^2) and the antrum capacitance (0.4 μF instead of 0.038 μF in [4]). Other values are close to those used by Puria & Allen [2] for the cat. The acoustic input impedance of the middle ear is approximately 3 dB too high, but similar in shape to that of published measurements [4] (not shown here).

4 Discussion

Unfortunately, due to 1/f-noise, up to now we have only very few data at frequencies below 3 kHz. This limitation makes phase unwrapping difficult. Generally speaking, it is to be expected that at low frequencies the tympanic membrane moves in unison as a single vibration mode and, therefore, efficiently produces a proportional volume velocity in the ear canal. Nonetheless, towards low frequencies the reverse transfer function must rise because of the free-field configuration – in other words, the radiation efficiency must drop. This can be seen only in the model. In the broad passbands of the measured reverse transfer function, the periphery of the membrane follows the umbo motion with increasing phase lag. Thus, the radiation efficiency of the membrane is reduced in the passband because tympanic-membrane motion breaks up into higher modes. At the frequency of the first notch, which occurs at 8-9 kHz, the umbo vibrates with less velocity amplitude compared with the frequency range of 1.5-6 kHz in order to produce the same sound pressure. Thus, the centre of the eardrum must be able to transfer the force optimally to a resonant vibration of higher amplitude in the periphery at this frequency. In the model, the notch frequency depends mainly on the delay parameter of the tympanic membrane and its characteristic impedance, which in turn is related to the distributed mass and stiffness of the membrane. The delay used in the model is 18 μs, which is considerably less than the values estimated for the cat tympanic membrane [2] and the overall middle-ear forward delay for gerbil [3]. The high Q of the notch at 8-9 kHz supports the idea that the normal tympanic membrane has very low internal friction [2].

Deficiencies of the model appear in the phase of the forward transfer function, as well as in the low-frequency range of the reverse transfer function. These might be specific for the guinea pig or might be general problems of the transmission-line representation of the tympanic membrane. The phase behaviour of the forward transfer function for instance could be explained if part of the delay were put "behind" the umbo. This could be possible because the handle of the malleus is also a distributed system. In addition, in the transmission-line model the whole stimulus is undergoing the same delay, which is physically unreasonable. On the other hand, it may be that the remarkable sudden widening of the ear canal in front of the tympanic membrane, which is characteristic in guinea pigs, plays a role which is not taken into account by the model.

In conclusion, it seems possible that the reverse transfer function can be used in the future, in order to aid fitting a middle-ear model to measured transfer functions of human subjects.

Acknowledgments

This work is supported by the German Research Council DFG Gu 194/8-1.

References

1. Dalhoff, E., Turcanu, D., Zenner, H. P., Gummer, A. W., 2007. Distortion product otoacoustic emissions measured as vibration on the eardrum of human subjects. Proc. Natl. Acad. Sci. U.S.A 104, 1546-1551.
2. Puria, S., Allen, J.B., 1998. Measurements and model of the cat middle ear: Evidence of tympanic membrane acoustic delay. J. Acoust. Soc. Am. 104, 3463-3481.
3. Dong, W., Olson, E.S., 2006. Middle ear forward and reverse transmission in gerbil. J. Neurophysiol. 95, 2951-2961.
4. Zwislocki, J., 1963. Analysis of the middle-ear function. Part II: Guinea pig ear. J. Acoust. Soc. Am. 35, 1034-1040.
5. Fernandez, C., 1952. Dimensions of the cochlea (guinea pig). J. Acoust. Soc. Am. 24, 519-523.
6. Ruggero, M.A., 1990. Middle-ear response in the chinchilla and its relationship to mechanics at the base of the cochlea. J. Acoust. Soc. Am. 87, 1612-1629.
7. Dancer, A., Franke, R., 1980. Intracochlear sound pressure measurements in guinea pigs. Hear. Res. 2, 191-205.
8. Wysocki, J., Sharifi, M., 2005. Measurements of selected parameters of the guinea pig temporal bone. Folia Morphol. (Warsz.) 64, 145-150.

DIFFERENTIAL INTRACOCHLEAR SOUND PRESSURE MEASUREMENTS IN NORMAL HUMAN TEMPORAL BONES

HIDEKO HEIDI NAKAJIMA[1,2], WEI DONG[4], ELIZABETH S. OLSON[4], SAUMIL N. MERCHANT[1,2], MICHAEL E. RAVICZ[2,3], JOHN J. ROSOWSKI[1,2,3]

[1]*Department of Otology and Laryngology, Harvard Medical School,*
[2]*Eaton-Peabody Lab., Mass. Eye and Ear Infirmary, 243Charles St., Boston, MA 02114, USA*
[3]*Resarch Lab. of Electronics, Mass. Institute of Technology, Cambridge, MA 02139, USA*
[4]*Columbia University, Department of Otolaryngology, Head and Neck Surgery, P & S 11-452, 630 W. 168th Street, New York, New York 10032*

We present the first simultaneous sound pressure measurements in scala vestibuli and scala tympani of the cochlea in human cadaveric temporal bones. Micro-scale fiberoptic pressure sensors enabled the study of differential sound pressure at the cochlear base. This differential pressure is the input to the cochlear partition, driving cochlear waves and auditory transduction. Results showed that: pressure of scala vestibuli was much greater than scala tympani except at low and high frequencies where scala tympani pressure affects the input to the cochlea; the differential pressure proved to be an excellent measure of normal ossicular transduction of sound (shown to decrease 30-50 dB with ossicular disarticulation, whereas the individual scala pressures were significantly affected by non-ossicular conduction of sound at high frequencies); the middle-ear gain and differential pressure were generally bandpass in frequency dependence; and the middle-ear delay in the human was over twice that of the gerbil. Concurrent stapes velocity measurements allowed determination of the differential impedance across the partition and round-window impedance. The differential impedance was generally resistive, while the round-window impedance was consistent with a compliance in conjunction with distributed inertia and damping. Our techniques can be used to study inner-ear conductive pathologies (e.g., semicircular dehiscence), as well as non-ossicular cochlear stimulation (e.g., round-window stimulation) – situations that cannot be completely quantified by measurements of stapes velocity or scala-vestibuli pressure by themselves.

1 Introduction

Traditional measurements of middle-ear function, such as stapes velocity, are useful for studying sounds transmitted by air conduction in normal ears and ears with middle-ear lesions. However, the study of inner-ear conductive pathologies (e.g., semicircular canal dehiscence, [1]) and the stimulation of the cochlea by non-ossicular means (e.g., round-window stimulation, [2]; or bone conduction) require a different approach: *differential pressure measurements*. We describe the first differential pressure measurements in human cadaveric temporal bones. These measurements are made possible by micro-scale pressure sensors developed by Olson [3] that enable simultaneous sound pressure measurements within scala vestibuli and scala tympani. The measurements reported in this study provide a baseline for future measurements to study various middle and inner-ear pathologies as well as the effect of non-ossicular cochlear stimulation.

The differential intracochlear pressure across the base of the cochlear partition is the input signal to the cochlea and starts the traveling wave moving toward the apex. The pressure difference thus drives auditory transduction. In this study, we measure the sound-evoked differential pressure at the base of the cochlea in human temporal bones with two micro-scale pressure sensors: one in scala vestibuli near the oval window and one in scala tympani near the round window. The sensitive tips of the pressure sensors are located within 200 μm of the bony wall of the cochlea and relatively distant from the cochlear partition. Hence, the pressures measured by the probes in our experiments are relatively unaffected by the near-field pressures that result from the motion of the basilar membrane.

Measurements of stapes velocity are made concurrently, allowing for the calculation of cochlear input impedance as well as the differential impedance across the partition and the round window impedance.

2 Methods

Six temporal bones, which did not appear to have intracochlear air and showed stable pressure probe calibrations, are presented. The specimens were stored in normal saline and refrigerated immediately upon removal. Several measures were taken to prevent introduction of air into the cochlea: the medial and posterior aspects of the petrous bone were sealed with dental cement; the temporal bone was positioned such that the locations of pressure-sensor insertions were uppermost relative to the rest of the bone to prevent intracochlear air leakage. Furthermore, cochleostomies just slightly larger than the pressure sensor diameter (with holes ~200 μm for the 167 μm sensors) were drilled by hand while the cochlea was immersed in saline solution to prevent introduction of air into the cochlea. The saline surrounding the cochlea was removed and the gaps between the pressure sensors and the bone were sealed with dental impression material, which dried to a rubbery consistency. A phase difference between the stapes and round window velocities of one-half cycle at frequencies below 500 Hz, before and after cochleostomies and pressure probe insertions, was used as confirmation that air was not introduced into the cochlea. Velocity measurements of the stapes and round window was made with a laser Doppler vibrometer in a similar manner to that of Nakajima et al. [4]. The fiberoptic pressure sensors were manufactured and calibrated in the manner described by Olson [5]. To ensure accuracy of the pressure sensors, the pressure sensors were calibrated in water just prior to insertion and immediately after removal from the scala in each sequence of measurements. Only measurements with pressure sensor calibrations that were stable

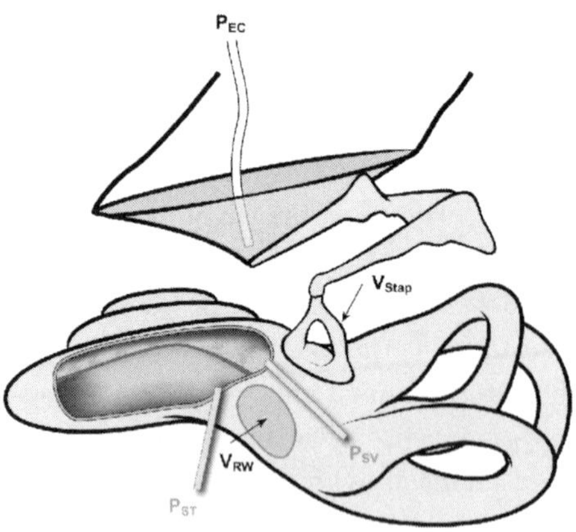

Figure 1. Illustration of the various types of recordings: pressure in scala vestibuli ($\mathbf{P_{SV}}$), pressure in scala tympani ($\mathbf{P_{ST}}$), pressure in the ear canal ($\mathbf{P_{EC}}$), velocity of the stapes ($\mathbf{V_{Stap}}$), velocity of the round window ($\mathbf{V_{RW}}$).

to within 2 dB are presented. At the end of the experiment, the positions of the pressure sensors were confirmed by opening the round window membrane and the surrounding bone over scala tympani and scala vestibuli. Figure 1 illustrates the locations of various measurements performed: the ear-canal pressure ($\mathbf{P}_{EC}$) was measured near the umbo (1-3 mm) with a probe tube, stapes velocity ($\mathbf{V}_{Stap}$) and round-window velocity ($\mathbf{V}_{RW}$) were measured on the posterior crus and center of the round window respectively, and the scala vestibuli pressure ($\mathbf{P}_{SV}$) and scala tympani pressure ($\mathbf{P}_{ST}$) were measured near the stapes and round window respectively.

3 Results and Discussion

3.1 Transfer Functions of Intracochlear Pressures

The relationship between the intracochlear pressures (both scala vestibuli and scala tympani) and the ear-canal pressures (between 80 to 130 dB SPL) were linear and stable with time. The average and standard deviation of the pressure transfer functions of scala vestibuli ($\mathbf{P}_{SV}/\mathbf{P}_{EC}$) and scala tympani ($\mathbf{P}_{ST}/\mathbf{P}_{EC}$) relative to the ear-canal pressure are plotted together in Fig. 2. The pressure in scala vestibuli is of a significantly higher magnitude than that in scala tympani for a wide range of frequencies, and the scala tympani pressure has a significant effect on the pressure difference only at low and high frequencies. For frequencies below 500 Hz, the phase of the scala tympani transfer function is slightly less than zero while the phase of scala vestibuli shows a lead of almost 1/8 cycle. Above 500 Hz, the phases of both scalae are generally similar. Similar findings were reported in cat [6] and the guinea pig [7].

3.2 Pressure Gain of the Middle Ear

The ratio of sound pressure in scala vestibuli to that of the ear canal ($\mathbf{P}_{SV}/\mathbf{P}_{EC}$) represents the pressure gain of the middle ear. Comparison between Aibara et al. [8] and the present study shows that for over most frequencies, the two studies are similar. The gain is bandpass in nature, rising at low frequencies and falling off above 3 kHz. The maximum middle-ear pressure gain is approximately 20 dB near 1 kHz.

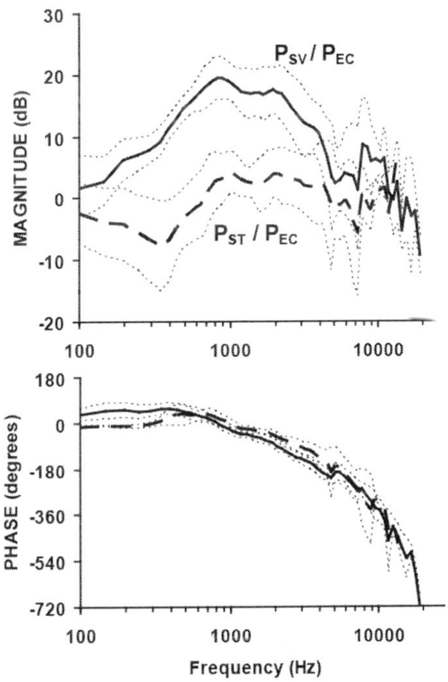

Figure 2. The mean and standard deviation of the pressure of scala vestibuli and scala tympani relative to the ear-canal pressure plotted together.

While the magnitude is bandpass, the phase falls proportionally with frequency above ~1 kHz, consistent with the middle ear having properties of a simple delay with an estimated group delay of 83 μs. This delay is more than twice the ~25–32 μs delay found in the gerbil [3], [9]. The increased delay observed in human temporal bones is likely related to inter-species differences in transduction from the ear canal to the inner ear.

3.3 Differential Pressure

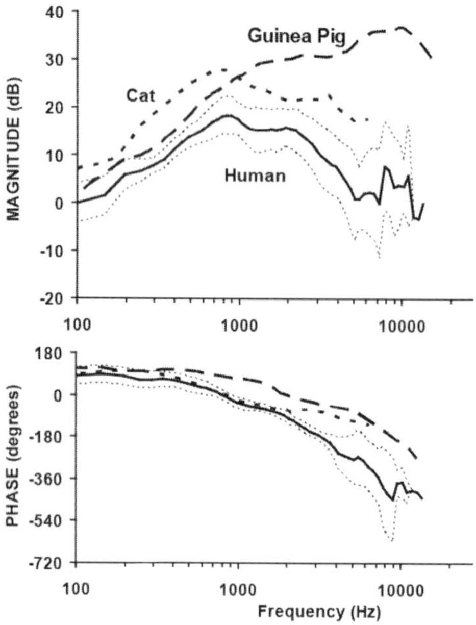

Figure 3. The differential pressure, $(\mathbf{P}_{SV}\text{-}\mathbf{P}_{ST})/\mathbf{P}_{EC}$, across the partition in humans (mean and standard deviation, n=6) is plotted with that of the cat (median, n=6, [6]) and the guinea pig (mean, n=5, [7]).

The intracochlear sound pressure difference across the organ of Corti $(\mathbf{P}_{SV}-\mathbf{P}_{ST})$ at the base of the cochlea is the input force that drives the motion of the cochlear partition, setting off the traveling wave. In the mid frequencies where $|\mathbf{P}_{SV}|>>|\mathbf{P}_{ST}|$ the differential pressure is dominated by $\mathbf{P}_{SV}$ (Fig. 2). At low and high frequencies, the differential input pressure is significantly affected by the scala tympani pressure because the latter is similar in magnitude to the scala vestibuli pressure.

The differential pressure at the cochlear base in humans (mean and standard deviation, n=6) from the present study is plotted in Fig. 3 with that of the cat (median, n=6, [6]) and the guinea pig (mean, n=5, [7]). Compared to other animals, the basal differential pressure of humans has a smaller magnitude with a high-frequency cutoff with increased phase lag at high frequencies. The cat and human data are similar, with a peak magnitude near 1 kHz but the magnitude of the pressure difference is larger in cat. The cat can hear up to ~ 60 kHz, yet the data available is limited to 7 kHz, therefore the high-frequency trend is unknown. The guinea pig differential pressure data is similar to the cat and human up to approximately 1 kHz, but continues to increase in magnitude with a peak near 10 kHz. These interspecies differences could be due to differences in the middle ear structures such as joint compliances or tympanic membranes.

3.4 Effect of Ossicular Discontinuity on Sound Transmission

The forward transduction of sound in the normal ear is assumed to be through the ear canal – tympanic membrane – ossicular chain – cochlea. To determine whether the measured intra-cochlear sound pressures are produced by ossicular transduction, we disrupted the ossicular chain. This manipulation was also a check for artifacts in our preparation, stimulus and recording system. When ossicular discontinuity occurs clinically, it results in a large (40 to 60 dB) conductive hearing loss throughout all frequencies.

Disarticulating the ossicular chain resulted in large (>20 dB) decreased sound pressure in both scala vestibuli and scala tympani for low frequencies (<5 kHz), but significant sound pressure was observed at higher frequencies in both scalae. Similar results were recorded by Puria et al. [10] in scala vestibuli in human temporal bones as well as by Nedzelnitsky [6] for both scalae in the cat. Close inspection of our data shows that this non-ossicularly conducted sound appears equal in scala vestibuli and scala tympani.

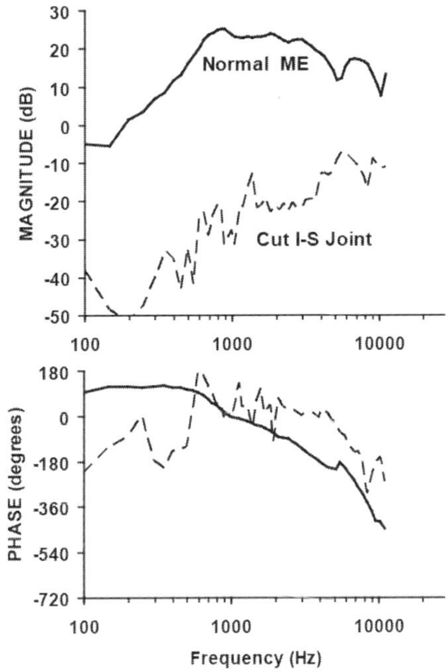

Figure 4. Effect of ossicular discontinuity on the normalized differential pressure $(\mathbf{P_{SV}}-\mathbf{P_{ST}})/\mathbf{P_{EC}}$. Disarticulation of the middle ear results in 30-50 dB decrease in the differential pressure.

Because the difference in pressure across the partition is the measure most relevant to cochlear stimulation, the effect of severing the ossicular chain on the differential input pressure was also quantified (Fig. 4). Ossicular discontinuity resulted in large (30-50 dB) decreases in the differential pressure for most frequencies, demonstrating the cancellation of similar non-ossicularly conducted sound signals in both scalae. Thus, the differential pressure measurement is a superior measure of the forward transduction of sound conducted by an intact ossicular chain. The large (30-50 dB) drop in differential pressure across a wide frequency range is consistent with the conductive hearing loss seen in patients as a result of ossicular discontinuity [11].

3.5 Cochlear Input Impedance (Z_C)

By measuring stapes velocity concurrently, the cochlear input impedance $\mathbf{Z_C}=\mathbf{P_{SV}}/\mathbf{U_{Stap}}$ was obtained (where $\mathbf{U_{Stap}}$ is the volume velocity of the stapes). $\mathbf{Z_C}$ fell between that observed by Aibara et al. [8] and Merchant et al. [12]. We also obtained the differential impedance across the partition, $\mathbf{Z_{Diff}}=(\mathbf{P_{SV}}-\mathbf{P_{ST}})/\mathbf{U_{Stap}}$, as well as the round window impedance, $\mathbf{Z_{RW}}=\mathbf{P_{ST}}/\mathbf{U_{Stap}}$, with the assumption that the stapes and round-window volume velocity are equal ($\mathbf{U_{Stap}}=\mathbf{U_{RW}}$). $\mathbf{Z_{Diff}}$ was generally resistive and $\mathbf{Z_{RW}}$ was dominated by compliance at low frequency (<300 Hz) and had distributed inertia and damping at high frequency (>500 Hz).

Acknowledgments

This work was carried out in part through the use of MIT's Microsystems Technology Laboratories and the help of Kurt Broderick. Discussions with William Peake were invaluable toward this study. We thank Diane Jones, Michaël Slama and Brian MacDonald for their generous efforts. Supported by grants R01 DC04798 and T32 DC00020 from the NIH. Wei Dong and Elizabeth Olson's time was supported by R01 DC003130. We also thank Mr. Axel Eliasen and Mr. Lakshmi Mittal for their support.

References

1. Merchant, S. N., Rosowski, J. J., 2008. Conductive hearing loss caused by third-window lesions of the inner ear. Otology & Neurotology, 29, 282-289.
2. Colletti, V., Soli, S. D., Carner, M., Colletti, L., 2006. Treatment of mixed hearing losses via implantation of a vibratory transducer on the round window. International Journal of Audiology, 45, 600-608.
3. E. S. Olson, E. S., 1999. Direct measurement of intra-cochlear pressure waves. Nature. 402, 526-529.
4. Nakajima, H. H., Ravicz, M. E., Merchant, S. N., Peake, W. T., Rosowski, J. J., 2005. Experimental ossicular fixations and the middle ear's response to sound: Evidence for a flexible ossicular chain. Hear. Res., 204, 60-77.
5. Olson, E. S., 1998. Observing middle and inner ear mechanics with novel intracochlear pressure sensors. J. Acoust. Soc. Am. 103, 3445-3463.
6. Nedzelnitsky, V., 1980. Sound pressure in the basal turn of the cat cochlea. J. Acoust. Soc. Am. 68, 1676-1680.
7. Dancer, A., Franke, R., 1980. Intracochelar sound pressure measurements in guinea pigs. Hear. Res. 2, 191-205.
8. Aibara, R., Welsh, J. T., Puria, S., Goode, R. L., 2001. Human middle-ear sound transfer function and cochlear input impedance. Hear. Res. 152, 100-109.
9. Dong, W., Olson, E. S., 2006. Middle ear forward and reverse transfer function. J. Neurophysiol. 95, 2951-2961.
10. Puria, S., Allen, J. B., 1998. Measurements and model of the cat middle ear: Evidence of tympanic membrane acoustic delay. J. Acoust. Soc. Am. 104, 3463-3481.
11. Merchant, S. N., Rososwski, J. J., 2003. Auditory physiology (middle-ear mechanics). In: Gulya, A.J., Glasscock, M.E. (Eds.), Surgery of the Ear. Hamilton Ontario, B.C. Decker, pp. 59-82.
12. Merchant, S. N.,. Ravicz, M. E, Rosowski, J. J., 1996. Acoustic input impedance of the stapes and cochlea in human temporal bones. Hear. Res. 97, 30-45.

COMPARING OTOACOUSTIC EMISSIONS AND BASILAR MEMBRANE MOTION IN INDIVIDUAL EARS

NIGEL P. COOPER

School of Life Sciences, Keele University, Keele, Staffordshire, ST5 5BG, UK

CHRISTOPHER A. SHERA

Eaton-Peabody Laboratory, Mass. Eye & Ear Infirmary, Harvard Medical School, 243 Charles St., Boston MA02401, USA

1 Summary

In order to distinguish between theories relating to the origin and existence of backward-traveling waves in the cochlea, basilar-membrane (BM) vibration patterns and stimulus-frequency and distortion-product otoacoustic emissions (SF and DP OAE) were recorded from individual, deeply-anesthetized guinea pigs. Methodological details are summarized

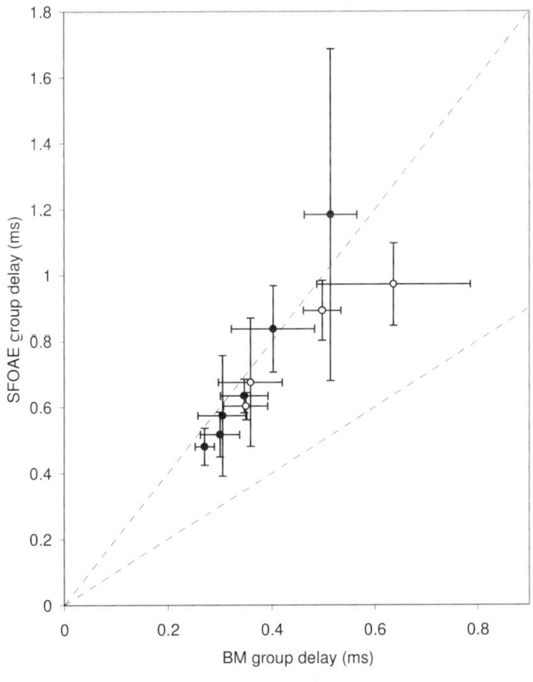

Figure 1. Basilar membrane (BM) and stimulus-frequency otoacoustic emission (SFOAE) group delays in 10 guinea-pig ears. SFOAE group delays were averaged across 1-2 kHz windows centered either at (filled symbols) or just below (open symbols) the characteristic frequency (CF) of each preparation's BM. BM group delays were averaged across 2 kHz windows centered on the BM's CF. Error bars indicate the standard deviations of the group delays within each window. Lower and upper dashed lines indicate theoretically predicted group delay (SFOAE/BM) ratios of 1 and 2, respectively.

elsewhere [1, 2]. Mechanical group delays associated with forward-traveling waves on the BM (at the BM site's characteristic-frequency or CF) varied from ~0.3 ms in the 30 kHz region of the cochlea to ~0.6 ms in the 14 kHz region. Group delays measured at corresponding frequencies from SF and DP OAE's (at f_1 and $2f_2$-f_1, respectively, where $f_2=f_1+100$ Hz, with $L_1=40$ dB SPL and $L_2=55$ dB SPL) varied from ~0.5 ms (at ~30 kHz) to ~1.2 ms (at ~15 kHz). Thus, within the estimated uncertainty of the measurements, the OAE group delays were roughly twice as long as the mechanical group delays measured on the BM at its CF (mean ratio = 1.85 +/-0.21 std. dev., n=10). These data are consistent with theories suggesting that SFOAEs and other reflection-source OAEs arise primarily via the coherent reflection of energy, in the form of backward-traveling waves, from regions of the BM where forward-traveling waves reach their peak amplitudes (e.g. [3]).

Analyses of BM responses to the two-tone stimuli used to evoke SF and DP OAEs revealed BM OAE-analogues that showed little accumulation of phase, relative to the f_1 primary, within ~0.25 octaves of CF. As f_1 and f_2 decreased further below CF, however, the BM OAE-analogues accumulated phase much more rapidly than the primary responses. These results are also consistent with the existence of waves which travel backwards, along the BM, from more apical regions of the cochlea. Similar results were observed using a 1-dimensional transmission line model of the cochlea's mechanics, in which the *only* mechanism available for OAE propagation was a backward-traveling wave.

Acknowledgments

Supported by Deafness Research UK, the MRC (UK), and the National Institutes of Health (USA).

References

1. Cooper, N.P., 1999. An improved heterodyne laser interferometer for use in studies of cochlear mechanics. J. Neurosci. Meth. 88(1), 93-102.
2. Shera, C.A., Guinan, J.J. Jr., 1999. Evoked otoacoustic emissions arise by two fundamentally different mechanisms: a taxonomy for mammalian OAEs. J. Acoust. Soc. Am. 105(2), 782-798
3. Zweig, G., Shera, C.A., 1995. The origin of periodicity in the spectrum of evoked otoacoustic emissions. J. Acoust. Soc. Am. 98(4), 2018-2047.

THE ROLE OF COMPRESSION AND TRAVELING WAVE PRESSURES IN THE TRANSMISSION OF SOUND OUT OF THE GERBIL COCHLEA

WEI DONG, ELIZABETH S. OLSON

Department of Otolaryngology / Head and Neck Surgery, Columbia University, New York, NY, 10032, USA

Cochlear emissions provide a noninvasive probe of cochlear mechanics, but their utility is hindered by incomplete understanding of their relationship to intracochlear activity. In particular, recent work has uncovered a question about the mode by which emissions travel out of the cochlea -- whether they emerge via a "fast" compression pressure or a "slow" traveling-wave pressure. We further probed this question with simultaneous measurements of intracochlear distortion products (DPs) at two well-separated locations and DP oto-acoustic emissions (DPOAEs). In the broad frequency range of the local best frequency (BF), the DP responses demonstrate the now well-known forward-traveling-wave character. However, at frequencies substantially lower than the BF, comparisons of both DPOAEs to DPs and of DPs at two locations support a reverse-traveling-wave. Finally, a compression pressure DP was observed when stimulating at high levels (90 dB) with frequencies that were well above the BF. Therefore, the compression / reverse-traveling-wave question appears to be a quantitative question of the relative size of these different pressure modes. In previous and present results we find that the reverse-traveling-wave mode can be dominant both within the cochlea and in the production of DPOAEs.

1 Introduction

In forward transmission, two modes - compression and traveling-wave - in the cochlea's response to stapes vibration were predicted theoretically over fifty years ago. The "fast" compression mode is supported by fluid inertia and compressibility (a sound wave); the "slow" traveling-wave mode is based on fluid inertia and partition stiffness [1]. Both modes are evident in intracochlear pressure responses to tone stimulation and can be decoupled by their particular characteristics as illustrated in Fig. 1 [2, 3]. The compression pressure, caused by the stapes vibration, almost instantaneously fills the cochlea and it is nearly spatially invariant. It grows linearly with sound pressure level, is in phase with stapes motion and dominates the pressure at frequencies above the local cutoff frequency (gray region in Fig. 1). The traveling-wave, caused by the pressure difference between the scala vestibuli (SV) and scala tympani (ST), travels much more slowly along the cochlear partition, is modified by the cochlear amplifier and peaks at a specific place (known as the BF place). The traveling-wave can lag the stapes motion by up to several cycles. It shows significant spatial variation around the BF (white region in Fig. 1); spatial pressure variations indicate fluid motion, and are substantial at locations close to the basilar membrane (BM) when BM motion is large. Therefore, the traveling-wave is especially pronounced at frequencies around the BF and close to the BM. Interference between the traveling-wave and compression pressure is apparent in the magnitude notches and corresponding phase steps (arrowheads in Fig. 1). At frequencies

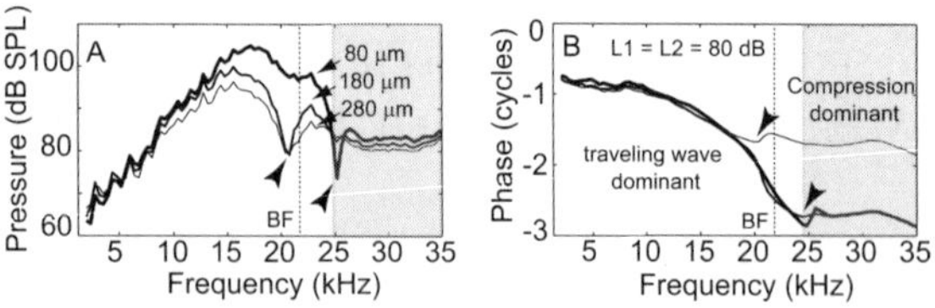

Figure 1. Spatial variation of primary f_2. (A) Magnitude of pressure. (B) Phase referenced to EC f_1 & f_2 (measured at 3 distances from the BM in turn-one with 22 kHz BF, wg131).

well below the BF, the small spatial variations in pressure are expected because the BM motion is small and the wavelength is long [4].

In contrast to this fairly coherent description of the basic intracochlear response to stapes stimulation, the manner in which sound travels out of cochlea is under active experimental and theoretical exploration. With regard to intracochlear DPs and their corresponding emissions, one group measured DPs at two longitudinal locations on the BM and observed a forward-traveling-wave, but no reverse-wave. Their interpretation of these findings was that the DPs, once generated, excited the stapes ~ instantaneously as a compression pressure [5, 6]. On the other side of the debate, detailed intracochlear DP pressure measurements gave evidence for a reverse-traveling-wave and set an upper limit to the contribution of the compression pressure to the emissions [7]. In the present study, the traveling-wave and compression pressure modes of DPs are investigated further: We find that both the traveling-wave and compression pressure are evident in intracochlear DP responses to two tone stimuli. The traveling-wave pressure dominates at frequencies around the BF, while the compression pressure emerges from the noise at frequencies well above the cutoff frequency with high intensity stimulation. However, the results support our previous conclusion that the major contributor to the emission is via a reverse-traveling-wave.

2 Methods

DPOAEs and intracochlear pressure DPs were simultaneously recorded in the ear canal (EC) and two positions in the basal ST (turn-one and very-basal) in deeply anesthetized young adult gerbil, using techniques and analyses described previously [3, 7]. Two equal - intensity primaries were used with fixed f_2/f_1 ratio, with f_1 & f_2 frequency swept from low to high. One micro-pressure sensor (turn-one) was positioned at a place with the BF ~ 20 kHz through a hand-drilled hole, and the 2nd sensor (very-basal) was positioned at a place with BF ~ 40 kHz through the round window opening after removing a portion of the round window membrane.

3 Results

In the following we show data found with f_2/f_1 ratio fixed at 1.05. With this low ratio, the results at $2f_1$-f_2 and $2f_2$-f_1 were similar on the points discussed below. We use $2f_2$-f_1 in the figures because the points are illustrated more clearly with this DP.

3.1 Traveling-wave is dominant at frequencies around the BF

DP pressure responses showed similar tuning and phase as the primaries [3, 7] when measured close to the BM at frequencies fairly close to the local BF (22 kHz) (thick line in Fig. 2A & B). Similar to its primary responses shown in Fig. 1, at frequencies around the BF, the DP exhibited significant spatial variations as the sensor moved away from the BM (Fig. 2A). The DP phase showed rapid phase accumulation even at location / frequency combinations for which the f_2 pressure was dominated by the compression pressure (gray in Fig. 2B and [3, 7]). Both the DP spatial variation and phase accumulation indicate that the DP pressure was dominated by the traveling-wave. Thus the compression pressure, which would exhibit almost no spatial variation and ~ invariant phase versus frequency response, was never big enough to dominate the DP responses at frequencies around the BF.

At frequencies below 14 kHz (arrowheads in Fig. 2), the DP magnitude was less smooth. Its phase varied more rapidly with frequency than the primary phase, indicating a longer delay time. These observations are consistent with previous observations [7] and accompanying interpretation that these DPs were likely generated at more apical positions and were detected on their way out. The lack of pressure spatial variation is as expected for a compression pressure but is also consistent with a traveling-wave mode, which theoretically does not change much at frequencies significantly lower than BF [4, 8].

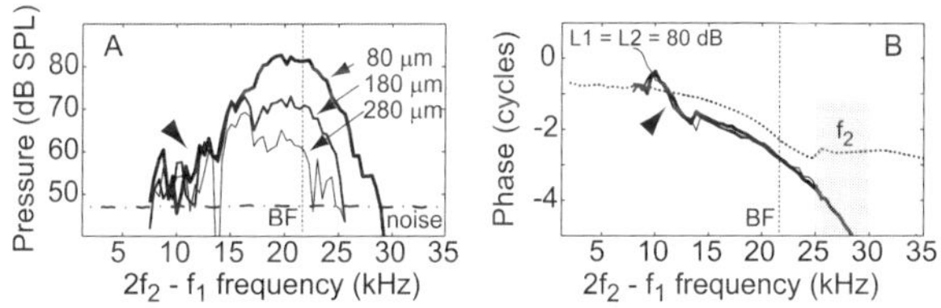

Figure 2. Spatial variation of $2f_2 - f_1$ DP. (A) Magnitude of pressure. (B) Phase referenced to EC f_1 & f_2. Dotted line is the f_2 primary phase (measured at 3 distances from the BM in turn-one with 22 kHz BF, wg131).

3.2 Compression pressure is apparent at frequencies well above the cutoff frequency

In the turn-one location, the DP was tuned to frequencies around the 20 kHz BF. The tuning became broader and shifted towards lower frequencies as the primary level was increased from 70 to 90 dB SPL (white region of Fig. 3A), similar to the known level-dependent tuning of single tone and primary responses. Its phase lined up with the single tone phase and changed little with the intensity of the primaries (white region of Fig. 3B). As in Fig. 2, both the tuning and phase suggested that the DP was dominated by either forward-traveling or locally generated distortion.

However, with 90 dB stimulation the DP response could be recorded up to 40 kHz, above the local cutoff frequency (gray region in Fig. 3). Interestingly, the phase in the high frequency region was similar to the single tone phase plateau, thus the high frequency response appears to be dominated by the compression pressure. In our

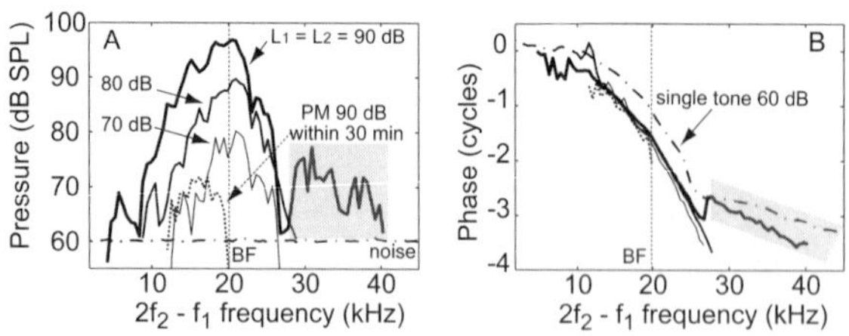

Figure 3. Level-dependence and physiological vulnerability of $2f_2 - f_1$ DP. (A) Magnitude of pressure. (B) Phase referenced to EC f_1 & f_2. Dot-dashed line is the single tone phase. Postmortem responses is plotted in dotted lines (sensor positioned at turn-one 10 μm from the BM, wg121).

interpretation of the observation, the supra-cutoff-frequency DP was generated in the overlapping region of the traveling-wave f_1 & f_2, basal to our turn-one observation point, and what we measured was the compression pressure component of this basally generated DP. Thirty minutes post-mortem the DP was substantially reduced, only measurable at frequencies from 12 to 20 kHz with 90 dB stimulation, illustrating that the DP was physiologically vulnerable and not an artifact.

3.3 Transmission of DPs inside the cochlea

The DPs shown in Fig. 3 were measured simultaneously, with the 1st sensor in the turn-one location (~20 kHz BF) and the 2nd sensor positioned at the very-basal (BF ~ 40 kHz) site. This simultaneous two-longitudinal-location measurement allowed us to directly explore the transmission of DPs in the cochlea. The 90 dB responses in Fig.3 are replotted in Fig. 4A & B together with the simultaneous measurement at the very-basal site. Only 90 dB SPL data are shown because the DPs were above the noise in a wide frequency range. When measured in the extreme base, the DPs were tuned to the local BF (Fig. 4C), similar to their behavior when measured in turn-one (Fig. 4A). In turn-one, the phase was similar to the single tone pressure response phase at frequencies up to

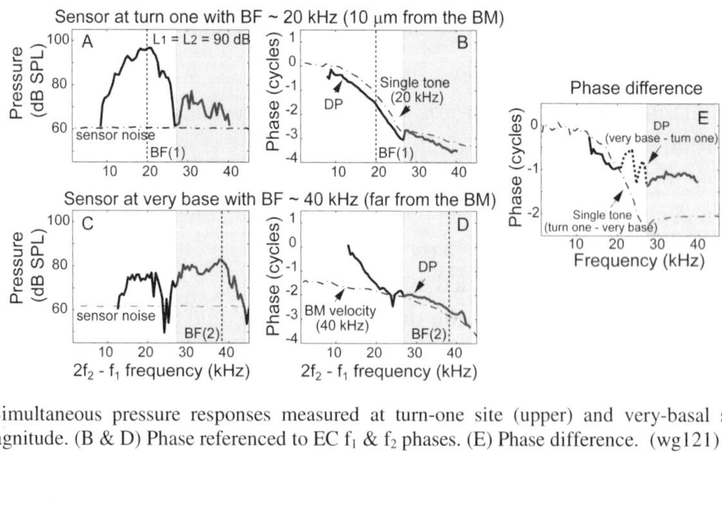

Figure 4. Simultaneous pressure responses measured at turn-one site (upper) and very-basal site (lower). (A & C) Magnitude. (B & D) Phase referenced to EC f_1 & f_2 phases. (E) Phase difference. (wg121)

27 kHz (Fig. 3B & 4B). The measurement at the very-basal location was not made very close to the BM and thus the single tone phase did not show a forward-traveling-wave to use for comparison (e.g., see spatial phase variations in Fig. 1 and [2]). However, a reasonable comparison phase can be found in the BM velocity phase measured with a Polytec vibrometer (different animal) and the DP phase was found to be similar (Fig. 4D). Thus, at frequencies around their BF the DPs at both locations were predominantly locally generated and/or forward-traveling. At the basal location (Fig. 4C) DPs in the region of the BF and DPs at lower frequencies were separated by a sharp notch at ~ 24 kHz, suggesting destructive interference between two modes – possibly locally generated and reverse-wave, or something different.

Comparison of DPs measured at the two longitudinal locations:

(i) DPs ~ 30 - 40 kHz (gray regions of Fig. 4). The DP extended to frequencies well above its cutoff frequency when measured in turn-one (Fig. 4A) and the phase was plateau-like (Fig. 4B), indicating that it was a compression pressure. This is supported by the fact these DPs were almost in phase with their counterparts measured at the basal location; this can be seen in the gray region of Fig. 4E, where the very-basal phase is referenced to the turn-one phase.

(ii) DPs ~ 12 - 20 kHz (within white regions of Fig. 4). As noted just above, at the turn-one location, these DPs appeared to be locally generated/forward-traveling (Fig. 4A). Measured at the basal location, the same DPs were smaller by ~ 20 dB (Fig. 4C) and the DP phase changed rapidly with frequency (Fig. 4D), departing from that of the forward-traveling phase (here indicated by the BM velocity phase). This is consistent with the picture that these DPs were measured on their way out of the cochlea by the very-basal sensor. (The rapid phase-frequency response is consistent with a "reflector" type emission [9]. The phase comparison of the two locations (basal – turn-one) in Fig. 4E was similar to that of the forward-traveling-wave phase in the 12 – 20 kHz frequency region, strongly supporting the dominance of the reverse-traveling-wave in transporting the signal from turn-one (near its generation point) to the more basal position.

(iii) DPs between 20 - 30 kHz. The phase of the DP at the very basal site is making a transition between steep (reflected) and primary-like (locally generated). The very-basal DP cannot be expected to be related to the turn-one DP in that case, and when they are related (dotted line in Fig. 4E) the result is not meaningful.

4 Discussion

In order to quantify the contribution of reverse-traveling-wave and compression pressure to the emissions, intracochlear pressure at DP frequencies was simultaneously measured at two longitudinal locations in the basal ST of the gerbil cochlea: one was at a turn-one place with BF of ~20 kHz, the other at a very-basal site with BF of ~40 kHz. The intracochlear DP pressure was shaped by cochlear filtering and tuned to the BF of its own place. In a wide frequency region near the BF, the DP appeared to be locally generated or forward-traveling [7]. With intense primary level, the DP could be measured at frequencies well above the local BF. In this case, the DP had the plateau-like phase expected for a compression pressure. In another data set, these high frequency / high level DPs did not vary with distance from the BM, also consistent with a compression pressure. As stated above, the DPs measured at high primary level are physiologically vulnerable, and thus are not due to system or middle ear distortion.

The micro-pressure-sensor measures the pressure in the cochlea. As illustrated in Fig. 5, the character of the measured DP is expected to depend on the relative positions of the f_1& f_2 overlapping region and the sensor location. *At frequencies well below the BF*, the DP was often variable in amplitude with a phase that departed from the forward-traveling-wave phase, consistent with it being generated apical to the measurement location and on its way out of the cochlea when measured. In support of this, the comparison of DPs measured at different cochlear locations (as shown in Fig. 4E), or of DPs to DPOAEs (as in [7]) indicates the existence of a reverse-traveling-wave. *At frequencies around the BF*, the local DP appeared to be locally generated / forward-traveling and it makes sense that, due to its forward-traveling character, this region cannot generally be examined for reverse-traveling-wave [7]. *At frequencies well above the local cutoff frequency*, a compression pressure DP was evident as shown in Figs. 3 and 4 (A & B). The high frequency DP was likely generated at a position basal to the turn-one sensor and the reasonable path by which this DP reached the sensor is through the fluid as a compression pressure. This DP might have arisen because at the high sound pressure level there was substantial nonlinear generation in a broad region where there was not much longitudinal phase change. Then the distortion would be in phase over a long region and be effective in generating a sizeable compression pressure. Thus, our measurements of intracochlear pressure clearly demonstrate the existence of compression pressure DPs.

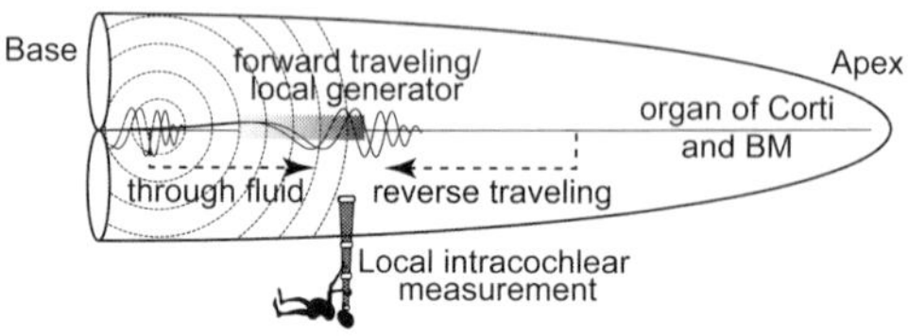

Figure 5. DP transmissions in the cochlea.

Nevertheless, the two-site intracochlear pressure measurement indicated that the reverse-traveling-wave was dominant in transporting DPs basal-ward. This result conflicts with the two-site BM velocity measurements of Ren and colleagues [5, 6]. In our measurements, the separation of the two sites (~ 1.44 mm [10]) was longer than the two-site BM velocity measurement, where the two sites were separated by 0.45 mm [6]. With the larger separation, the more basal site was not within the generation region of the < 20 kHz DPs, and this absence of locally generated distortion made it possible to detect the reverse-wave.

Acknowledgments

Work supported by grant NIH R03DC009080.

References

1. Peterson, L.C., Bogert, B.P., 1950. A dynamical theory of the cochlea. J. Acoust. Soc. Am. 22, 369-381.
2. Olson, E.S., 1998. Observing middle and inner ear mechanics with novel intracochlear pressure sensors. J. Acoust. Soc. Am. 103, 3445-63.
3. Dong, W., Olson, E.S., 2005. Two-tone distortion in intracochlear pressure. J. Acoust. Soc. Am. 117, 2999-3015.
4. Taber, L.A., Steele, C.R., 1981. Cochlear model including three-dimensional fluid and four modes of partition flexibility. J. Acoust. Soc. Am. 70, 426-36.
5. Ren, T., 2004. Reverse propagation of sound in the gerbil cochlea. Nat. Neurosci. 7, 333-4.
6. He, W., Nuttall, A.L., Ren, T., 2007. Two-tone distortion at different longitudinal locations on the basilar membrane. Hear. Res. 228, 112-22.
7. Dong, W., Olson, E.S., 2008. Supporting evidence for reverse cochlear traveling waves. J. Acoust. Soc. Am. 123, 222-40.
8. Yoon, Y.J., Puria, S., Steele, C.R., 2007. Intracochlear pressure and derived quantities from a three-dimensional model. J. Acoust. Soc. Am. 122, 952-66.
9. Shera, C.A., Guinan, J.J., Jr., 1999. Evoked otoacoustic emissions arise by two fundamentally different mechanisms: a taxonomy for mammalian OAEs. J. Acoust. Soc. Am. 105, 782-98.
10. Muller, M., 1996. The cochlear place-frequency map of the adult and developing Mongolian gerbil. Hear. Res. 94, 148-56.

Comments & Discussion

Puria: Thank you for the presentation and analysis of your great measurements. You did a nice job analyzing the measurements in the different frequency regions. However, you did not discuss your results below about 10 kHz. Here we see that the total pressure is below the compression pressure and decreases with frequency. This is unexpected but also seen in our 3D cochlear model calculations of pressure (Yoon et al. 2007). There at least two different explanations. First is that the fast wave and slow wave pressures are approximately equal and opposite and thus tend to cancel. The second explanation is that the low impedance boundary condition imposes a zero pressure condition.

Olson: The comment concerns the response at the primary frequency shown in Fig. 1. The explanation is valid - the very flexible round window is expected to impose a boundary condition where the fluid pressure must be equal to atmospheric pressure ("zero"). For this to happen in a model with two pressure modes (fast/compressional and slow/traveling), the fast and slow wave pressure modes must be equal and opposite at the round window. The measurement is made a bit downstream from the round window, but still close enough to be influenced by the boundary condition. In addition to the Yoon et al. paper, Peterson and Bogert (1950, J. Acoust. Soc. Am. 22) is a good reference for these ideas.

OBVIOUS AND 'HIDDEN' WAVES IN THE COCHLEA

EGBERT DE BOER
Room D2-226, Academic Medical Centre, University of Amsterdam, Meibergdreef 9, 1105 AZ, Amsterdam, The Netherlands

ALFRED L. NUTTALL
Oregon Hearing Research Centre, NRC04, Oregon Health & Science University, 3181 SW Sam Jackson Park Road, Portland, Oregon 97239-3098, USA; Kresge Hearing Research Institute, University of Michigan, 1301 E. Ann Street, Ann Arbor, Michigan 48109-0506, USA

We have tried to shed more light on the amplification process in the cochlea by measuring properties of 'hidden waves', i.e. waves produced by the assembly of OHC-generated pressure sources and not by stimulation at the stapes. From experiments we have confirmed that hidden waves have properties that cannot always be explained by a 'classical' model of the cochlea – a model in which BM dynamics is described by a point-impedance function. Therefore, we have considered a non-classical model with feed-forward, where excitation of each OHC comes from a somewhat more basal location.

1 Introduction – 'hidden waves'

We consider 'hidden waves' to be waves in a mechanical model of the cochlea that are not directly elicited by imposed signals at the stapes. There are at least three categories:

a) For pure-tone stimulation, the elements responsible for amplification (assumed to be Outer Hair Cells, OHCs) generate a hidden wave of their own. It has been shown in [1] that including that wave in the wave produced by stapes excitation in the dead cochlea accurately reproduces the total wave in the live and active cochlea.
b) When two tones are presented, nonlinear distortion in the OHCs gives rise to the production of Distortion Products (DPs). The distortion components in the output of the OHCs produce a hidden wave starting in the region where the two excitation patterns overlap. One component of that wave travels toward the stapes; another component travels towards the helicotrema and produces a local maximum at the location corresponding to its frequency (this applies to lower-sideband DPs). Experiments tell a different story: much of the former DP wave is observed to travel in the forward direction, towards the helicotrema (e.g. [2], [3], [4]) instead of the other way around.
c) A third type of hidden wave can be distinguished: the compression wave. If we assume the fluid in the model to be ideal (incompressible, inviscid and linear), the pressure associated with this wave is the same everywhere. That wave turns into a more realistic wave with a finite propagation speed once the fluid is assumed to be compressible.

The difference between theory and experiment mentioned under b) has been attributed to the compression wave (e.g., in [2] and [3]). In some more detail: the two primary tones produce local DPs, a compression wave arises, that wave reaches the stapes and creates there a travelling wave, still with the DP frequency, which travels in the direction of the helicotrema. That travelling DP wave would then be the forward wave that is observed in experiments. This notion has been severely challenged (e.g. in [5]). In particular, the amplitude of the so-induced DP wave would not decrease enough when the frequency ratio approaches 1.0 (as has repeatedly been observed in DP Otoacoustic Emissions, DPOAEs).

In a recent study ([6]) the DP travelling wave resulting from a compression wave has been discussed. A brief account of the underlying computation follows. First, the individual components of the DP pressure generated by all OHCs are computed. Next, assuming an ideal fluid, the sum (integral) of these pressure components is determined. This is the compression-wave pressure which instantly invades all of the fluid and all soft tissues in the model. At the stapes end the compression wave finds two membranes with different mechanical impedances. These two membranes will vibrate with different amplitudes and will thus set up a "normal" travelling wave. The result of the computation was that the forward wave induced by the compression wave does not dominate the DP wave, at least in the 'classical' model computed according to [1]. Therefore, the compression-wave theory does not apply. We are left with an enigma: why does the cochlear DP wave almost entirely consist of a forward wave, and why is there so little evidence about its reverse-travelling component?

2 Classical and non-classical models

The computations described in our earlier publications were all executed in a "classical" cochlear model. In such a model the mechanics of the organ of Corti is lumped into the concept of a mechanical impedance that has the character of a point-impedance. In more accessible language this means that the local velocity of the basilar membrane (BM) only depends on the *local* pressure difference across it, and not on influences from elsewhere, from other locations. In view of the results reported above it seems that we have to consider non-classical models ([7]) to describe and explain the data. It is difficult to generalize, hence we will mention a few types of such models. The first type is the multi-compartment model, of which we know two examples: [8], and [9] (the sandwich model). Only a limited amount of general theorems are available for this type of model ([10]). The second type is the travelling-wave amplifier, of this type only one example is cited in the literature ([11]), and little or no general theory is available. The third type is the model with feed-forward and/or feed-backward. This type of model has been described and analyzed a few times ([12], [13], [14]).

Of the multi-compartment models we first briefly consider the sandwich model. As described elsewhere ([10]), that model has two fundamentally distinct modes of wave propagation. Unfortunately, we have not (yet) been able to realize in the sandwich model a combination of these modes that would explain a forward DP wave in the two-tone case.

This leaves us with the feed-forward model and its variants. It has been shown that this type of model has quite unusual properties with respect to reverse and forward waves [15]. In particular, when the feed-forward mechanism is adjusted to provide *amplification* for forward waves, it will show *attenuation* for reverse waves. The same is true for feed-backward. A combination of these two models would be advantageous because each of those, when considered by itself, tends to produce deviant phase shifts that are difficult to reconcile with data (see further below).

3 An example: Distortion-Product (DP) waves in a feed-forward model

Analysis of the feed-forward model follows the same pattern as that of the classical model, with one major difference. In the classical model the basic equation to handle contains two matrices, the Green's function matrix and the impedance matrix, plus a vector, called the "stapes propagator". The Green's function matrix describes the hydrodynamics of the fluid; the impedance matrix is a diagonal matrix of which the main diagonal contains the BM impedance function Z_{BM} – as the descriptor of BM mechanics. The same formulation is used for the inverse solution and for resynthesis.

In a feed-forward model the OHC contribution from the BM velocity at location x results in an added pressure at location $(x+\Delta x)$. In that model the impedance matrix contains a transfer impedance function X_{ffw} in a sub-diagonal matrix $\mathbf{X}_{\mathrm{ffw}}$, that sub-diagonal is distant from the main diagonal by a number of steps that corresponds to Δx. When $\mathbf{X}_{\mathrm{ffw}}$ contains elements below the main diagonal, we have *feed-forward*, when it contains elements above the main diagonal, we have *feed-backward*; any combination of these two is also possible. Provided the correct indexing of that sub-diagonal is used, the inverse solution as well as the resynthesis of the model response proceed in the normal manner. In passing, it should be noted that the 'passive' BM impedance component is excluded from the feed-forward procedure, it must be treated in the normal manner and placed in the main diagonal of the matrix $\mathbf{X}_{\mathrm{ffw}}$. The result of resynthesis is, in general, excellent, the resynthesized response accurately matches the original response, in amplitude as well as in phase. Note that, while the feed-forward model has a tendency to produce excessive phase shifts [12, 13], that tendency is absent when the transfer function X_{ffw} is derived from a measured response.

Figure 1 shows representative results, derived from responses of four animals. In each case a feed-forward model has been constructed fitting the experimental data, and it is stimulated with two tones with frequencies f_1 and f_2. Thin dashed curves indicate the amplitudes of the cochlear patterns evoked by the two tones. Shading is applied as a form of embossment. The (hidden) DP wave with frequency $2f_1-f_2$ is computed according to the process outlined above (based on the generation scheme outlined in [1]). Amplitude and phase of the DP wave are drawn to stick out from the figures. For

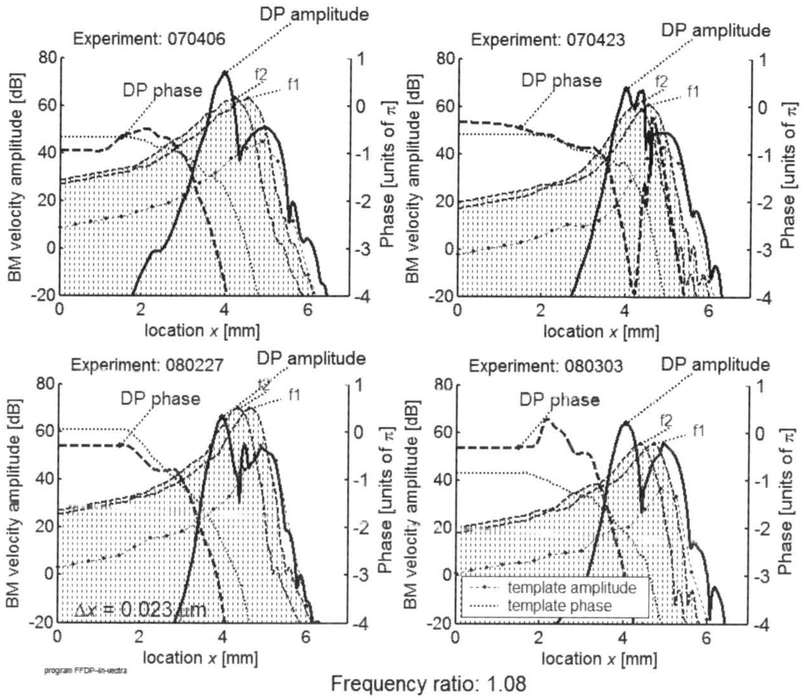

Figure 1. Distortion-Product (DP) waves generated in a stylized three-dimensional model with feed-forward. Length of the model: 12 mm, divided into 1024 segments. Parameters of model are derived from four experiments. Two tones of 80 dB SPL with frequencies f_1 and f_2 are used as stimuli. The ensuing amplitude excitation patterns are outlined by thin dashed curves (labelled: 'f1' and 'f2'), not drawn to actual level, and suggested by shading. Source of the DP wave with frequency $2f_1 - f_2$ estimated as described in [1]. Computation of the ensuing DP wave with the modified off-diagonal matrix method as described in the text. Reflection of the DP wave at the stapes suppressed. Solid thick curve: amplitude of the DP wave. Over-all amplitude level has been normalized for best visibility. Ordinate scale on the left. Coarsely dashed curve: the associated phase response. Curves have been displaced by multiples of 2π radians for optimal visibility. Ordinate scale on the right. For comparison two 'template' curves are added. The 'template' is the response to a fictitious low-level tone with the DP frequency starting at the stapes.

comparison two more finely dotted curves show amplitude and phase of the ('template') pattern produced by a fictitious single tone with the DP frequency (with arbitrary amplitude and zero phase) presented at the stapes. See [1] for more details on the layout of the figure, the genesis of distortion products and suppression of reflection at the stapes. Pay particular attention to the curve labelled 'DP phase'. The slope of that curve is mainly negative to the left of the region where the f1 and f2 patterns overlap. This means that in that region a *forward* DP wave is generated. Note, in addition, that in the same region the amplitude of the generated DP wave – 'DP amplitude' – is generally smaller than that of the stapes-generated 'template' wave – which was drawn with approximately the same maximum amplitude; this is in agreement with the theoretical prediction in [15].

We should note at this point that the forward wave in that region is definitely not due to imperfect reflection of a – very weak – reverse wave at the stapes. This view is further confirmed by the finding that removing the condition of minimal reflection at the stapes brings about only minor variations in the DP response.

It should be clear that the dominance of the forward wave in this type of model would provide a good explanation for the observations on the (forward) direction of DP waves [2, 3, 4]. It should be remembered, though, that in each figure panel the two phase curves drawn, that of the generated DP (coarsely dashed) and the comparison curve (finely dotted), are not parallel; the data ([4] and later extensions of that study) suggest that the phase courses should be almost parallel over most of their courses.

In addition, there exists a fundamental problem with, for instance, the feed-forward model. Suppose the model is adjusted to have amplification for forward waves; it will then show attenuation of reverse waves. What, then, about coherent reflection, and near-periodicity of spontaneous otoacoustic emissions (SOAEs)? The basic theory of coherent reflection [16] is well established, and it has been demonstrated that the principle works well to describe the distribution of spurious peaks in cochlear responses [17]. If reverse waves are severely attenuated so that only a forward DP wave remains, local coherent reflection would still be possible but it could not lead to peaks in the resulting response, let alone to cause spontaneous oscillation. The wave reflected from the stapes would be too weak. That means we have to search among the possible models one that does attenuate reverse waves but not too much, a kind of "combi" model in which a fraction of 'activity' is carried out by feed-forward (or a combination of feed-forward and feed-backward). Further study has to reveal whether such a model is feasible.

Acknowledgments

All of the model computations used in our work are directly based on data from measurements. The authors gratefully acknowledge Jiefu Zheng for his invaluable contributions to the experiments and Edward Porsov for the development of the new software and his contributions to many discussions. The authors also want to thank Renato Nobili and Fabio Mammano who kindly put their Green's function program code at our disposal. This study received support from NIH NIDCD, R01 DC 00141.

References

1. de Boer, E., Nuttall, A.L., Shera, C.A., 2007. Wave propagation patterns in a classical three-dimensional model of the cochlea. J. Acoust. Soc. Am. 121, 352-362.
2. He, W., Nuttall, A.L., 2006. Backward propagation of otoacoustic emissions in the cochlea. In: Nuttall, A.L., Ren, T., Gillespie, P., Grosh, P., de Boer, E. (Eds.), Auditory Mechanisms: Processes and Models. World Scientific, Singapore, pp. 79-85.
3. He, W., Fridberger, A., Porsov, E., Grosh, K., Ren, T., 2008. Reverse wave propagation in the cochlea. Proc. Natl. Acad. Sci. USA 105, 2729-2733.

4. de Boer, E., Zheng, J., Porsov, E., Nuttall, A.L., 2008. Inverted Direction of Wave Propagation (IDWP) in the Cochlea. J. Acoust. Soc. Am. 123, 1513-1521.
5. Shera, C.A., Tubis, A., Talmadge, C.L., de Boer, E., Fahey, P.A., Guinan, J.J., 2007. Allen–Fahey and related experiments support the predominance of cochlear slow-wave otoacoustic emissions. J. Acoust. Soc. Amer. 121, 1564-1575.
6. de Boer, E. (----) Inverse solution method for a class of non-classical cochlear models. J. Acoust. Soc. Am., submitted.
7. de Boer, E., 1997. Classical and non-classical models of the cochlea. J. Acoust. Soc. Am. 101, 2148-2150.
8. Chadwick, R.S., Dimitriades, E.K., Iwasa, K.H., 1996. Active control of waves in a cochlear model with subpartitions. Proc. Natl. Acad. Sci. USA 93, 2564-2569.
9. Hubbard, A.E., Yang, Z., Shatz, L., Mountain, D.C., 2000. Multi-mode cochlear models. In: Wada, H., Takasaka, T., Ikeda, K., Ohyama, K., Koike, T. (Eds.), Recent developments in auditory mechanics. World Scientific, Singapore, pp. 167-173.
10. de Boer, E., 1990. Wave-propagation modes and boundary conditions for the Ulfendahl-Flock-Khanna (UFK) preparation. In: Dallos, P., Geisler, C.D., Matthews, J.W., Ruggero, M.A., Steele, C.R. (Eds.), Mechanics and Biophysics of Hearing, Springer, Berlin, pp. 333-339.
11. Hubbard, A.E., 1993. A traveling wave-amplifier model of the cochlea. Science 259, 68-71.
12. Steele, C.R., Baker, G., Tolomeo, J., Zetes, D., 1993. Electro-mechanical models of the outer hair cell. In: Duifhuis, H., Horst, J.W., van Dijk, P., van Netten, S.M. (Eds.), Biophysics of Hair-Cell Sensory Systems, World Scientific, Singapore, pp. 207-214.
13. Geisler, C.D., Sang, C., 1995. A cochlear model using feed-forward outer-hair-cell forces. Hear. Res. 86, 132-146.
14. Wen, B., Boahen, K., 2003. A linear cochlear model with active bi-directional coupling. In: Proceedings of the 25th Annual International Conference of the IEEE Engineering in Medicine and Biology Society, Vol. 3, Cancun, Mexico, pp. 2013-2016.
15. de Boer, E., 2007. Forward and reverse waves in nonclassical models of the cochlea. J. Acoust. Soc. Am. 121, 2819-2821.
16. Zweig, G., Shera, C.A., 1995. The origin of periodicity in the spectrum of evoked otoacoustic emissions. J. Acoust. Soc. Am. 98, 2018-2047.
17. de Boer, E., Nuttall, A.L., 2006. Spontaneous Basilar-Membrane Oscillation (SBMO) and coherent reflection. J. Assoc. Res. Otolaryng. 7, 26-37.

Comments & Discussion

Chadwick: The width of the generation area of DP waves seems to be a critical factor in such studies. Can this be determined by solving an inverse problem?

Shera: Your analysis assumed a certain mechanism and distribution for generating the DPs - what do you think about the validity of that assumption and, in particular, do you think that the DP generation region might extend more basally than we all have been assuming?

de Boer: The width of the generation area of DP waves seems to be the prime factor. We are working on that topic and hope to submit a paper soon.

van der Heijden: A non-symmetrical impedance matrix violates the principle of reciprocity and cannot be realized by a circuit containing only passive elements. For example, take a transmission line which supports ingoing waves but not outgoing ones. Connect heat reservoirs to its ends, both at the same temperature. The directionality will cause energy to be transported only one way, heating up one reservoir and cooling the other - in clear violation of the Second Law of Thermodynamics.

de Boer: The principle of reciprocity cannot be applied here, because the system is locally active, it has a power source built-in.

DP PHASES IN MAMMALIAN COCHLEAE, PREDICTED FROM LIQUID-SURFACE-WAVE FORMULAS

REINHART FROSCH

PSI Villigen and ETH Zurich (retired)
Sommerhaldenstrasse 5B, CH-5200 Brugg, Switzerland

Hydrodynamic formulas for mass-loaded spring-driven liquid-surface waves, together with cochlear parameters derived from experimental literature values of single-tone BM-vibration phases, are used to predict the phases of cubic DP's and DPOAE's. These predictions depend on the primary-tone level and on the category (low- or high-side) of the distortion; they agree [disagree] with recent fixed- f_2 low-side DPOAE experiments by two different groups in Mongolian gerbil if the backward cochlear DP-wave is assumed to be a slow surface wave [a fast compression wave]. The weakness of DPOAE's, particularly of high-side ones, is argued to be due to destructive interference among slow backward DP-TW's from different parts of the DP generation zone.

1 Introduction

In Fig. 3 of T. Ren et al. [1], the measured phases (re stapes) of the BM vibration at the 17-kHz low-level best place ($x \approx 2.4$ mm) of a sensitive Mongolian-gerbil cochlea at any given single-tone frequency up to ~21 kHz (i.e., up to ~1.2 CF, where CF = characteristic frequency ≈ 17 kHz) are shown to vary by less than ~1 rad among the 9 levels 5, 15, ... , 85 dB (SPL). In the present study it was attempted to predict these phases, and also those of cochlear travelling waves (TW's) generated by cubic distortion-products (DP's), from formulas for un-damped mass-loaded spring-driven surface waves.

2 Methods

2.1 Surface-Wave Formulas

For a conventional box-shaped channel with x- and y-independent properties the following relation between the wave number $k = 2\pi / \lambda$ and the angular frequency $\omega = 2\pi \cdot f$ of a single-tone TW is obtained [2]:

$$\omega^2 = \frac{S}{M + 2\rho /[k \cdot \tanh(k \cdot H)]} . \tag{1}$$

Here S is the BM stiffness per square mm, M is the BM mass per square mm, and ρ is the liquid density. In the first 2.5 mm of the gerbil cochlea the effective half-channel height H amounts to several mm (see Section 2.3), so that the *short-wave approximation* [3], $k \cdot H >> 1$, is valid even for fairly low frequencies; for instance, $\tanh(k \cdot H)$ is greater than 0.964 if the short-wave condition $k \cdot H > 2$ holds. Replacement of the hyperbolic tangent in Eq. (1) by 1 and solving for k yields

$$k \approx 2\rho \cdot \omega^2 /[S - M \cdot \omega^2] . \tag{2}$$

Now the BM stiffness S is assumed to depend on the longitudinal coordinate x [3]:

$$S = S_0 \cdot \exp(-\alpha \cdot x)\,. \tag{3}$$

According to Eq. (23) of W. M. Siebert [4], the function $k(x)$ defined by Eqs. (2) and (3) represents a local wave number, i.e., $k(x) = -\partial\phi / \partial x$. If the phase ϕ of a TW having an angular frequency ω (high enough for the short-wave approximation to be valid) is defined to vanish at the stapes (i.e., at $x = 0$), then integration of Eq. (2) yields

$$\phi = -\int_0^x k(x')\cdot dx' = -[2\rho/(\alpha\cdot M)]\cdot[\,\ln(1-\eta) - \ln(e^{-\alpha\cdot x} - \eta) - \alpha\cdot x\,]\,. \tag{4}$$

The dimension-less quantity η on the right-hand side of Eq. (4) is defined as follows:

$$\eta = M \cdot \omega^2 / S_0\;. \tag{5}$$

A function $\phi(\omega)$ according to Eqs. (4) and (5) is shown in Fig. 1A below.

2.2 Cochlear Parameters for Gerbil

The parameter α in Eq. (3) above was derived from Fig. 4C of T. Ren [5], where x = 2.30 and 2.75 mm are seen to be the best 60-dB places of ~17.6 and ~13.3 kHz. That corresponds to $\Delta x_8 = 1.1$ mm/octave. The experiment of M. Müller [6] yields, for the basal half of the gerbil cochlea, $\Delta x_8 = 1.5$ mm/octave, but is consistent with 1.1 mm/octave for $x < 3$ mm. The assumption that Δx_8 = 1.1 mm/octave holds also for the BM mechanical resonance frequency f_{res} leads to the following parameter α:

$$f_{res} = [1/(2\pi)]\cdot\sqrt{S_0\cdot[\exp(-\alpha\cdot x)]/M}\;;\; \alpha = \ln(4)/\Delta x_8 = 1260\,\mathrm{m}^{-1}\,. \tag{6}$$

The parameters S_0 and M were optimised so that, at x = 2.4 mm, Eqs. (4) and (5) yield the phase $\phi(17\mathrm{kHz}) = -10$ rad read off in Fig. 3B of T. Ren et al. [1], and that Eqs. (2) and (3) agree optimally with the wave numbers $k = -(\partial\phi/\partial x)$ defined by the experimental single-tone phase-versus-x curves in Fig. 1 of T. Ren [7]; see Eq. (7).

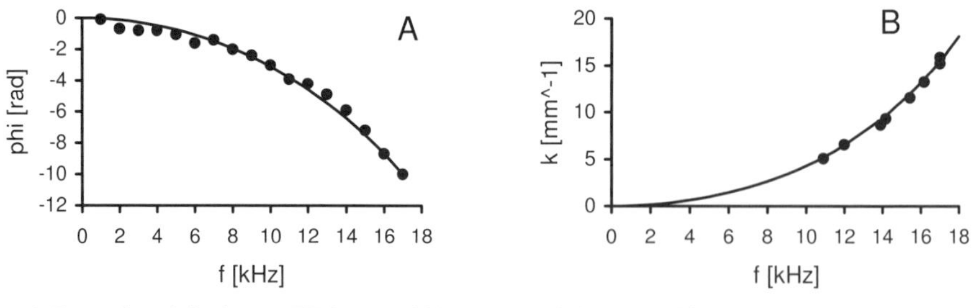

Figure 1. Properties of single-tone TW's on gerbil BM at x = 2.4 mm. A) Phase ϕ versus frequency f; filled circles: experimental 85-dB phases taken from Fig. 3B of [1]; curve: phases calculated from Eqs. (4), (5), (6), and (7). B) Local wave number k versus frequency f; filled circles: experimental wave numbers $k = -(\partial\phi/\partial x)$ taken from Fig. 1 of [7]; curve: wave numbers calculated from Eqs. (2), (3), (6), and (7).

$$S_0 = 4.16 \cdot 10^{10}\ \mathrm{kg/(m^2 s^2)};\quad M = 0.0478\ \mathrm{kg/m^2}\ ; \tag{7}$$

Fig. 1 shows the good agreement of these parameters with the experiments [1], [7].

2.3 Validity of the Short-Wave Approximation

For the first 2.5 mm of the gerbil cochlea, an order-of-magnitude estimate yields half-channel cross sections of 2 and 0.5 mm^2 and the corresponding effective half-channel cross section $q = 2 \cdot 2 \cdot 0.5/(2+0.5) = 0.8\ \mathrm{mm}^2$. Since the BM width w is ~0.1 mm, the effective half-channel height according, e.g., to Section 4.2 of [3] is $H \approx q/w \approx 8\ \mathrm{mm}$. A straight box with that H-value fulfils the short-wave condition $k \cdot H > 2$ if the wave number k exceeds $2/H \approx 0.25\ \mathrm{mm}^{-1}$. With Eqs. (2), (3), (6), (7) one finds that at x = 2.4 mm the short-wave condition is met at frequencies above 2.5 kHz. The phase shown by the curve in Fig. 1A, on the other hand, is calculated from Eq. (4), and so is influenced also by wave numbers at x = 0, where the just mentioned calculation yields a minimal short-wave frequency as high as 11 kHz. A phase-versus-frequency curve for all wavelengths, obtained by numerical integration of the function $k(x)$ defined by Eqs. (1) and (3), however, is very close to the curve in Fig. 1A, because the extreme base (due to the smallness of the local wave number below 11 kHz) makes only a small contribution to the integral $\phi = \int_0^x k(x') \cdot dx'$; thus the short-wave approximation is found to yield good phase estimates in the most basal 2.5 mm of the gerbil cochlea even at frequencies below the above-mentioned value of 11 kHz.

2.4 Locations of DP Generation

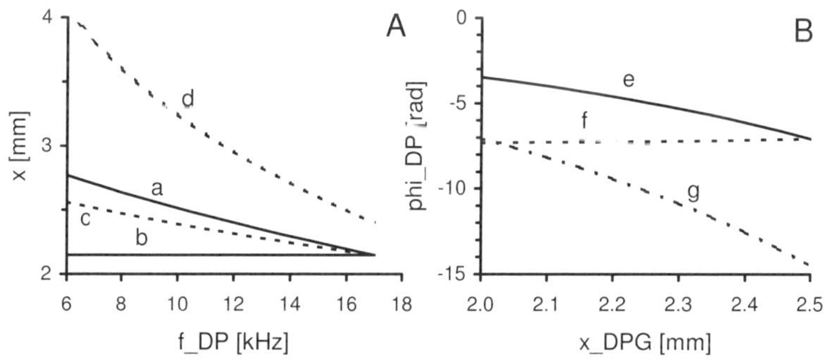

Figure 2. Predictions for the fixed-f_2 low-side DP experiment yielding Figs. 2D and 2E of [1] (f_2 = 17 kHz; $L_2 = 70$ dB ; $f_1 = (f_{DP} + f_2)/2$; $L_1 = 75$ dB). A) Locations x of four peaks in the gerbil cochlea; solid curve a: single tone at f_1 ; solid curve b: single tone at f_2 ; dashed curve c: low-side DP generation peak; dashed curve d: low-level active best place at f_{DP} . B) Phases ϕ_{DP} for $f_1 = 15.5$ kHz, $f_2 = 17.0$ kHz, $f_{DP} = 14.0$ kHz; solid curve e: ϕ_{DP} at the DP-generation location x_{DPG} ; dashed curve f: ϕ_{DP} at x = 2.5 mm of a forward DP-TW generated at x_{DPG} ; dashed curve g: ϕ_{DP} at $x = 0$ of a backward DP-TW generated at x_{DPG} .

The sources of low-side ($f_{DP} = 2f_1 - f_2$) and high-side ($f_{DP} = 2f_2 - f_1$) distortion products are spatially distributed over the region where the BM excitations due to the f_1-and the f_2-waves overlap. According to the Helmholtz combination-tone theory [8], the location of strongest sources coincides with that of the largest function $a_1^2 a_2$ (low-side) or $a_2^2 a_1$ (high-side) of BM vibration amplitudes and so depends on the primary-tone levels; at levels below ~40 dB, the single-tone peaks occur at the best "active" places, but at higher levels they are more basal. Fig. 2 above shows our predictions relevant for the experimental DP-phases shown in Figs. 2D and 2E of Ren et al. [1]. Solid curves b and a: estimates of the BM vibration peaks for single tones at $f_2 =$ 17 kHz [70 dB] and $f_1 = (f_{DP} + f_2)/2$ [75 dB], based on the low-level active best place at 17 kHz (x = 2.4 mm), on $\Delta x_8 = 1.1\,\text{mm}$, and on the peak shift from 17 kHz to 14.5 kHz (corresponding to 0.23 octave and to 0.25 mm) between 35 and 75 dB in Fig. 3A of [1]; that experimental diagram also yielded an estimate of 0.8 mm for the (–3dB) full widths of the peaks represented by curves a and b. If these peaks are assumed to be of gaussian shape, then [8] yields that the DP generation peak (curve c) is gaussian too, is located at $x = (2x_{f_1} + x_{f_2})/3$, and has a (–3dB) full width of $0.8/\sqrt{3} \approx 0.5$ mm. For example, at $f_{DP} = 14\,\text{kHz}$, the just mentioned full width ranges from ~2.0 to ~2.5 mm. For that DP-generation zone, curve e in Fig. 2B represents the phase $\phi_{DP} = 2\phi_1 - \phi_2$ [calculated from Eqs. (4), (5), (6), (7)] of the DP at its generation place x_{DPG}. Curve g, representing ϕ_{DP} at x = 0 of a backward DP-TW generated at x_{DPG}, indicates that the phases of the backward DP-TW's from x = 2.0 mm (basal end of DP generation zone) and from x = 2.5 mm (apical end) differ by as much as 7.4 radians; this implies strong destructive interference among backward DP-TW's from different parts of the DP generation zone. A calculation at $f_{DP} = 11\,\text{kHz}$ (i.e., $f_1 = 14\,\text{kHz}$, DP generation zone from x = 2.1 mm to x = 2.6 mm) yields a corresponding phase difference of 4.1 radians only, so that less destructive interference is predicted. Indeed the experimental DP stapes velocity in Fig. 2C of [1] is greater at $f_{DP} = 11\,\text{kHz}$ (i.e., $f_1 = 14\,\text{kHz}$) than at $f_{DP} = 14\,\text{kHz}$ ($f_1 = 15.5$ kHz), although the primary-wave peak overlap is weaker in the $f_{DP} = 11\,\text{kHz}$ case. Curve f in Fig. 2B, representing ϕ_{DP} at x = 2.5 mm of a forward DP-TW generated at x_{DPG}, indicates that the *forward* DP-TW's from the various parts of the DP-generation zone interfere constructively. At 75 dB, however, the amplification of that wave between curves c and d (best low-level f_{DP} place) in Fig. 2A is fairly weak, as confirmed by an earlier experiment of T. Ren [7]: even at a primary-wave level as low as 60 dB, the experimental DP BM vibration magnitude at the best low-level f_{DP} place in Fig. 1a of [7] (i.e., at $x \approx 2.7$ mm; $f_{DP} = 13.91$ kHz, $f_1 = 15.45\,\text{kHz}$) is not greater than that in the DP-generation zone (2.2 mm < x < 2.5 mm). In the high-level cases to be presented below, the backward DP-TW's from the DP generation zone are therefore assumed to be stronger than the backward DP-TW's due to reflection at the best low-level f_{DP} place.

3 Results

In Fig. 3 experimental literature values of DP phases are shown by filled circles. A (Fig. 2E of [1]) and B (Fig. 11B of [9]): DPOAE phases after cochlear round-trip; fixed $f_2 = 17$ kHz (A) or 20 kHz (B, C). Solid curves in A, B: our slow-backward DP-TW calculations; dashed curves in A, B: our fast compression backward DP-wave calculations. The slow backward DP-TW curves in A, B are seen to agree with the measurements, in contrast to the conclusion of the authors of [1], but in agreement with the general conclusion in [9]. For the pro-slow verdict in B it was important that Fig. 1B of [9] implies an 80-dB single-tone BM vibration peak basal of the low-level active peak by ~0.5 octave (i.e., 0.55 mm).

Fig. 3C, circles: experimental curve labelled DP in Fig. 11B of [9]. Our solid curve in Fig. 3C represents the phase at $x \approx 2.1$ mm of a forward DP-TW generated at the DP-generation peak location (basal of curve c in our Fig. 2A because of the higher f_2 and the just mentioned 0.55-mm shift between low-level and 80-dB BM vibration peaks).

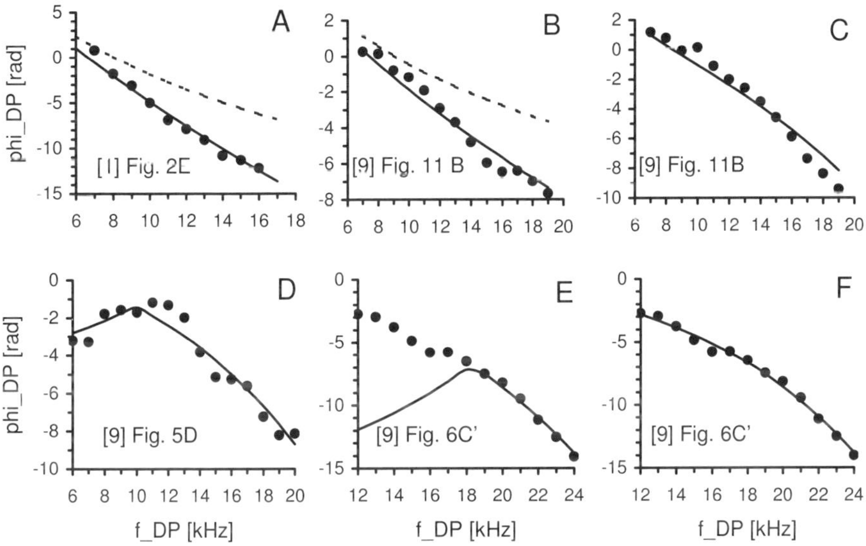

Figure 3. Comparison of our predictions of gerbil DP phases with experimental literature values (shown by filled circles; references are given in the diagrams); A, B, C, D: low-side DP; E, F: high-side DP; in all six cases experimental phases have been shifted by a frequency-independent amount and adapted so that both primary-wave phases vanish at the stapes. Solid curves in A-E are based on slow forward and backward DP-TW's. Dashed curves in A, B are for fast compression backward DP waves; further details are given in the text.

Fig. 3D, circles: experimental low-side DP phases from Fig. 5D of [9] (fixed $f_2 / f_1 = 1.25$; $L_1 = L_2 = 80\,\text{dB SPL}$; phases measured at $x \approx 2.1$ mm). At $f_{DP} \approx 10\,\text{kHz}$ ($f_1 = 13.3$ kHz, $f_2 = 16.7\,\text{kHz}$) the DP generation peak is at $x \approx 2.1$ mm; at $f_{DP} < 10\,\text{kHz}$ the measured phase is dominated by a slow backward DP-TW generated at the DP generation peak ($x > 2.1$ mm); at $f_{DP} > 10\,\text{kHz}$ the measured phase is dominated by a forward DP-TW generated at the DP generation peak ($x < 2.1$ mm).

Fig. 3E, circles: experimental high-side DP phases from Fig. 6C' of [9] (fixed $f_2 / f_1 = 1.25$; $L_1 = L_2 = 80\,\text{dB SPL}$; phases measured at $x \approx 2.1$ mm). At $f_{DP} \approx 18$ kHz ($f_1 = 12$ kHz, $f_2 = 15\,\text{kHz}$) the DP generation peak is at $x \approx 2.1$ mm. Solid curve: at $f_{DP} < 18$ kHz the measured phase was assumed to be dominated by a slow backward

DP-TW generated at the DP generation peak, similarly to Fig. 3D; that assumption is seen to disagree with the measurements.

Fig. 3F: same measurements as in Fig. 3E. Calculations similar to those for Fig. 2B above yielded that, for $12\,\text{kHz} < f_{\text{DP}} < 18\,\text{kHz}$, the phases at x = 2.1 mm of backward DP-TW's generated at the two ends of the DP generation zone differed by ~9 radians, implying strong destructive interference among the TW's from the different parts of that zone. The assumption that in this case the DP's generated locally at x = 2.1 mm are stronger than the just mentioned backward DP-TW yields the solid curve in Fig. 3F which agrees with the measurements. Calculations of the corresponding backward-DP-TW phase difference for the low-side case of Fig. 3D, however, yielded differences of <2 radians, implying little destructive interference.

4 Discussion

Figs. 1 and 3 imply that the short-wave equations (2)-(5) give a realistic description of the ~6–24-kHz TW phases in the first millimetres of the gerbil cochlear channel. The cochlear-round-trip phase curves and data in Figs. 3A and 3B indicate that low-side cubic DPOAE's are mainly due to slow backward DP-TW's. The experimentally observed weakness of high-side cubic DPOAE's is attributed to destructive interference among the slow backward DP-TW's from different parts of the DP generation zone.

Acknowledgments

The author wishes to thank N. P. Cooper, W. Dong, H. Duifhuis, G. R. Long, E. S. Olson, T. Ren, and C. A. Shera for their kind and helpful comments.

References

1. Ren, T., He, W., Scott, M., Nuttall, A.L., 2006. Group delay of acoustic emissions in the ear. J. Neurophysiol. 96, 2785-2791.
2. Patuzzi, R., 1996. Cochlear micromechanics and macromechanics. In: Dallos, P., Popper, A.N., Fay, R.R. (Eds.), The Cochlea. Springer, New York, pp. 186-257.
3. de Boer, E., 1996. Mechanics of the cochlea: modeling efforts. In: Dallos, P., Popper, A.N., Fay, R.R. (Eds.), The Cochlea. Springer, New York, pp. 258-317.
4. Siebert, W.M., 1974. Ranke revisited – a simple short-wave cochlear model. J. Acoust. Soc. Am. 56, 594-600.
5. Ren, T., 2002. Longitudinal pattern of basilar membrane vibration in the sensitive cochlea. Proc. Natl. Acad. Sc. 99, 17101-17106.
6. Müller, M., 1996. The cochlear place-frequency map of the adult and developing mongolian gerbil. Hear. Res. 94, 148-156.
7. Ren, T., 2004. Reverse propagation of sound in the gerbil cochlea. Nature Neuroscience 7, 333-334.
8. von Helmholtz, H., 1954. Theory of combinational tones. In: von Helmholtz, H., On the Sensations of Tone. Dover, New York, pp. 411-413.
9. Dong, W., Olson, E.S., 2008. Supporting evidence for reverse cochlear traveling waves. J. Acoust. Soc. Am. 123, 222-240.

Comments and Discussion

de Boer: I am always amazed at how much of the functioning of the cochlea can be derived from (what I call) short-wave behaviour. Your paper is another convincing example.

Frosch: Thank you, Egbert. May I add that the success of the short-wave approximation (valid at large effective half-channel heights H) at distances x from the base less than a few mm appears understandable if it is taken into account that (in humans, at least) inside of the oval window is the wide-open vestibulum. The cross-section of the scala tympani is large at small x, too, because it has to accomodate the round window with its diameter of typically 1 mm. The BM starts between the two windows and has the shape of a hook, I believe. In that basal region, in travelling waves at frequencies above about 10 kHz, the liquid motion is confined to two semicircular regions just above and just below the (narrow) BM, so the roof (vestibule) and the floor (round window) of the cochlear channel are not involved.

DISTORTION PRODUCT EMISSIONS: WHERE DO THEY COME FROM?

XUEDONG ZHANG, DAVID C. MOUNTAIN
Department of Biomedical Engineering, Hearing Research Center, Boston University, Boston, MA 02215 USA

The location from which distortion product otoacoustic emissions (DPOAE) originate and the means by which they propagate back to the ear canal have been a rich source of discussion and controversy since DPOAEs were discovered. In order to shed light on the origins and means of propagation we show through modeling studies that data thought to support different hypotheses can be reconciled and emissions can originate from a wide range of cochlear locations. The contribution pattern of these distributed components at the stapes, however, can be complex and change with level and f_2/f_1 ratios. We also show that using basilar membrane group delay to estimate travel time from the generation site back to the stapes can lead to incorrect conclusions.

1 Introduction

Distortion-product-otoacoustic-emissions (DPOAEs) are pressure signals at combination-tone frequencies (e.g., $f_{dp}=2f_1 - f_2$) evoked when the cochlear is excited with two primary tones at frequencies f_1 and f_2 ($f_1 < f_2$). The DPOAEs measured at ear canal [1-3] suggest that the generated distortion product (DP) travels back to the stapes as a transverse wave that is similar to the forward traveling wave. Mountain et al [4] estimated the forward travel times from the group delay for the cochlear microphonic, reverse travel times from the electrically-evoked otoacoustic emissions generated using the same electrode, and round trip travel times from DPOAEs. Their experimental results directly support the idea that the delay of DPOAEs at ear canal is roughly twice the time the forward traveling wave needed to reach the generation site.

By measuring the DPOAEs at stapes and the basilar membrane (BM) motion in the cochlea simultaneously, Ren et al [5, 6] recently challenged the idea that the emissions propagate via a backward-traveling wave. They found the DPOAE group delay measured at the stapes was much smaller than the twice of the group delay for BM vibration at the putative emission-generation site, the forward delay. They also found that the BM vibration phase measured at the DP frequency for a basal location leads the phase measured at a more apical location. They suggest that these data imply that the DP travels back via compressive fast wave rather than the slow transverse-traveling wave.

In this study, we use a simple nonlinear traveling wave model to study how emissions originate and propagate. We show that the Ren et al data can be replicated with this model without the mechanism of compressive fast waves. We also show that a wide range of cochlear locations can contribute to the DPOAEs measured at stapes, and the traditional DP propagation theory and BM vibration data can be reconciled because of the distributed nature of DPOAE generation. The simulation results also explain the variation observed in experimental data at different f_2/f_1 ratios and presentation levels.

2 Methods

2.1 The classical one dimension nonlinear model

A simple one dimensional classical cochlea model [7] was used to generate the simulation results. The discretized equivalent acoustic impedance model is shown in Fig. 1. Only the cross-partition pressure difference $P(i)$ is driving the system in the model, which means that we assume the sound speed in fluid is infinite and there is no compressive wave.

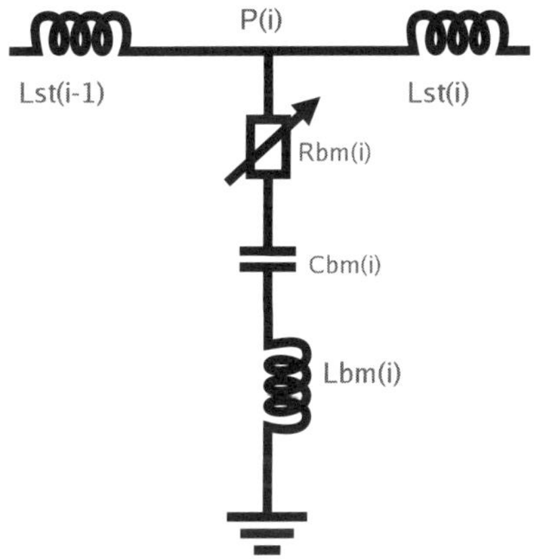

Figure 1. The equivalent circuit realization of one section of the classical model.

The BM compliance map was derived from measurements by Naidu and Mountain [8], and equivalent mass and resistance is then computed based on the Greenwood frequency map and the a tuning factor. The overall partition resistance contains a nonlinear resistor whose value varies with BM velocity (I_{BM}):

$$R_{BM}[i]=R_0[i]+R_1[i]*I_{BM}(t)^2 \quad (1)$$

When two tones are presented simultaneously, the nonlinear resistor will generate the distortion component $2f_1$-f_2 which then propagates along the cochlea in both directions.

A pressure source P_{stim} is added at the first section of the model together with the a matching impedance R_{foot} to simulate the acoustic stimulation system and middle ear. The helicotrema is modeled as a resistance R_h and is connected to the last section of the model. Table 1, at the end of this paper, lists the parameters used in the model. The simulation was done in time domain using an ODE solver (SUNDIALS).

2.2 DPOAE responses of the nonlinear cochlea model

When two tones are presented together to the model, many nonlinear components along the cochlea will generate distortion products (or emissions). These emissions will propagate along the cochlea and interact with each other in the nonlinear model, thus the final response at stapes is not a linear summation of the responses from each generation site calculated independently.

In order to understand how distortion components from different generation sites contributed to the total DPOAE, we computed the results from the nonlinear model for each simulation, and then replaced the nonlinear resistor in each section with an active pressure source which equals to the value of the pressure measured across that resistor. The new system is thus a linear system which have the exactly same responses as the original system for this particular simulation. The value of the active pressure sources in each section already include the nonlinear interaction between different sections, and the responses of these active sources at stapes can be added linearly in the new system and studied individually.

We call these active sources derived from the nonlinear resistance at DP frequency the DPOAE generator sources: $P_a^i(\omega_{dp})$. The generator contribution $P_{stapes}^i(\omega_{dp})$ is defined as the response at the stapes for each generator source calculated in the new system. The DPOAEs measured at stapes is then a summation of contributions from these distributed generators:

$$P_{stapes}^{dp}(\omega_{dp}) = \sum P_{stapes}^{i}(\omega_{dp}) \quad (2)$$

3 Results

3.1 Comparison to experimental data

Figure 2 shows the DPOAE responses from the model measured at ear canal for two sound levels. At high level (dashed lines), notches apear in the magnitude response and the phase of DPOAE responses flattens, especially when the f_2/f_1 ratio is small.

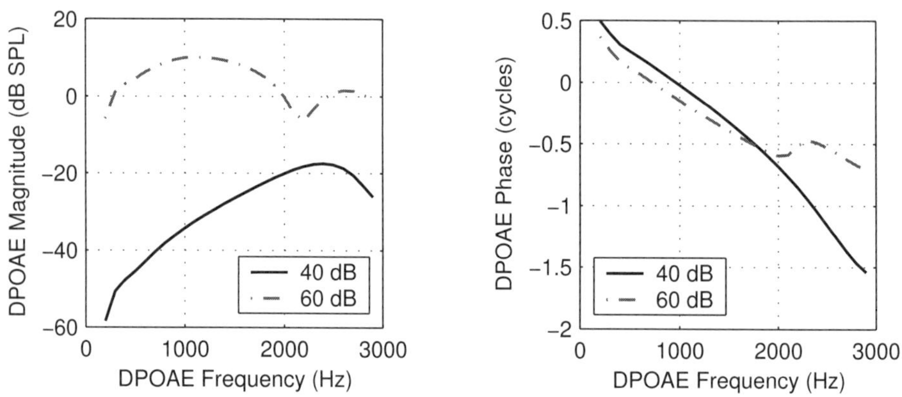

Figure 2. DPOAE Responses from model simulation. The f_2 is fixed at 3000 Hz with f_1 sweeping from 1500 Hz to 3000 Hz. Both tones were presented at 40 or 60 dB SPL.

Figure 3 and Figure 4 compare the model simulation results (left panel) of DPOAEs with experimental data reported by Ren et al [6, 7] (right panel). In Figure 3 the group delay of BM vibration measured at the 14 kHz location was compared with the DPOAE round trip delay measured at stapes. The BM group delay is much larger than half of DPOAE delay as the frequency approaches the CF. Figure 4 shows the phase delay between two locations (12 kHz and 15 kHz) at DP frequency for fixed f_2/f_1 ratio experiments, the positive value means the phase of BM vibration at basal location leads that at the apical location. And the delay increased with the DP frequency. Both simulation results is consistent with the experimental data.

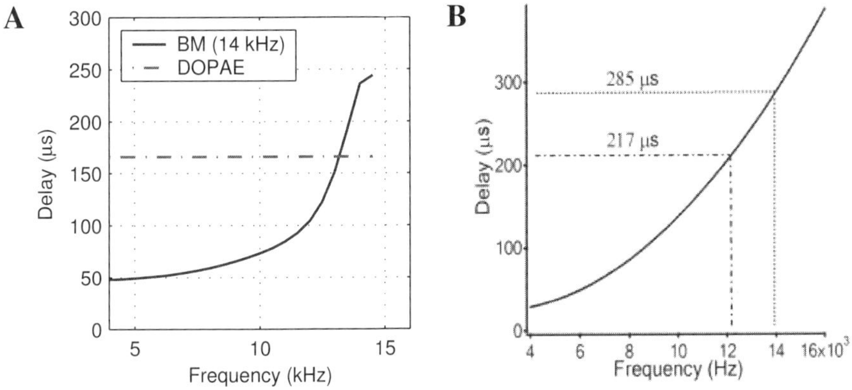

Figure 3. (A) Two-tone simulation with F1 sweeping and F2 fixed at 14 kHz (both at 40 dB SPL). Solid line is BM vibration group delay measured at 14 kHz location, dashed line is the group delay of DPOAEs at stapes. (B) data from Ren et al [6, Fig. 4F], The DPOAE round trip delay is 217 μs, and BM vibration phase data was measured at 14 kHz location.

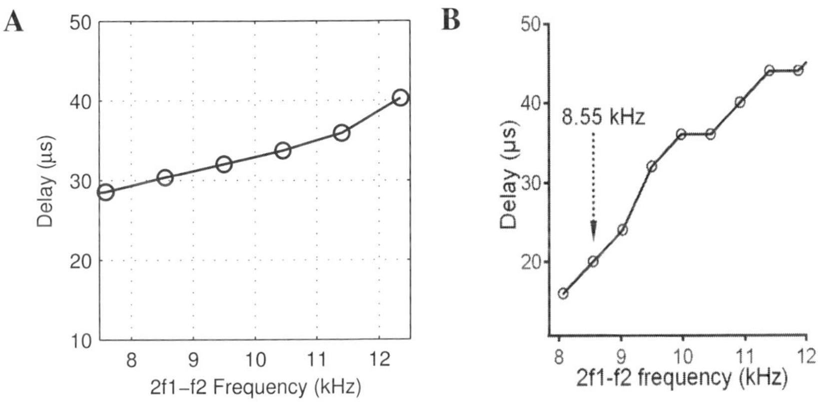

Figure 4. (A) Phase delay between two locations (see text) at DP frequencies. The two-tone simulation has a fixed f2/f1 ratio at 1.05, with f2 presented at 60 dB SPL and f1 at 55 dB SPL. The paradigms and parameters are very similar to those used in experimental studies by Ren et al (right panel). (B) Data from Ren et al [7, Fig. 1C].

3.2 Distortion product origins

To locate the region from where the largest DPOAE components originate, we add the individual contributions starting from the apex and find the location where the cumulative sum is 10 dB below the total contribution, and treat this point as the apical edge of the DP generation region. We can do the similar cumulation starting from the base to find the basal edge of the DP generation region.

Figure 5 shows the boundary lines for DPOAE generation region for different f_2/f_1 ratios. The normalized location for peak f_2 response is about 0.59 (3 kHz). The generation region tends to be wider for small f_2/f_1 ratios. At high levels (right panel), the DPOAE contains components contributed by generators far basal to the peak f_2 response region.

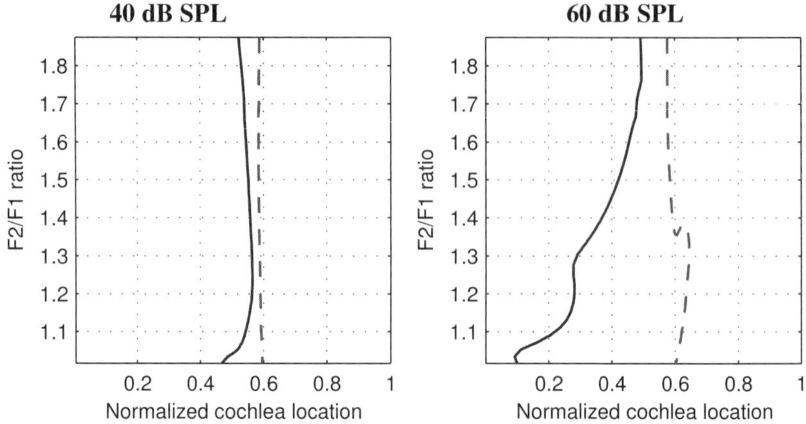

Figure 5. The right dashed lines represent the apical boundaries, and the left solid lines represent the basal boundaries of the DP generation region. The y-axis is for different f2/f1 ratios, and the f2 is always fixed at 3 kHz in the simulation. Both tones were presented at either 40 dB SPL (left panel) and 60 dB SPL (right panel).

4 Discussion

Using a simple classical nonlinear model, we show that we can produce realistic DPOAE responses at ear canal as well as the results comparable with Ren et al [6, 7]. Since this model only supports slow traveling waves, our results suggest that the Ren et al data and the idea that DPOAEs propagate to the stapes via slow backward traveling waves can be reconciled. The reconciliation came from the fact that the DPOAE generation region is very wide at low f_2/f_1 ratio and at moderate to high levels. For the simulation results presented in Fig. 4, we compared the BM vibrations caused by the generators basal to the measurement site (12 kHz) with that by the apical generators. We found the responses from basal is about twenty times larger of that from apical. This suggests that the measurement of BM vibration in Fig. 4 is dominated by these forward traveling waves from basal generators, and that these generators made significant contributions to the DPOAEs at stapes.

Strictly, the delay of traveling wave between two locations is defined by the accumulated phase difference divided by radial frequency. The traveling wave delay can be approximated by the group delay only when all the frequencies travel at similar speed. Since the traveling wave slows down dramatically near the characteristic region, the travel times for single tones to the region increase rapidly as the frequency of the tone approaches the CF (so called "dispersion"). The BM vibration group delay near CF is actually dominated by the dispersion and is much larger than the true traveling wave delay. Thus the BM vibration group delay around CF should not be used as an estimate of the forward traveling time.

TABLE I. Model parameters

Parameter	Value	Description and unit
N	512	Number of discretized sections
L	1.21	Cochlea length (cm)
R_h	1.38e6	Helicotrema (dyn cm^{-3} s)
R_{foot}	1.74e7	Matching impedance (dyn cm^{-3} s)
W_{BM}	0.0175	BM width (cm)
CF[i]	398 * ($10^{2.2*(N-1-I)/N}$ - 0.631)	Characteristic frequency (Hz)
C_{BM}[i]	$e^{30.9+4.04*L*(i+1)/N}/A_{BM}[i]$	BM compliance (cm^3 dyn^{-1})
L_{BM}[i]	$(2\pi CF[i])^2/C_{BM}[i]$	BM effective mass (g cm^{-2})
R_0[i]	sqrt(L_{BM}[i]/C_{BM}[i]) / Q, Q=10	Resting BM Resistance (dyn cm^{-3} s)
R_1[i]	R0[i]*1e10/$(W_{BM}*L/N)^2$	Nonlinear BM resistance coeff. (dyn cm^{-3} s)
As[i]	0.01 (1-3*i/N) for i = 1 to 102; 0.00352*$e^{(-8.514*(i-102)/N)}$ + 0.0005 for others	Scala cross section area (cm^2)
Ls[i]	$2\rho\Delta$/As[i], ρ = 1.02 g m^{-3}	Fluid mass (g cm^{-2})

In conclusion, our model suggests that the DPOAEs are generated in a distributed region and propagate back in the same way as forward traveling waves. The generation region varies as stimulus parameters change, becomes much larger at high level and small f_2/f_1 ratios, and these variations explain the DPOAE data observed in different experiments.

Acknowledgments

Supported by NIDCD.

References

1. Mahoney, C. F., Kemp, D. T., 1995. Distortion product otoacoustic emission delay measurement in human ears. J. Acoust. Soc. Am. 97(6), 3721-3735.
2. Mills, D. M., Rubel, E. W., 1997. Development of distortion product emissions in the gerbil: "filter" response and signal delay. J. Acoust. Soc. Am. 101(1), 395-411.
3. Shera, C. A., Guinan, J. J., 1999. Evoked otoacoustic emissions arise by two fundamentally different mechanisms: a taxonomy for mammalian OAEs. J. Acoust. Soc. Am. 105(2 Pt 1), 782-798.
4. Mountain, D.C., Nakajima, H.H., Rafee,S., Hubbard, A.E., 2000. Forward and reverse traveling waves in the gerbil cochlea. In: Wada, H., Takasaka, T., Ikeda, K., Ohyama, K., Koike, T. (Eds). Recent Developments in Auditory Mechanics, World Scientific Publishing. pp. 102-108.
5. Ren, T., He, W., Scott, M. Nuttall, A. L., 2006. Group Delay of Acoustic Emissions in the Ear. J. Neurophysiol. 96, 2785-2791.

6. He, W., Fridberger, A., Porsov, E., Grosh, K., Ren, T., 2008. Reverse wave propagation in the cochlea. PNAS 105 (7), 2729-2733.
7. de Boer, E., 1995. Mechanics of the Cochlea: Modeling Efforts. In: Dallos, P., Popper, A. N., Fay, R.R. (Eds). The Cochlea, Springer-Verlag New York, pp. 258 - 317.
8. Naidu, R. C., Mountain, D. C. 1998. Measurements of the stiffness map challenge a basic tenet of cochlea theories. Hear. Res. 124, 124-131.

Comments and Discussion

de Boer: You report no "widening" of the source area of DP waves at primary-tone levels of 40 dB. Yet Ren specifically showed that the "Ren paradox" exists not only at high but also at low levels, as low as 40 dB.

Mountain: The sound level at which the model shows "widening" of the DP source area is very sensitive to model parameters. For the parameters that we used, the "widening" is just beginning to happen at 40 dB SPL for very closely space primaries. This model is quite generic and is used to show that a simple model without any compressional waves can produce the types of responses that have been interpreted to mean that DPs do propagate via compressional waves.

RETROGRADE PROPAGATION OF COCHLEAR DISTORTION

STEPHEN T. NEELY, YI-WEN LIU
Boys Town National Research Hospital, 555 North 30th Street
Omaha, Nebraska 68131, USA

In 1992, Allen and Fahey suggested a method of estimating "cochlear amplifier" (CA) gain by employing simultaneous measurement of auditory nerve fibers and acoustic distortion products (DP). Their conclusion that CA gain must be less than 6 dB, based on their measurements in a cat, has been controversial. We have simulated the retrograde propagation of distortion in a nonlinear, active, one-dimensional model of cochlear mechanics. We conclude, based on results of this model, that the Allen-Fahey paradigm does not provide accurate estimates of CA gain.

1 Introduction – The Allen-Fahey Experiment

The experimental paradigm suggested by Allen and Fahey [1] requires simultaneous measurement of (1) sound pressure level in the ear canal and (2) spike discharge of a single auditory nerve fiber (ANF). The primary frequencies of a two-tone stimulus are selected to produce intermodulation distortion ($2f_1$-f_2) at the characteristic frequency (CF) of the ANF. The two tones are equal in level and this level is adjusted so that the $2f_1$-f_2 distortion product (DP) produces threshold excitation of the ANF. If we assume (1) that the DP is generated near the f_2 place within the cochlea and (2) that the cochlear amplifier (CA) [2] achieves gain by means of a negative damping region (NDR), then the effective gain applied to the DP, as it travels toward its characteristic place (CP), will vary depending on the location of f_2 place relative to the NDR. If the forward component travels through less than the entire NDR, then its amplification will be less than the maximum possible CA gain. For the same reason, the DP-component generated at the f_2 place that travels back toward the stapes will also receive varying amounts of gain from the NDR. As f_2 is swept toward CF from one-octave above CF, the forward DP gain will decrease and the reverse DP gain will increase. Because the stimulus level must increase to compensate for reduced forward DP gain, the net increase in the backward DP gain is effectively doubled. As the f_2 place is swept through the NDR, the DP level observed in the ear canal should increase as much as twice the maximum forward DP gain.

The Allen-Fahey experiment has been replicated by Shera and Guinan [10] and has been extended by de Boer et al. [3] and Shera and Guinan [10] with similar results. In general, the DP level observed in the ear canal decreases as f_2 approaches CF, contrary to what would be expected if there were any CA gain. Shera [8] has suggested that beam-forming in DP generation may invalidate a key assumption of Allen-Fahey experiment, which assumes that the relative size of the forward and backward DP components does not change as f_2 is swept. Although beam-forming must certainly be a factor in DP generation and is consistent other DPOAE measurements, uncertainties in spatial properties of DP generation make it difficult to quantify how the relative size of forward and reverse components change as a function of f_2.

The importance of the Allen-Fahey experiments extends beyond estimates of CA gain. The Allen-Fahey experiment has been cited by Ren et al. [7] as evidence to support the existence of compression waves in the cochlea; however, this interpretation has been disputed by Shera et al. [11].

2 Cochlear Model

There are many ways to model the cochlea. The present model is based on physical variables and approximates Greenwood's frequency-place map [4], but the emphasis is on representing signal processing characteristics.

2.1 *Macro-mechanics*

Consider a model of cochlear macro-mechanics with one spatial dimension that consists of the following differential equations:

$$\partial_x P = -M_f \ddot{\xi}_f \tag{1}$$

$$\partial_x \xi_f = w_p \dot{\xi}_r \tag{2}$$

In these equations, ∂_x represents a spatial derivative, the over-dot represents a time derivative, P represents the pressure difference across the basilar membrane (BM), $\dot{\xi}_f$ represents longitudinal volume-velocity of the scala fluid, ξ_r represents transverse displacement of the reticular lamina (RL), $M_f = 2\rho / A_s$ is the inertial component of the scala fluid impedance (per unit length), w_p is the effective width of the BM, and A_s is the cross-sectional area of the scala.

2.2 *Micro-mechanics*

To complete the cochlear model description, we represent cochlear *micro*-mechanics (at each point on the BM) by the following equations:

$$\xi_b = \xi_r + \xi_o \tag{3}$$

$$M_b \ddot{\xi}_b + R_b \dot{\xi}_b + K_b \xi_b = \left(-P/w_p\right) \tag{4}$$

$$M_o \ddot{\xi}_o + R_o \dot{\xi}_o + K_o \xi_o = \gamma \cdot \left(R_r \dot{\xi}_r + K_r \xi_r\right) \tag{5}$$

In these equations, ξ_b represents BM displacement and ξ_o represents outer hair cell (OHC) contraction, which reduces the separation between BM and RL. RL displacement ξ_r is assumed to provide the relevant input to the inner hair cell, which converts this mechanical signal into a neural signal. Simultaneous solution of Eqs. 1-5, with appropriate boundaries conditions, quantifies RL displacement for a stimulus specified in terms of stapes displacement.

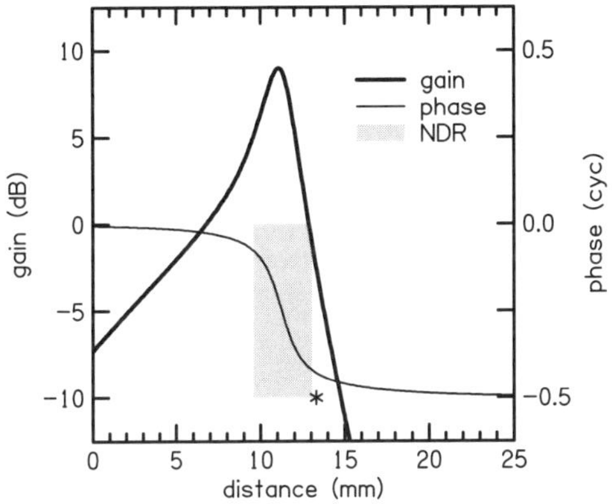

Figure 1. OHC contraction relative to RL displacement at 4 kHz as a function of distance from the stapes when $\gamma = 1$. The gray bar indicates the extent of the NDR produced when this OHC gain is included in the cochlear model. The asterisk indicates the characteristic place for a 4-kHz signal.

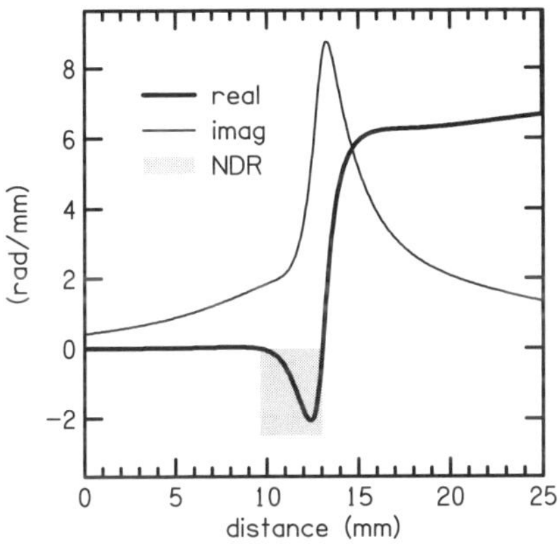

Figure 2. Propagation function at 4 kHz as a function of distance from the stapes. The real part (thick line) determines the slope of the pressure log-magnitude, while the imaginary part (thin line) determines the slope of the pressure phase. The gray bar indicates the extent of the NDR.

2.3 Outer hair cell

The ability of an OHC to change its length in response to changes in membrane potential gives the OHC the role of a controlled source of mechanical energy in a model of cochlear mechanics. For simplicity, OHC motility is represented in the present model only in terms of its input and output, which hides the details of mechano-electric and electro-mechanic transductions.

The OHC is tuned in this model by its mechanical load. This tuning is independent of basilar membrane tuning and may be demonstrated by looking at the transfer function between OHC input, which is hair-bundle (HB) displacement, and output, which is contraction of the length of the cell body. In the frequency domain, OHC contraction relative to RL displacement is a low-pass filter:

$$H_o \equiv \left(\frac{\xi_o}{\xi_r}\right) = \gamma \cdot \left(\frac{R_r + K_r/s}{sM_o + R_o + K_o/s}\right). \tag{6}$$

In this equation s is the Laplace transform frequency variable. The parameter γ allows the model to represent a transition from completely passive ($\gamma = 0$) to fully active ($\gamma = 1$).

The OHC transfer function H_o is shown in Fig. 1 at 1, 4, and 16 kHz for model parameters selected to represent a cat cochlea (in particular, $R_r = 0$). When OHC gain at 4 kHz is maximum, the OHC contraction is about 2.8 times the RL displacement. OHC contraction is in-phase with RL displacement at low frequencies, lags RL displacement by ¼ cycle when OHC gain is maximum, and lags RL displacement by ½ cycle at high frequencies.

2.4 *Propagation function*

In the frequency-domain, the propagation function $\kappa \equiv \sqrt{Z_f Y_p}$ describes wave propagation in the cochlea, where Z_f is the impedance (per unit length) of the scala fluid and Y_p is the admittance (per unit length) of the cochlear partition:

$$Z_f \equiv \left(\frac{\partial_x P}{-\dot{\xi}_f}\right) = sM_f \tag{7}$$

$$Y_p \equiv \left(\frac{\partial_x \dot{\xi}_f}{-P}\right) = \frac{1}{(sM_b + R_b + K_b/s) \cdot (1 + H_o)}. \tag{8}$$

The propagation function is shown in Fig. 2 at 1, 4, and 16 kHz for model parameters selected to represent a cat cochlea. The propagation functions shown in Fig. 2 are similar in shape and amplitude to the propagation functions and wave numbers derived from cat and chinchilla tuning-curve data by Shera [9].

3 Model Results

Iso-displacement tuning curves are shown in Fig. 3 for model parameters selected to represent a cat cochlea. These tuning curves were derived from impulse responses in a linear version of the model. The entire length of the BM is represented in this model by 500 points, which is 20 points/mm for the typical 25-mm length of a cat cochlea.

Cochlear nonlinearity is introduced by γ a function of RL displacement:

$$\gamma(\xi_r) \equiv \frac{1}{1 + |\xi_r / \xi_h|^{0.2}}, \tag{9}$$

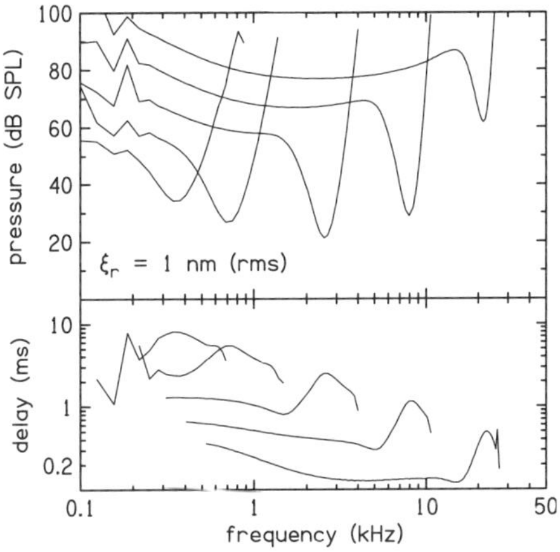

Figure 3. Iso-displacement curves at locations 20, 40, 60, 80, and 90% of the distance from the stapes. These transfer functions were obtained from impulse responses in a linear version of the time-domain model.

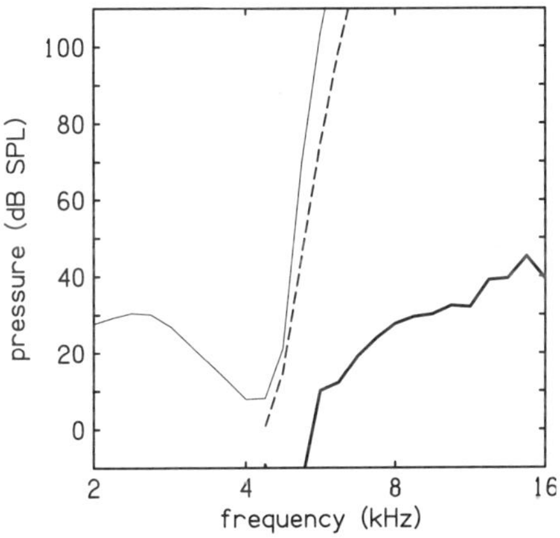

Figure 4. DPOAE level (thick line) as a function of f_1. For comparison, the two-tone stimulus level (dashed line) and single-tone threshold level (thin line) are also shown. The DPOAE level shows no indication of the CA gain present in the model.

where the parameter ξ_h specifies the RL displacement at which $\gamma = 1$. For smaller displacements, γ approaches one and CA gain is maximum. For larger displacements, γ approaches zero and CA gain is eliminated.

To simulate the Allen-Fahey experiment, the 4-kHz place in the cochlea was selected to represent the location at which excitation was kept constant under all stimulus conditions. Prior to generating DPOAEs, a threshold-tuning curve (TTC) was obtained by presenting single-tone stimuli over the range of frequencies from 2 to 8 kHz and varying the level of these tones to maintain constant displacement at the 4-kHz place. The single-tone stimulus levels determined by this procedure are shown in Fig. 4 as a thick-solid (blue) line.

DPOAEs were generated in the model by presenting two-tone stimuli with f_1 decreasing over the range from 16 to 4 kHz and f_2 selected to make the $2f_1$-f_2 distortion product always at 4 kHz. The stimulus tones always had the same level L_1=L_2, which was adjusted to maintain the same displacement at the 4 kHz place in the cochlea that was used to produce the TTC. The two-tone stimulus level is shown in Fig. 4 at the f_1 frequency as a dashed line. The thick line in Fig 4 indicates the corresponding DPOAE level at the eardrum in response to this two-tone stimulus. Although there is fine-structure in the DPOAE level across frequency that suggests partial cancellation due to multiple sources, there is no evidence of amplification as f_1 approaches 4 kHz. Thus, the model results are consistent with results observed by Allen and Fahey in that the DPOAE shows no evidence of cochlear amplifier gain. However, examination of RL displacement and power flow reveals that the model provides 12 dB of power gain to a 4-kHz tone at 20 dB SPL and 47 dB of tip-to-tail displacement gain at 0 dB SPL. We must conclude that the Allen-Fahey paradigm does not provide an accurate indication of CA gain in this model.

4 Discussion

The Allen-Fahey paradigm reveals important information about cochlear mechanics; however, due to the inherent limitations of this paradigm, estimates of cochlear amplifier gain obtained by this method may underestimate both displacement gain and power gain present within the cochlea. Two-tone neural responses, recorded from hundreds of single auditory-nerve fibers [6] suggest a spatial distribution of distortion generation that is less localized than assumed by Allen and Fahey [1]. Our time-domain model results are consistent with the quasi-linear examination of the Allen-Fahey paradigm in frequency-domain model by Kanis and de Boer [5], which also found DPOAE level to be an inaccurate indication of CA gain.

Acknowledgments

This work was supported by a grant from NIH-NIDCD (R01-DC8318).

References

1. Allen, J.B., Fahey, P.F., 1992. Using acoustic distortion products to measure the cochlear amplifier gain on the basilar membrane. J. Acoust. Soc. Am. 92, 178-188.
2. Davis, H., 1983. An active process in cochlear mechanics. Hear. Res. 9, 79-90.
3. de Boer, E., Nuttall, A.L., Hu, N., Zou, Y., Zheng, J., 2005. J. Acoust. Soc. Am. 117, 1260-1266.
4. Greenwood, D.D., 1990. A cochlear frequency-position function for several species—29 years later. J. Acoust. Soc. Am. 87, 2592-2605.
5. Kanis, L.J., de Boer, E., 1993. The emperor's new clothes: DP emissions in a locally-active nonlinear model of the cochlea. In: H. Diufhuis, J.W. Horst, P. van

Dijk, S.M., van Netten (Eds.), Biophysics of Hair Cell Sensory Systems. World Scientific, Singapore, pp. 304-314.
6. Kim, D.O., Molnar, C.E., Matthews, J.W., 1980. Cochlear mechanics: nonlinear behavior in two-tone responses as reflected in cochlear-nerve fiber responses and ear canal pressure. J. Acoust. Soc. Am. 67, 1704-1721.
7. Ren, T., Nuttall, A.L., 2006. Cochlear compression wave: An implication of the Allen-Fahey experiment. J. Acoust. Soc. Am. 119, 1940-1942.
8. Shera, C.A., 2003. Wave interference in the generation of reflection- and distortion-source emissions In: A.W. Gummer (Ed.), Biophysics of the Cochlea: From Molecules to Models. World Scientific, Singapore, pp. 439-449.
9. Shera, C.A., 2007. Laser amplification with a twist: Traveling-wave propagation and gain functions from throughout the cochlea. J. Acoust. Soc. Am. 122, 2738–2758.
10. Shera, C.A. Guinan, J.J., 2007. Cochlear traveling-wave amplification, suppression, and beamforming probed using noninvasive calibration of intracochlear distortion sources. J. Acoust. Soc. Am. 121, 1003-1016.
11. Shera, C.A., Tubis, A., Talmadge, C.L., de Boer, E., Fahey, P., Guinan, J.J., 2007. Allen–Fahey and related experiments support the predominance of cochlear slow-wave otoacoustic emissions. J. Acoust. Soc. Am. 121, 1564-1575.

Comments and Discussion

de Boer: What is new here? I am sorry but cannot accept precise conclusions from model computations that include a non-physiological type of nonlinearity. Can you clarify that problem?

Neely and Liu: Our goal was to present a "counter-example" to Allen and Fahey's assertion that cochlear-amplifier gain must be evident in DPOAE measurements. We have demonstrated that it is possible for power amplification to exist but be not observable by the Allen-Fahey paradigm. The Allen-Fahey paradigm has not previously been simulated in an active, nonlinear, time-domain model of cochlear mechanics.

Although the nonlinearity in our model may be non-physiological, the time-domain simulation is robust in the sense that, when the system is linear, it produces frequency responses that are identical to frequency-domain simulation. For this reason, we are confident that the results of the time-domain simulation described here will generalize to more physiological representations of OHC electro-motility, which is the focus of our current efforts. The essential feature that determines the results of Allen-Fahey experiment is the spatial distribution the generated distortion. The exact manner by which this distortion was generated is less important.

RETROGRADE WAVES IN THE COCHLEA

S. T. NEELY

Boys Town National Research Hospital,
Omaha, Nebraska 68131, USA

J. B. ALLEN

University of Illinois at Urbana-Champaign,
Urbana, Illinois 61801, USA

Retrograde waves in the cochlea are important because they provide information about cochlear mechanics that may be measured in the external ear canal as otoacoustic emissions. By extending traditional wave-variable methods, an explicit equation is derived for the forward-traveling wave in a one-dimensional cochlear model that is equivalent to the WKB approximation. The corresponding equation for the retrograde waves requires no additional approximation and provides a simple characterization of cochlear reflectance.

1. Introduction

The importance of retrograde waves was not fully appreciated until the discovery of sounds measured in the ear canal that originate within the cochlea [1]. These otoacoustic emissions (OAE) provide a noninvasive means to acquire information about cochlear function.

Coherent reflection theory is the accepted explanation for stimulus-frequency OAEs (SFOAE) [2]. Zweig and Shera use the WKB approximation [3, 4] to obtain a solution for the fluid pressure in a cochlear transmission line. Their *coherent reflection* theory provides a theoretical explanation for observed amplitude and phase characteristics of SFOAEs.

In this paper, wave-variable methods that have traditionally been applied to uniform transmission lines are extended to obtain the WKB formula described by [4]. This formulation leads to a simple equation for reflectance looking into the cochlea from the base.

2. Cochlear model

The equation for a one-dimensional long-wave transmission-line model of the cochlea[3] is

$$\frac{d}{dx}\begin{bmatrix} P(x,f) \\ U(x,f) \end{bmatrix} = -\begin{bmatrix} 0 & Z_s(x,s) \\ Y_b(x,s) & 0 \end{bmatrix}\begin{bmatrix} P(x,f) \\ U(x,f) \end{bmatrix}, \qquad (1)$$

where $P(x,f)$ is the scala fluid pressure, $U(x,f)$ is the longitudinal volume velocity of the scala fluid, $Z_s(x,s)$ is the scala impedance (per unit length) and $Y_b(x,s)$ is the BM admittance (per unit length). The Laplace frequency s is used for expressions of frequency, such as impedances, that are causal functions in the time domain. The real frequency f (and $\omega \equiv 2\pi f$) are restricted to the Fourier transform of noncausal functions. These model variables are shown in Fig. 1 in the context of a single section of a transmission line.

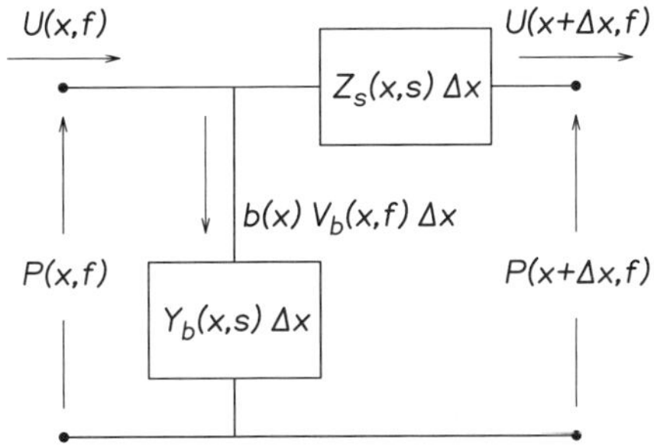

Fig. 1. Transmission line model for a section of the cochlea of length Δx. The series-impedance $Z_s(x,s)\Delta x$ limits fluid translational volume velocity $U(x,f)$, while the shunt-admittance $Y_b(x,s)\Delta x$ limits the basilar membrane volume velocity.

Scala impedance depends on fluid density ρ_0 and cross-sectional area $A(x)$.

$$Z_s(x,s) = s\rho_0 / A_s(x). \tag{2}$$

BM velocity is proportional to the gradient of the scala volume velocity:

$$\frac{dU}{dx}(x,f) = -Y_b(x,s) \cdot P(x,f) = b \cdot V_b(x,f), \tag{3}$$

where b is the effective BM width.

3. Wave variables

The wave *propagation function* $\kappa(x,s)$ and the *characteristic impedance* $z_0(x,s)$ are determined entirely by the medium, and therefore only depend on the per unit length series impedance $Z(x,s)$ and shunt admittance $Y_b(x,s)$ [5]:

$$\kappa(x,s) \equiv \sqrt{Z_s(x,s) \cdot Y_b(x,s)} \tag{4}$$

$$z_0(x,s) \equiv \sqrt{Z_s(x,s)/Y_b(x,s)}. \tag{5}$$

The pressure *wave* variables P_+ and P_- are defined by

$$\begin{bmatrix} P_+ \\ P_- \end{bmatrix} \equiv \frac{1}{2} \begin{bmatrix} 1 & z_0 \\ 1 & -z_0 \end{bmatrix} \begin{bmatrix} P \\ U \end{bmatrix} \tag{6}$$

We invert Eq. 6

$$\begin{bmatrix} P \\ U \end{bmatrix} = \begin{bmatrix} 1 & 1 \\ y_0 & -y_0 \end{bmatrix} \begin{bmatrix} P_+ \\ P_- \end{bmatrix}, \tag{7}$$

where $y_0 = 1/z_0$, and substitute Eq. 7 into Eq. 1:

$$\frac{d}{dx} \begin{bmatrix} P_+ \\ P_- \end{bmatrix} = -\frac{1}{2} \begin{bmatrix} y_0 Z_s + z_0 Y_b + z_0 y' & -y_0 Z_s + z_0 Y_b - z_0 y' \\ y_0 Z_s - z_0 Y_b - z_0 y' & -y_0 Z_s - z_0 Y_b + z_0 y' \end{bmatrix} \begin{bmatrix} P_+ \\ P_- \end{bmatrix}. \tag{8}$$

In this equation $y_0' = dy_0/dx$. The equivalence of $y_0 Z_s = z_0 Y_b = \kappa$ allows Eq. 8 to be written as

$$\frac{d}{dx}\begin{bmatrix} P_+ \\ P_- \end{bmatrix} = \begin{bmatrix} -\kappa+\varepsilon & -\varepsilon \\ -\varepsilon & \kappa+\varepsilon \end{bmatrix}\begin{bmatrix} P_+ \\ P_- \end{bmatrix} \tag{9}$$

with

$$\varepsilon(x,f) \equiv \frac{1}{2}\frac{d}{dx}\ln z_0(x,s). \tag{10}$$

Note that when $z_0(x,s)$ is independent of x, ε is equal to zero, which decouples $P_+(x,f)$ and $P_-(x,f)$ in Eq. 9. When $z_0(x,s)$ varies slowly with x, $|\varepsilon/\kappa|$ is small, which may be exploited to obtain approximate model solutions.

Figure 2 shows examples of P_+ and P_- that demonstrate the importance of BM roughness in generating retrograde waves. Model parameters were selected to

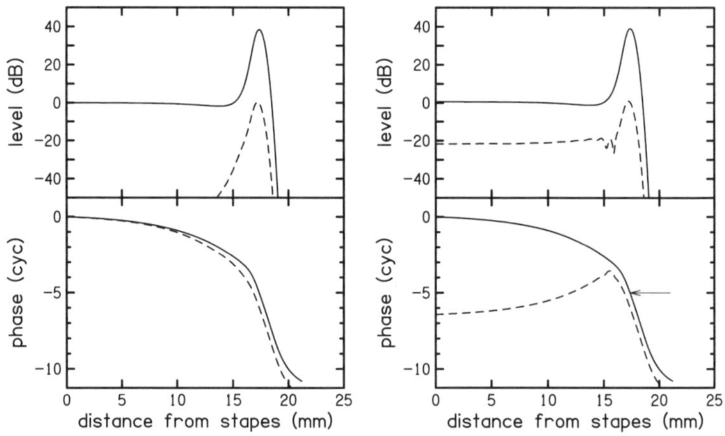

Fig. 2. Forward-traveling (solid) and retrograde (dashed) pressure components for the approximate model. The magnitude and phase are shown relative to pressure at the stapes. The results on the left side are for smooth BM impedance. The results on the right side are for rough BM impedance. The arrow indicates the phase at the characteristic place.

maintain nearly uniform characteristic impedance by making the scala area decrease with x at the same rate as BM stiffness [4]. The "tall broad peaks" of the BM excitation patterns were created by specifying a negative damping region at a location just basal to the tonotopic place. The negative damping was sufficient to boost the tip-to-tail ratio of the BM excitation patterns by about 60 dB. The model results shown in Fig. 2 were obtained by finite-difference solution of Eq. 1 and are called *exact* to contrast them with approximate results described below.

4. Approximate wave variables

The top row of Eq. 9 provides a differential equation for P_+ coupled to P_-

$$\frac{d}{dx}P_+ = (\varepsilon - \kappa)\,P_+ - \varepsilon P_-. \tag{11}$$

When $y_0(x,s)$ varies slowly with x, we can obtain an approximate equation for P_+ that is independent of P_- by assuming that $|(\kappa+\varepsilon)\,P_+| \gg |\varepsilon\,P_-|$. With this approximation, Eq. 11 reduces to

$$\frac{d}{dx}\widetilde{P}_+ = (\varepsilon - \kappa)\,\widetilde{P}_+ \tag{12}$$

which can be integrated to obtain

$$\widetilde{P}_+(x,f) = P_+(0,f)\cdot\exp\left[\int_0^x \varepsilon(x_1,f) - \kappa(x_1,f)dx_1\right], \tag{13}$$

giving an approximate expression for P_+. Surprisingly this expression is the WKB approximation for the forward-traveling pressure wave [6, 3, 7].

The bottom row of Eq. 9 gives us a differential equation for P_- that is coupled to P_+,

$$\frac{d}{dx}P_- = (\kappa + \varepsilon)\,P_- - \varepsilon P_+. \tag{14}$$

No further approximations are needed to obtain a solution for the retrograde wave. Substitute Eq. 13 into Eq. 14 to obtain

$$\frac{d}{dx}\widetilde{P}_- - (\kappa+\varepsilon)\,\widetilde{P}_- = -\varepsilon P_+(0,f)\cdot\exp\left[-\int_0^x (\kappa-\varepsilon)\,dx_1\right]. \tag{15}$$

The integration of this equation requires a boundary condition. If we assume that $\widetilde{P}_-(L,f) \approx 0$, we obtain

$$\widetilde{P}_-(x,f) = P_+(0,f)\cdot\exp\left[\int_0^x (\kappa+\varepsilon)\,dx_1\right]\cdot\int_x^L \varepsilon\exp\left[-2\int_0^{x_2}\kappa dx_3\right]dx_2, \tag{16}$$

where x_2 and x_3 are additional integration variables for x. The sum of Eqs. 13 and 16 provides an approximate solution for the total pressure $\widetilde{P}(x,f)$. The accuracy of this approximation is demonstrated in Fig. 3 by comparing exact and approximate solutions of peak pressure (at the characteristic place) relative to stapes pressure for versions of the cochlear model with smooth and rough BM. The exact solutions required about 20 times longer to compute.

We use Eqs. 13 and 16 to write an approximate equation for reflectance

$$\widetilde{\mathcal{R}}(x,f) = \exp\left[2\int_0^x \kappa(x_1)dx_1\right]\cdot\int_x^L \varepsilon(x_2)\exp\left[-2\int_0^{x_2}\kappa(x_3)dx_3\right]dx_2, \tag{17}$$

which at the stapes reduces to

$$\widetilde{\mathcal{R}}(0,f) = \int_0^L \varepsilon(x_2)\exp\left[-2\int_0^{x_2}\kappa(x_3)dx_3\right]dx_2. \tag{18}$$

Comparison between exact and approximate reflectance (not shown) demonstrate excellent agreement.

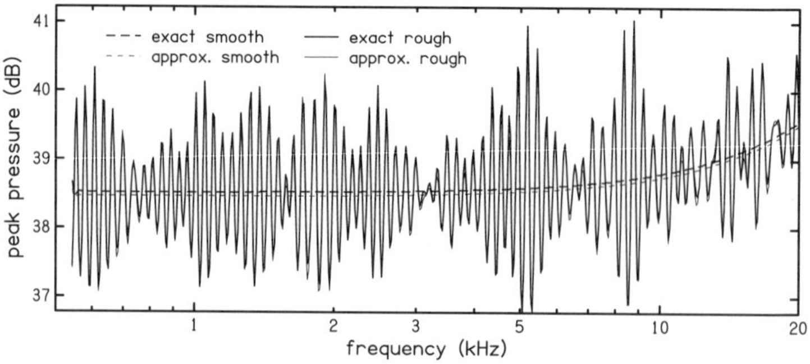

Fig. 3. Peak pressure (re stapes pressure) as a function of frequency. The peak pressure is defined as the maximum pressure magnitude over the entire length of the BM at each frequency. The four curves represent different model conditions: (1) exact-smooth (thick dashed); (2) exact-rough (thick solid); (3) approximate-smooth (thin dashed); (4) approximate-rough (thin solid). The approximate results (thin) are barely visible in this figure because they are covered by the exact results (thick).

5. Discussion

Figure 2 shows exact model results for forward-traveling pressure P_+, and retrograde pressure P_-, at one frequency. The results on the left of Fig. 2 are for a smooth BM stiffness. The relatively small P_- level at the stapes indicates that reflection at the tonotopic place is insignificant. The results on the right are for a rough BM stiffness. The rough stiffness had a small, random increase added to the smooth stiffness at each place along the BM. The random stiffness increase was uniformly distributed between 0 and 0.001%. This small amount of roughness was sufficient to produce significant reflection of the forward-traveling wave in the vicinity of the tonotopic place. The positive slope of the phase of the P_- component indicates that retrograde energy is being propagated. Another interesting feature demonstrated by the cochlear model results in Fig. 2 is that the round trip delay of an SFOAE may be less than twice the forward delay to the characteristic place.

In Fig. 3, we see that the pressure difference between the smooth and rough models (i.e., the peak pressure fine structure) is the same for the exact and approximate results. In other words, the fine structure of the approximate pressure is the same as the fine structure of the exact pressure. Model peak-pressure fine structure represents *threshold* fine-structure rather than *SFOAE* fine-structure [8].

In Eq. 18, the accumulated phase of the incident wave $\widetilde{P}_+$ at location x comes primarily from the integral of κ. In Eq. 18, note that the contribution to the stapes reflectance $\widetilde{R}$ of the retrograde wave from any location x has twice the phase accumulation of the incident wave to that location. This doubling of phase is consistent with the mechanism of reflection being *place-fixed*. Experimental observations of OAE phase as a function of frequency suggest that some types of emissions are generated by a place-fixed mechanism [9].

6. Conclusions

The wave variable decomposition traditionally applied to uniform transmission lines may be extended to the case of nonuniform transmission lines, which may be used to represent one-dimensional models of cochlear mechanics. When the characteristic impedance z_0 varies slowly, the equations for $\widetilde{P}_+$, $\widetilde{P}_-$ and $\widetilde{\mathcal{R}}$ provide useful approximations for solutions of cochlear models with "tall broad peaks."

Acknowledgments

This work was partially supported by a grant from NIH-NIDCD (R01-DC8318).

References

1. Kemp, D., 1980. Stimulated acoustic emissions from within the human auditory system. J. Acoust. Soc. Am. 5, 1386-1391.
2. Zweig, G., Shera, C.A., 1995. The origin of periodicity in the spectrum of evoked otoacoustic emissions. J. Acoust. Soc. Am. 98, 2018-2047.
3. Zweig, G., Lipes, R., Pierce, J.R., 1976. The cochlear compromise. J. Acoust. Soc. Am. 59, 975-982.
4. Shera, C.A., Zweig, G., 1991. Reflection of retrograde waves within the cochlea. J. Acoust. Soc. Am. 89, 1290-1305.
5. Brillouin, L, 1953. Wave propagation in periodic structures. Dover, London, second edition.
6. Durney, C.H. Johnson, C.C., 1969. Introduction to modern electromagnetics. McGraw-Hill, New York.
7. Viergever, M.A., de Boer, E., 1987. Matching impedance of a nonuniform transmission line: Application to cochlear modeling. J. Acoust. Soc. Am. 81, 184-186.
8. Choi, Y.-S., Lee, S.-Y., Parham, K., Neely, S., Kim, D., 2008. Stimulus-frequency otoacoustic emission: Measurements in humans and simulations with an active cochlear model. J. Acoust. Soc. Am. 123, 2651-2669.
9. Shera, C.A., Guinan, J.J., 1999. Evoked otoacoustic emissions arise by two fundamentally different mechanisms: A taxonomy for mammalian OAEs. J. Acoust Soc. Am. 105, 782-798.

Comments and Discussion

Shera: With appropriate allowance for changes in notation, Eq. (18) for the reflectance is identical to Eq. (11) derived by Viergever (1986) using the "Bremmer series" (1951). The relationship between reflectances based on Bremmer and WKB series is discussed in Ref. 4, above.

Bremmer, H., 1951. The W.K.B. approximation as the first term of a geometric-optical series. Commun. Pure Appl. Math. 4, 105-115.

Viergever M.A, 1986. Asymmetry in reflection of cochlear waves. In Auditory Frequency Selectivity, edited by B.C.J. Moore and R.D. Patterson. (New York, Plenum), pp. 31-38.

COCHLEAR REFLECTIVITY AND TEOAE TRANSFER FUNCTION

RENATA SISTO

Occupational Hygiene Department, ISPESL,
Via di Fontana Candida,1, Monte Porzio Catone (RM), 00040 Italy

ARTURO MOLETI

Physics Department, University of Roma Tor Vergata,
Via della Ricerca Scientifica, 1,Roma, 00133, Italy

FILIPPO SANJUST

Occupational Hygiene Department, ISPESL,
Via di Fontana Candida,1, Monte Porzio Catone (RM), 00040 Italy

Cochlear reflectivity has been theoretically estimated for different otoacoustic emission generation mechanisms, using a standard perturbative approach based on the WKB approximation and the osculating parameters technique. The predictions of theoretical models have been compared with experimental measurements of the TEOAE transfer function at different stimulus levels, between 60 and 90 dB pSPL. The phase-gradient delay of the TEOAE response has also been compared to the estimates of the cochlear transmission delay obtained with time-frequency analysis of the same TEOAE waveforms. Place-fixed and wave-fixed OAE generation mechanisms are discussed in the light of the results.

1 Introduction

Measuring the transfer function of transient evoked otoacoustic emission (TEOAEs), provides a fast estimate of the cochlear reflectivity function, defined as the ratio between the spectral amplitudes of the backward and forward traveling waves, measured at the cochlear base [1]. Several assumptions have to be made; in fact, the actually measured quantity is the ratio between the OAE and click stimulus spectral amplitudes, which cannot be considered the frequency response of the system, due to cochlear nonlinearity. Moreover, this function of frequency is dependent on the middle ear forward and backward transmission (defined as a frequency dependent input/output matrix) [2]. Analytical estimates of the cochlear reflectivity have been obtained for linearized cochlear transmission line models, whose validity is not guaranteed at high stimulus (or, more precisely, response) levels. Different OAE generation mechanisms have been proposed, and, among them, the most popular are nonlinear distortion and linear reflection [3]. The standard perturbative technique that we have followed in this study is based on the osculating parameter technique, applied to the WKB approximated solutions of the linearized cochlear 1-d transmission line equations [1]. Reflectivity is computed as the ratio between the amplitudes of the backward and forward components of the traveling wave at the cochlear base. The overall reflectivity is given by a cochlear path integral of a "scattering potential" associated with the difference between the true cochlear equation and that whose the WKB functions are the exact solutions. For a smooth linear cochlea, a null result is obtained, but introducing a perturbation associated either to weak nonlinearity or to roughness, significant reflectivity levels can be obtained, capable of explaining the measured OAE levels. Examining the theoretical reflectivity estimates, it turns out that they crucially depend on the width of the cochlear activity pattern, which is a function of the cochlear tuning.

2 Model

In the frequency domain, the linearized cochlear model equations have the form:

$$\frac{\partial^2 P_d(x,\omega)}{\partial x^2} + k^2(x,\omega) P_d(x,\omega) = 0, \tag{1}$$

$$\ddot{\xi}(x,\omega) + \gamma\dot{\xi}(x,\omega) + \omega_0^2 \xi(x,\omega) = \frac{P_d(x,\omega)}{\sigma_{bm}\omega^2} \tag{2}$$

where x is the longitudinal position measured along the BM starting from the base, ω is the angular frequency, k is the wave vector, P_d is the differential pressure applied to the BM. Active terms can be inserted either in the damping term or in the elastic term of Eq.(2), in the form of explicit anti-damping terms or as time-delayed stiffness terms [1].

The wave vector is a tonotopically resonant function:

$$k^2(x,\omega) = \frac{k_0^2\omega^2}{\Delta(x,\omega)}, \tag{3}$$

where k_0 is a constant dependent on geometrical and structural cochlear properties, and Δ is a tonotopically resonant denominator:

$$\Delta(x,\omega) = \omega_0^2 - \omega^2 + i\omega\gamma \tag{4}$$

The local resonance frequency is related to cochlear position by the Greenwood map:

$$\omega_0(x) = \omega_{max} e^{-k_\omega x} + \omega_l \tag{5}$$

where k_ω is the cochlear inverse scale length, and ω_l a small scale-invariance violating term.

The WKB solution is a linear combination of basis functions of the type:

$$\beta_\pm(x,\omega) \approx \frac{1}{\sqrt{k}} e^{\mp j\int_0^x k(x',\omega)dx'}, \tag{6}$$

representing forward and backward traveling waves on the BM. These functions are good approximations of the exact solution as long as the wave vector changes slowly with x, and the group velocity is different from zero [1]. These two conditions are not verified near the resonant place, particularly for sharp, high-Q, resonances.

The WKB basis functions, $\beta_\pm$, are the exact solutions of the equation:

$$\frac{\partial^2 \beta_\pm}{\partial x^2} + k^2(1+\varepsilon_\pm)\beta_\pm(x,\omega) = 0\,, \tag{7}$$

which differs from Eq.(1) by a relative error function ε. If ε is sufficiently small, a perturbative regime holds. The exact solution can be developed as:

$$P_d = P_+ + P_- = \psi_+\beta_+ + \psi_-\beta_-, \tag{8}$$

where P_+ represents a progressive wave traveling toward the apex, whereas P_- represents a regressive wave traveling back towards the cochlear base. The amplitude at the base ($x=0$) for a regressive wave generated by the reflection of a progressive wave of unitary amplitude (basal reflection coefficient) is:

$$\psi_- \cong \int_0^\infty dx' \beta_+ \sigma_+ \beta_+\,, \tag{9}$$

where the perturbative potential σ_+ is a local reflection source.

The sharp variation of the wave vector of each frequency component of the forward traveling wave near its tonotopic place could be considered as a possible source of reflection. The osculating parameter technique gives in this case:

$$\sigma_+ = k^2\gamma\varepsilon_+,\ \gamma = \frac{1}{2\,jk(0)},\ \text{and}\ \varepsilon_\pm = -\frac{1}{k^2}\sqrt{k}\,\frac{\partial^2}{\partial x^2}\frac{1}{\sqrt{k}}\,. \tag{10}$$

Although this "backscattering potential" is not negligible, the integral of Eq.(9) is null, if extended up to the resonant place, due to exact cancellation of opposite sign contributions in the last part of the path. This null result could be questioned, because the last part of the path is that in which the validity of the WKB approximation itself is not guaranteed. Neglecting this possibility, we restrict to considering reflectivity associated with perturbations that remove the exact symmetry responsible for the above-mentioned cancellation, namely cochlear roughness and weak nonlinearity.

We consider a saturated regime in which the BM gain is given by:

$$\hat{\chi} = \eta\left(\frac{p}{p_0}\right)^a \frac{\hat{k}^{3/2}}{k^{3/2}(0)} = \eta\left(\frac{p}{p_0}\right)^a \left(\frac{k_0\sqrt{Q}}{k(0)}\right)^{3/2}, \tag{11}$$

where p/p_0 is the stimulus pressure in units of the reference pressure $p_0 = 20\ \mu$Pa. The scaling with a power a of the stimulus amplitude keeps into account the fact that the cochlear amplifier gain is partly saturated. For $\eta = 35$ and $a = -1/3$, we obtain reasonable values of the BM gain $\hat{\chi}$ between 40 and 60 dB, decreasing with stimulus level.

We express the activity pattern width σ_x in terms of the cochlear tuning Q:

$$\sigma_x = \frac{\alpha}{k_\omega Q(f)}, \tag{12}$$

where $\alpha = 1$ corresponds to a simple passive resonator, and using the approximate estimate of the wave vector amplitude at the resonant place:

$$\hat{k} = k_0 \sqrt{Q(f)}\,. \tag{13}$$

The wave-fixed OAE sources considered in the literature are associated with nonlinear distortion. The result [1] is:

$$\psi_{nl} \cong -j\gamma \frac{\varepsilon_\gamma \omega^4 k_0{}^2 |b_+|^2 |\hat{\chi}|^4}{b^2{}_{nl}\omega^4{}_{\max}} e^{-4j\pi N_\lambda} e^{-j\pi} e^{\frac{j\hat{k}k_\omega \sigma_x^2}{2}} \sigma_x \sqrt{\frac{\pi}{2}} \exp\left[-(\hat{k}^2 - \frac{k^2{}_\omega}{4})\frac{\sigma_x^2}{2}\right]. \tag{14}$$

Here, b_{nl}^2 is a cochlear displacement saturation level, σ_x is the width of the cochlear activity pattern, N_λ is the number of wavelengths from the base to the tonotopic place, b_+ is the basal cochlear displacement, and a hat indicates the resonant place.

Using Eqs.(11-13), the amplitude of the nonlinear distortion reflectivity becomes:

$$|\psi_{nl}| \cong \sqrt{\frac{\pi}{2}} \frac{k_0}{k_\omega} \frac{\varepsilon_\gamma a_+^2}{2b^2{}_{nl}\sigma_{bm}^2 \omega_{\max}} \eta^4 \left(\frac{p}{p_0}\right)^{4a} \alpha \frac{Q^2}{\omega^3} \exp\left[-\frac{\alpha^2 k_0^2}{2Qk_\omega^2}\right], \tag{15}$$

where a_+ is the differential pressure amplitude at the base, assumed proportional to the stimulus amplitude.

The reflection sources responsible for the rapid-rotating phase component of the OAE response are assumed to be associated with small irregularities randomly distributed along the BM. The computation of the reflectivity contribution due to roughness was given by Talmadge [1]:

$$\psi_r = -\frac{\omega^4 k_0{}^2 \hat{\chi}^2 2r_0}{2jk(0)\omega_0^4} e^{-4j\pi N_\lambda} e^{-j\pi} \sigma_x e^{k_\omega^2 \sigma_x^2} e^{2j\hat{k}\hat{x}} \sqrt{\pi}\, \tilde{r}_f(\hat{x} - k_\omega \sigma_x^2) \tag{16}$$

Using Eqs.(11-13), the amplitude of the roughness reflectivity function becomes:

$$|\psi_r| = \sqrt{\pi}\, r_0 |\tilde{r}_f| \eta^2 \alpha \frac{k_0}{k_\omega} \left(\frac{p}{p_0}\right)^{2a} Q^{\frac{1}{2}} e^{\frac{\alpha^2}{Q^2}}, \tag{17}$$

where $r_0|\tilde{r}_f|$ parameterizes the cochlear roughness amplitude.

3 Methods

In this study we have analyzed TEOAE data recorded in the nonlinear acquisition mode from 28 ears of young subjects, at stimulus levels ranging from 60 to 90 dB pSPL. Fourier analysis provided estimates of the response level and of the phase-gradient delay.

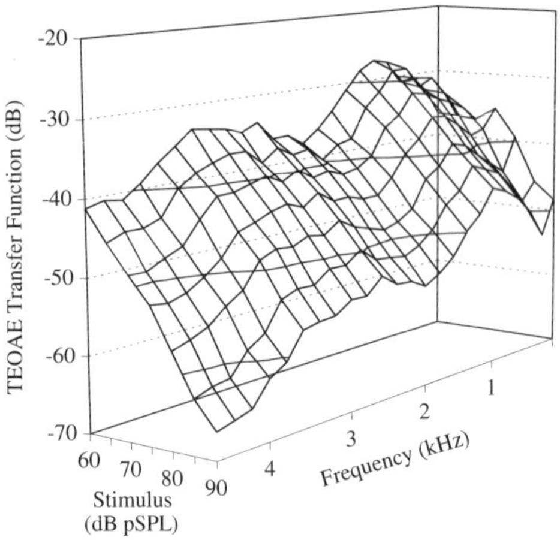

Figure 1. TEOAE transfer function. Note the rather regular dependence on stimulus level throughout the whole frequency range.

Wavelet analysis was used to get direct time-domain estimates of the TEOAE latency, associated with the round-trip cochlear delay. The dependence of cochlear tuning Q on the stimulus level was derived from TEOAE latency [4].

4 Results and Discussion

As shown in Fig.1, a clear dependence of the average transfer function on stimulus level is observed in the whole TEOAE frequency range. For each frequency, we may analyze more quantitatively the dependence of the TF on stimulus level. In Fig.2 the reflectivity at 2 kHz, estimated from the TEOAE TF, considering the middle ear round-trip gain [2] is plotted as a function of the stimulus level. The best-fit curves obtained fitting the data to the theoretical predictions for roughness only and for nonlinear distortion only are shown. The fit shown in Fig.2 can be obtained for roughness using the same fit parameters at all frequencies, whereas, for nonlinear distortion, different fine tuning of the fit parameters at different frequencies. A good fit to cochlear reflectivity assuming nonlinear distortion source only would also require $\alpha<1$ (in this case $\alpha=0.4$), i.e., narrow activity patterns, which is in disagreement with measurements of the phase of the BM transfer function [5]. These measurements led Zweig [6] to insert time-delayed stiffness terms in the BM equations to obtain sufficiently tall and broad activity patterns.

The relation between the wavelet latency and the phase-gradient delay for four stimulus level between 60 and 90 dB pSPL is shown in Fig.3. The agreement between these two characteristic times is rather good, as expected for place-fixed OAE generation. In fact, wave-fixed generation mechanism should exhibit a null phase-gradient delay (in the scale-invariant limit) [1].

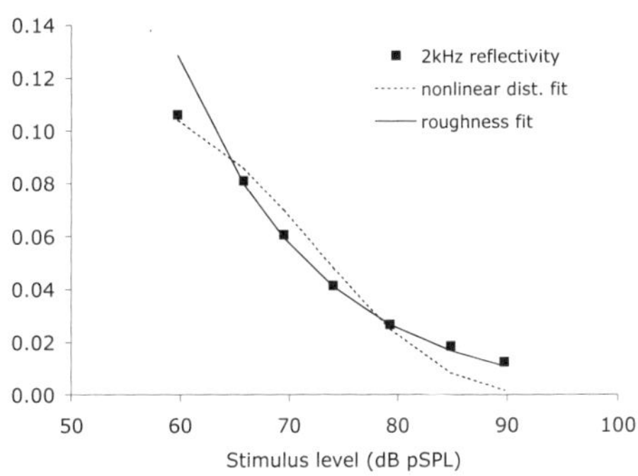

Figure 2. Cochlear reflectivity estimate from TEOAE I/O function at 2 kHz, as a function of stimulus level, and best fit to theoretical models for OAE generation.

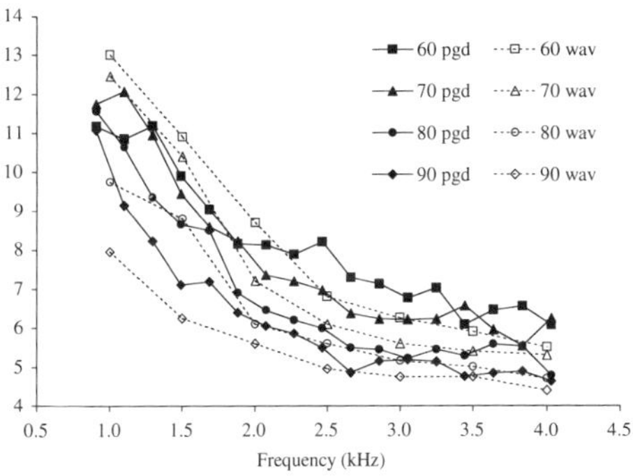

Figure 3. TEOAE phase gradient delay and wavelet latency at different stimulus levels. Note the rather good agreement between the two characteristic times.

5 Conclusions

Summarizing, the analysis of the TEOAE TF data suggests that a scenario in which place-fixed linear reflection from cochlear roughness is the main TEOAE source is in good agreement with the experimental evidence, but other mechanisms cannot be ruled out. In fact, several open theoretical problems suggest keeping a cautious approach in the interpretation of the experimental data.

References

1. Talmadge, C.L., Tubis, A., Long, G.R., Tong, C., 2000. Modeling the combined effects of basilar membrane nonlinearity and roughness on stimulus frequency otoacoustic emission fine structure. J. Acoust. Soc. Am. 108, 2911-2932.
2. Puria, S., 2003. Measurements of human middle ear forward and reverse acoustics: Implications for otoacoustic emissions. J. Acoust. Soc. Am. 113, 2773-2789.
3. Shera, C. A., Guinan, J. J. Jr., 1999. Evoked otoacoustic emissions arise from two fundamentally different mechanisms: A taxonomy for mammalian OAEs. J. Acoust. Soc. Am. 105, 782-798.
4. Sisto, R., Moleti, A., 2007. Transient evoked otoacoustic emission latency and cochlear tuning at different stimulus levels. J. Acoust. Soc. Am. 122, 2183-2190.
5. Rhode, W.S., 1971. Observations of the Vibration of the Basilar Membrane in Squirrel Monkeys using the Mössbauer Technique. J. Acoust. Soc. Am. 49, 1218-1231.
6. Zweig, G., 1991. Finding the impedance of the organ of Corti. J. Acoust. Soc. Am. 89, 1229-1254.

TIME DOMAIN MODEL OF A NONLINEAR INHOMOGENEOUS COCHLEA

STEPHEN J. ELLIOTT, EMERY M. KU, BEN LINETON
Institute of Sound and Vibration Research, University of Southampton, Southampton SO17 1BJ, UK

A state space model of the cochlea has been developed that represents its internal dynamics in the time domain. It can be used to investigate the variation of the linear frequency response at low levels with different spatial inhomogeneities in the cochlea and also simulate the nonlinear time domain response at higher levels, or for unstable systems. The model combines a macromechanical formulation for fluid coupling with a discrete distribution of nonlinear, active micromechanical elements. The model seems to provide a convenient framework for the quantitative investigation of several types of OAE within a common formulation.

1 Introduction

The cochlea is a distributed and dispersive system, whose complicated behaviour can be difficult to understand because it is, in general, also nonlinear and inhomogeneous [1]. A state space representation of the cochlea [2] was originally developed to enable the poles of a linear model of the cochlea to be calculated, which allowed its stability to be investigated for various types of inhomogeneity. The state space model is fundamentally a time domain representation, however, and it quickly became apparent that direct time domain simulation, using well developed solvers, allowed the nonlinear behaviour to be incorporated and provided results that gave considerable insight into the interaction between the different physical processes within the cochlea.

In this paper the formulation of the state space model is briefly outlined and results are presented for perturbations about a particular set of parameters, which approximate the behaviour of the human cochlea. The effect of small inhomogeneities, which do not cause the linear model to become unstable, on the frequency response is first investigated. Larger inhomogeneities make the linear model unstable and nonlinearities in the cochlear amplifier need to be introduced to illustrate the transient behaviour in time domain simulations.

2 Formulation of the State Space Model

The main assumption of the model presented here is that radial distribution of the basilar membrane (BM) motion, in response to the pressure difference in the fluid chambers, is reasonably independent of excitation frequency and level. A single variable can then be used to describe the BM motion at each point along the cochlea. The component of the differential pressure distribution in the fluid chambers that drives the BM motion can then also be represented by a single variable at each point on the cochlea. More complicated motion across the BM could be accounted for by increasing the number of variables, but it is felt that this "one-dimensional" model captures most of the important behaviour.

The BM is now assumed to be divided into a finite number of elements, N, and the dynamic behaviour of the cochlea is split into that due to the fluid coupling along the BM, and that due to the isolated, uncoupled, behaviour of each BM element, as illustrated in Figure 1. If the set of N pressure difference waveforms at each element is represented by the vector $\mathbf{p}(t)$ and the set of BM accelerations at each element is represented as $\ddot{\mathbf{w}}(t)$, then the fluid coupling can be written as in Figure 1, where $\mathbf{G}$ and $\mathbf{H}$ are matrices corresponding to the sampled Green's function for the fluid coupling and $\dot{\mathbf{q}}(t)$ is the time derivation of the driving velocity at the stapes. Note that even though the BM response is set out as a one-dimensional array, a full three-dimensional fluid coupling model could be represented by $\mathbf{G}$ and $\mathbf{H}$, since only the single pressure variable that affects the BM is represented in $\mathbf{p}(t)$.

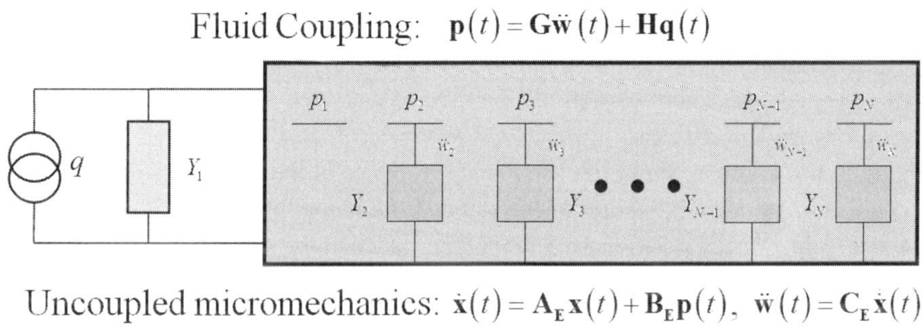

Figure 1. The discrete model of the cochlea in which displacements of the individual BM elements, $w_2\, w_3 \ldots w_N$, are related to the corresponding pressure differences $p_1\, p_2 \ldots p_N$, by individual micromechanical models $Y_2\, Y_3 \ldots Y_{N-1}$, including the effects of the middle ear, Y_1 and helicotrema, Y_N.

The dynamics of each of the individual isolated elements of the BM can be represented by finite-dimensional micromechanical models, which can be written as a set of first order differential equations in terms of their internal states [3]. Combining together all the internal states for each of these models into an overall state vector, $\mathbf{x}(t)$, all these uncoupled models can be represented by a set of state space equations in which $\mathbf{A}_E$, $\mathbf{B}_E$ and $\mathbf{C}_E$ are matrices that combine the dynamics for each isolated element. Note that if the elements are assumed to only react locally, then $\mathbf{A}_E$ and $\mathbf{B}_E$ are block diagonal [2] but, in general, longitudinal coupling along the cochlea could be incorporated by allowing off-diagonal blocks.

Combining these equations then gives a state space model for the coupled cochlea that can be written in the standard form [3]

$$\dot{\mathbf{x}}(t) = \mathbf{A}\,\mathbf{x}(t) + \mathbf{B}\,\mathbf{u}(t), \tag{1}$$

where $\mathbf{A} = [\mathbf{I} - \mathbf{B}_E \mathbf{G} \mathbf{C}_E]^{-1} \mathbf{A}_E$, $\mathbf{B} = [\mathbf{I} - \mathbf{B}_E \mathbf{G} \mathbf{C}_E]^{-1} \mathbf{B}_E$ and $\mathbf{u}(t) = \mathbf{H}\,\mathbf{q}(t)$. The waveforms of all the parameters within the model, including those of the pressure differences at each location, can be inferred from the state vector.

If the system is linear, then its stability can be determined by examining the eigenvalues of the system matrix, **A**, which are equal to the poles of the coupled system. The system is unstable if the real parts of any of these poles have a positive value.

In the simulation results presented below, the fluid coupling model makes the long wavelength assumption [1] and the BM elements are assumed to be uncoupled longitudinally and to behave as active two degree of freedom systems [4], The parameters are adjusted to provide distributions of characteristic frequency, active enhancement and wavelength that approximate those in the human cochlea [5] and have an individual value of active gain, γ, for each element along the cochlea.

3 The Effect of Inhomogeneities in the Linear Cochlea

When excited at low levels, the cochlea is active, but behaves almost linearly [1]. Provided the system is stable, which can be established from the calculated position of the poles, the state space formulation can be used to calculate the frequency response at any position within the cochlea.

Figure 2, for example, shows the frequency response of the pressure at the basal end of the cochlea to velocity excitation of the stapes for a model having a 2% step decrease in the active gain, γ, at a position 16.3 mm along the cochlea. Also shown in Figure 2 is the distribution of complex pole positions, plotted with the imaginary part, proportional to frequency, along the horizontal axis and the real part, σ, proportional to the transient divergence rate, plotted along the vertical axis. Since there are 500 elements, and each element has 4 states, there are about 2,000 poles in this model, of which only the poles with positive frequencies and σ greater than -1,000 s^{-1} are plotted in Figure 2. It can be seen that in this case all the poles have a negative real part and are thus stable. The main effect of the step in γ is to perturb a few of the poles close to the characteristic frequency for the step position. The poles closest to the imaginary axis have the smallest damping ratio and give rise to the peaks in the computed frequency response. If the size of the step becomes larger, the dominant pole position has a positive real part. This indicates that the system is unstable and, although a finite frequency response can still be calculated, it does not represent the response of the physical system because it ignores the exponentially diverging transient due to the unstable pole.

Figure 2 also shows the corresponding distribution of pole positions and basal pressure frequency response for a model in which the distribution of cochlear gain, γ, is randomly perturbed by about 0.1% from its nominal value. Such random perturbations would be expected in any biological system and their average behaviour are represented by a wavenumber spectrum [2,5]. It can be seen from the pole positions that the system is stable, but that the positions of a number of poles have been perturbed so that the real parts are less negative. These poles give rise to peaks in the frequency response of the pressure at the basal end of the cochlea, and so will also give rise to peaks in the measured pressure in the outer ear that are thought to be the source of stimulus frequency otoacoustic emissions, SFOAE. Although the individual frequencies of these peaks depends on the particular form of the random perturbation, their average spacing is predicted to only depend on the length scales and wavelength in the model [6,7]. The state space model allows these predictions to be tested [8].

If the BM frequency response at any position along the cochlea is computed with this model, these will, to a greater or lesser extent, also exhibit peaks at the same frequencies as those at the basal end. These peaks are thought to give rise to the fine structure in the hearing threshold, which is thus intimately connected to the generation of SFOAEs [6,7].

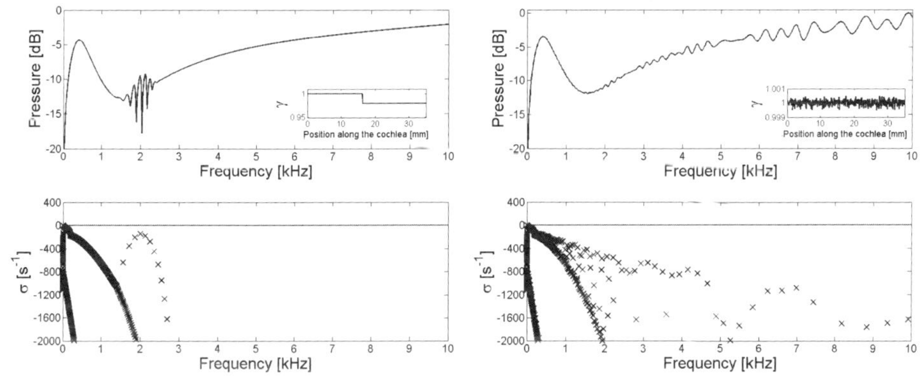

Figure 2. The calculated frequency response of the pressure difference at the basal end of the cochlea (upper plots) for a model with a small step (left) or a spatially random perturbation (right) in the cochlea gain, together with the distribution of poles for these models.

4 Instabilities in the Nonlinear Inhomogeneous Cochlea

Once the inhomogeneities within the linear cochlear model become large and/or abrupt enough, the model has poles with positive real parts and is thus unstable. If a time domain simulation were performed of such a linear but unstable cochlea, we would see exponentially diverging behaviour at the frequency of the unstable poles, which would grow without limit. In the real cochlea the amplitude is constrained by the saturating nonlinearity of the outer hair cells [1]. This can be incorporated into the state space model by making the appropriate elements of the system matrix, **A**, dependent in the relevant states. Recent work has shown the importance of the correct form and position of this nonlinearity in accurately predicting harmonic distortion in the cochlea [9]. Well-developed solvers can then be used to simulate the time domain behaviour of this nonlinear model, such as *ode45* in MATLAB, which uses a fourth order Runge-Kutta algorithm with variable internal step size.

Figure 3 shows a colour-coded plot of the transient response of a model with a 12% step decrease in active gain 16.3 mm along the cochlea to an impulsive excitation at the stapes. This discontinuity generates three unstable poles. The velocity is shown as a function of time at each location along the cochlea, where red denotes a positive velocity and dark blue denotes a negative velocity. The main curved part of the picture corresponds to the transient response of the linear cochlea, which slows and broadens as it propagates along the cochlea. It is also just possible to see evidence of a backward travelling wave, originating at the position of the step after about 3 ms, which then reflects off the stapes and reinforces the oscillation that occurs at this position after about 9 ms. The frequency of this oscillation is about 2 kHz, corresponding to the characteristic frequency at the step position and the peaks in the frequency response in Figure 2.

The transient response can be represented in two-dimensional form as in Figure 3, but greater insight can be obtained from viewing the BM velocity along the cochlea as a movie [10]. Although only the physical behaviour of the individual parts of the BM and the fluid coupling have been used to construct the state space model, it is clear that wave behaviour naturally emerges from these simulations, as predicted by WKB analysis [1,6,7]. The reflection of the incoming wave by the step is also apparent, although the subsequent reflection of the backward wave by the stapes is less clear, because the wavespeed is high and the amplitude is low. These simulations do, however, clearly show the 'active laser' mechanism of oscillation within the cochlea, and subsequent generation of spontaneous otoacoustic emissions (SOAEs), as predicted by Shera [11].

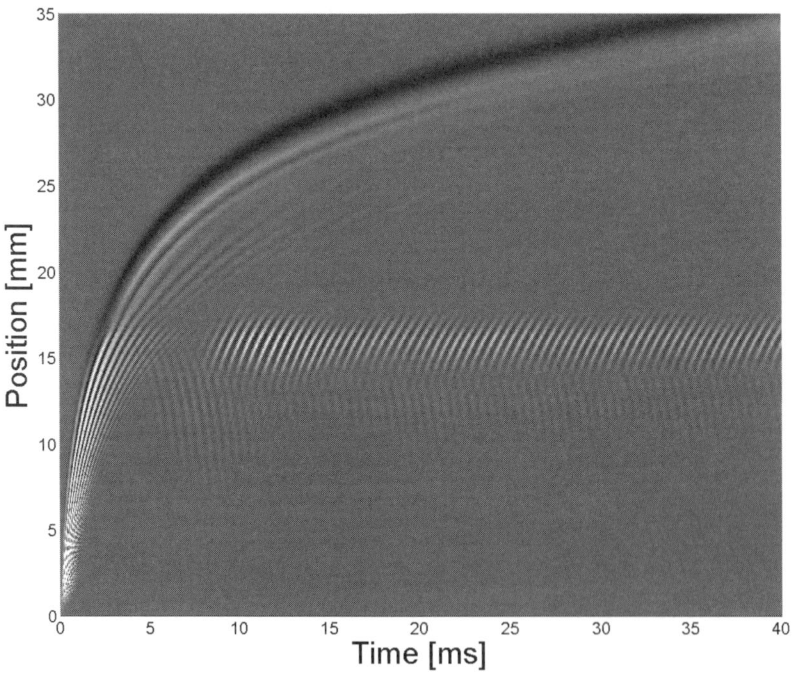

Figure 3. The transient response at each point along a nonlinear model with a step in active gain 16.3 mm along the cochlea.

Figure 4 shows the corresponding transient response for a model in which the perturbation in γ is random, but large enough to generate 28 unstable poles. It is clear that a number of different frequency oscillations exist at various places along the cochlea, which interact in rather subtle ways. The frequencies of the instabilities are reasonability well predicted from the imaginary parts of the unstable poles in this case, and their average spacing is about the same as those in the stable case above.

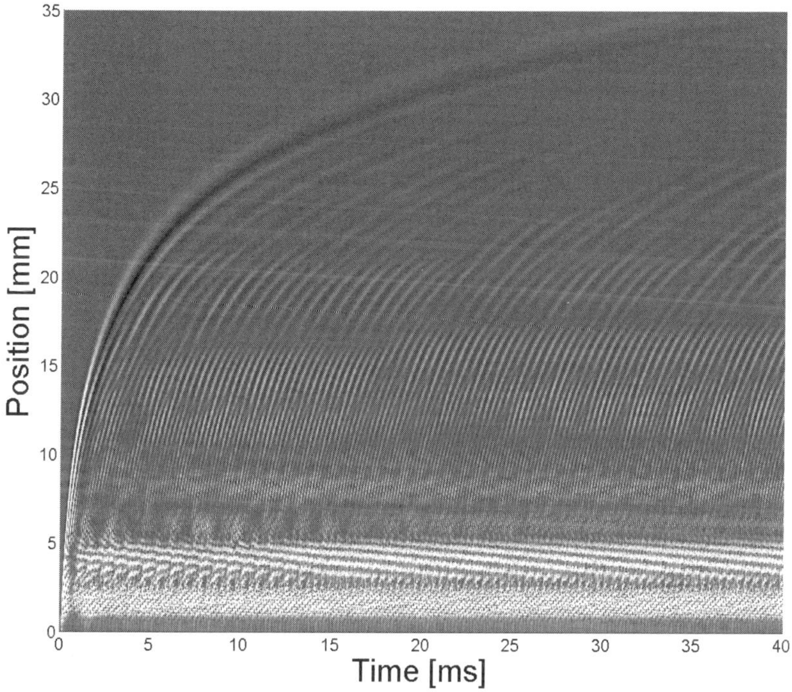

Figure 4. The transient response at each point along a nonlinear model with a spatially random variation in active gain, 10% peak-to-peak change in γ.

5 Conclusions

A state space representation of the cochlea has been presented that allows the stability of linear models and the time domain responses of nonlinear models to be easily calculated.

Small spatial inhomogeneities in the cochlear dynamics are shown, on average, to modulate the SFOAE and the hearing threshold in the way predicted by previous authors [6,7], but the detailed behaviour for particular forms of perturbation can now be investigated. Larger spatial inhomogeneities cause the linear model to become unstable, but the incorporation of saturation into the cochlea amplifier model allows time domain simulations to be performed that predict the stable limit cycle oscillations that are thought to generate SOAEs [11].

Future work will exploit the ability of the model to simultaneously represent the effects of nonlinearities and inhomogeneities to investigate the interaction between wave-fixed and place-fixed features in otoacoustic emissions [7].

References

1. E. de Boer, 1996. Mechanics of the cochlea; modeling efforts. In: P. Dallos, A.N. Popper and R.R. Fay (Eds) The Cochlea, Springer Verlag, pp258-317 .
2. S.J. Elliott, E.M. Ku and B. Lineton, 2007. A state space model for cochlear dynamics. J. Acoust. Soc. Am. 122(5), 2759-2771.
3. G.F. Franklin et al, 2005. Feedback Control of Dynamic Systems. Addiston Wesley.
4. S.T. Neely and D.O. Kim, 1986. A model for active elements in cochlear biomechanics. J. Acoust. Soc. Am. 79(5), 14721-1480.
5. E.M. Ku, S.J. Elliott and B. Lineton, 2008. Statistics of instabilities in a state space model of the human cochlea. Accepted for publication in J. Acoust. Soc. Am.
6. G. Zweig and C.A. Shera, 1995. The origin of periodicity in the spectrum of evoked otoacoustic emissions. J. Acoust. Soc. Am 98(4), 2018-2047.
7. C.L. Talmadge et al. 1998. Modelling otoacoustic emissions and hearing fine structures. J. Acoust. Soc. Am. 104(3), 1517-1543.
8. E.M. Ku, S.J. Elliott and B. Lineton, 2008. Periodicity in the spectrum of modelled spontaneous otoacoustic emissions. This volume.
9. J. How, S.J. Elliott and B. Lineton, 2008.The influence on predicted harmonic generation of the position of the nonlinearity within micromechanical models. This volume.
10. www.cochlearmodelling.isvr.soton.ac.uk
11. C.A. Shera, 2003. Mammalian spontaneous otoacoustic emissions are amplitude-stabilised cochlear standing waves. J. Acoust. Soc. Am. 114(1), 244-262.

Comments and Discussion

Sisto: I would like to know how the nonlinearity is included in your state space model formulation in the time domain. It seems to me that the matrix A and B only act linearly, so how do you model the effect of nonlinear saturation in the damping term? In your 2007 article the state variable of each micromechanical element is made by four degrees of freedom associated to the position and velocity of two coupled oscillators. Are they necessary to model an active feedback mechanism? It seems to me that it is a linear model in any case. Could I compare this model with the action of delayed stiffness terms introduced by Zweig to get a broad activity pattern profile? What happens if in your model the state variable x was made of only two degrees of freedom, the transversal displacement of the BM and its first derivative?

Elliott: If the A and B matrices are constant, the state space model is linear. Nonlinearity is introduced by making some of the elements of the A matrix dependent on the states. Specifically, the instantaneous active gain in each micromechanical model in each micromechanical element is dependent on the instantaneous relative motion between the

BM and the TM, which is scaled by a saturating Boltzmann function. A single degree of freedom model could be used for each micromechanical element, reducing the number of state variables from four to two per element as suggested. Nonlinearity could still be included, for example by making the stiffness and/or damping of each of thee mass-spring-damper elements dependent on the amplitude of its vibration, thus implementing a Van der Pol oscillator for example. Unfortunately, however, it is difficult to accurately represent a pure delay in a set of continuous-time differential equations, so Zweig's delayed stiffness term would have to be approximated by a model in which the stiffness was dependent on, for example, a low pass filtered version of the state variables.

Van Schaik: From simulating and building silicon cochleae with a 2D network to model fluid coupling, I have learned that reflections depend strongly on the amount of overlap in the band-pass characteristics of neighbouring sections. For a cochlea tuned to have higher Q values for the BM resonators, a larger density (resonators/octave) is needed to ensure stability. The instability otherwise obtained results from forward and backward reflections at frequencies that lack a low impedance path to ground, even in a perfectly linear cochlea.
You claim that your model allows investigation of the detailed behaviour for particular forms of perturbation resulting in SFOAEs. What is unclear to me is if the results of these investigations depend on the number of elements per octave in your model. The results shouldn't depend on this, but I suspect they might, particularly when simulating low SPLs. Could you please verify this?

Elliott: Thank you for your interesting question. We have briefly investigated the effect of the number of elements on the response and stability results, and find that these change very little if the number of elements is doubled, from 500 to 1000. If, however, the number of elements is reduced to 300 the predicted response is very different. This suggests that the number of elements used is sufficient for convergence. Also, the model is stable, with a uniform distribution of feedback gains, up to sufficiently high values of this gain that it is able to provide levels of active enhancement that are larger than those which have been measured in the physical cochlea.
The model becomes unstable, for lower and more representative average feedback gains, however, if random inhomogeneities of the order of 1% are introduced. This indicates that these inhomogeneities (which we want to model) are much larger than the inhomogenieties generated by spatial quantisation (which we do not want to model), thus also suggesting that our results are reasonably reliable, I think.

de Boer: When the only communication between a model and the "outer world" is possible via the cochlear windows, there is a much simpler way of finding out whether the model is stable or not. All one has to do is to compute the input impedance.

PERIODICITY IN THE SPECTRUM OF MODELLED SPONTANEOUS OTOACOUSTIC EMISSIONS

EMERY M. KU, STEPHEN J. ELLIOTT, BEN LINETON
Institute of Sound and Vibration Research, University of Southampton, Southampton SO17 1BJ, UK

1 Summary

A state space model of the cochlea [1, 2] is used to study the distribution of frequencies at which a linear, inhomogeneous cochlea will become unstable. This is of interest as an unstable cochlea may generate spontaneous otoacoustic emissions (SOAEs). The salient features of the multiple-reflection theory of SOAE generation of [3, 4] are observed in this model. Specifically, the travelling wave can experience multiple reflections between the middle ear boundary and inhomogeneities along the basilar membrane (BM) [5, 6]; only waves that undergo an integer number of cycles of phase change between the oval window and their best place reflection sites interfere constructively across multiple reflections, thus potentially becoming unstable. Furthermore, the response must peak in the region of the cochlear inhomogeneity such that enough energy is reflected to overcome the damping between the reflection sites. It is also found that the average distance along the BM between the best places of unstable frequencies is approximately one half the wavelength of the travelling wave at its peak:

$$\overline{\Delta x}_{SOAE} \approx \frac{1}{2}\lambda_{peak}. \tag{1}$$

This is observed if and only if the cochlear reflection sites are densely spaced such that the wavenumber corresponding to $\frac{1}{2}\lambda_{peak}$ is well-represented in the spatial distribution of the inhomogeneities [2], as predicted by [3]. A distribution of inhomogeneities that is not well-represented at $\kappa = \frac{1}{2}\lambda_{peak}$ is referred to as 'sparsely spaced' in this investigation.

In order to disrupt the smooth variation of BM impedance along the cochlea, the distribution of micromechanical feedback gain along the cochlea is randomly perturbed with band-passed Gaussian white noise. The spacings between unstable frequencies can be log-normalised [3]:

$$f/\Delta f = \frac{\sqrt{f_a f_b}}{|f_a - f_b|}, \tag{2}$$

where f_a and f_b are two neighbouring unstable frequencies. This spacing is predicted to be related to the wavelength of the travelling wave in its peak, λ_{peak}, by approximately

$$f/\Delta f \approx 2l/\lambda_{peak}, \tag{3}$$

where l is the distance along the BM by which the best frequency changes by a factor of e [3]. Both λ_{peak} and l are roughly constant as a function of position in the model, and give rise to a predicted $f/\Delta f$ of approximately 15. The most commonly measured spacing between SOAE frequencies in humans is also ~15 when expressed in terms of $f/\Delta f$ [4].

Figure 1 shows stability data corresponding to models with sparsely- (left) and densely-spaced (right) reflection sites. Note that the bandwidth of unstable frequencies is wider than those of commonly detected SOAEs; this is believed to be due to the bandpass-like nature of the reverse-transmission characteristics of the middle ear which precludes the conduction of high-frequency oscillations. The bottom plots of Figure 1 each summarize the averaged results of 200 stability plots by plotting all adjacent unstable frequencies in terms of $f/\Delta f$, arranged in log bins. A clear peak is visible at $f/\Delta f \approx 15$ for the cochlear models with densely-spaced reflection sites (right), but is not correctly located for the sparsely-spaced cochleae (left); this finding agrees well with predictions.

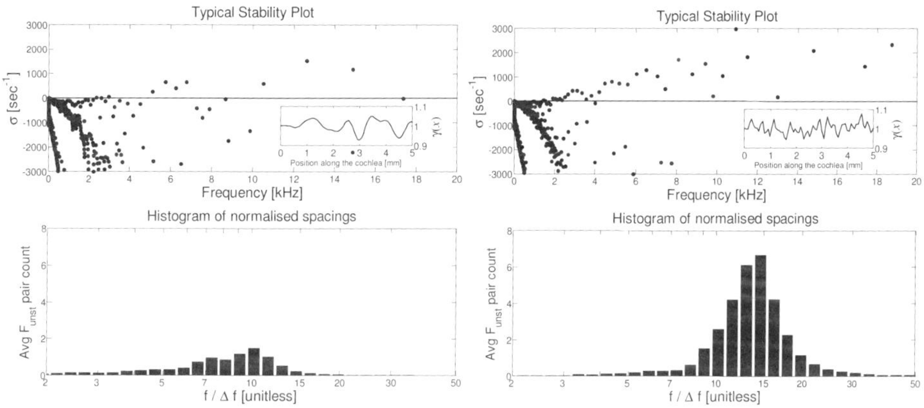

Figure 1. (left) Results of models with sparsely spaced reflection sites. (right) Results of models with densely spaced reflection sites. (top) Stability plot with sample distribution of gains inset; note that only the first 5 mm of this $\gamma(x)$ are shown for clarity. Poles with $\sigma > 0$ represent unstable frequencies. (bottom) Averaged histogram of the normalised spacings for 200 cochlear models arranged in log-bins.

A more detailed explanation of the findings outlined in this paper can be found in [2].

Acknowledgments

This work was partially supported by the US-UK Fulbright Commission.

References

1. Elliott, S.J., Ku, E.M., Lineton, B., 2007. A state space model for cochlear mechanics. J. Acoust. Soc. Am. 122(5), 2759-2771.
2. Ku, E.M., Elliott, S.J., Lineton, B. 2008. Statistics of instabilities in a state space model of the human cochlea. Accepted for publication to the J. Acoust. Soc. Am.
3. Zweig, G., Shera, C.A. 1995. The origin of periodicity in the spectrum of evoked otoacoustic emissions. J. Acoust. Soc. Am. 98, 2018-2047.

4. Shera, C.A. 2003. Mammalian otoacoustic emissions are amplitude-stabilized cochlear standing waves. J. Acoust. Soc. Am. 114(1), 244-262.
5. Elliott, S.J., Ku, E.M., Lineton, B. Time domain model of a nonlinear inhomogeneous cochlea. *This volume*.
6. www.cochlearmodelling.isvr.soton.ac.uk

MODELING STIMULUS-FREQUENCY OTOACOUSTIC EMISSIONS IN THE GECKO

CHRISTOPHER BERGEVIN

Department of Mathematics, University of Arizona, Tucson, Arizona 85705 USA

CHRISTOPHER A. SHERA

Eaton-Peabody Laboratory, Massachusetts Eye & Ear Infirmary, Boston, Massachusetts 02114 USA

Although lizards lack the basilar-membrane traveling waves evident in mammals, their ears produce stimulus-frequency otoacoustic emissions (SFOAEs) with latencies comparable to those measured in many mammals (1–2 ms or greater). To probe the origin of these relatively long OAE delays, we developed a model of SFOAE generation in the gecko. The model inner ear comprises a collection of linear, coupled oscillators (hair bundles and associated tectorium) whose effective damping manifests a small degree of irregularity. The model reproduces the major qualitative features of gecko SFOAEs, including their substantial delays. The SFOAE delays predicted by the model increase with the assumed sharpness of tuning, reflecting the build-up time associated with mechanical resonance.

1. Introduction

Stimulus-frequency otoacoustic emissions (SFOAEs) exhibit significant phase-gradient delays in both mammalian [1] and non-mammalian ears [2]. Although mammalian SFOAE delays are often attributed to basilar-membrane (BM) traveling waves, the origin of the delay in non-mammals, especially those lacking BM waves, remains unclear. Here we explore the origin of OAE delay in the gecko, a species in which SFOAE phase-gradient delays are 1–2 ms or greater, comparable to those measured in many mammals. Lizard ears lack BM traveling waves [3], and their tuning and tonotopy are thought to arise micromechanically at the level of the hair-cell bundle [4,5]. Our goal is to determine whether a simple model of the gecko inner ear—a collection of coupled oscillators representing the hair bundles, associated structures, and adjacent fluid—can account for measured SFOAE delays.

2. The Model

We simplify the gecko inner ear morphology as shown in Fig. 1. Middle-ear delays are assumed negligible [6]. The model consist of two rigid-walled cavities (the scala vestibuli and tympani) filled with incompressible fluid and separated by the papilla. The papilla is assumed to vibrate as a rigid body in the transverse direction only, driven by the pressure difference ($P_{\mathrm{v}} - P_{\mathrm{t}}$) between the two scalae. The scalae pressures P_{v} and P_{t} are assumed uniform at the frequencies of interest. The papilla and the hair cells on its surface are represented by linear oscillators (i.e., masses,

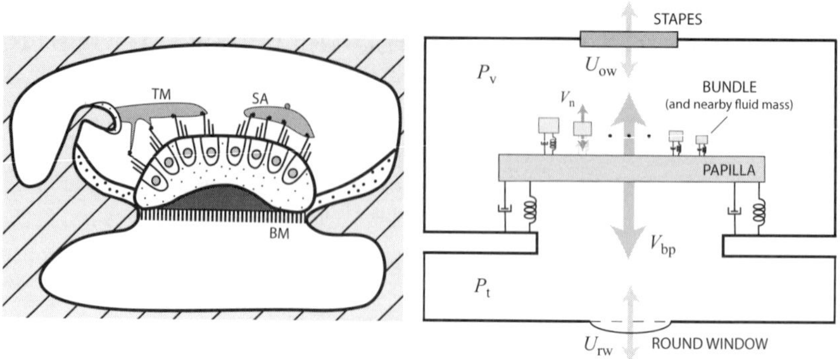

Fig. 1. *Left:* Simplified transverse-radial cross-sectional schematic of the gecko inner-ear anatomy showing hair cells embedded in the papilla. The papilla has an effective area A_{bp} and length L ($\sim$ 1 cm). The gecko is one of many lizard species with sallets, discretized sections of tectorium that are thought to behave as resonant filters. White regions are fluid-filled, gray region represent the overlying tectorium, gray striped areas represent bone, and stippled areas are supporting cellular structures. Abbreviations: BM–basilar membrane, TM–tectorial membrane, SA–sallet. *Right:* Longitudinal-transverse cross-section of the model, consisting of a collection of n linear oscillators coupled only by the motion of the basilar papilla.

springs, and dashpots). We assume harmonic time dependence and represent dynamical variables by Fourier coefficients at angular (driving) frequency ω.

By analogy with models of mammalian OAEs, SFOAEs are produced by introducing micromechanical irregularity [7]. Studies of lizard inner-ear anatomy [8,9] show irregular variations in features such as the width and thickness of the BM and basilar papilla, the number of hair cells in a given radial cross-section, and the coupling between the hair bundles and overlying tectorium. We quantify the emission as the pressure $\Delta P(\omega)$, defined by

$$\Delta P(\omega) \equiv \left[\tilde{Z}(\omega) - Z(\omega)\right] U_{\text{ow}} , \tag{1}$$

where U_{ow} is the stapes volume velocity, $\tilde{Z}(\omega)$ and $Z(\omega)$ are the cochlear input impedances $[(P_{\text{v}} - P_{\text{t}})/U_{\text{ow}}]$ computed with and without micromechanical irregularity, respectively.

We compute the input impedance by solving the equations of motion for the system. Newton's second law implies that

$$Z_{\text{bp}} V_{\text{bp}} = A_{\text{bp}}(P_{\text{v}} - P_{\text{t}}) + \sum_n \frac{k_n}{i\omega}(V_n - V_{\text{bp}}) . \tag{2}$$

In this equation, V_{bp} is the transverse velocity of the papilla, and A_{bp} and Z_{bp} are its effective cross-sectional area and mechanical impedance, respectively. V_n and k_n represent the velocity and effective stiffness of the nth hair bundle. The individual bundles are coupled only through the motion of the underlying papilla and satisfy the equation

$$i\omega m_n V_n = -r_n V_n - \frac{k_n}{i\omega}(V_n - V_{\text{bp}}) , \tag{3}$$

where m_n, r_n, and k_n are the effective mass, damping, and stiffness of bundle n. Thus, each bundle exerts a force on the papilla proportional to the difference in transverse displacement between the bundle and the papilla. Solving Eq. (3) for the velocity V_n of bundle n yields

$$V_n = \frac{1}{1-\beta_n^2 + i\beta_n/Q_n} V_{\rm bp} \,, \tag{4}$$

where $\beta_n \equiv \omega/\omega_n$ is normalized frequency, with ω_n and Q_n representing the bundle's resonant frequency and quality factor ($\omega_n \equiv \sqrt{k_n/m_n}$ and $Q_n \equiv k_n/r_n\omega_n$). Substituting Eq. (4) into Eq. (2) and using conservation of mass ($U_{\rm ow} = A_{\rm bp}V_{\rm bp}$) to solve for the input impedance yields

$$Z(\omega) \equiv \frac{P_{\rm v} - P_{\rm t}}{U_{\rm ow}} = \frac{1}{i\omega A_{\rm bp}^2}\left[Z_{\rm bp} + \sum_n k_n \frac{-\beta_n^2 + i\beta_n/Q_n}{1-\beta_n^2 + i\beta_n/Q_n}\right] . \tag{5}$$

We introduced micromechanical irregularity into the model by including a small stochastic element in the effective damping of each oscillator (Q_n). Specifically, we took $\tilde{Q}_n = Q_n(1+\epsilon_n)$, where Q_n is the unperturbed value (see below) and ϵ_n is a small random value close to zero.

3. Analytic Approximation for the Delay

To derive an approximate analytic expression for the emission phase-gradient delay we make several simplifying assumptions (validated using numerical analysis). First, we convert the sum to an integral and neglect spatial variations in bundle stiffness ($k = k_{\rm o}$). Expressing the integral in terms of β yields

$$\Delta P \approx \frac{U_{\rm ow}k_{\rm o}\ell}{\omega A_{\rm bp}^2} \int_{\beta_0}^{\beta_L} (\epsilon/Q)(Ae^{i\theta})^2 \, d\beta \,, \tag{6}$$

where ℓ is the exponential space constant of the papilla tonotopic map, $\beta_L \equiv \omega/\omega_{\rm min}$, $\beta_0 \equiv \omega/\omega_{\rm max}$, and $A(\beta)e^{i\theta(\beta)} \equiv 1/(1-\beta^2+i\beta/Q)$ is the transfer function of the harmonic oscillator. Following the analysis of the coherent-reflection model [7], we decompose the irregularity function, $\epsilon(x)$, into spatial-frequency components and assume that the value of the integral is dominated by spatial frequencies in a (relatively) narrow range about some "optimal" value $\kappa_{\rm opt}$. The integral then becomes

$$\Delta P \propto e^{-i\kappa_{\rm opt}\ell \ln \beta_0} \int_{\beta_0}^{\beta_L} \frac{A^2(\beta)}{Q(\beta)} e^{i[\kappa_{\rm opt}\ell \ln \beta + 2\theta(\beta)]} d\beta \,, \tag{7}$$

(constant terms contributing only to the magnitude have been dropped to simplify the notation). Because the amplitude of the integrand is sharply peaked, the value of the integral does not depend strongly on frequency except near frequencies close to the ends of the tonotopic map. The principal frequency dependence of the emission phase ($\phi \equiv \angle \Delta P$) therefore arises from the argument of the exponential outside the integral and depends on the value of $\kappa_{\rm opt}$. We determine $\kappa_{\rm opt}$ by applying the principle of stationary phase and requiring that the phase of the integrand be constant

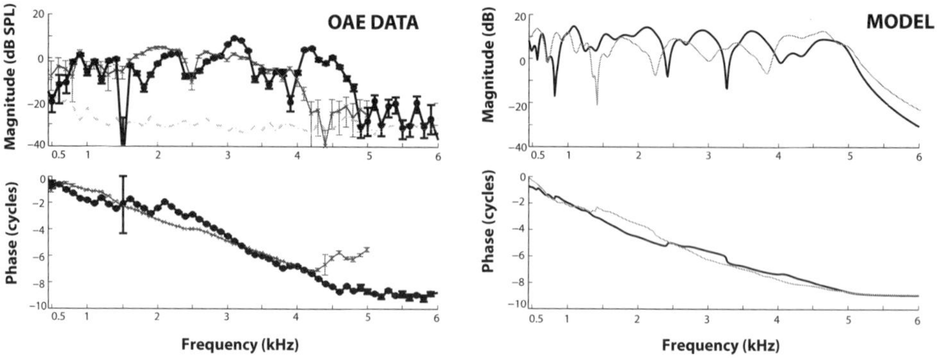

Fig. 2. *Left:* Representative SFOAE data at probe levels of 20 dB SPL from both a leopard gecko and tokay gecko [2]. The dashed line shows the acoustic noise floor. *Right:* Model values of $\Delta P(\omega)$ (simulated SFOAEs) based on random micromechanical irregularities. The model papilla comprised 150 bundles with CFs logarithmically distributed from 0.2 to 5 kHz. A roughness factor of 3% was used. Model results are shown for two different "ears" (i.e., different irregularity patterns). The overall model SFOAE magnitude was scaled to approximate that of the measurements.

near its magnitude peak (i.e., near $\beta \approx 1$). This yields

$$\frac{\partial}{\partial \beta}\left[\kappa_{\text{opt}}\ell \ln \beta + 2\theta(\beta)\right]\Big|_{\beta=1} = 0 \qquad \Longrightarrow \kappa_{\text{opt}} = -\frac{2}{\ell}\frac{\partial \theta}{\partial \beta}\Big|_{\beta=1} . \tag{8}$$

For the harmonic oscillator, one can show that $\frac{\partial \theta}{\partial \beta}|_{\beta=1} = -Q/\pi$. Combining our expressions to compute N_{SFOAE}, the emission phase-gradient delay expressed in stimulus periods $(-\frac{\omega}{2\pi}\frac{\partial \phi}{\partial \omega})$, yields the approximation

$$N_{\text{SFOAE}} \approx \frac{2Q}{\pi} . \tag{9}$$

According to this analysis, the emission phase-gradient delay (in periods) is proportional to the sharpness of tuning.

4. Comparison with Experiment

Figure 2 compares model simulations with published SFOAEs measured in the gecko [2]. The figure shows the magnitude and phase of ΔP computed from Eq. (1) using Eq. (5) for $Z(\omega)$. Parameter values were obtained by assuming that bundle resonant (or characteristic) frequencies (CF) are distributed logarithmically along the papilla [4] with a constant density of oscillators per octave. In the absence of irregularity, the quality factors Q_n were assumed to vary with CF according to the power law $Q_n = 3Q_{10}(\text{CF}_n)$ where $Q_{10}(\text{CF}) = 2.3[\text{CF}/(\text{kHz})]^{0.355}$ is a power-law fit to tuning data obtained from gecko auditory-nerve fibers [4].[a] ϵ_n was randomly sampled from a Gaussian distribution with zero-mean and standard deviation 0.03.

[a] We used the data of Manley et al. [4] but note that gecko ANF Q_{10} values vary considerably across studies. The factor of 3 used to convert to Q_{10} to Q_n is an approximation valid for second-order filters.

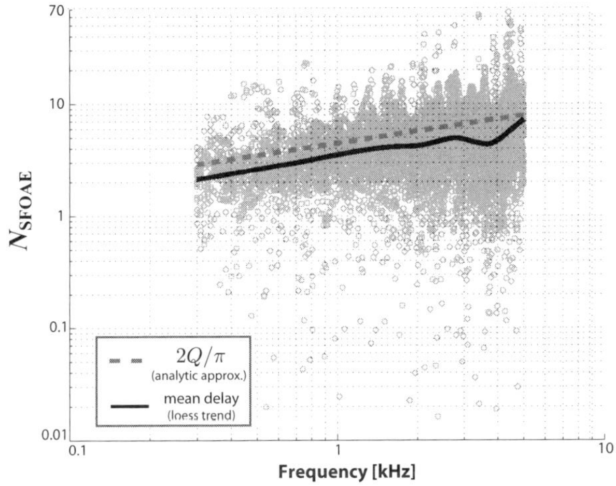

Fig. 3. Model SFOAE phase-gradient delays expressed in stimulus periods. The solid curve is a trend line computed from the results for 20 different simulated ears (grey circles) and trend line (solid line). The analytic approximation given by Eq. (9) is also plotted (dashed line).

As shown in Fig. 2, the model captures many of the qualitative features seen in the emission data. For example, the model reproduces the frequency-dependent variations in SFOAE magnitude, such as spectral notches unique to a particular ear and the high frequency roll-off at frequencies outside the range of the tonotopic map. In addition, both the simulated emissions and the SFOAE data exhibit substantial phase accumulation. These qualitative features are relatively insensitive to model parameters such as the bundle stiffness and the total number of bundles.

When the model Q values are chosen to match ANF responses, the model predicts generally realistic OAE delays. Below 1 kHz, however, the predicted delays are longer than those measured. This low-frequency deviation is similar to that observed in mammals between measures of Q and N_{SFOAE} and may stem from interference between multiple emission source mechanisms. Whereas only one mechanism is represented in the current model, the gecko shows evidence for at least two different OAE generation mechanisms [2]. In both the model and the lizard, substantial SFOAE delays occur despite the absence of BM traveling waves.

Figure 3 demonstrates that the mean phase-gradient delay is well approximated by the analytic expression derived in Eq. (9). Variations in the value of Q along the length of the papilla produce corresponding changes in N_{SFOAE}. These results indicate that the emission phase-gradient delay (N_{SFOAE}) is proportional to the sharpness of tuning (Q).

5. Summary

We have described a simple model for the gecko inner ear in order to predict SFOAE magnitude and phase. When the sharpness of tuning of the model resonators is chosen to match ANF responses, the model captures many of the qualitative features of gecko SFOAEs. In particular, the model reproduces the substantial phase-gradient delays in spite of the absence of a tuned BM or traveling waves. The model predicts that SFOAE phase-gradient delays are proportional to the sharpness of tuning of the resonators inside the ear (i.e., the hair cells and associated tectorium). Despite the absence of traveling waves, many mechanisms in the model are qualitatively similar to those of coherent reflection filtering [7] in the mammalian cochlea (e.g., the role of a dominant spatial frequency in determining SFOAE delay). The oscillators used here are presumably too simple (e.g. passive, linear, no fluid coupling between adjacent bundles). Nevertheless, we conjecture that the proportionality between SFOAE delays and sharpness of tuning described here holds in more realistic models.

Acknowledgments

Supported by Howard Hughes Medical Institute (grant 52003749), the National Science Foundation Division of Mathematical Sciences (grant 0602173), and the National Institutes of Health (R01 DC003687).

References

1. Shera, C.A., Guinan, J.J. Jr., Oxenham, A.J., 2002. Revised estimates of human cochlear tuning from otoacoustic and behavioral measurements. Proc. Natl. Acad. Sci. 99, 3318–3323.
2. Bergevin, C., Freeman, D.M., Saunders, J.C., Shera, C.A., 2008. Otoacoustic emissions in humans, birds, lizards, and frogs: Evidence for multiple generation mechanisms. J. Comp. Physiol. A 194, 665–683.
3. Peake, P.T., Ling, A. Jr., 1980. Basilar-membrane motion in the alligator lizard: its relation to tonotopic organization and frequency selectivity. J. Acoust. Soc. Am. 67, 1736–1745.
4. Manley, G.A., Koppl, C., Sneary, M., 1999. Reversed tonotopic map of the basilar papilla in Gekko gecko. Hear. Res. 131, 107–116.
5. Aranyosi, A.J., Freeman, D.M., 2004. Sound-induced motions of individual cochlear hair bundles. Biophys. J. 87, 3536–3546.
6. Rosowski, J.J., Peake, W.T., Lynch, T.J., Leong, R., Weiss, T.F., 1985. A model for signal transmission in an ear having hair cells with free-standing sterocilia. II. Macromechanical stage. Hear. Res. 20, 139–155.
7. Zweig, G., Shera, C.A., 1995. The origin of periodicity in the spectrum of evoked otoacoustic emissions. J. Acoust. Soc. Am. 98, 2018–2047.
8. Miller, M.R., 1973. A scanning electron microscope study of the Papilla basilaris of *Gekko gecko*. Z. Zellforsch. 136, 307–328.
9. Wever, E.G., 1978. The Reptile Ear. Princeton University Press.
10. Shera, C.A., Guinan, J.J. Jr., 2003. Stimulus-frequency-emission group delay: A test of coherent reflection filtering and a window on cochlear tuning. J. Acoust. Soc. Am. 113(5), 2762–2772.

SECTION II

COCHLEAR AMPLIFICATION: CHARACTERISTICS, MODULATION AND CONTROL

A. Fridberger and R. Frosch

W. Dong and A.L. Nuttall

NONLINEAR COCHLEAR SIGNAL PROCESSING AND PHONEME PERCEPTION

JONT B. ALLEN, MARION RÉGNIER, SANDEEP PHATAK, FEIPENG LI
University of Illinois, Urbana IL, USA

The most important communication signal is human speech. It is helpful to think of speech communication in terms of Claude Shannon's information theory channel model. When thus viewed, it immediately becomes clear that the most complex part of speech communication channel is in auditory system (the receiver). In my opinion, even after years of work, relatively little is know about how the human auditory system decodes speech. Given cochlear damaged, speech scores are greatly reduced, even with tiny amounts of noise. The exact reasons for this SNR-loss presently remain unclear, but I speculate that the source of this must be cochlear outer hair cell temporal processing, not central processing. Specifically, "temporal edge enhancement" of the speech signal and forward masking could easily be modified in such ears, leading to SNR-Loss. What ever the reason, SNR-Loss is the key problem that needs to be fully researched.

1. Introduction

A fundamental problem in auditory science is the perceptual basis of speech, that is, phoneme decoding. How the ear decodes basic speech sounds is important for both hearing aid and cochlear implant signal processing, both in quiet and in noise. To address these issues, we need a theory of speech perception. Other than Claude Shannon's theory of information, depicted in Fig. 1, such theories are limited.

Consonant speech sounds are typically described in terms of production concepts, such as voicing, manner and place [1], however these categories tell us very little about the perception of speech, and nothing about the effect of masking noise on the received signal. It has not proved to be possible to generalize from copious examples, or the problem would have proved to be easy. Easy is not a word we may associate with this decoding problem.

It is clear from decades of research that the state of the cochlea is an important variable in speech perception studies. For example, *auditory masking* is critical to our understanding of speech and music processing. Furthermore, once the organ is damaged, our ability to process speech in noise is seriously impaired. The reasons for this impairment are not known, but it seems possible, or even likely, that such impairments are related to outer hair cell (OHC) processing in the cochlea.

The goal of this paper is to outline a theory of speech processing and to isolate the features in speech. We would like to answer questions as *What separates /t/ from /d/ or /t/ from /k/ and /p/?* and *Can we quantify the role of NL cochlear processing in this classification task?* We shall show that across-frequency onsets define the plosive consonants, while bandwidth and duration define fricative consonants. Finally we shall speculate on the role of the OHC processing in speech perception.

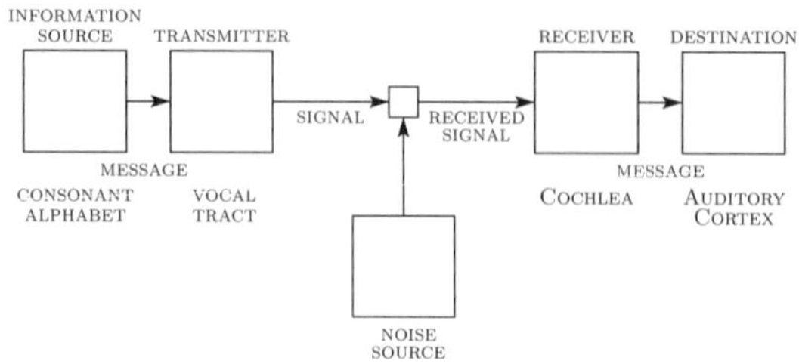

Fig. 1. Shannon's model of information transmission.

2. Key studies

The first speech studies were done in by Rayleigh (1908) [2], following telephone commercialization. Within a few years, Western-Electric's George Campbell (1910) [3] developed the electrical wave filter to high and low-pass speech signals, as well as probabilistic models of speech perception such as the *confusion matrix method* of analysis. With these tools established, by 1921 Harvey Fletcher was deeply into similar studies [4]. Fletcher soon discovered that by breaking the speech into bands having equal scores, he could formulate a rule relating the errors in each band to the wide-band error. This method became known as the *articulation index method.* Today we now know that it is closely related to information theory, introduced many years later by Claude Shannon (1948) [5], as summarized in Fig. 1.

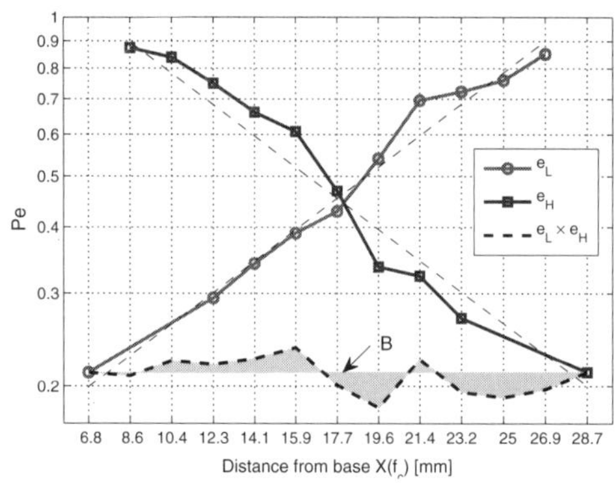

Fig. 2. This figure shows the core concept behind the *articulation index*. Speech is high and low-pass filtered to cutoff f_c, and two errors $e_L(f_c)$ and $e_H(f_c)$ are plotted on a log-error scale and cross near 1.5 kHz, at 17.7 mm along the cochlea, the abscissa scale. Each curve has been fit with a linear regressions having equal slopes but with opposite signs. Thus the product of the high and low-bands obeys Eq. 1.

In Fig. 2 we see a recent version of Fletcher's results, measured in my lab. Two complimentary filters are used, a high and low-pass with cutoff frequency f_c, with a 12 dB SNR masker. As the cutoff is varied, the average speech phone error is determined, as shown in the figure, where the probability of error $P_e(f_c)$ is on a log scale, as a function of position along the cochlea $X(f_c)$. While this data has been collected in our lab in 2006, it is similar in many ways to Fletcher's 1921 results. Fletcher demonstrated that the product of the low and high band errors is a constant, namely that $e_{total} = e_L(f_c) \times e_H(f_c)$ is a constant. The dashed line

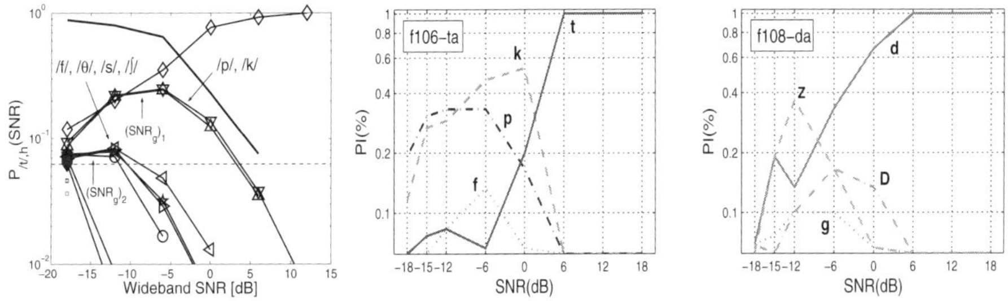

Fig. 3. Confusion patterns are defined by a row of the articulation matrix $\mathcal{A}(h|s)$, defined as the probability of hear sound h after speaking sound s, as a function of the signal to noise ratio (SNR), in dB. In each panel, the curve that rises to 1 is the diagonal element of the matrix, namely the score of /t/ given /t/ for the left and middle curves, and /d/ for the curves on the right. For the left panel, at -6 dB SNR, the *average* confusions with /p/ and /k/ are nearly equal to the score of /t/. Other confusions appear at even lower SNR, with /f/ for example. The other two panels are CPs for a specific /t/ (middle) and /d/ (right), with white noise maskers. Each of these sounds *morphs* (heard as another consonant) below 6 dB SNR. For example, for the middle panel at 0 dB, /k/ is reported 55% of the time while /t/ and /p/ are reported 20% and 18% of the time.

along the bottom, having an average error of 21%, has small fluctuations, labeled B (Bias), shaded in green. The average phone error $s = 1 - e_{total}$, may be computed from the articulation confusion matrix $\mathcal{A}_{h|s}(SNR)$, defined as the probability of hearing sound h after speaking sound s. As an example, confusions for the case of $s \equiv$ /t/ are shown in Fig. 3, left.

The *Confusion patterns* (CP) shown in Fig. 3 allow one to determine the precise nature of the confusions of each sound. The confusion set, and their dependence on SNR is not predictable without running a masking experiment. These confusions, and their masked dependence, are important because they reveal the mix of underlying perceptual features, or *events*.

It is easy to create a sound that *primes*, meaning that it can be heard as any of several sounds, depending on one's state of mind. In this case the confusion patterns show subject responses that are equal (the curves cross each other), similar to the CP of Fig. 3 (left) at -10 dB, where one naturally primes /p/, /t/ and /k/, and on the right at -9 dB where /z/ and /d/ prime.

Fletcher found that the log-errors $e_L(f_c, SNR)$ and $e_H(f_c, SNR)$ are linear on a cochlea place scale $X(f_c)$ [6]. The implication is the total error may be generalized

$$e_{total} = e_1 e_2 \cdots e_{20}, \tag{1}$$

where e_k is the error contributing to the speech score due to cochlear band, indexed by integer k. This relationship is the key to the *articulation index* method, as reviewed in many places [6, 7, 8, 9, 10], and summarized in Fig. 4.

Along the top of the figure the response-measure is shown, such as the output of the cochlea or the phone scores. Just below the block diagram the mathematical model measure is displayed. For example, the output of the cochlea defines the signal to noise ratio in cochlear bands, as specified by the band articulation index AI_k for band k. Along the very bottom the figure indicates which are physical Φ

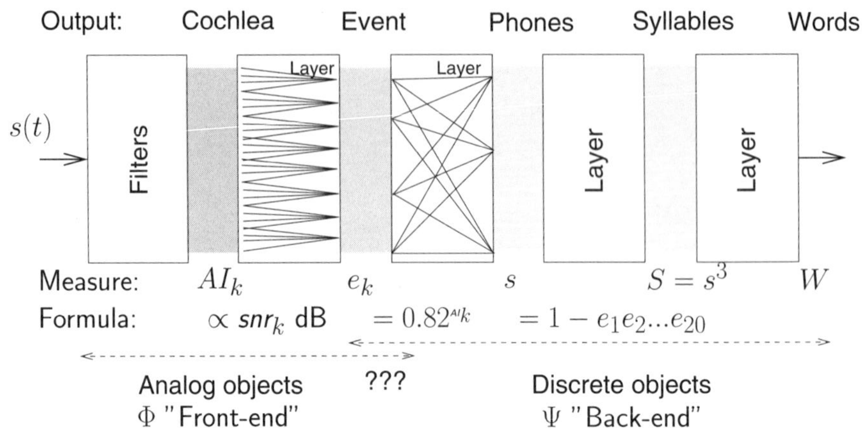

Fig. 4. Outline for the theory of speech perception. The output of the cochlea defines the signal to noise ratio in cochlear bands, as specified by the band articulation index AI_k for band k. The *event band error* is given by e_k. The maximum entropy average phone error is defined as $s(AI)$, also known as the *nonsense phone error*. The consonant-vowel (CV) syllable score S_{cv} is the square of the phone error while the CVC error is the cube (i.e., $S_{cvc} = s^3$).

measures and which are psychophysical Ψ discrete objects. The critical transition from Φ to Ψ is presumed to happen at the event level [6]. Once a speech event is quantal, central processing is assumed to be error free.

2.1. *Identifying events*

Two methods have been established for precisely identifying *events* (perceptual features). The first method is outlined in a recent paper by Régnier and Allen (2008) [11]. Rather than reviewing this method here, since it is so recently published, we present a second, perhaps more general method, as yet unpublished.

2.1.1. *Speech-Plosive events*

In Fig. 5 there are six sets of 4 panels, as described in the caption. Each of the six sets corresponds to a specific consonant, labeled by a character string that defines the gender (m,f), subject ID, consonant and SNR for the display. For example, in the upper left 4 panels we see the analysis of /ta/ for female talker 105 (`f105ta0dB`) at 0 dB. Along the top are unvoiced plosives /t/, /k/ and /p/ while along the bottom are voiced plosives /d/, /g/ and /b/. Data from the same talker was not available in our database, so three different talkers have been used in this analysis.

Each sound was first time-truncated from the onset to a given time, in 10 ms steps [12], and played back in random order to 14 listeners, who indicated what they heard by clicking an icon on the screen. Noise was added to the truncated sound at an SNR of 12 dB. The results of the truncation experiment (TR07) are presented in the upper-left panel with the title TR07 at the top. Each curve is the probability of what was reported, and is labeled with the identified consonant. In a second experiment (MN16R) the same sound was subjected to a variable signal to

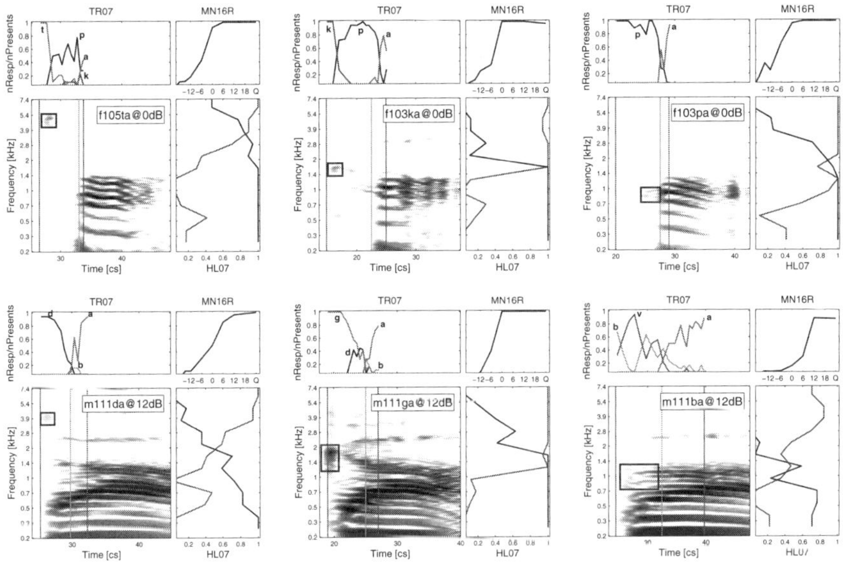

Fig. 5. Identification of features by time and frequency bisection. Along the top we have unvoiced consonants /t/, /k/ and /p/, while along the bottom, the corresponding voiced consonants /d/, /g/ and /b/. Each of the six sounds consists of 4 sub-panels. For example, for /t/ in the upper left wee see four panels consisting of the time-truncation confusions (upper-left), the score vs. SNR (upper-right), the AI-gram (lower-left) and the score as a function of low and high-pass filtering. There are six such groups, one for each of the six consonants displayed.

noise ratio, from -12 dB SNR to quiet, and the average score was measured across the 23 listeners [13]. Finally the same sample was high and low-pass filtered to a variable cutoff frequency (experiment HL07), as indicated on the frequency axis.

A summary of the audible sound features at the threshold of masking, are shown by the AI-gram [11], as exampled in the lower-left panel of Fig. 5. This plot is similar to a spectrogram, but differs in several important ways. First the AI-gram is normalized to the noise floor. This is similar to the cochlea which dynamically adapts to the noise floor due to OHC NL processing [14, 15], as discussed in Section 3. Second, unlike a fixed-bandwidth spectrogram, the AI-gram uses a cochlear filter bank, with bandwidths given by Fletcher critical bands (ERBs) [8]. Finally the intensity scale in the plot is proportional to the signal-to-noise ratio, in dB, in each critical band, as in AI-band densities $AI_k(SNR)$ described in Fig. 4.

At the present time the AI-gram is imperfect in that it contains no forward masking, no upward-spread of masking, and no neural masking components. Much work remains to be done on time-domain NL cochlear models of speech.

One may identify the speech event from these displays. For example, the feature that labels the sound (e.g., /t/) is indicated by the red-square in the lower-left panel of each of the six sounds (e.g., next to the descriptor of the sound (e.g. `105ta@0dB`) there is a red box showing the burst of energy that defines the /t/ sound). When this burst is truncated, as in the TR07 experiment, the /t/ morphs to /p/. When

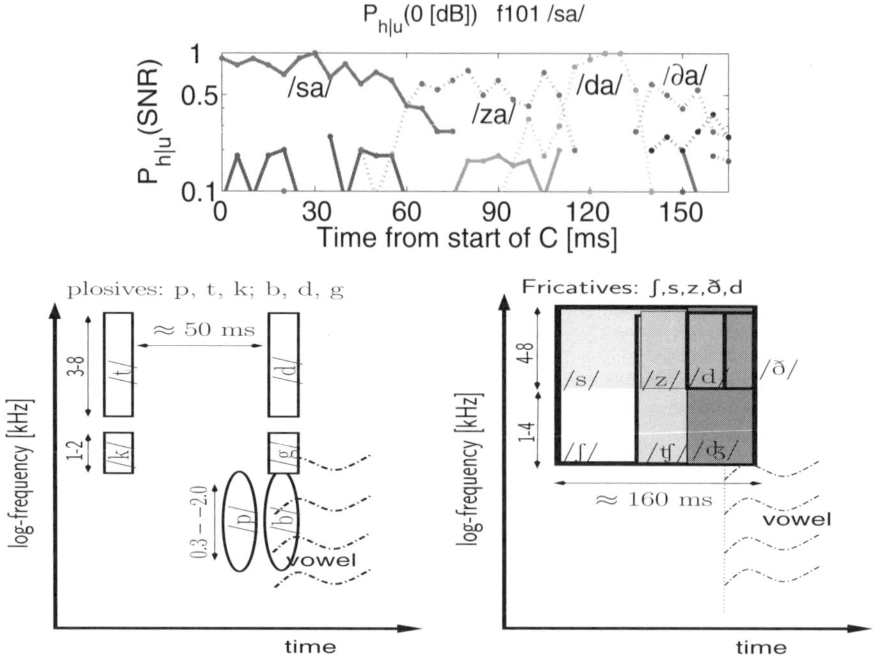

Fig. 6. **Upper panel:** Frication sound female 101 saying /sa/. As the sound is truncated from the onset, the /s/ is heard as /z/, then /d/ and finally /ð/. Each time the conversion happens at about a factor of two in frication duration. **Lower two panels:** Structure of the plosives and the fricatives, in terms of time-frequency allocation. Mapping these regions into events requires extensive perceptual experiments. But once the sounds have been evaluated, it is possible to prove where the key noise-robust events live in perceptual space.

masking noise is added to the sound, such that it masks the boxed region, the percept of /t/ is lost. When the high and low-pass filters remove the frequency of the /t/ burst, again the consonant is lost. Thus the three experiments are in agreement, and they uniquely isolate the location of the event responsible for /t/. This nicely generalizes to the other plosive consonants shown (i.e. voiced /t/, /k/, /p/, and unvoiced /d/, /g/, /b/).

From such data we see that /t/ is labeled by a 4 kHz burst of energy ≈50ms before the vowel, whereas /k/ is defined as a 1-2 kHz burst, also 50 ms or so before the vowel. A burst of energy leading the vowel at 0.3-2 kHz defines /p/. The three voiced sounds /d/, /g/ and /b/ have similar frequencies but onset with the vowel .

The two high-frequency sounds (top and bottom left) are /t/ and /d/, each produced with the tongue tip on the roof of the mouth slightly behind the teeth. The two mid-frequency sounds, /k/ and /g/ are produced with the back of the tongue, labeled in the frequency domain as bursts between 1-2 kHz, for the examples shown. Finally low-frequency /p/ and /b/ are produced with the release of the lips. These two sounds produce a low frequency 0.2-2 kHz burst.

We have analyzed all the sounds in our consonant database, and similar results have been found. Thus we are confident that these tags of energy label the identity of these consonant. The distributions of the burst frequencies, durations and delays to [1] voicing needs more study, as does the relationship between tongue place and burst frequency.

2.1.2. *Fricative sounds*

Not surprisingly, the events associated with fricative sounds are quite different from the plosives. Obviously timing and bandwidth remain important variables. For the fricative sounds, a swath of bandwidth of fixed duration and intensity is used to indicate the sound, as shown in Fig. 6.

Using a time-truncation experiment similar to Furui (1986) [12], as disclosed by Régnier and Allen (2008) [11], we see the importance of duration to these consonants. In the top panel of Fig. 6, a /sa/, spoken by female talker 101 and presented at 0 dB, was truncated in 10 ms steps. After about 60 ms of truncation from the onset of the sound, our pool of subjects reported /za/ instead of /sa/. After 30 additional ms of truncation, /d/ was heard. Finally at the shortest duration /ða/ was reported. A related experimental result found ʃa → ʧ→ ʤ→ d. At the end of this chain is the plosive. Thus the fricatives and the voiced-plosives seem to form a natural continuum, in the limit of very-short duration sounds.

2.2. *Verification methods*

To further verify all these results we have developed a method to modify the speech sounds using *short-time Fourier transform* (STFT) methods [16, 17], to attenuate and amplify these Φ bursts of energy underlying the Ψ speech-events. These unpublished studies have confirmed that the narrow band bursts of energy shown in Fig. 5 are both necessary and sufficient to robustly label the plosive consonants. Above the feature's masked threshold, the event is independent of SNR [11].

Two STFT modifications are exampled in Fig. 7. For this case the /g/ event has been removed and the /d/ event enhanced, resulting in the morph /ga/ → /da/.

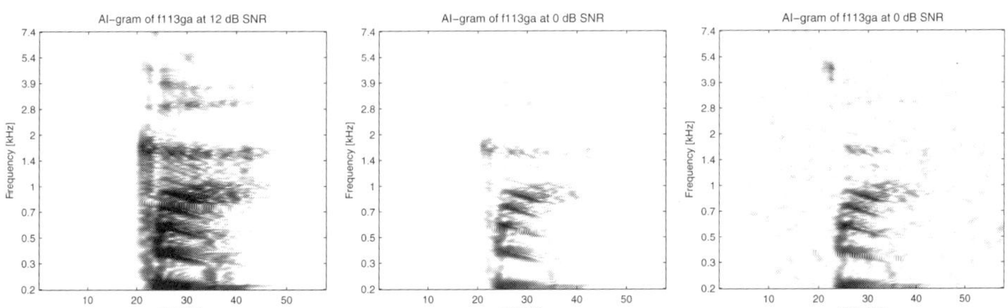

Fig. 7. On the left we see an AI-gram of the original sound f113ga at 12 dB SNR, and in the middle, at 0 dB. The score is 100% at and above 0 dB, and 90% at -6 dB. At -12 dB it is 30%. On the right is an AI-gram of the sound after modification by the STFT method, where the mid-frequency burst at [20 cs, 1.5 kHz] was removed, along with remnants of the pre-vocalic burst, and a 12 dB of gain was applied at 20 cs between 3.9-5.4 kHz, creating a burst of energy seen in the right panel. These two changes resulted in the sound being reported as /da/.

3. Nonlinear cochlear speech processing

The discussion next focuses on NL cochlear processing. Understanding and modeling NL OHC processing seems key to many speech processing applications. It seems under-appreciated that NL OHC processing (i.e., dynamic masking) is largely responsible for *forward masking* (post-stimulus masking), which results in very large effects over long time scales. For example OHC effects (FM/USM/2TS) can be as large as 50 dB, with a FM "latency" (return to base line) of up to 200 ms [18, 19, 15]. *Forward masking* (FM) and NL OHC *signal onset enhancement* are important to the detection and identification of perceptual features of a speech signal. In contrast, some studies have concluded that forward masking is not related to OHC processing [20, 21], so the topic remains controversial.

3.1. *Function of the Inner Ear in speech perception*

One key goal of cochlear modeling is to refine our understanding of how speech signals are processed [15]. The two main roles of the cochlea are to separate the input acoustic signal into overlapping frequency bands, and to compress the large acoustic intensity range into the much smaller mechanical and electrical dynamic range of the inner hair cell, synapse and neuron. This is a basic issue of signal, noise and information processing by the ear. The eye plays a similar role as a peripheral organ. It breaks the light image into rod and cone sized pixels, as it compresses the dynamic range of the visual signal. Based on the intensity JND, the corresponding visual dynamic range is about 9 to 10 orders of magnitude of intensity [22, 23], while the ear has about 11 to 12. The visual and auditory stimulus has a relatively high information rate compared to the low bandwidth of neural channels. The eye and the ear must cope with this problem by reducing the stimulus to a large number of low bandwidth signals. It is the job of the cortex to piece these pixelated signals back together, to reconstruct the world as we hear and see it.

Thus in general terms, the role of the cochlea is to convert sound at the eardrum into ≈30,000 neural pulse patterns in the human auditory (VIIIth) nerve. After being filtered by the cochlea, a low-level pure tone has a narrow spread of excitation which excites the cilia of about 40 contiguous inner hair cells [24, 8, 25]. The IHC excitation signal is narrow band with a center frequency that depends on the inner hair cell's location along the basilar membrane.

The prevailing and popular "cochlear amplifier" (CA) view is that the OHC provide *cochlear sensitivity* and *frequency selectivity* [25, 26, 27, 28]. The alternative view, argued here, is that the OHC compresses the excitation to the inner hair cell, thereby providing dynamic range expansion [29, 15].

There are key differences between these two views. The CA view deemphasizes the role of the OHC in providing dynamic range control (the OHC's role is to improve sensitivity and selectivity), and assumes that the NL effects result from OHC saturation. Such a simple model fails many comparisons to neural data. The *NL-compression* view places the dynamic range problem as the top priority. It assumes that the sole purpose of the OHC nonlinearity is to provide dynamic range compression, and that the OHC plays no role in either sensitivity or selectivity, which are treated as important, but independent issues.[a]

3.2. *The dynamic range problem*

The question of how the large (120 dB) dynamic range of the auditory system is attained has been a long standing problem which remains fundamentally incomplete. Based on a simple noise analysis of the IHC membrane voltage, one may prove that the dynamic range of the IHC must be less than 65 dB [30]. In fact it is widely accepted that IHC dynamic range is less than 50 dB. The obvious question arises: *How can the basic cochlear detectors (the IHCs) have a dynamic range of less than 50 dB (a factor of* 0.3×10^2*), and yet the auditory system has a dynamic range of up to 120 dB (a factor of* 10^6*)?* This discrepancy in dynamic range forms a basic paradox. This would seem to be the necessary condition for the dynamic range compression, provided by OHC processing, that we are looking for. Indirect evidence has shown that this increased dynamic range results from mechanical NL signal compression provided by outer hair cells. This dynamic range compression shows up in auditory psychophysics and in cochlear physiology in many ways.

For example, *recruitment*, the most common symptom of neurosensory hearing loss, is best characterized as the loss of dynamic range [31, 8, 32, 14]. Recruitment results from outer hair cell damage [33]. To successfully design hearing aids that deal with the problem of recruitment, we need models that improve our understanding of *how* the cochlea achieves its dynamic range. Given the observations shown here on speech events, we need to extend our primitive understanding of *wide-dynamic range compression* [14] into the time domain.

As a second example, explaining the proportionality between the neural threshold in dB and the linear membrane voltage, is also key Sewell (1984) [34] has demonstrated that as the EP voltage driving the hair cells changes, the neural gain in dB at CF changes proportionally by 1 dB/mV. It is not yet known why the dB gain is proportional to the voltage, however Sewell's observation might explain why cochlear forward masking decays exponentially in dB with time, after a strong excitation. Sewell's result implies an exponential decay with time of the neural sensitivity, *in decibels*. In other words, the log-decibel sensitivity should decay linearly in time. If $v_m(t) \propto e^{-t/\tau_m}$, then the dB value with change by a factor of $1/\mathrm{e} = 1/2.7$ ≈ 8.7 [dB/dB] in τ [ms]. Given Sewell's result, a plot of the dB change in forward masking on a log scale would be linear. Just this relationship has been demonstrated by Duifhuis (1973) [18], who shows a slope of $\approx$30 dB in 100 ms. From this one would predict an OHC recovery time constant of 30/100=0.3 [dB/ms]. Thus these estimates define a release time constant for the OHC of the right form (linear in log-dB).

[a]Of course other views besides these two are possible.

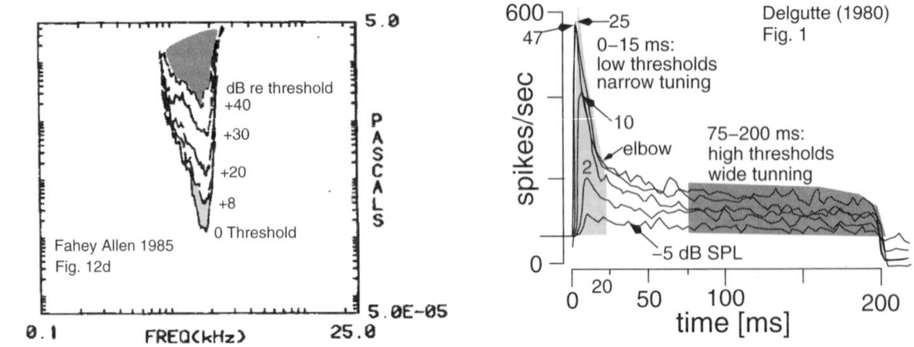

Fig. 8. On the left we see a typical 2 kHz tuning curve in various states of two-tone suppression, due to a low-frequency masker. These tuning curves are similar to what one would expect from driving the neuron with noise at various levels above threshold. On the right we see a PSTH to a 180 ms CF tone. The first 20 ms (to the point labeled elbow) shows a strong adaptation effect, due to the onset. At least some of this onset must be due to the high sensitivity of the neuron when a burst comes on, and reflects the time-course for it to reach a new state of tuning. It appears that this time is longer that one would expect from basilar membrane time constants, which are more like 100 μs (1/2 cycle of BM response). This effect will strongly enhance mid to high frequency transient onsets, such as those seen in speech with /t/, /k/, /g/, etc.

In Fig. 8 we relate the NL upward spread of masking seen in a typical 2 kHz neural tuning curve [35] to the neural rate-based *peristimulus time histogram* (PSTH) [19]. If we assume some level of noise is present, then the cochlear sensitivity must adapt to the level of noise. By way of an example, assume that the noise level is +20 dB re threshold (0 dB in Fig. 8). If a /t/ or /k/ burst then drives the nerve fiber, the rate will jump up, as shown in the right panel (adapted from Delgutte (1980) [19]). According to the PSTH response, the neural sensitivity will remain in the low-threshold, high-rate state, for up to 20 ms (elbow). Other data in [19] shows that this adaptation can forward mask (rate-suppress) a response up to 50 ms.

In summary, onsets will be enhanced by OHC processing, due to the overshoot seen in the auditory nerve PSTH functions of Fig. 8. In the hearing impaired ear, such enhancements would be gone, and this extra kick of response would not be available in those ears.

Summary

This article has reviewed some of what we have recently learned about speech perception of consonants, and how this might impact our understanding of NL cochlea speech processing of speech.

The application of NL OHC processing in speech processing is still an underdeveloped application area [15]. The key open problem here is "How does the auditory system (e.g., the NL cochlea and the auditory cortex) processes human speech?" There are many applications of these results including speech coding, speech recognition in noise, hearing aids, cochlear implants, as well as language learning and reading disorders in children. If we can solve the *robust phone decoding problem,* we will fundamentally change the effectiveness of human-machine interactions. For example, the ultimate hearing aid is the hearing aid with built in robust speech feature detection and phone recognition. While we have no idea when this will come to be, and it is undoubtedly many years off, when it happens there will be a technology revolution that will change human communications. The speech perception results shown here are relevant to this application.

Outer hair cells provide intensity dynamic range control and are responsible for the NL cochlear processing of speech. The OHCs are the one common element that link all the NL data previously observed, and a missing piece of the puzzle that most needs to be understood before any model can hope to succeed in predicting basilar membrane, hair cell, and neural tuning, and NL compression. Understanding the outer hair cell's two-way mechanical transduction is viewed as the key to solving the problem of the cochlea's dynamic range. The OHCs membrane's τ_m RC recover time constant (defined as the τ_m with an OHC cilia admittance of zero) is a determining factor in the cochlear response recover time.

Acknowledgments

We gratefully acknowledge Phonak and the ECE department of UIUC for their generous support of this research.

References

1. Olive, J., Greenwood, A., and Coleman, J., 1993. Acoustics of American English Speech: A dynamic approach. Springer-Verlag, New York, Berlin, Heidelberg.
2. Rayleigh, Lord, 1908. Acoustical notes – viii. Philosophical Magazine, 16(6), 235–46.
3. Campbell, G. A., 1910. Telephonic intelligibility. Phil. Mag., 19(6), 152–9.
4. Fletcher, H., 1921. An empirical theory of telephone quality. AT&T Internal Memorandum, 101(6).
5. Shannon, C. E., 1948. The mathematical theory of communication. Bell System Tech. Jol., 27, 379–423 (parts I, II), 623–56 (part III).
6. Allen, J. B., 2005. Articulation and Intelligibility. Morgan and Claypool, 3401 Buckskin Trail, LaPorte, CO 80535. ISBN: 1598290088.
7. Allen, J. B., 1994. How do humans process and recognize speech? IEEE Transactions on speech and audio, 2(4), 567–77.
8. Allen, J. B., 1996. Harvey Fletcher's role in the creation of communication acoustics. J. Acoust. Soc. Am., 99(4), 1825–39.
9. Allen, J. B., 2005. Consonant recognition and the articulation index. J. Acoust. Soc. Am., 117(4), 2212–23.
10. Phatak, S. and Allen, J. B., 2007. Consonant and vowel confusions in speech-weighted noise. J. Acoust. Soc. Am., 121(4), 2312–26.
11. Regnier, M. S. and Allen, J. B., 2007. A method to identify noise-robust perceptual features: application for consonant /t/. J. Acoust. Soc. Am., 123(5), 2801–14.
12. Furui S., 1986. On the role of spectral transition for speech perception. J. Acoust. Soc. Am., 80(4), 1016–25.
13. Phatak, S., Lovitt, A., and Allen, J. B., 2008. Consonant confusions in white noise. J. Acoust. Soc. Am., 124(2), 1220–33.

14. Allen, J. B., 2003. Amplitude compression in hearing aids. In R. Kent, editor, MIT Encyclopedia of Communication Disorders, chapter Part IV, 413–23. MIT Press, MIT, Boston Ma.
15. Allen, J. B., 2007. Nonlinear cochlear signal processing and masking in speech perception. In Jacob Benesty and Mohan Sondhi, editors, Springer Handbook on speech processing and speech communication, chapter 3, 1–36. Springer, Heidelberg Germany.
16. Allen, J. B., 1977. Short time spectral analysis, synthesis, and modification by discrete Fourier transform. IEEE Trans. Acoust. Speech and Sig. Processing, 25, 235–38.
17. Allen J. B., and Rabiner, L. R., 1977. A unified approach to short-time Fourier analysis and synthesis. Proc. IEEE, 65(11), 1558–64.
18. Duifhuis, H., 1973. Consequences of peripheral frequency selectivity for nonsimultaneous masking. J. Acoust. Soc. Am., 54(6), 1472–88.
19. Delgutte, B., 1980. Representation of speech-like sounds in the discharge patterns of auditory-nerve fibers. J. Acoust. Soc. Am., 63(3), 843–57.
20. Relkin, E.M., and Turner, C.W., 1988. A reexamination of forward masking in the auditory nerve. J. Acoust. Soc. Am., 84(2), 584–91.
21. Hewitt, M.J. and Meddis, R., 1991. An evaluation of eight computer models of mammalian inner hair-cell function. J. Acoust. Soc. Am., 90(2), 904–17.
22. Hecht, S., 1934. Vision II. The nature of the photoreceptor process. In C. Murchison, editor, Handbook of General Experimental Psychology. Clark University Press, Worcester, MA.
23. Gescheider, G.A., 1997. Psychophysics: The Fundamentals, 3d edition. Lawrence Erlbaum Associates, Mahwah, NJ; London.
24. Allen, J. B. and Neely, S.T., 1992. Micromechanical models of the cochlea. Physics Today, 45(7), 40–47.
25. Dallos, P., 1996. Cochlear neurobiology. In P. Dallos, A.N. Popper, and R.R. Fay, editors, The cochlea, 186–257, Springer, New York.
26. Narayan, S.S., Temchin, A.N., Recio, A. and Ruggero, M.A., 1998. Frequency tuning of basilar membrane and auditory nerve fibers in the same cochleae. Science, 282, 1882–4.
27. deBoer, E. , 1996. Mechanics of the cochlea: modeling efforts. In The cochlea, 258–317 Springer, New York.
28. Geisler, D. C., 1998. From Sound to Synapse: Physiology of the Mammalian Ear. Oxford University Press.
29. Duifhuis, H., 1992. Cochlear modelling and physiology. In M. E. H. Schouten, editor, The Auditory Processing of Speech, 15–27, Mouton de Gruyter, Berlin.
30. Allen, J. B., 2001. Nonlinear cochlear signal processing. In A.F. Jahn and J. Santos-Sacchi, editors, Physiology of the Ear, Second Edition, chapter 19, 393–442. Singular Thomson Learning, 401 West A Street, Suite 325 San Diego, CA 92101.
31. Steinberg, J.C., and Gardner, M.B., 1937. Dependence of hearing impairment on sound intensity. Journal of the Acoustical Society of America, 9, 11–23.
32. Neely, S. T. and Allen, J. B., 1997. Relation between the rate of growth of loudness and the intensity DL. In W. Jesteadt and et al., editors, Modeling Sensorineural Hearing Loss, 213–222. Lawrence Erlbaum Assoc., Mahwah, NJ.
33. Carver, W.F., 1978. Loudness balance procedures. In Jack Katz, editor, Handbook of Clinical Audiology, 2^{d} edition, chapter 15, 164–178. Williams and Wilkins, Baltimore MD.
34. Sewell, W.F., 1984. The effects of furosemide on the endocochlear potential and auditory-nerve fiber tuning curves in cats. Hearing Res., 14, 305–14.
35. Fahey, P. F., and J. B., Allen, 1985. Nonlinear phenomena as observed in the ear canal, and at the auditory nerve. J. Acoust. Soc. Am., 77(2), 599–612.

Comments and Discussion

van Schaik (and others): The adaptation shown in Figure 1 of Delgutte (1980) in response to tone bursts is generally attributed to adaptation of neurotransmitter release in the synaptic cleft (see for instance the Meddis IHC model). If this is correct, surely the adaptation has nothing to do with a change in tuning bandwidth nor does it allow one to estimate the time constant of the change in bandwidth?

The adaptation at the IHC does indeed enhance onsets, but it is not clear to me if it necessarily follows that hearing impaired ears would not have this adaptation. If the mechanism for neurotransmitter release is normal in a hearing impaired ear, then the same onset enhancement would be available.

Allen: I agree (See section 3, page 100) that the neural adaptation model you quote (Hewitt and Meddis [21]) is the widely held view. It is exactly for the reason you quote, that neural adaption would not lead to SNR-Loss, that I am suggesting there must be another mechanism. How else can we explain the widely observed SNR-Loss? The adaptation model has other serious flaws:

- It is widely believed that hearing impaired ears loose a natural robustness to noise, an effect sometimes called *SNR-Loss*. This cannot be accounted for by neurotransmitter adaptation, since there is no reason to believe that the synapse would be modified in the HI ears (as you point out in your question). Rather it is the OHC (e.g., their cilia) that are different.
- Forward masking data shows a "return to baseline" of 200 ms, with a linear in log-dB (a double log). How could such properties come from a simple synapse and its adaptation?
- Sewell found a 1 dB/mV dependence between the EP and the tip of the tuning curve, which seems indicative of a mechanical induced transformation, and it seems highly inconsistent with the adaptation model.

Thus we need something more than such a simple neurotransmitter adaptation to account for these other observations.

In my view, we are a long way from fully understanding nonlinear processing in the cochlea. Many people make a sweeping assumption that the CA explains all the the things they don't understand. Here are a few things that I believe: The first spike at the speech onset is highly significant (Heil). The role of the OHC is for signal dynamic range control, and as I have said many times, the evidence for the OHC's role in providing significant power gain (i.e., cycle-by-cycle) on the BM is quite limited. The dynamics of OHC processing are still not fully understood. As of yet, there are no forward masking data measured on the BM. Please reconsider my explanation of how Sewell's 1 dB/mV might come about [15].

DETERMINING THE IDENTITY OF THE COCHLEAR AMPLIFIER: ELECTRICAL STIMULATION OF THE *TECTA* MOUSE COCHLEA

MARCIA M. MELLADO LAGARDE, MARKUS DREXL, VICTORIA A. LUKASHKINA, ANDREI N. LUKASHKIN, IAN J. RUSSELL

School of Life Sciences, University of Sussex, Falmer, Brighton, BN1 9QG, UK.

The sensitivity, large dynamic range and narrow frequency tuning of the mammalian cochlea is determined by the passive mechanical properties of the basilar membrane (BM) and active feedback from the outer hair cells (OHCs). Two mechanisms have been proposed to provide amplification: Hair bundle motility, and OHC somatic-motility. Acoustically- and electrically-elicited mechanical responses were measured from the BMs of the cochleae of wild type and genetically modified mice where the hair bundles are freestanding and cannot react against the tectorial membrane (TM) to contribute to amplification. We found the electrically elicited responses in mutant mice, where only somatic motility can provide amplification, to be remarkably similar to acoustical and electrical responses in the wild type animals. We, therefore, conclude that somatic, not stereocilia motility is the basis of the cochlear amplifier.

1 Introduction

The resulting flow of current into the stereocilia of the sensory motor OHCs when the hair bundles are deflected through sound-induced shear displacements between the reticular lamina and TM initiates active mechanical forces. These forces constitute the cochlear amplifier, amplifying low-level and compressing high-level BM displacements [1, 2]. Effective amplification around the characteristic frequency of each location along the BM, the cochlear amplifier provides the basis for the cochlea's exquisite, frequency tuning, sensitivity and enormous dynamic range [3].

The cochlear amplifier has been attributed to voltage-dependent somatic motility of the OHCs [4] due to the activity of the motor protein prestin [5] and to hair bundle motility driven by calcium currents [6, 7]. Although prestin is unique to OHCs and is essential for cochlear frequency tuning and sensitivity [8], it has yet, however, to be demonstrated as the unequivocal source of cochlear amplification. Calcium-dependent hair bundle motility is a ubiquitous feature of hair cells but its role as the source of amplification in the mammalian cochlea is predicted so far only from *in vitro* studies [9, 10].

Here we present data on the relative contributions of hair bundle and somatic motility in amplifying and tuning the BM responses to acoustical and electrical cochlear stimulation in wild type ($Tecta^{+/+}$) and mutant ($Tecta^{\Delta ENT/\Delta ENT}$) mice [11]. The OHC hair bundles of $Tecta^{+/+}$ mice are coupled mechanically to the TM but those of $Tecta^{\Delta ENT/\Delta ENT}$ mice are freestanding and not attached to the TM, which is vestigial and not associated with the organ of Corti in $Tecta^{\Delta ENT/\Delta ENT}$ mice. BM responses to acoustic stimulation depend on sensory transduction mediated via displacements of the OHC hair bundles. Electrical stimulation of the cochlea drives both OHC somatic and hair bundle motility directly [6, 9]. In *Tecta*$^{+/+}$ mice however, the presence of the TM will permit electrically elicited hair bundle movements to interact with the TM, an opportunity precluded to the hair bundles of *Tecta*$^{\Delta ENT/\Delta ENT}$ mice at low to moderate levels of stimulation [11].

2 Methods

Experiments were performed in *Tecta*$^{+/+}$ and *Tecta*$^{\Delta ENT/\Delta ENT}$ mice [11], 4–6 weeks old, on a C57B6/J or CBA /J background that were bred and maintained at the University of Sussex. Animals were initially anaesthetized (i.p.) with Fentanyl (0.05 mg/ kg bodyweight,), Midazolam (5mg/ kg bodyweight) and Medetomidine (0.5 mg/ kg bodyweight). Anaesthesia was maintained throughout the experiments, at the end of which the mice were overdosed with pentobarbital sodium (150 mg/ kg bodyweight i.p.). Carbogen (95% O_2, 5% CO_2, flow rate ~ 0.3 l/ min) was provided via a head mask. The heartbeat was monitored with skin electrodes and core temperature maintained at 38°C.

The right auditory bulla was opened with a ventrolateral approach to gain access to the round window. The pinna was removed and the sound system was placed into the meatus ~1mm from the eardrum [11]. We adopted Nuttall and Ren's [12] technique to deliver extracochlear electrical stimulation using either silver, or tungsten electrodes placed on the round window and a Ag/ AgCl reference electrode in the neck tissue. Sodium salicylate was applied as a crystal on the round window membrane for 5 minutes. This technique, and not cochlear perfusion, was employed to avoid a reduction of cochlear sensitivity when the round window is breached. We were therefore unable to quantify the concentration of salicylate in the perilymph. Enhancement of the CM due to increases in the basolateral and not the mechanoelectrical conductance of the OHCs, which occur when the level of salicylate in the perilymph exceeds ~ 2mM [13], were not observed however.

The sound system, its calibration in dB SPL re 2 x 10-5 Pa, and the method of generating computer-controlled command voltages, have been described [11, 14], but here we used a custom-built condenser loudspeaker [15]. For electrical stimulation the loudspeaker was unplugged and the output signal from the GPIB-controlled attenuator was used as the command voltage for a custom-built current-pump with a sensitivity of 100 μA/V. The current pump delivered through the round window electrode sinusoidal current of constant amplitude corresponding to the applied sinusoidal command voltage. BM displacements were measured by focusing the beam of a self-mixing, laser-diode interferometer [14] through the round window membrane to form an ~5 μm spot on the centre of the BM in the 60–65 kHz region of the cochlea. Experimental control, data acquisition, and data analysis were performed using programs written in TestPoint (CEC). All procedures involving animals were performed in accordance with UK Home Office regulations with approval from the local ethics committee.

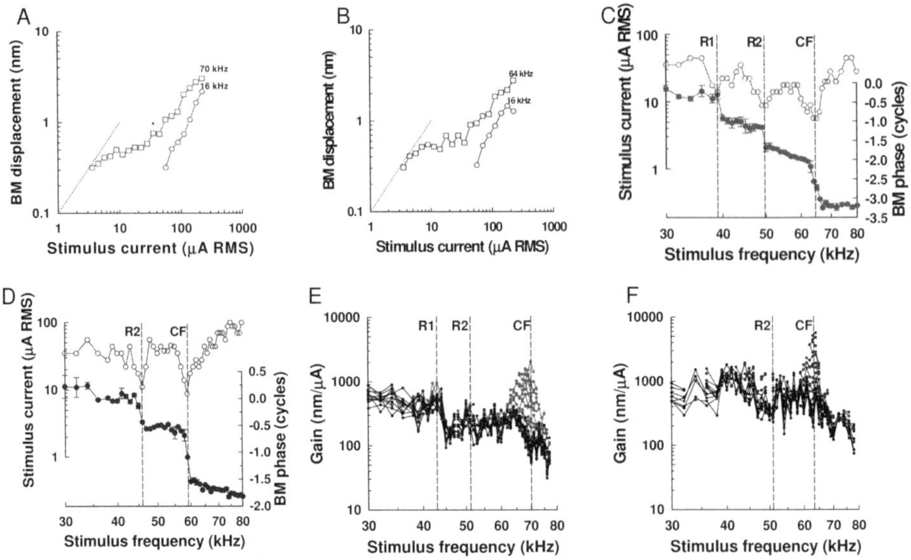

Figure 1. Basilar membrane responses to electrical stimulation of the mouse cochlea. (a, b) BM displacement as a function of stimulating current (μA RMS, root mean square) measured at (a) the CF (70 kHz, squares) and 16 kHz (circles) from the 70 kHz location in the cochlea of a *Tecta*$^{+/+}$ mouse and (b) the CF (64 kHz, squares) and 16 kHz (circles) from the 64 kHz location in the cochlea of a *Tecta*$^{\Delta ENT/\Delta ENT}$ mouse. Dotted line represents a slope of 1. (c, d) Threshold frequency tuning curves (open symbols) measured from the basal turn BM in the cochleae of (c) a *Tecta*$^{+/+}$ and (d) a *Tecta*$^{\Delta ENT/\Delta ENT}$ mouse. The phase of BM displacement (solid symbols) is plotted for a stimulus current of 15 dB above threshold for both genotypes. (e, f) Gain of BM displacement with respect to stimulating current in 2 dB steps (2.2 - 223.6 μA RMS) vs. stimulus frequency measured from (e) a *Tecta*$^{+/+}$ mouse and (f) a *Tecta*$^{\Delta ENT/\Delta ENT}$ mouse. Intense traces are for stimulus currents between 35.4 – 223.6 μA RMS. Vertical dashed lines indicate CF, R2, and R1 frequencies. The data in each figure are from different *Tecta*$^{+/+}$ and *Tecta*$^{\Delta ENT/\Delta ENT}$ mice. Modified with permission from [16]

3 Results

Near-threshold electrically elicited displacements measured from the BM of both *Tecta*$^{+/+}$ and *Tecta*$^{\Delta ENT/\Delta ENT}$ mice in response to stimulation via an electrode on the round window membrane are very similar to those measured from *Tecta*$^{+/+}$ mice in response to acoustic stimulation [11]. Growth of electrically elicited BM displacements for near-threshold responses at or near the characteristic frequency (CF) are nonlinear and compressive (Figure 1a, b). BM threshold tuning curves for both *Tecta*$^{+/+}$ and *Tecta*$^{\Delta ENT/\Delta ENT}$ mice are very similar in form and characteristics to those obtained in response to acoustic stimulation in *Tecta*$^{+/+}$ mice (Figure 1c, d, Table 1). A novel resonance (R2) [16] is present in the acoustic tuning curves of *Tecta*$^{+/+}$ mice [16] and in the electrical tuning curves of both *Tecta*$^{+/+}$ and *Tecta*$^{\Delta ENT/\Delta ENT}$ mice, but R1, which has been attributed to the TM [11], is absent from tuning curves measured from *Tecta*$^{\Delta ENT/\Delta ENT}$ mice. The BM responses to electrical stimulation are presumably due to the simultaneous excitation by the stimulating current of OHCs in the basal turn of the cochlea. There are, however

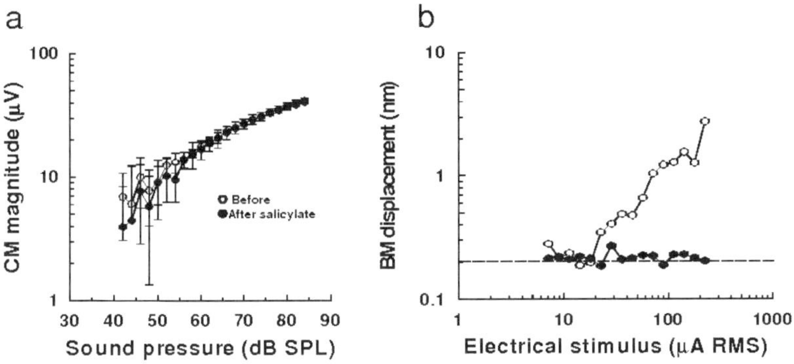

Figure 2. Salicylate blocks electrically evoked BM displacements but not acoustically evoked cochlear microphonic potentials in *Tecta* mice. (a) Means ± s.d. (n = 5) of cochlear microphonic potentials as functions of sound pressure level recorded before and after salicylate treatment. (b) Basilar membrane displacement as a function of electrical stimulus level recorded before and after placing a crystal of salicylate on the round window membrane. This is a typical example of the data obtained in *Tecta*$^{\Delta ENT/\Delta ENT}$ mice, the data in *Tecta*$^{+/+}$ is similar. Dashed line: measurement noise floor (modified with permission from [16]).

Table 1. Characteristics of acoustically- and electrically-elicited BM tuning curves. Measurements were made in the basal turn of the cochleae of *Tecta*$^{+/+}$ and *Tecta*$^{\Delta ENT/\Delta ENT}$ mice. Oct<CF: octaves below the CF of the location on the BM where the laser beam was focused. For acoustic stimulation, the threshold gain at the CF is compared with that post mortem. For electrical stimulation the threshold gain at the CF is compared with that due to stimulating currents 30 dB above threshold (from [16].

Acoustic	R1 Oct<CF	R2 Oct<CF	CF (kHz)	CF Threshold (dB SPL)	Q_{10dB}	Gain (dB)	n
Tecta$^{+/+}$	0.61 ± 0.07	0.27 ± 0.04	64.5 ± 2.5	19 ± 3	10.0 ± 1.4	27.5 ± 4.9	5
Tecta$^{\Delta ENT/\Delta ENT}$	-	0.34 ± 0.08	62.2 ± 4.9	55 ± 8	-	6.7 ± 2.7	5
Electrical				Threshold (μA RMS)			
Tecta$^{+/+}$	0.60 ± 0.08	0.33 ± 0.06	65.2 ± 3.4	6.75 ± 1.75	9.5 ± 1.6	29.9 ± 5.4	5
Tecta$^{\Delta ENT/\Delta ENT}$	-	0.37 ± 0.04	63.2 ± 5.2	6.41 ± 1.63	9.6 ± 0.4	28.5 ± 2.4	9

phase jumps of ~180° associated with R2 and R1 and an accumulating phase lag of ~ 450° associated with the CF, which are similar to those obtained with acoustic stimulation.

The gains of electrically elicited vibrations of the BM were obtained by comparing the sensitivities of BM displacements (nm / μA) for threshold currents with those at high levels (224 μA RMS). The gains were found to be very similar for both *Tecta*$^{+/+}$ and *Tecta*$^{\Delta ENT/\Delta ENT}$ mice and not significantly different from those of the acoustically stimulated cochleae of *Tecta*$^{+/+}$ mice [11, 16] (Table 1).

Electrically evoked BM vibrations, but not the sound-evoked receptor potentials (cochlear microphonic potentials), in *Tecta*$^{+/+}$ and *Tecta*$^{\Delta ENT/\Delta ENT}$ mice are completely blocked by sodium salicylate within the range of currents used in these experiments

(Figure 2), presumably by blocking the functioning of the motor protein prestin and therefore OHC somatic motility [14, 17, 18, 19, 20].

4 Discussion

Near threshold, when BM mechanical responses are dominated by contributions from OHCs, frequency tuning measured from *Tecta*$^{+/+}$ mice in response to acoustic stimulation and from both *Tecta*$^{+/+}$ *and Tecta*$^{\Delta ENT/\Delta ENT}$ *mice* in response to extracochlear electrical stimulation are very similar. Both prestin- and calcium-based motility have the potential to provide cochlea amplification in *Tecta*$^{+/+}$ mice because both are excited by electrical stimulation of the cochlea [9, 10]. For calcium-based motility to be the source of cochlear amplification, OHC hair bundles must interact with the TM if they are to exert forces on the BM [9]. Electrically elicited calcium-based motility cannot, therefore, amplify BM responses measured from *Tecta*$^{\Delta ENT/\Delta ENT}$ mice at low stimulus levels because the sensory bundles of the OHCs are freestanding and are not able to react against the TM [11]. For this reason, amplification and compression of electrically elicited BM responses from the cochleae of *Tecta*$^{\Delta ENT/\Delta ENT}$ mice, which are similar in all respects to those of acoustically and electrically elicited responses from *Tecta*$^{+/+}$ mice, cannot be due to hair bundle motility. We therefore propose that, at levels close to the threshold of hearing, it is not necessary to invoke a role for hair bundle motility in amplifying the vibrations of the BM. Indeed, electrically evoked BM vibrations in both *Tecta*$^{+/+}$ and *Tecta*$^{\Delta ENT/\Delta ENT}$ mice are completely suppressed by salicylate. Salicylate has been shown to block prestin-based OHC somatic motility [19, 20], but has no direct action on the hair cell mechano-electrical transducer channels [21, 22]. We are thus lead to attribute amplification in the basal turn of the cochlea to the prestin-based somatic motility of OHCs.

Acknowledgments

This work was supported by the Medical Research Council, M.M. was supported by a FENS – IBRO Fellowship, M.D. was supported by a DFG Fellowship. We thank Guy Richardson for making available the *Tecta* mice. We thank James Hartley for technical assistance.

References

1. Dallos P., 1992. The active cochlea. J. Neurosci. 12, 4575-4585.
2. Lukashkin A.N., Walling M.N., Russell I.J., 2007. Power amplification in the mammalian cochlea. Curr. Biol. 17, 1340-1344.
3. Robles L., Ruggero M.A., 2001. Mechanics of the mammalian cochlea. Physiol. Rev. 81, 1305-1352.
4. Brownell W.E., Bader C.R., Bertrand D., de Ribaupierre Y., 1985. Evoked mechanical responses of isolated cochlear outer hair cells. Science 227, 194-196.
5. Zheng J., Shen W., He D.Z., Long K.B., Madison L.D., Dallos P., 2000. Prestin is the motor protein of cochlea outer hair cells. Nature 405, 149-155.

6. Fettiplace, R. and Hackney, C. M., 2006. The sensory and motor roles of auditory hair cells. Nat. Rev. Neurosci. 7, 19-29.
7. Chan, D. K. and Hudspeth, A. J., 2005. Ca2+ current-driven nonlinear amplification by the mammalian cochlea in vitro. Nat. Neurosci. 8, 149-55.
8. Liberman M.C., Gao J., He H.Z., Wu X., Jia S., Zuo J., 2002. Prestin is required for electromotility of the outer hair cell and for the cochlear amplifier. Nature 419, 300-304.
9. Chan, D.K. and Hudspeth, A, J., 2005. Mechanical responses of the organ of corti to acoustic and electrical stimulation in vitro. Biophys J. 89, 4382-95.
10. Kennedy, H. J., Crawford, A. C. and Fettiplace, R. *Nature* **433**, 880-3 (2005).
11. Legan P.K., Lukashkina V.A., Goodyear R.J., Kossl M., Russell I.J., Richardson G.P., 2000. A targeted deletion in alpha-tectorin reveals that the tectorial membrane is required for the gain and timing of cochlear feedback. Neuron 28, 273-285.
12. Nuttall, A. L. and Ren, T., 1995. Electromotile hearing: evidence from basilar membrane motion and otoacoustic emissions. Hear. Res. 92, 170–177 (1995).
13. Fitzgerald, J. J., Robertson, D. and Johnstone, B.M., 1993. Effects of intra-cochlear perfusion of salicylates on cochlear microphonic and other auditory responses in the guinea pig. Hear. Res. 67, 147-56.
14. Lukashkin A.N., Bashtanov M.E., Russell I.J., 2005. A self-mixing laser-diode interferometer for measuring basilar membrane vibrations without opening the cochlea. J. Neurosci. Methods 148, 122-129.Schuller G., 1997. A cheap earphone for small animals with good frequency response in the ultrasonic frequency range. J. Neurosci. Methods 71, 187-190.
15. Mellado Lagarde, M.M., Drexl, M., Lukashkina, V. A., Lukashkin, A. N., Russell, I. J., 2008. Outer hair cell somatic, not hair bundle, motility is the basis of the cochlear amplifier. Nature Neurosci. 11, 746-8.
16. Murugasu E., Russell I.J., 1995. Salicylate ototoxicity: the effects on basilar membrane displacement, cochlear microphonics, and neural responses in the basal turn of the guinea pig cochlea. Aud. Neurosci. 1, 139-150.
17. Zheng, J., Deo, N., Zou, Y., Grosh, K. and Nuttall, A. L., 2007. Chlorpromazine alters cochlear mechanics and amplification: in vivo evidence for a role of stiffness modulation in the organ of corti. J. Neurophysiol. 97, 994-1004.
18. Oliver D., He D.Z., Klocker N., Ludwig J., Schulte U., Waldegger S., Ruppersberg J.P., Dallos P., Fakler B., 2001. Intracellular anions as the voltage sensor of prestin, the outer hair cell motor protein. Science 292, 2340-2343.
19. Santos-Sacchi, J.,Song, L., Zheng J. and Nuttall, A. L. J., 2006. Control of mammalian cochlear amplification by chloride anions. Neurosci. 26, 3992-8.
20. Kimitsuki, T., Nakagawa, T., Hisashi, K., Komune, S. and Uemura, T., 1994. The effects of ototoxic drugs on mechano-electric transduction channels in chick cochlear hair cells. Eur. Arch. Otorhinolaryngol. 251, S53–S56.
21. Kennedy, H. J., Evans, M. G., Crawford, A. C. and Fettiplace, R. 2006. Depolarization of cochlear outer hair cells evokes active hair bundle motion by two mechanisms. J. Neurosci. 26:2757–2766.

Comments and Discussion

Braun: The reported results show that somatic OHC motility at low sound levels is visible through BM motion, even when the TM is uncoupled from the OHCs. From these findings, however, it is not possible to draw conclusions on the role of hair bundle motility. "Cochlear amplifier" means “amplifier of forces that excite the IHCs". It does not mean “BM amplifier”. Nowotny et al. (2006) have shown that OHCs can excite IHCs independently from any BM motion. Chen et al. (2008) have shown that the reticular lamina (RL), and thus both OHCs and IHCs, vibrate independently from the basilar membrane (BM).

Nowotny M, Gummer AW, 2006. Nanomechanics of the subtectorial space caused by electromechanics of cochlear outer hair cells. Proc. Natl. Acad. Sci. USA 103, 2120-2125.

Chen F, Zheng J, Choudhury N, Jaques S, Nuttall AL, 2008. Organ of Corti micromechanics with local electrical stimulation. This volume.

DIFFERENTIAL MEASUREMENT OF BASILAR MEMBRANE VIBRATION IN SENSITIVE GERBIL COCHLEAE

TIANYING REN[1,2], WENXUAN HE[1]

[1]*Oregon Hearing Research Center, Oregon Health & Science University, 3181 SW Sam Jackson Park Road, NRC04, Portland, Oregon 97239-3098, USA;* [2]*Department of Physiology, Xi'an Jiaotong University, Xi'an, Shaanxi 710061, P.R. China;*

Basilar membrane vibrations in the transverse direction were measured at two longitudinal locations using a laser interferometer. The magnitude transfer function of the cochlear partition between the two locations was obtained by calculating the ratio of the vibration magnitude at the apical location to that at the basal location as a function of the stimulus frequencies. Measured differential transfer function showed a response peak at a frequency below the best frequency of the apical location with a magnitude >1, and the minimum magnitude near the best frequency of the basal location with an amplitude as small as 0.01. These data indicate that the cochlea achieves its high sensitivity, sharp tuning, and nonlinearity through amplification and attenuation of basilar membrane vibration.

1 Introduction

Current knowledge of cochlear mechanics in sensitive mammalian ears is mainly from the measurements of basilar membrane vibration in the transverse direction at a single longitudinal location. To quantify the mechanical properties of the cochlear partition, basilar membrane vibration is often measured as a function of the stimulus frequency at different sound pressure levels. Magnitude transfer functions of the basilar membrane are commonly presented as the ratio of the basilar membrane to stapes vibration as a function of frequency. The phase transfer function is obtained by subtracting stapes vibration phase from that of the basilar membrane [1]. Basilar membrane response to a low level sound at the best frequency (BF) can be 1,000 fold more than stapes vibration [2-6]. As the stimulus frequency moves away from the BF, the vibration magnitude decreases quickly and forms a sharp response peak. The magnitude near the BF decreases with the sound level, indicating a compressive nonlinearity. It is a common belief that cochlear high sensitivity, sharp tuning, and nonlinearity result from the activity of the cochlear amplifier, an outer hair cell-based local feedback mechanism, which uses metabolic energy to amplify the low level sound-induced basilar membrane vibration [7]. This mechanical power amplification is believed to occur over an area on the cochlear partition basal to the BF location [8]. The conventional transfer function, however, characterizes the basilar membrane from the base to the BF site as a whole and does not provide information about the function of a specific length of the basilar membrane. The aim of this preliminary experiment is to quantify the local function of the basilar membrane, where cochlear amplification is supposed to take place, by measuring basilar membrane vibrations at two longitudinal locations using a heterodyne laser interferometer.

2 Materials and Methods

2.1 Animals and surgical procedures

Young healthy Mongolian gerbils (40-80 g) were used in this study. The animal preparation and surgical approaches for opening the auditory bulla were the same as in a previous study [9]. Anesthesia was achieved by intraperitoneal injection of 30 mg/kg ketamine followed by intramuscular injection of 5 mg/kg xylazine. The animal's head was firmly attached to a head-holder with x-y-z translation and rotation capability. A tracheotomy was performed and a ventilation tube was inserted into the trachea to ensure natural breathing. Body temperature was monitored using a rectal thermal probe and maintained at 38±1°C with a servo-regulated heating blanket. Animal procedures were approved by the Oregon Health & Science University Institutional Animal Care and Use Committee.

A round window approach was used to expose the basilar membrane of the first cochlear turn [9]. The round window membrane was removed using a micro-hook, and the bony round window was enlarged anteriorly. The animal's head was positioned so that bone chips from the edge of the round window would not fall onto the cochlear partition. A few gold-coated glass beads with diameter of ~20 μm were placed on the basilar membrane with longitudinal separation of 150 to 400 μm. The opened round window was partially covered using a clear thin glass cover slip to avoid optical distortion and maintain low impedance of the round window. The distance between beads was calculated from x, y, and z coordinates of several reference points measured using a positioning system.

2.2 Measurement of conventional basilar membrane transfer functions

The laser beam from an optical head of a laser interferometer (Polytec OFV 301) was focused on a bead at the apical or basal location. The magnitude and phase of the basilar membrane vibration velocity in response to tones at different frequencies were measured. The stimulus frequency was varied from 500 Hz to 25 kHz in 250 Hz steps, and the measurements were taken at sound pressure levels of 40 and 90 dB SPL. To obtain conventional transfer functions of the basilar membrane, the vibration of the stapes was also measured. The magnitude transfer function was obtained by plotting the ratio of the basilar membrane velocity to that of the stapes as a function of frequency. The phase transfer function was measured as phase difference between basilar membrane and stapes vibrations as a function of frequency. The BF was indicated by the frequency with maximum magnitude on the magnitude transfer function at 40 dB SPL.

2.3 Measurement of local basilar membrane transfer functions

The local basilar membrane transfer function was obtained by plotting the ratio of the vibration velocity at the apical location to that at the more basal site against the stimulus frequency. The corresponding phase transfer function was presented as the phase difference between two measured locations calculated by subtracting the basal phase from the apical phase.

3 Results

3.1 Conventional basilar membrane transfer functions

Based on the data measured at the basilar membrane and the stapes, the conventional basilar membrane transfer functions at the two longitudinal locations were calculated and presented in Fig. 1. Data in Fig. 1A and B were collected at the apical location, and those in Fig. 1C and D were from the basal location. The distance between the two measured locations is ~292 μm. The magnitude transfer function at 40 dB SPL (solid line in Fig. 1A) shows that the magnitude increases with the stimulus frequency and reaches the maximum at ~14 kHz, i.e., BF. As the stimulus frequency becomes higher than the BF, the magnitude decreases quickly and forms a sharp peak centered at the BF. The magnitude at the BF is up to ~1,000. The response at 90 dB SPL (dashed line in Fig. 1A) shows a broad peak at a frequency below 14 kHz. The peak magnitude of the high-level

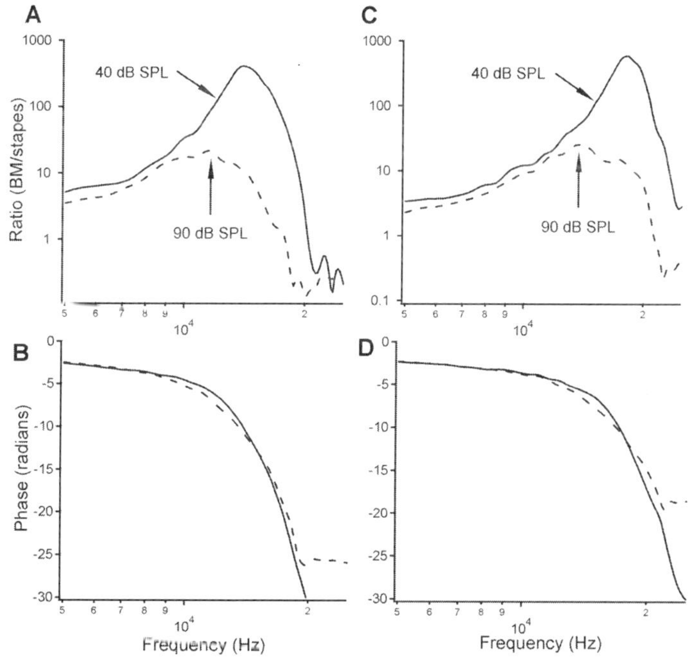

Figure 1. Conventional transfer functions of basilar membrane vibrations at two measured locations. A and B, magnitudes and phases measured at the more apical location. Panel A shows a sharp peak at ~14 kHz at 40 dB SPL. As sound level increases from 40 to 90 dB SPL, the magnitude at ~14kHz decreases by ~30 dB, and the response peak broadens and shifts toward ~11 kHz. Phase transfer functions (B) show that phase lag progressively increases with frequency and slightly decreases with sound level. The data in C and D measured at the basal location are similar to those in A and B except for a higher BF at ~18 kHz.

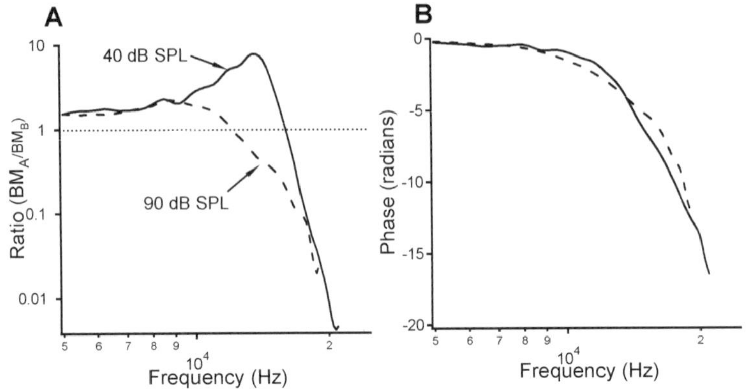

Figure 2. Magnitude (A) and phase (B) of local transfer functions. Response magnitude below 10 kHz is approximately independent of frequency or sound level. At low sound level, the response peak is at ~13.7 kHz, which decreases with sound level. The smallest magnitude is ~0.01 near 20 kHz. BM_A/BM_B: the ratio of basilar membrane vibration at the apical location to that at the basal location.

response is ~30 dB below that of the low-level data, indicating a compressive nonlinear growth. Thus, Fig. 1A shows high sensitivity, sharp tuning, and nonlinearity of the basilar membrane response, indicating a relatively healthy cochlear condition. The phase transfer functions in Fig. 1B show a progressive lag, indicating that traveling speed of basilar membrane vibration becomes slower with the frequency. The phase slope becomes flatter at high sound level at frequencies above the BF, indicating that basilar membrane vibration travels faster.

The conventional transfer functions measured at the more basal location are presented in Fig. 1C and D. Besides a relatively high BF, the data in Fig. 1C and D are very similar to those in Fig. 1A and B.

3.2 Local basilar membrane transfer functions

Local basilar membrane transfer functions were derived from data measured at the apical and basal locations and presented in Fig. 2A and B. The magnitude responses in panel A are presented by the ratio of the vibration amplitude at the apical location to that at the basal location as a function of frequency. At low-sound level, the magnitude increases with frequency and reaches the maximum at a frequency below the BF of the apical location. The magnitude above the dotted line in Fig. 2A indicates amplification of basilar membrane vibration. The magnitude of the peak response is positively related to the distance between the two measured locations. Although there is variation, the pattern of the local magnitude transfer function is consistent across animals. As the frequency continuously increases, the magnitude decreases quickly and becomes as small as 0.01 near the BF of the basal location, indicating strong attenuation of the basilar membrane vibration. The phase change over the distance between the two measured locations is characterized by Fig. 2B, which shows a pattern similar to that in Fig. 1B and D except

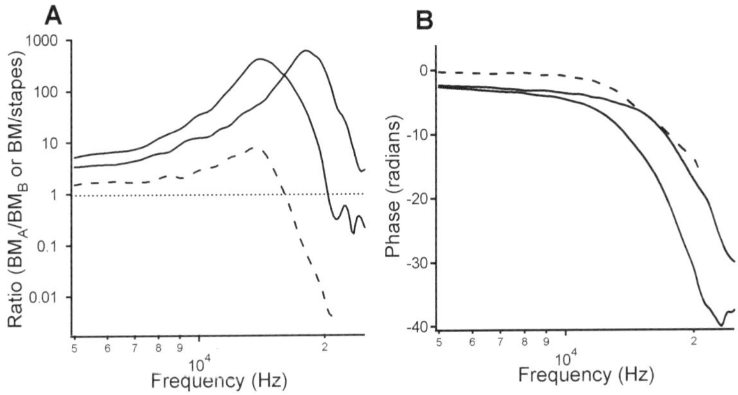

Figure 3. Magnitude (A) and phase (B) of the conventional (solid lines) and differential transfer functions (dashed line) at 40 dB SPL. Magnitude transfer functions are presented by the ratio of basilar membrane vibration at the apical location (BM_A) to that at the basal location (BM_B) or the ratio of the basilar membrane to stapes vibration (BM/stapes) as a function of frequency.

for a smaller phase lag. At the high sound level, the response peak becomes smaller or disappears and the phase lag becomes slightly smaller.

3.3 *Comparison of conventional and local basilar membrane transfer functions*

In order to show their differences, local and conventional transfer functions are plotted together in Fig. 3. Fig. 3A shows that the peak-response magnitude of the local transfer function (dashed line) is >100-fold smaller than that of conventional transfer functions (solid lines). Also, the minimum magnitude of the local transfer function is as small as 0.01 while the minimum of the conventional transfer function is ~1.0.

4 Discussion

A transfer function is a mathematical representation of the relation between the input and output of a system, which is commonly measured as the ratio of the output and input amplitudes of a single-input and single-output system. The term is often exclusively referred to a linear, time-invariant system. There was concern whether the term "transfer function" or "describing function" should be used to describe basilar membrane responses. In spite of this discussion, the ratio of the basilar membrane vibration amplitude to that of the stapes as a function of frequency has been used as the transfer function of the basilar membrane in the literature. In order to remain consistent with previous studies, we use the term "transfer function" to describe the current data.

The principle of the local transfer function measurement is the same as that of the conventional transfer function. The only difference is that one refers to stapes vibration and the other to basilar membrane vibration at a more basal location. The rationale for the use of basilar membrane vibration at a basal location as the input include the fact that sound-induced cochlear partition vibration is a forward traveling wave from base to apex, and the vibration at the apical location results from the vibration at the basal locations [10]. In addition, the basilar membrane vibration at a single longitudinal location has been routinely used as the output for the measurement of the conventional basilar membrane transfer function [1]. In contrast to stapes vibration, the basilar membrane vibration at a basal location is more comparable to the vibration at the apical location due to similar physical properties of the two vibration sites.

Conventional transfer functions in previous studies demonstrated high sensitivity, sharp tuning, and nonlinearity of basilar membrane responses to low-level tones. These cochlear functions have been attributed to the cochlear amplifier. This amplification has been commonly believed to occur over an area basal to the peak-response place. Since vibrations were measured at two longitudinal locations in this study, the local transfer function should reveal some functions of the proposed cochlear amplifier, which is responsible for the vibration at the apical location. The local transfer function demonstrates not only amplification but also attenuation of the basilar membrane vibration. Observed amplification is qualitatively consistent with the cochlear amplifier theory. The fact that the minimal response is ~100-fold smaller than 1 shows a strong attenuation near the BF of the basal location. In fact, this attenuation is as critical as the amplification because the peak response at the basal location at high frequencies needs to be attenuated to achieve the sharp tuning at the apical site. Thus, the data indicate that the cochlea achieves its high sensitivity, sharp tuning, and nonlinearity through amplification and attenuation of the incoming vibration.

Acknowledgments

We thank A.L. Nuttall and other colleagues at the Oregon Hearing Research Center for valuable discussions. This work was supported by the National Institute of Deafness and Other Communication Disorders grant R01 DC004554.

References

1. Robles, L., Ruggero, M.A., 2001. Mechanics of the mammalian cochlea. Physiol. Rev., 81, 1305-52.
2. Rhode, W.S., 1984. Cochlear mechanics. Ann. Rev. Physiol., 46, 231-46.
3. Ruggero, M.A., Rich, N.C., 1991. Application of a commercially-manufactured Doppler-shift laser velocimeter to the measurement of basilar-membrane vibration. Hear. Res., 51, 215-30.
4. Nuttall, A.L., Dolan, D.F., Avinash, G., 1991. Laser Doppler velocimetry of basilar membrane vibration. Hear. Res., 51, 203-13.
5. Cooper, N.P., 1998. Harmonic distortion on the basilar membrane in the basal turn of the guinea-pig cochlea. J. Physiol. (Lond.), 509, 277-88.
6. Russell, I.J., Nilsen, K.E., 1997. The location of the cochlear amplifier: spatial representation of a single tone on the guinea pig basilar membrane. Proc. Natl. Acad. Sci. U S A, 94, 2660-4.
7. Davis, H., 1983. An active process in cochlear mechanics. Hear Res, 9, 79-90.

8. de Boer, E., 1983. No sharpening? a challenge for cochlear mechanics. J. Acoust. Soc. Am., 73, 567-73.
9. Ren, T., 2002. Longitudinal pattern of basilar membrane vibration in the sensitive cochlea. Proc. Natl. Acad. Sci. U S A, 99, 17101-6.
10. von Bekesy, G., 1970. Travelling waves as frequency analysers in the cochlea. Nature, 225, 1207-9.

Comments and Discussion

Shera: One would find qualitatively similar "local" transfer functions (i.e., with magnitudes that increase to a maximum below BF and then decline sharply) in a purely passive cochlear model. So it's not really correct to say that the "magnitude above the dotted line in Fig. 2A indicates amplification of basilar membrane vibration".

Ren: The transfer function amplitude >1 means that the vibration at the apical location is greater than that at the more basal location. Since the apical vibration (output) results from the vibration at the basal location (input) according to the cochlear traveling wave theory the magnitude above the dotted line in Fig. 2A indicates amplification of basilar membrane vibration. This experimental result does not necessarily conflict with the common belief that there is no power amplification in a purely passive cochlear model because the vibration was measured only in the transverse direction not in power. In addition, the result is consistent with broad tuning of the basilar membrane vibration in a passive cochlea.

Shera: I agree with your first sentence. However your use of the word "amplification" is problematic. For example, one reason that the vibration may be greater at the more apical location is because the BM stiffness is smaller there. Or (if one considers a model where the BM response peak arises through passive "resonance") because the stimulus frequency is closer to the BF of the apical location than it is to the BF of the basal location. Neither of these reasons has anything to do with power amplification.
Also, even in 1D models the motion of the BM at a basal location is not "the input" that drives the motion of more apical locations. If anything at the basal location could be called the input it would be the volume velocity of the fluids in the scalae. But the truth (in models, at least) is that the motion of everything affects the motion of everything else.

Ren: The experimental data show that the basilar membrane vibration in the transverse direction increases as waves propagate from the basal location to the more apical site. This observation reveals amplitude amplification of the basilar membrane vibration in the transverse direction. Measured amplitude amplification in one direction is different from assumed power amplification.

According to the cochlear traveling wave theory, the basilar membrane vibration at an apical location occurs after the vibration at a more basal location. Moreover, the basilar membrane vibration measured at a single longitudinal location has been commonly used as the output to calculate the basilar membrane transfer function. Since the criteria for the output is the same as those for the input, the basilar membrane vibration at a basal location was used as the input in this study.

Braun: Figs. 1 & 2 clearly show the difference between the passive and the active cochlea by presenting data for both 90 and 40 dB SPL. Most qualities of this difference are not new. What is new, however, is that a response peak can disappear, totally, at 90 dB SPL. This result indicates that the passive mechanics of the basilar membrane may be poorly equipped, or even unable, to function as a bandpass filter. Thus we are now confronted with a further reminder that bandpass filtering in hearing, [even] in mammals, is likely to be found elsewhere.

Ren: Thank you for your comment. You pointed out a fundamental issue for cochlear theories.

Gummer: Could you please explain the added value of using the "local" transfer function of apical to basal responses? For example, is not the "attenuation" of this function simply an expression of a passive hydrodynamic property of the system; namely, that of the true resonant frequency of the BM response?

Ren: In contrast to the conventional transfer function, the value of using the local or differential transfer function is to quantify the function of a segment of the basilar membrane between the two measured locations where the hypothetical cochlear amplifier resides. The results show that the observed region increases basilar membrane vibration in the transverse direction at frequencies near and below the best frequency of the apical location, and decreases vibrations at higher frequencies. Observed magnitude amplification is conceptually consistent with the cochlear amplifier theory. Since observed attenuation in vibration amplitude does not depend on the stimulus level and cochlear sensitivity, we agree with you that it is likely an expression of a passive hydrodynamic property of the system. This property is one of functions of the observed region of the basilar membrane, which cannot be revealed by the conventional transfer function. The passive resonance mechanism certainly contributes to the local transfer function, particularly at frequencies above the characteristic frequency of the more basal location.

de Boer: As I have always stressed, amplification in the cochlea is local. Where amplification ceases, damping takes over and the wave amplitude starts to decrease. To give this effect a special name, attenuation, and to treat it as something peculiar is, in my opinion, superfluous.

Ren: We are very open about the name and the proposed mechanism of observed magnitude attenuation of basilar membrane vibration. The main purpose of the current paper is to present experimental data showing that the speculated cochlear-amplifier region of the basilar membrane not only increases but also decreases the vibration amplitude. The reason for stressing the magnitude attenuation is that this phenomenon has not been revealed by the conventional single-point measurement.

de Boer: You quite rightly say: 'the speculated cochlear-amplifier ....'. It is a hypothesis. In the case of a stable system or model, stability implies automatically that any power generated locally will be dissipated elsewhere. In my opinion, this is more a matter of interpretation than of experiment.

Ren: Your clarification that the cochlear amplifier is a hypothesis is significant because the cochlear amplifier has often been treated as a fact rather than a hypothesis. As a hypothesis, it remains to be tested and may coexist with other possible mechanisms*.
* Ren, T. and P. G. Gillespie (2007). "A mechanism for active hearing." Curr. Opin. Neurobiol. 17(4): 498-503.

Frosch: I would like to add three remarks to the preceding comments:
1) BM-velocity increase with distance-from-base x (panoramic view, fixed frequency f) or with f (frequency domain, fixed x) is possible without power gain because the group velocity decreases if the low-level best x, or the best f, is approached from below.
2) In Fig. 1A (1C) the 90-dB velocity ratio BM/stapes has a clear peak at 11 kHz (14 kHz). In Fig. 2A there is no high 90-dB peak because the distance between the two observed locations is only 0.292 mm.
3) In agreement with de Boer's comment: At fixed f, in the x-region of the basal slope of the low-level BM-velocity peak the OHC's feed energy into the travelling wave.

AMPLIFICATION IN THE COCHLEAR APEX

ANDERS FRIDBERGER, STEFAN JACOB

Center for Hearing and Communication Research, Karolinska Institutet, M1 Karolinska University Hospital, SE-17176 Stockholm, Sweden

It is well established that an active process, the 'cochlear amplifier' increases auditory sensitivity and frequency resolution in the basal, high frequency regions of the inner ear. The amplifier depends critically on the endocochlear potential, the value of which will greatly influence hair cell transduction and amplification. In the apical, low frequency regions of the cochlea, the situation is much less clear. We therefore preformed a series of experiments on cochlear explants. A glass microelectrode inserted through Reissner's membrane allowed rapid changes in the endocochlear potential during simultaneous sound stimulation. Changes in cochlear vibrations were assessed with laser interferometry and resampled confocal imaging. At low sound pressures, displacements were in the range 1 – 10 nm. Positive current injected through the pipette resulted in substantial enhancement of vibration amplitudes, an effect that vanished at higher stimulus levels. However, the increase in amplitude was not accompanied by substantial changes in frequency tuning or response phase.

1 Introduction

1.1 Amplification in the apex

Recent studies suggest that transduction mechanisms in the apex differ from those found in the base [1,2]. Whereas a compressive nonlinearity is present in the base, expansive nonlinearity may dominate in the apex. Following the death of the animal, apical vibration amplitudes generally increase, suggesting that a negative feedback mechanism may shape apical responses at low and moderate stimulus levels. However, other reports describe a compressive nonlinearity sharing some characteristics with that found in the base [*e.g.* 3]. In an attempt to clarify the situation, we injected currents in scala media while recording sound evoked vibration.

2 Methods

Temporal bones were isolated from young guinea pigs as previously described [4]. In brief, the animal was anaesthetized and decapitated using procedures approved by the local ethics committee. The excised temporal bone was glued to a rubber holder, the external auditory meatus facing a tube connected to a loudspeaker. The preparation was immersed in tissue culture medium and an opening made in scala tympani in the basal turn. This opening provided access for a thin piece of plastic tubing that ensured a continuous supply of oxygenated tissue culture medium. A second opening at the apex served as the outlet and also allowed visualization using water immersion optics.

2.1 *Interferometry*

A custom heterodyne laser interferometer was used for measuring cochlear vibrations. The design is loosely based on those published by Cooper [5] and by Willemin et al. [6] Important modifications include the use of a 'fake-object' beam providing a reference signal for the demodulation; the use of software digital demodulation based on direct sampling of the carrier waveform; and integration with a laser scanning confocal microscope (LSM Pascal, Zeiss Inc., Germany). Calibrations with a feedback-controlled piezo stack confirmed linearity over the range 0.05 – 1000 nm, with a noise floor on the order of 0.5 Å / √Hz with carrier powers resembling those typically achieved during experiments on the cochlea.

2.2 *Resampled confocal imaging*

As an adjunct to the interferometric measurements, we also measured two-dimensional vibrations using resampled confocal imaging [4, 7]. For these measurements, the hearing organ was loaded with fluorescent dyes through the perfusion system and placed on the stage of a Zeiss LSM 510 confocal microscope. The acquisition system produced image sequences showing the sound-evoked motion of the inner ear, and also allowed static images to be acquire during a variety of conditions, as described below.

2.3 *Current injection and recording of cochlear microphonics*

Borosilicate glass capillaries with tip size around 1 μm and impedance 5 – 10 MΩ when filled with 3M potassium chloride were used for recording cochlear microphonics and for injecting current. Later experiments were performed with pipettes filled with artificial endolymph. For recording of cochlear microphonic potentials in scala media, the pipette was connected to the headstage of an amplifier. Current injections were driven by a WPI constant current stimulator.

3 Results

3.1 *Interferometry*

Figure 1 shows a representative recording. The sound stimulus was a multi-tone complex with 6 components spaced around the expected best frequency of the recording location, 200 Hz. At the onset of the recording, vibration amplitudes were small. The main peak was located at 200 Hz where the amplitude was 4 nm. The horizontal black bar marks the timing of the current injection, 5 μA DC injected through the glass pipette into scala media. A 100-nm shift in the static position of the organ of Corti resulted from the current. At the end of the positive current step, the polarity of the current was switched. The hearing organ then resumed its original position. A small step is evident near the end, when the negative current was turned off.

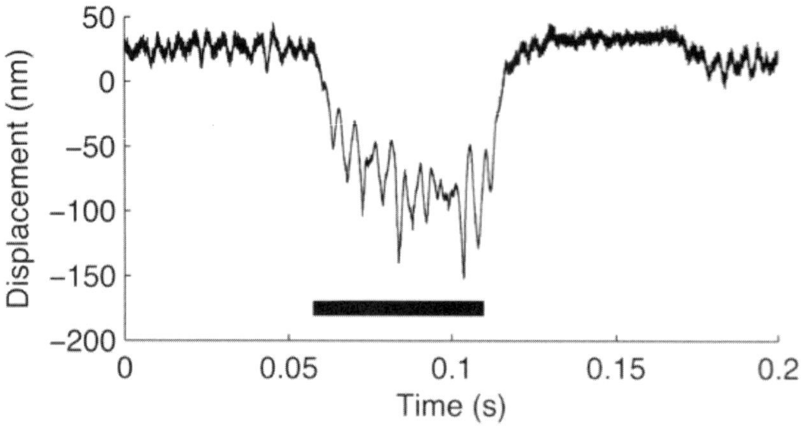

Figure 1. Sound-evoked displacement before, during and following current injection. Black bar denotes timing of positive current. Sound pressure, 60 dB SPL at the tympanic membrane.

During positive current, a substantial enhancement of sound-evoked vibration was observed. Displacements increased from 4 to 14 nm at the characteristic frequency. Negative current led to an immediate decrease of vibration amplitudes. When compared against the negative current amplitude, the enhancement during positive current was approximately 22 dB.

3.2 Resampled confocal microscopy

Two-dimensional vibration measurements were performed with resampled confocal microscopy. This method is less sensitive than interferometry and it is therefore necessary to use higher sound pressures during data acquisition. Nonetheless, positive current did produce enhancement of cochlear vibrations. The maximum increase was 6 dB. This increase of vibration amplitude occurred without significant alterations in the shape of the motion trajectories. As in previous studies, trajectories were almost rectilinear at low frequencies, radial components becoming more significant at higher frequencies. Current did not alter the relation between transverse and radial vibration components, and trajectories therefore retained the same overall shape.

4 Discussion

The present results can be summarized as follows:

1. Positive current injection in scala media enhances sound-evoked vibration.
2. Despite the change in amplitude, there was no significant change in either frequency tuning or response phase.

These results demonstrate that cochlear amplification in the apex is positive, just as in the base. In the base, amplification is coupled to an increased sharpness of tuning and a shift in the best frequency toward higher frequencies. These effects were absent in our recordings, suggesting that transduction mechanisms in the apex may still harbor significant differences from those of the base, despite the similarity in the sign of the feedback. These results are in accordance with some previous *in vivo* studies [*e.g.* 3] that found compressive nonlinearity in the apex, but differ from the findings of more recent studies [1,2]. The reason for the discrepancy may be related to the choice of control condition. In both of the latter studies, vibration amplitudes *in vivo* were compared to those measured after the death of the animal. After death, vibration amplitudes increased, suggesting that negative feedback may be important in the low-frequency areas of the cochlea. Although death certainly does reduce the endocochlear potential, thereby eliminating the driving force for the amplifier, death induces a complicated series of events that may profoundly affect the system. Therefore, increases of vibration amplitude may be possible even if feedback is positive, as demonstrated here.

Acknowledgments

Supported by the Swedish Research Council, the Tysta Skolan foundation, Hörselskadades Riksförbund and the Wallenberg foundation. We thank Martin Pienkowski for performing preliminary experiments.

References

1. Zinn, C., Maier, H., Zenner, H.-P., Gummer, A.W., 2000. Evidence for active, nonlinear, negative feedback in the vibration response of the apical region of the in-vivo guinea-pig cochlea. Hear. Res. 142, 159-183.
2. Khanna, S.M., Hao, L.F., 2000. Amplification in the apical turn of the cochlea with negative feedback. Hear. Res. 149, 55-76.
3. Cooper, N.P., Rhode, W.S., 1997. Mechanical responses to two-tone distortion products in the apical and basal turns of the mammalian cochlea. J. Neurophysiol. 78, 261-70.
4. Fridberger, A., Tomo, I., Ulfendahl, M., Boutet de Monvel, J., 2006. Imaging hair cell transduction at the speed of sound: dynamic behavior of mammalian stereocilia. Proc. Natl. Acad. Sci. (USA) 103, 1918-1926.
5. Cooper, N.P., 1999. An improved heterodyne laser interferometer for use in studies of cochlear mechanics. J. Neurosci. Meth. 88, 93-102.
6. Willemin, J.F., Khanna, S.M., Dändliker, R., 1989. Heterodyne interferometer for cellular vibration measurement. Acta Otolaryngol Suppl. 467, 35-40.
7. Jacob, S., Tomo, I., Fridberger, A., Boutet de Monvel, J., Ulfendahl, M., 2007. Rapid confocal imaging for measuring sound-induced motion of the hearing organ in the apical region. J. Biomed. Opt. 12, 021005.

Comments and Discussion

Shera: It may be worth noting that an analysis of auditory-nerve fiber responses using an inverse method (Shera, 2007) reveals traveling-wave amplification throughout the cochlea, including the apex.

Shera, C.A, 2007. Laser amplification with a twist: Traveling-wave propagation and gain functions from throughout the cochlea, J. Acoust. Soc. Am. 122, 2738–2758.

Braun: Analyses of auditory-nerve fiber responses by inverse methods reveal local amplification, yes. But they do not reveal information about traveling waves of any kind. The phase and delay data that have been derived in such analyses are fully compatible with the frequency specific resonance build-up of local point amplifiers. Ruggero and Temchin (2007) have demonstrated that frequency specific delays of neural responses are remarkably similar across tetrapod classes, irrespective of the occurrence of traveling waves.

Ruggero MA, Temchin AN (2007) Similarity of traveling-wave delays in the hearing organs of humans and other tetrapods. J. Assoc. Res. Otolaryngol. 8, 153-166.

Gummer: First, I do not agree that your current injection experiment has shown that amplification in this extreme apical region is "positive". Instead, you have shown that cochlea vibration is increased by the current injection. In Zinn et al. (2000) we suggested that the active attenuation was possibly due to a change of the phase relation, relative to that found in the base, between electromechanical OHC (somatic) force and the parallel impedance of stereocilia stiffness and tectorial-membrane inertia. This mechanical system in the apical region of the cochlca is nevertheless an active one – in the sense that energy is inserted from the endolymphatic potential – but because of the phase relationship the OHC force attenuates rather than amplifies. If this phase relation is irreversibly disturbed by the post-mortem conditions of the present experiment, then the attenuation found in vivo is unlikely to be restored by attempting to restore "endolymphatic potential".

Second, one must be careful when comparing data from a 1-kHz region (Cooper and Rhode, 1997) with the data (Khanna and Hao, 2000; Zinn et al., 2000) from the extreme apical region, where the best frequency is two or three octaves below 1 kHz. I suspect that mechanical conditions are quite different in this extreme apical region of the cochlea.

Fridberger: I think that we have quite convincingly shown that current does increase sound-evoked vibration. To me, that is the essence of "positive" feedback (amplification). In the case of "negative" feedback, as in Zinn et al (2000), the OHC would still be producing force, so yes, I definitely agree that such a cochlea is active.

As regards the phase relations, you seem to suggest that the relation between OHC force and the motion of the TM/stereocilia complex was altered in our experiments. In my opinion, such a change could only occur in cases of gross damage. Such damage was not seen (assessed by confocal microscopy) and the presence of cochlear microphonic potentials with respectable amplitudes suggest that forward transduction is working. Also, we previously measured stereocilia deflections in our preparations, those measurements show the expected polarity of stereocilia deflection for the OHC. A 180-degree phase flip in the OHC force would therefore require a profound change in the relation between voltage and OHC force. This does not seem very likely, given that our current-evoked response polarities are the same as the ones reported by others.

Your second point highlights the fact that we really don't know a whole lot about apical mechanics! All we can say is that more work is needed.

Braun: The finding that amplification of motion of the organ of Corti was not accompanied by macromechanical effects such as shifts of characteristic frequency or phase, or changes of motion trajectories, indicates that amplification was inherent in micromechanical local resonators, such as hair cells or complex structures with hair cells as central elements. This is a crucial observation, which is fully in line with the report by Chen et al. (2008).

Chen F, Zheng J, Choudhury N, Jaques S, Nuttall AL, 2008. Organ of Corti micromechanics with local electrical stimulation. This volume.

QUANTIFYING THE PASSIVE SUBSTRATE FOR ACTIVE COCHLEAR TUNING

ELIZABETH S. OLSON

Depts. of Otolaryngology, Head & Neck Surgery and Biomedical Engineering, Columbia University, New York, NY 10032, USA

OMBELINE DE LA ROCHEFOUCAULD, WEI DONG

Dept. of Otolaryngology, Head & Neck Surgery, Columbia University, New York, NY 10032, USA

Two aspects of passive mechanics that are relevant to the integration of active mechanics were examined:

(i) Role of organ of Corti complex (OCC) mass in frequency tuning: Some cochlear models employ significant OCC mass to produce tuning while other models get by with zero OCC mass. We used closely spaced measurements of traveling wave phase, coupled to a cochlear model, to measure OCC mass and found it was significant, at least in the cochlear base.

(ii) OCC stiffness and resistance: Active cochlear mechanics is thought to work as a negative resistance that is large enough to overcome passive damping, and the size and frequency dependence of the OCC acoustic stiffness are fundamental to traveling wave mechanics. By simultaneously measuring cochlear pressure and basilar membrane (BM) velocity in the cochlear base (BF ~ 40 kHz), the frequency-dependent OCC impedance was measured from 4 to 20 kHz. The imaginary part of the impedance was negative (stiffness dominated) with a size that decreased with frequency. The real part (resistance) often appeared negative at frequencies below ~ 8 kHz, an unexpected finding. At ~10 kHz, the real part flattened out to a positive value that remained smaller than the imaginary part.

1 Introduction

Since the discovery of active cochlear nonlinearity, much experimental activity has been devoted to its characterization [1] and to studies of the cellular mechanics that likely contribute to the observed nonlinearity [2]. Cochlear models complement the experimental studies by exploring possible mechanisms by which the amplifying element is integrated into passive mechanics. Several themes have emerged from these models. For example, in many models the amplifier works within a framework of a passively resonant OCC (defined to include the organ of Corti plus basilar and tectorial membranes). As another example, based on the level dependence of BM motion amplitude and phase, the amplifier is thought to work as a negative resistance that is large enough to more than compensate for passive resistance [3]. Whether OCC mass is big enough to be involved in local resonance, and the size of passive resistance are vital unknowns of cochlear mechanics. Their exploration was undertaken in the studies here. Mass (M_{OCC}), resistance (R_{OCC}) and stiffness (S_{OCC}) are the elements of OCC impedance (Z_{OCC}), as $Z_{OCC} = R_{OCC} + i[M_{OCC}\omega - S_{OCC}/\omega]$ (where ω is radial frequency). Fundamentally, these studies are in vivo measurements of the passive Z_{OCC}.

1.1 Background to M_{OCC} + local resonance

Both active and passive cochlear models often employ substantial M_{OCC}, which serves to decrease the impedance of the organ of Corti via local resonance, enhancing the peak of the response [e.g. 4]. However, other cochlear models operate without (or without significant) M_{OCC} [e.g. 5]. (In this thinking, the cells of the OCC are malleable enough to be mechanically continuous with the fluid.) With such a simple and robust concept as mechanical resonance, it seemed that the answer, "yes" or "no" should be within reach. However, the wave nature of cochlear mechanics makes the probing of local resonance elusive: The cochlear traveling wave is based on the interplay of fluid mass and OCC stiffness, with wave speed decreasing with stiffness decrease. OCC mass will serve to reduce the "effective" OCC stiffness: $S_{OCC}' = S_{OCC} - M_{OCC}\omega^2$. At the frequency and location for which the OCC mass and stiffness terms are equal, $S_{OCC}' = 0$. This occurs at the resonance frequency familiar from simple spring/mass systems: $\omega = \sqrt{S_{OCC}/M_{OCC}}$. However, when S_{OCC}' is 0 the wave speed is zero, and at near-by slightly basal (stiffer) locations the speed is very small, and with even a small amount of damping the wave never makes it to the resonance location. Lighthill [4], who argued that the *approach to* local resonance was essential to the (passive) tuning peak, tied his argument to the observed rapidly decreasing wavelength in the region of the peak. In a recent study [6] we applied this strategy by analyzing closely-longitudinally-spaced BM motion measurements in combination with predictions from a 3D cochlear model by deBoer [7]. This allowed us to "measure" M_{OCC} and S_{OCC} (both assumed frequency independent in this case). This methodology is most robust in the frequency region of the peak, where the traveling wave phase change is rapid.

1.2 Background to resistive (real) and reactive (imaginary) parts of Z_{OCC}

Similar to local resonance, R_{OCC} also has a large effect on the amplitude of the response peak. For example, a passive model [8] showed a substantial increase in the peak when the basilar membrane resistance was reduced from a nominal value (corresponding to 5% viscosity in the elastic modulus) to zero. Active cochlear models typically incorporate cochlear amplification via a negative resistance that is large enough to more than compensate for the positive resistance [3]. This model for amplification has the effect, compatible with observations, of greatly increasing the response amplitude with relatively small effect on the response phase. So, how big is R_{OCC}? In the literature, the resistive and reactive parts of the impedance of the OCC and isolated OCC elements have been probed through a wide frequency range by finding the force required for localized mechanical motion [9,10]. In our studies simultaneous measurements of the velocity of the OCC at the BM and the pressure difference across the OCC (ΔP) upon acoustic stimulation were used to find the OCC specific acoustic impedance, $Z_{OCC} = \Delta P/v$ (v = BM velocity). The interpretation of these measurements is relatively direct compared to the locally forced impedance measurements, in that we measure the natural intracochlear pressure and motion. However, with this directness came a limitation to frequencies about an octave beneath the BF, as discussed in methods below.

2 Methods

Experiments were performed in-vivo in deeply anesthetized young adult gerbils, with methods approved by Columbia University's Institute Animal Care and Use Committee.

Sound stimulation was delivered closed-field and monitored in the ear canal with a probe-tube microphone. Due to the fragility of the cochlear base, the preparations did not show non-linear scaling to pure tone stimulation. Because these studies were designed to probe passive cochlear mechanics, this was alright. In order to attain high signal:noise ratio the results presented here were done at fairly high sound pressure level (SPL) (80 dB for the M_{OCC} studies, 90 dB for the Z_{OCC} study). Data gathered at lower SPL were similar, consistent with linear preparations.

For the studies on M_{OCC}, BM velocity was measured with a confocal interferometer [11] through the round window (RW) opening, in some experiments at a very basal site (best frequency (BF) ~ 35-40 kHz) and in others at a turn-one site (BF ~ 20-25 kHz). In each experiment motion measurements were made at several very closely spaced longitudinal locations and the data used to find the traveling wave wavenumber (k). k is defined as the derivative of the BM motion phase, ϕ, with respect to longitudinal distance. Experimentally, k_{exp} (subscript identifies the experimental k) was found as $\Delta\phi/\Delta$, from two phase measurements spaced by a known small distance, Δ, of ~ 200 μm. k_{exp}-vs-frequency is the experimental curve to be compared with model predictions. The cochlea in deBoer's 3D model was box-like, with stapes and RW at one end of the box, and separated by a divider that included the OCC. The model characterized the cochlear fluid as an effective fluid mass (M_{fluid}). The expression for M_{fluid} is derived from fundamental fluid mechanics and depends on the box dimensions and OCC width (known quantities) and the wavenumber, k. In cochlear modeling, M_{fluid} is then related to the Z_{OCC} by force balance. The resulting expression, termed the dispersion relationship, is in [6]. Most importantly for us, the dispersion relationship relates k to Z_{OCC}. Using the relationship, k_{exp} can be fit with S_{OCC}, M_{OCC} and R_{OCC} as free parameters. Because response phase (and thus k_{exp}) is not very sensitive to R_{OCC}, R_{OCC} was neglected (but see [6]). S_{OCC} and M_{OCC} were assumed frequency independent (but see [6]). Because our main objective was to answer the question as to whether M_{OCC}, and thus local resonance was significant to passive cochlear tuning, we were foremost interested in whether a fit would work with $M_{OCC} = 0$. Beyond that, the size of the best-fit S_{OCC} and M_{OCC} were of interest.

For the direct measurements of Z_{OCC}, ΔP should be found as the difference between pressure in scala media (SM) close to the tectorial membrane and pressure in scala tympani (ST) exactly on the other side and close to the BM. Inserting a sensor into SM proved impractical. A practical approximate-ΔP measurement was accomplished by measuring scala vestibuli (SV) pressure in the very base and keeping the fluid level low in the RW opening, so that the ST pressure was ~ 0. Previous measurements [12] justified this approximation for ΔP for frequencies an octave below the local BF. Therefore, we can measure Z_{OCC} to ~ 20 kHz. BM velocity was measured without reflecting beads in the very base through the RW opening, using a Polytec vibrometer. Below ~ 4 kHz the signal often went into the noise, setting the low frequency limit. Pressure was measured with tiny intracochlear pressure sensors [12]. Phase was critical, and we took steps to measure it very accurately. As a result, in the 4 - 20 kHz frequency range we correct for a sensor phase lag that was ~ unchanging in frequency with a mean value of 7° and a standard deviation of 7°. The contribution to Z_{OCC} by M_{OCC} is small at frequencies < ½ BF, so this study probed S_{OCC} and R_{OCC} – their size and frequency dependence.

3 Results

3.1 M_{OCC} and S_{OCC} found with BM velocity measurements + model

In Fig. 1 we show BM velocity measurements from two experiments, and k_{exp} derived from each. Panels A and B (left) are from expt. 18, in which the very basal region was examined, panels C and D (right) from the turn-one location of expt. 8. In each of the top plots, two BM velocity-vs-frequency measurements are shown. In A these measurements were separated by a distance, Δ, of 116 µm. In C Δ was 270 µm. The corresponding, $k_{exp} = \Delta\phi/\Delta$ curves are in the bottom plots. Also shown are theoretical predictions of k-vs-frequency (solid, dashed and dotted curves). The dashed and dotted curves are the best fits that could be obtained when M_{OCC} was taken as zero. The solid lines considered M_{OCC} to be a free parameter. At both locations the fit was poor with $M_{OCC} = 0$, and quite good when M_{OCC} was included. The best-fit values for M_{OCC} corresponded to OCC

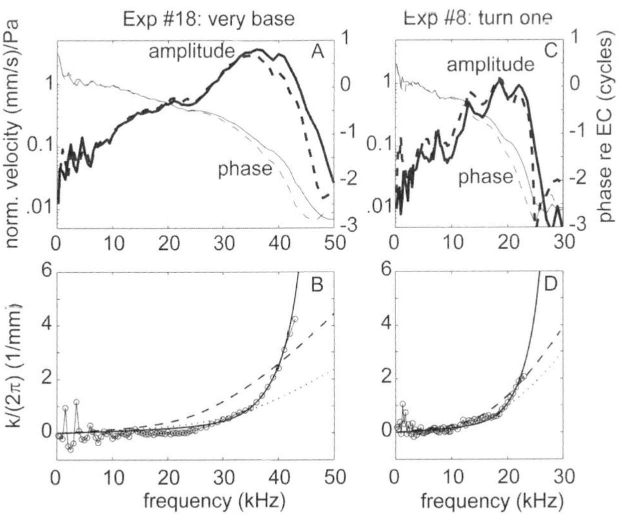

Fig. 1. Top panels: BM velocity at very-basal (A) and turn-one (C) sites. At each site, velocity was measured at several closely spaced locations, with results shown from two locations at each site. Bottom panels show wavenumber, k_{exp}, from very-basal (B) and turn-one (D) sites, and fitted k curves.

S_{OCC} and M_{OCC} values are as follows: (S_{OCC} in Pa/(m/s), M_{OCC} in kg/m^2.)

B:

solid: $S_{OCC}=20\times10^9$, $M_{OCC} = 0.22$

dashed: $S_{OCC}=6\times10^9$, $M_{OCC} = 0$

dotted: $S_{OCC}=10\times10^9$, $M_{OCC} = 0$

D:

solid: $S_{OCC}=5.2\times10^9$, $M_{OCC} = 0.15$

dashed: $S_{OCC}=2.4\times10^9$, $M_{OCC} = 0$

dotted: $S_{OCC}=3\times10^9$, $M_{OCC} = 0$

heights of 220 and 150 μm, reasonably in line with the anatomical height of the OCC. This correspondence with anatomical height only applied in the very cochlear base; when regions slightly more apical were considered, a good fit still required M_{OCC} but it could be substantially smaller [6].

3.2 Direct measurements of resistive and reactive components of Z_{OCC}

In Fig. 2 we show the SV pressure and BM velocity from one preparation, results that were similar with findings from six animals [13].

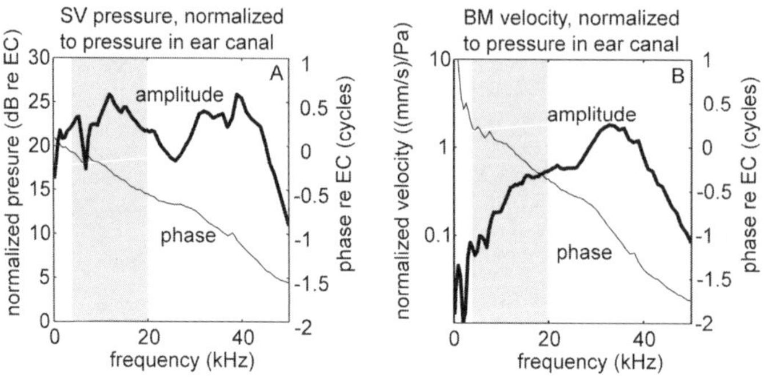

Fig. 2. (A) SV pressure near the stapes. (B) BM velocity at the very basal site, adjacent to the stapes. The data in the gray shaded region can be used to calculate Z_{OCC}. (wg120 run35, 90 dB SPL)

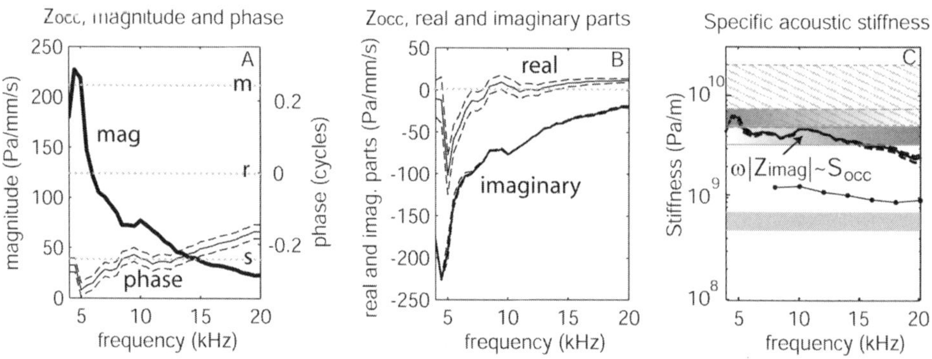

Figure 3. Z_{OCC} from Fig. 2 data. (A) Magnitude and phase, with phase uncertainty discussed in methods. (B) Real and imaginary parts, showing uncertainty. (C) The imaginary part of Zocc recast as S_{OCC}: The thick black line is $|Z_{OCC\_imag}|\omega$, corresponding to $S_{OCC}-M_{OCC}\omega^2$. For the analysis of this study, frequency < ½ BF and in that case $S_{OCC}-M_{OCC}\omega^2 \sim S_{OCC}$. Basal BM stiffness data from other gerbil studies are also plotted: The diagonal-patterned region shows the range of stiffness values from Fig. 1, derived from BM motion and a cochlear model; line with dots was a result derived from pressure measurements alone [12]. Graded gray tones and gray band near the bottom are ranges of specific acoustic stiffness derived from point stiffness values [6]. Light-to-dark from [14], dark-to-light from [15] and solid gray bar from [16].

Fig. 3 shows the Z_{OCC} results from the Fig. 2 data. From A, the phase was much like that of a stiffness (stiffness, mass and resistance phases are indicated at the right). The increasing phase above 15 kHz might be due to the approximations that limited the analysis to frequencies ~ an octave below BF, so those results are less robust. When shown as real and imaginary parts in panel B, the imaginary (reactive) part was negative, thus stiffness dominated, with a size that decreased steadily from ~ 200 Pa/(mm/s) at 4 kHz to ~ 20 Pa/(mm/s) at 20 kHz. Below ~ 8 kHz, the real part (resistive) was negative, an unexpected but common finding in this study, which was present even post-mortem. Potential sources of artifact (RW motion, background vibration) have been eliminated as possibilities and this aspect of the findings needs further exploration. Above 8 kHz the real part varied non-monotonically between values of 0 and 12 Pa/(mm/s), thus was always smaller than the imaginary part, with the difference decreasing as frequency increased. In C we show the calculated specific acoustic stiffness (bold line). Up to 15 kHz, the stiffness was nearly constant, ranging from ~ $4 - 6 \times 10^9$ Pa/m. Also in panel C are acoustic stiffness values from the gerbil base from a variety of sources. There is a clustering of values at ~ 5×10^9 Pa/m, based on direct pressure and motion measurements (this study), the results from the study presented above (Fig. 1 C & D) and point stiffness measurements from the literature [15, 16]. The highest 20×10^9, value is from the very basal measurements of Fig. 1 A & B and might reflect an inability of a box model to well-represent the region of the cochlea right at the stapes and RW. Some of the overall range is attributable to longitudinal variations in stiffness within the basal region.

4 Discussion

Passive cochlear mechanics forms the substrate upon which active mechanics operates. In this contribution we discuss fundamental questions of passive OCC mechanics: Is M_{OCC} large enough that the approach to local resonance is a relevant concept? What are the sizes and frequency dependencies of S_{OCC} and R_{OCC}? In our results, M_{OCC} was significant at the very base of the cochlea. S_{OCC} was ~ frequency independent, with a size consistent with values found with several different experimental techniques. R_{OCC} was small compared to reactive impedance, an expected result. Other characteristics of R_{OCC} were unexpected. The size and shape of the elements of Z_{OCC} described here will be useful for the further development of realistic models of cochlear mechanics.

Acknowledgments

This work was supported by the NIDCD and the Emil Capita Foundation

References

1. Robles, L., Ruggero, M.A., 2001. Mechanics of the mammalian cochlea. Physiol. Rev 81, 1305-1352.
2. Fettiplace, R., Hackney, C.M., 2006. The sensory and motor roles of auditory hair cells. Nat. Neuroscience Rev. 7, 19-29.
3. Kolston, P.J., 2000. The importance of phase data and model dimensionality to cochlear mechanics. Hear. Res. 145, 25-36.
4. Lighthill, J., 1981. Energy flow in the cochlea. J. Fluid Mech. 106, 149-213.

5. Lim, K.M., Steele, C.R., 2002. A three-dimensional nonlinear active cochlear model analyzed by the WKB-numeric method. Hear. Res. 170, 190-205.
6. de La Rochefoucauld, O., Olson, E.S., 2007. The role of organ of Corti mass in passive cochlear tuning. Biophys. J. 93, 3434-3450.
7. deBoer, E., van Bienema, E., 1982. Solving cochlear mechanics problems with higher-order differential equations. J. Acoust. Soc. Am. 72, 1427-1434.
8. Steele, C.R., Taber, L.A., 1981. Three dimensional model calculations for guinea pig cochlea. J. Acoust. Soc. Am. 69, 1107-1111.
9. Freeman, D.M. Abnet, C.C., Hemmert, W., Tsai, B.S., Weiss, T.F., 2003. Dynamic material properties of the tectorial membrane: a summary. Hear. Res. 180, 1-10.
10. Scherer, M.P., Gummer, A.W., 2004. Impedance analysis of the organ of Corti with magnetically actuated probes. Biophys. J. 87, 1378-1391.
11. Khanna, S.M., Koester, C.J., Willemin, J.F., Dandliker, R., Rosskothen, H., 1996. A noninvasive optical system for the study of the function of inner ear in living animals. SPIE 2732, 64-81.
12. Olson, E.S., 1998. Observing middle and inner ear mechanics with novel intracochlear pressure sensors. J. Acoust. Soc. Am. 103, 3445-3463.
13. Dong, W., Olson, E.S., 2008. In vivo impedance of gerbil organ of Corti from 4 – 20 kHz. Assoc. Res. Otolaryngol. Abstract # 178.
14. Naidu, R.C., Mountain, D.C., 1998. Measurements of the stiffness map challenge a basic tenet of cochlear mechanics. Hear. Res. 124, 124-131.
15. Olson, E.S., Mountain, D.C., 1994. Mapping the cochlear partition's stiffness to its cellular architecture. J. Acoust. Soc. Am. 95, 395-400.
16. Emadi, G., Richter, C.P., Dallos, P., 2004. Stiffness of the gerbil basilar membrane: radial and longitudinal variations. J. Neurophysiol. 91, 474-488.

Comments and Discussion

Shera: It would be interesting to compare your effective OCC height with the value of 1/k, where k is 2pi/wavelength at the peak.

Olson: The question is based on the understanding that in a 2D model, the effective mass of the fluid is ~ density/k. In our work, which uses a 3D model by deBoer, the fluid mass is a slightly more complicated function of density, k and cochlear dimensions, but the 2D model gives the basic idea. However, the OCC mass we are using our study to "measure" is by definition a mass that is not k dependent. You could think of the study as one that asks whether a transverse section of the OCC moves as a "rigid" body ("transverse section of the" needed to not get into the issue of longitudinal coupling here!) In our fig.1, k/2pi changes a lot (from ~ 0 to ~ 4 mm^-1) but the point of the study was to see if a non-k dependent mass – *in addition to* the k-dependent mass that is ascribed to the fluid – was needed to fit the 3D model to the observations.

ORGAN OF CORTI MICROMECHANICS WITH LOCAL ELECTRICAL STIMULATION

FANGYI CHEN, JIEFU ZHENG

Oregon Hearing Research Center, NRC04, Oregon Health & Science University, 3181 SW Sam Jackson Park Road, Portland, Oregon 97239-3098, USA

NILOY CHOUDHURY, STEVE JAQUES

Department of Biomedical Engineering, Oregon Health & Science University, 3181 SW Sam Jackson Park Road, Portland, Oregon 97239-3098, USA

ALFRED L. NUTTALL

Oregon Hearing Research Center, NRC04, Oregon Health & Science University, 3181 SW Sam Jackson Park Road, Portland, Oregon 97239-3098, USA; Kresge Hearing Research Institute, University of Michigan, 1301 E. Ann Street, Ann Arbor, Michigan 48109-0506, USA

Optical low coherence interferometry is able to both image and measure the vibration of the cellular and non-cellular structures of the organ of Corti *in vivo*. In this study we applied an electric current to the basal turn from a pair of electrodes, one in scala tympani and the other in scala vestibuli, at the location corresponding to ~17 kHz when interferometry measurements were made. The coherence gate of the interferometer was positioned 1) at the basilar membrane (BM) near the radial location of the outer hair cells (OHCs) (approximately ½ the width of the BM) and 2) at the reticular lamina (RL) where the OHCs are located. We confirmed that electrical stimulation with a frequency sweep (12 kHz –25 kHz) caused a mechanical BM displacement with a peak and a traveling wave-like phase delay as we reported previously using laser Doppler velocimetry and reflective beads on the BM. Reflective beads were not used in the current study. The vibration of the RL had little or no phase delay that would characterize a traveling wave. These data suggest a very high compliance system for the electrically activated cellular structure of the organ.

1 Introduction

To understand the cochlear micromechanics, it is critical to measure and compare the vibration at two locations of the organ of Corti: the reticular lamina (RL) and basilar membrane (BM). These locations are the places where the hydrodynamic forces from acoustic stimulation of the cochlea are applied to the organ. The outer hair cells (OHCs) which can also produce force are also acting on both locations. Electrical stimulation of the OHCs can potentially reveal the local 'micromechanical' behavior of the organ. A new instrument employing optical low-coherence interferometry (OCT) was developed to achieve this long-desired goal. Here we report the OCT technique and a study on the vibration of the RL and BM under electrical stimulation.

2 Methods

2.1 OCT

Optical low-coherence interferometry utilizes a wide-band light source to achieve the axial resolution, which is normally determined by the numerical aperture in an optical microscope system. In the traditional in vivo vibration measurement using laser Doppler vibrometer (LDV), the practical numerical aperture is often limited by the size of the measurement window opened on the cochlear bony wall for accessing the BM. A longer wavelength (1310 nm in current system vs. 633 nm in the typical LDV) of the light source also improves the tissue penetrating capability of the system. Both wider bandwidth and longer wavelength help achieve good optical sectioning with low coherence interferometry, allowing measurement at different location inside the organ of Corti [1, 2]. In the present study, we focus on the top and bottom surface of the organ, BM and RL.

2.2 Animal preparation

Albino guinea pigs weighing 250-350g were used. Experimental protocols were approved by the Institutional Animal Care and Use Committee, Oregon Health & Science University. The animals were anesthetized using both ketamine (40 mg/kg, i.m.) and xylazine (10 mg/kg, i.m.). A ventral approach was used to open the left auditory bulla and expose the cochlea. The middle ear muscle tendons were sectioned. Cochlear sensitivity was monitored by measuring the sound-evoked compound action potential (CAP) of the auditory nerve via a round window electrode. A plastic coupler with two speakers (made of 1/2" B&K microphones) mounted inside was fitted to the ear canal to deliver acoustic stimuli. To measure BM motion, a small fenestra (approx. 0.3×0.4 mm) was created in the first-turn scala tympani bony wall of the cochlea. A cover slip was used to close the hole to prevent perilymph surface vibration and to correct the optical distortion of imaging through a fluid meniscus. Two wire electrodes (Pt–Ir, 50 μm in diameter) were inserted into the scala vestibuli (SV) and scala tympani (ST) respectively in the basal turn, forming a bipolar pair at the BM measurement location across the cochlear duct. Sinusoidal current (100~200 μArms) from a constant-current stimulator was delivered through the electrodes. Voltage control to the constant-current stimulator was from the oscillator output of a digital lock-in amplifier (Stanford Research Systems SR830). Signals from the laser interferometer were recorded by the same lock-in amplifier and displayed in terms of amplitude and phase.

3 Results

3.1 OCT images

Different from an LDV system, the OCT system can work as an imager as well as an interferometer. Under the imaging mode, our system generates an X-Z (Radial-Axial)

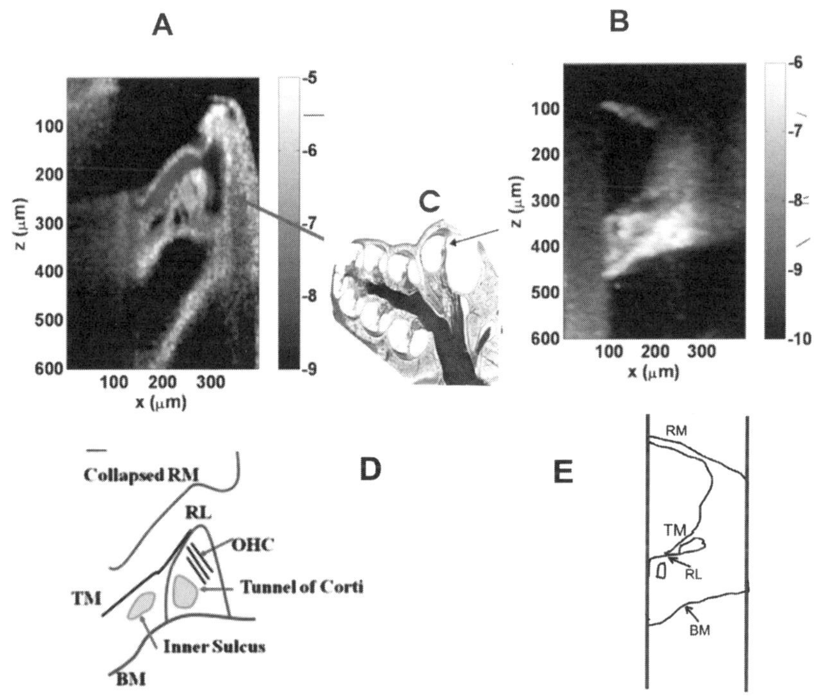

Figure 1. OCT images taken through holes in the cochlear bony wall. The gray scale represents the logarithms of the normalized tissue reflectivity, where the mirror reflectivity is defined as 1. (A) An *in vivo* image taken at the cochlear apex as illustrated by cochlear cross-section (C). A line drawing on the lower-right side (D) depicts the profile of structures of the organ of Corti, where the BM, RL, Reissner's membrane (RM), inner sulcus, tunnel of Corti, tectorial membrane and outer hair cells can be identified. (C) An *in vivo* image taken at the basal turn. This image is not as clear as the one taken at the apex. However, the positions, BM, RL, TM and RM are all discernable as illustrated in the drawing (E). The vertical lines in E represent the imaging boundaries established by the size of the opening in the cochlear bone.

scan image. This image gives a cross-section of the tissue morphology and is used to localize and identify the cellular structure under measurement by positioning the scanning apparatus electronically. Figure 1 shows two OCT images of the organ of Corti taken from (a) the apical turn through an opening at the apex of the cochlea and (b) the basal turn through an opening made in the bony wall of the cochlea.

3.2 BM responses at 3 longitudinal positions under RW stimulation

The OCT interferometer does not require reflective objects to be placed onto the BM. Thus it is possible to image and measure vibration of the BM through the round window membrane (RW). In Figure 2 we illustrate this method and the result from three longitudinal positions along the basilar membrane. The magnitude responses show the expected shifts in the best frequency (BF) with the longitudinal location, and phase data reveal a location-related progressive phase lag with frequency. The BF and phase slopes of responses decrease with the distance from the cochlear base.

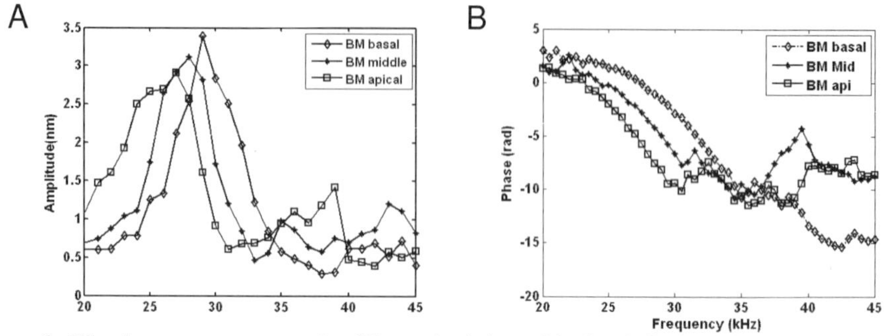

Figure 2. Vibration measurement on the GP round window with electrical stimulation. (A) The amplitude responses. The peak frequency shifts from 28 kHz to 24 kHz from a more basal to a more apical location. (C) The phase responses. The steep phase slope of the more apical response indicates a greater delay.

3.3 BM and RL phase difference under electrical stimulation

Figure 3 shows the phase difference between the RL and BM vibration. The BM had a large monotonic delay that is characteristic of the expected travelling wave. The RL phase is different from that of BM vibration, and had a shallow negative slope with increasing frequency over the observed frequency range. This example is typical in the current experiment.

4 Discussion

The measurement of organ of Corti motion without reflective objects has been accomplished in a number of studies [3, 4, 5]. Ren and Nilsen and Russell [4, 5] have been able to show the longitudinal patterns of BM motion that occur and exploit the freedom provided when not confined to the random locations of beads placed on the BM.

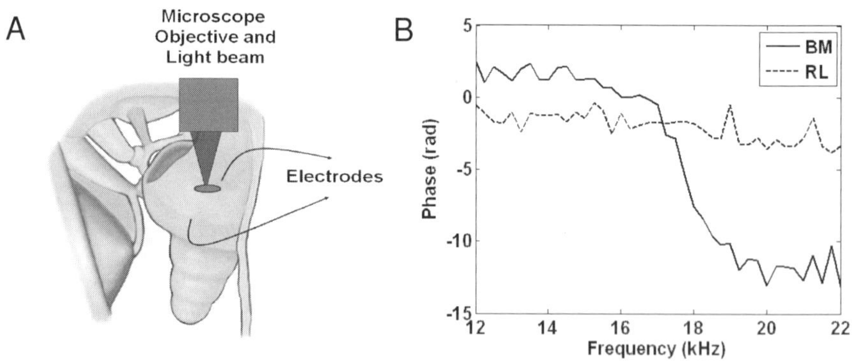

Figure 3. A comparison of BM and RL phase under electrical stimulation. A) The measurement is made of the organ of Corti at about the 18kHz location via an opening in the cochlea. Electrodes are placed so as to pass current across the cochlear duct. One electrode is in the scala tympani and the other is in the scala vestibuli. B) The blue line is the phase responses measured on BM. It shows a sharp decrease at the frequency range of 17 kHz to 19 kHz, which is the peak region. This phase curve is characteristic of a travelling wave. However, the RL phase responses (dashed line) is almost flat over the same frequency range.

Data in Figure 2 show that the OCT method can provide a similar level of freedom and confirm the expected longitudinal patterns of BM motion in the hook area of the guinea pig cochlea.

The measurement of the RL and BM at the same radial location across the width of the organ of Corti has not been accomplished *in vivo*. Nowotny et al. [6] measured the electrically evoked responses of the tectorial membrane/organ of Corti *in vitro*. Tomo et al. [7] have measured the Henson's cells *in situ* in the apical turn where OCT method also shows great possibility as a new method (Figure 1). The interferometer microscope developed by Khanna has accomplished *in situ* measurements of scala media surface of the organ of Corti in the apex [8].

Mammano and Ashmore [9] have measured both the BM and the RL under electrical stimulation for an *in vitro* preparation. It is their work that is most relevant to the current study. They found that the motion of the RL was about 15 times greater than the BM (BM vibration estimated from motion of the scala media aspect of the Claudius cells) and that the phase difference was 180 degrees when the organ was stimulated with rectangular current pulses. We confirm that the RL vibrates at a greater magnitude than the BM (about 3 times, data not shown) but we find *in vivo* that the RL vibrates independently of the BM for near BF frequencies. This suggests a weak mechanical coupling within the organ of Corti and a likely significant mechanical influence of the tectorial membrane.

Acknowledgments

This study received support from NIH NIDCD R01 grants DC 00141 and DC 006273.

References

1. Choudhury, N., Chen, F., Matthews, S.K., Tschinkel, T., Zheng, J., Jackques, S., Nuttall, A.L. 2006. Low coherence interferometry of cochlear partition. Hear. Res. 220(1-2), 1-9.
2. Chen, F., Choudhury, N., Zheng, J., Matthews, S.K., Nuttall, A.L., Jacques, S.L. 2007. *In vivo* imaging and low-coherence interferometry of organ of Corti vibration. J. Biomed. Opt. 12, 021006.
3. Cooper, N.P. 1999. Vibration of beads placed on the basilar membrane in the basal turn of the cochlea. J. Acoust. Soc. Am. 106, L59-L64.
4. Nilsen, K.E., Russell, I.J. 2000. The spatial and temporal representation of a tone on the guinea pig basilar membrane. Proc Natl Acad Sci U S A 97, 11751-8.
5. Ren, T. 2002. Longitudinal pattern of basilar membrane vibration in the sensitive cochlea. Proc Natl Acad Sci U S A 99, 17101-17106.
6. Nowotny, M., Gummer, A.W. 2006. Nanomechanics of the subtectorial space caused by electromechanics of cochlear outer hair cells. Proceedings of the National Academy of Sciences 103, 2120-2125.
7. Tomo, I., Boutet de Monvel, J., Fridberger, A. 2007. Sound-Evoked Radial Strain in the Hearing Organ. Biophys. J. 93, 3279-3284.

8. Ulfendahl, M., Flock, A., Khanna, S.M. 1989. A temporal bone preparation for the study of cochlear micromechanics at the cellular level. Hear Res 40, 55-64.
9. Mammano, F., Ashmore, J.F. 1993. Reverse transduction measured in the isolated cochlea by laser Michelson interferometry. Nature 365, 838-41.

Comments and Discussion

Cooper: In your talk you showed tuning curves derived from recording sites on the BM & RL in the same preparation. You reported that these sites had slightly different characteristic frequencies as well as two-to-three-fold different sensitivities. This seems likely to be a very important finding for our understanding of the cochlea. However, the modelling community will need to know whether the two sites were in the same transverse, anatomical cross-section, or whether your recordings were made at a longitudinally oblique angle. Also, were the recordings made through a cover-slip, to stabilise and control the perilymphatic meniscus?

Nuttall: The transverse locations of the basilar membrane and reticular lamina measurements were the same. That is, the angle of a line through the organ of Corti was 90 degrees with respect to the basilar membrane in the longitudinal direction. We checked this by acquiring cross-sectional images of the organ at three longitudinal positions; one at the place of measurement and one each to either side of that place (200 micrometers to either side). With OCT imaging, these images have an absolute (constant distance) relationship such that by the amount of mis-registration of the images one can find the angle of the BM along the longitudinal direction. The images fully registered indicating no tilt. Also all measurements with the OCT system were made using a cover slip.

Braun: The finding of independent vibration of the reticular lamina (RL) and the basilar membrane (BM) in vivo, as shown in Fig. 3B, definitely is an important one. It is in conflict with certain model hypotheses. But it is consistent with our knowledge on inner ear function in non-mammalian vertebrates, where we are generally confronted with the feature of auditory frequency selectivity by means of local resonators, such as hair cells or complex structures with hair cells as central elements.

Nuttall: Thank you for your comment and we agree that these data support a concept of multiple resonances for the organ of Corti.

DOES THE COCHLEA COMPROMISE ON SENSITIVITY AND FREQUENCY SELECTIVITY?

A. N. LUKASHKIN, V. A. LUKASHKINA, G. P. RICHARDSON, I. J. RUSSELL
School of Life Sciences, University of Sussex, Falmer, Brighton, BN1 9QG, UK

A tenet of cochlear physiology is that sharp tuning and sensitivity are directly interrelated. Here we show a reciprocal interdependence between tuning and sensitivity in the mammalian cochlea from measurements of basilar membrane (BM) mechanical tuning and neural suppression tuning curves of wild-type (*Tectb*$^{+/+}$) and β-tectorin mutant (*Tectb*$^{-/-}$) mice. The tectorial membrane (TM) of the mutants lacks striated-sheet matrix, which is likely to decrease longitudinal elastic coupling. Mechanical and neural tuning curves recorded in mutants are slightly less sensitive, although more sharply tuned. The inverse relationship between sensitivity and tuning observed in the mutants could be attributed to smaller numbers of the outer hair cells responding in synchrony due to reduced longitudinal coupling in the TM. We suggest that frequency tuning and high sensitivity are not necessarily concomitant but reciprocal properties of the cochlea.

1 Introduction

The majority of cochlear models assume that elastic coupling along the TM is negligible [see 1 for review]. In vitro measurements of the spread of excitation in the TM, however, demonstrate significant elastic coupling within the TM in the longitudinal direction [2-4]. These findings support the contention that the TM contributes towards elastic coupling between adjacent sections of the cochlea [5]. Furthermore, Ghaffari et al. [4] demonstrated that the isolated TM behaves as an elongated elastic structure that is capable of supporting travelling wave propagation. The extent of longitudinal coupling is determined by two opposing restrictions. It cannot be too low as amplification efficiency would be impaired due to a reduction in the number of OHCs working in synchrony. Longitudinal coupling cannot be too high because frequency resolution would deteriorate due to the greater extent of the cochlear partition responding to the same pure tone. Hence, for a cochlea of a given design, there must be an optimum value for the elastic coupling (TM space constant) which is based on a trade-off between frequency tuning and sensitivity. This work demonstrates the inter-dependence of tuning and sensitivity in recordings of BM mechanical tuning and neural suppression tuning made in wild-type and *Tectb*$^{-/-}$ mutant mice.

2 Methods

Detailed description of the methods can be found elsewhere [6]. Briefly. BM displacements, extracellular receptor potentials and compound action potentials of the auditory nerve (CAPs) were recorded from deeply anaesthetized mice. CAPs were measured from the round window membrane using pipettes with tip diameters of 50–100 μm (recording bandwidth 430 kHz) filled with artificial perilymph. Extracellular receptor potentials were recorded from the organ of Corti with pipettes pulled from thin-walled quartz tubing, filled with 3 M KCl, resistance 15–20 MΩ, bandwidth 12–18 kHz with capacitance compensation. Signals were amplified with a recording bandwidth of 0–100 kHz.

Tone-evoked BM displacements were measured by focusing the beam of a self-mixing, laser-diode interferometer through the round window membrane to form a spot on the center of the BM. Voltage responses from the interferometer were measured with a two-channel lock-in amplifier (Princeton 5210) and digitized at 250 kHz.

The sound system was calibrated in situ for frequencies between 1 and 70 kHz in dB SPL. White noise and tone pulses with rise-fall times of 0.2 ms were synthesized by a DT 3010 board, attenuated and used for sound system calibration and the measurement of electrical and acoustical cochlear responses.

CAP tuning curves were derived from simultaneous tone-on-tone masking [7] using a 10-ms probe tone centered on a 40-ms masker tone. The probe tone was set to a level where a stable CAP appeared just above the recording noise floor. The frequency of the masker was set and its attenuation was adjusted until the probe tone CAP was suppressed. The masker frequency and level was noted, a new masker frequency was set and the process was repeated. In a second group of experiments, the probe level was set to increasingly higher values to see how the CAP tuning curves varied with probe tone level.

All procedures involving animals were performed in accordance with UK Home Office regulations, with approval from the local ethics committee.

3 Results

3.1 BM tuning is sharper in Tectb$^{-/-}$ mice

The TM of the *Tectb*$^{-/-}$ mice lacks the highly organised striated-sheet matrix. Hence, elastic coupling along the TM in *Tectb*$^{-/-}$ mice may be decreased. CAP thresholds recorded from the round windows revealed a distinct low-frequency hearing loss in the *Tectb*$^{-/-}$ mutant mice for frequencies below 20 kHz, presumably due to the distinct enlargement and swelling of the TM seen in the apical end of the cochlea. For frequencies above ~20 kHz, however, the TM dimensions and neural sensitivities in *Tectb*$^{+/+}$ and *Tectb*$^{-/-}$ mice were not significantly different.

BM threshold tuning curves (0.2-nm criterion) (Fig. 1A) were measured from the cochleae of eight *Tectb*$^{+/+}$ and eight *Tectb*$^{-/-}$ mice at the 53-kHz place the basal turn. The thresholds at the tips of the tuning curves measured in *Tectb*$^{+/+}$ and *Tectb*$^{-/-}$ mice were 22.33 ± 6.23 dB SPL and 33.25 ± 4.72 dB SPL, respectively. In *Tectb*$^{-/-}$ mutant mice, BM vibrations at the characteristic frequency of the recording site were therefore about 10 dB less sensitive than those of wild-type mice. The bandwidths of tuning curves measured from *Tectb*$^{-/-}$ mutants were, however, less broad with a mean Q_{10dB} (characteristic frequency/bandwidth 10 dB from tip) of 18.9 ± 2.6 (mean ± SD) compared with that of

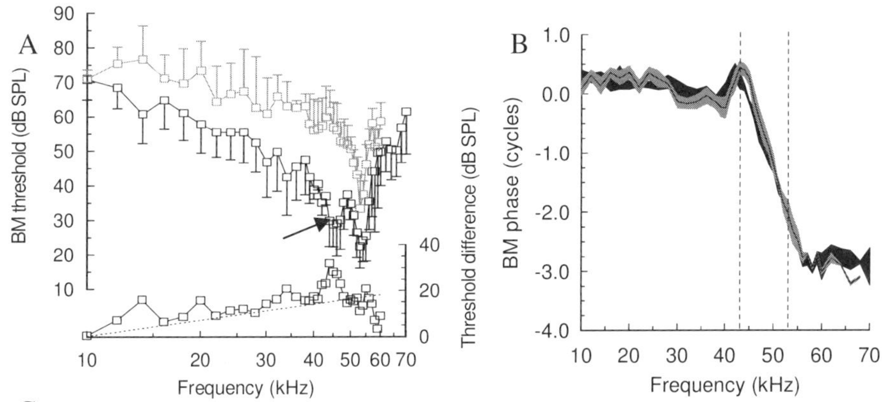

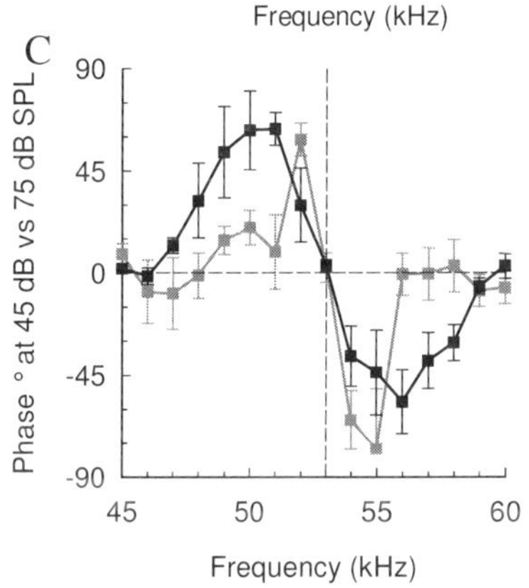

Figure 1. BM responses from the cochleae of *Tectb*$^{+/+}$ (black) and *Tectb*$^{-/-}$ (grey) mice. (A) Isoresponse displacement frequency tuning curves (mean ± SD, n=8 for each genotype; response criterion 0.2 nm, 53-kHz location). Arrow, low-frequency resonance. Lowest curve shows difference between wild-type and *Tectb*$^{-/-}$ tuning curves. Dotted line, 6 dB per octave. (B) The phases of BM responses (mean ± SD, n=4 for each genotype) to 70-dB SPL tones as a function of frequency. Vertical dashed lines indicate frequencies of the low-frequency resonance (43 kHz) and the characteristic frequency (53 kHz). (C) Phase difference (in degrees, mean ± SD, n=4 for each genotype) between BM displacements at 45 and 75 dB SPL. Zero crossing slopes, wild-type 33.28° kHz^{-1}, *Tectb*$^{-/-}$ mice 61.751° kHz^{-1}.

9.6 ± 3.3 for wild-type mice. The low-frequency resonance (Fig. 1A; arrow) that has been observed in previous measurements from the mouse BM and attributed to the TM [2] was seen only in the tuning curves of wild-type mice. The low-frequency tail of the tuning curve from *Tectb*$^{-/-}$ mutants was, on average, less sensitive than that of wild-type mice and decreased in sensitivity with a mean slope of 8.2 ± 1.6 dB per octave, compared with 14.5 ± 1.4 dB per octave for wild-type mice ($P < 0.001$), a difference of 6.3 dB per octave from 10 to 53 kHz (Fig. 1A; lowest curve).

Plots of BM phase (mean ± SD) relative to that of the malleus, measured at 70 dB SPL and as a function of stimulus frequency, taken from the 53-kHz place of four *Tectb*$^{+/+}$ and four *Tectb*$^{-/-}$ mice are shown in Fig. 1B. The plots were similar in overall phase delay to phase curves measured from guinea pigs, chinchillae and gerbils [3], but they differed in form, with a phase lead of about 270° in the region of the low frequency resonance. In wild-type mice this phase lead occurred at a frequency that is 3 kHz lower than in *Tectb*$^{-/-}$ mutants. The slope of the phase curve in *Tectb*$^{-/-}$ mice (0.2904 ± 0.0077 octave per kHz) was steeper than that of wild-type mice (0.2171 ± 0.0151 octave per kHz; $P < 0.001$). The curves were, however, coincident close to the characteristic frequency of the measurement place. Differences in the phase of BM displacement measured at moderate levels (45 dB SPL, when BM vibrations are actively amplified) and at high levels (75 dB SPL, when BM vibrations are largely passive) [4] revealed that BM phase changed more rapidly with frequency about the characteristic frequency in the *Tectb*$^{-/-}$ mutants (Fig. 1C), as would be expected from a system with actively enhanced frequency tuning.

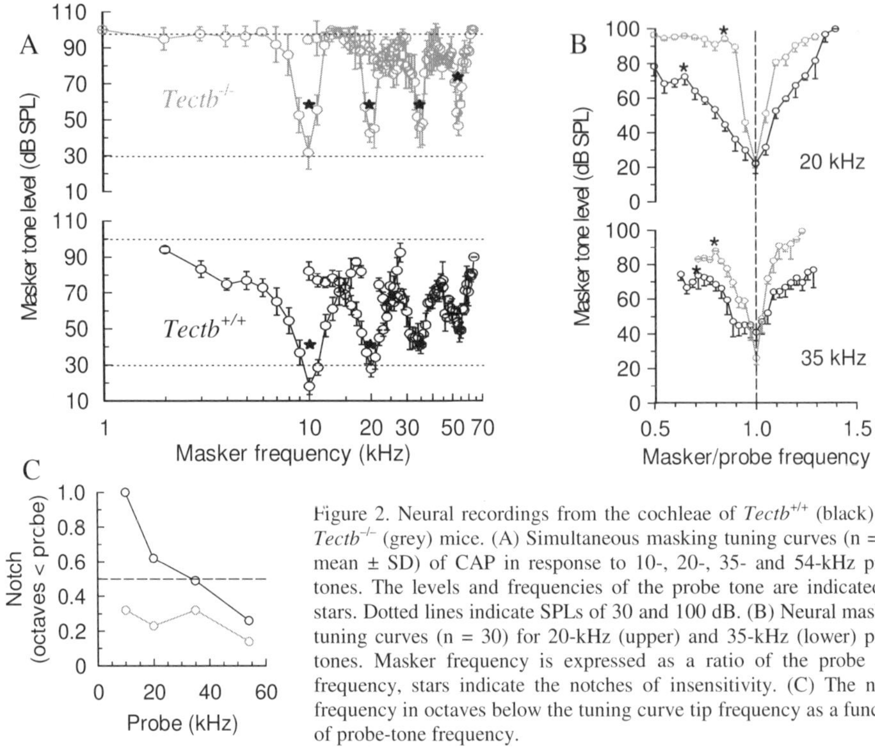

Figure 2. Neural recordings from the cochleae of *Tectb*$^{+/+}$ (black) and *Tectb*$^{-/-}$ (grey) mice. (A) Simultaneous masking tuning curves (n = 30, mean ± SD) of CAP in response to 10-, 20-, 35- and 54-kHz probe tones. The levels and frequencies of the probe tone are indicated by stars. Dotted lines indicate SPLs of 30 and 100 dB. (B) Neural masking tuning curves (n = 30) for 20-kHz (upper) and 35-kHz (lower) probe tones. Masker frequency is expressed as a ratio of the probe tone frequency, stars indicate the notches of insensitivity. (C) The notch frequency in octaves below the tuning curve tip frequency as a function of probe-tone frequency.

3.2 Neural masking tuning is sharper in Tectb$^{-/-}$ *mice*

Simultaneous masking neural tuning curves [5] closely resemble the tuning properties of single auditory nerve fibers [6] and are more sharply tuned in *Tectb*$^{-/-}$ mutants than in wild-type mice (Fig. 2A and Table 1). The low-frequency slope of the tuning curves was almost three times steeper in *Tectb*$^{-/-}$ mice and the high-frequency slope was almost twice as steep as that of *Tectb*$^{+/+}$ mice.

For frequencies within the tuning curve tip, the neural masking and the mechanical tuning curves of wild-type mice had similar Q_{10dB}. The neural responses of *Tectb*$^{-/-}$ mice were, however, more sharply tuned than the mechanical responses at the tip of the tuning curves and the low-frequency shoulder was very insensitive (Fig. 2A,B and Table 1). A notch of insensitivity is seen in the neural masking frequency tuning curves at the frequency where the tip of the tuning curve meets the tail [8]. In *Tectb*$^{+/+}$ mice, the notch occurred between 1 and 0.26 octaves below the probe frequency. In *Tectb*$^{-/-}$ mice the notch occurred between 0.32 and 0.14 octaves below the probe frequency (Fig. 2C).

Table 1. Neural tuning characteristics of *Tectb*$^{+/+}$ and *Tectb*$^{-/-}$ mice.

Probe, kHz	Q_{10dB} *Tectb*$^{+/+}$	Q_{10dB} *Tectb*$^{-/-}$	Threshold, dB SPL *Tectb*$^{+/+}$	Threshold, dB SPL *Tectb*$^{-/-}$
10	6.25	12.5	18.1 ± 4.5	31.6 ± 9.1
20	8.3	12.5	27.8 ± 4.5	42.6 ± 5.2
36	5.8	15.9	47.1 ± 7.1	45.7 ± 11.3
54	7.7	25.7	49.2 ± 4.5	46.6 ± 5.5

4 Discussion

Two mechanisms could be responsible for sharpening of the mechanical and neural suppression responses in *Tectb*$^{-/-}$ mice. It is likely that the absence of the striated-sheet matrix in the TM of mutant mice [6] leads to a decrease in the longitudinal elastic coupling along the TM. If the TM space constant is reduced then a smaller number of the outer hair cells will cooperate in synchrony to amplify the cochlear partition vibrations at any given place. This is expected to produce responses that are less sensitive but more sharply tuned. Reduction in the frequency separation between the local notch of insensitivity, which corresponds to the TM resonance frequency [6,8,9], and the tuning curve tip of the neural masking curves (Fig. 2C) also supports the conclusion that a smaller TM region contributes towards the local cochlear response. Figure 3 illustrates that this frequency separation should reduce in this case as a result of a reduction in the number of outer-hair-cell hair bundles effectively contributing towards the elastic coupling between the BM and TM. Disappearance of the BM low-frequency resonance (arrow in Fig. 1A) is also expected in this case.

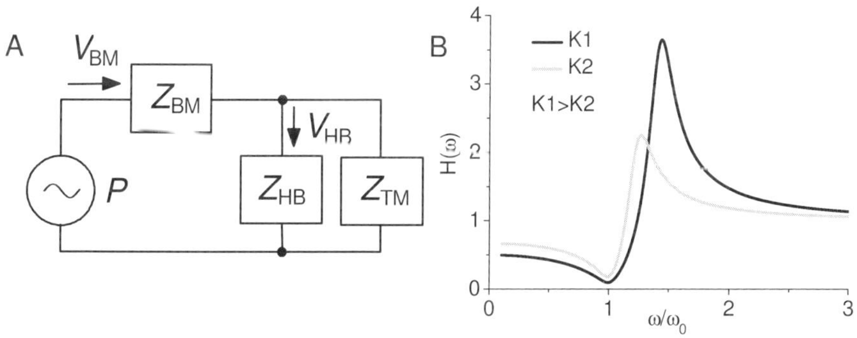

Figure 3. Equivalent diagram (A) of the TM resonance [8,9] and the hair bundle transfer function (B) calculated for two values of the bundle stiffness. (A) Z_{BM}, Z_{TM} and Z_{HB} are BM, TM and hair bundle impedances respectively. The hair bundles are assumed to have no mass. Pressure source P drives the BM and hair bundles with velocities V_{BM} and V_{HB} respectively. (B) Dependence of the hair bundle transfer function $H=V_{HB}/V_{BM}=Z_{TM}/(Z_{TM}+Z_{HB})$ on the driving frequency ω for two values of the hair bundle stiffness K1 and K2. ω_0 is the TM resonance frequency.

Changes in interaction between the BM and TM travelling waves [4,5,8] should also contribute towards sharpening of the responses in *Tectb*$^{-/-}$ mice. Effective amplification of the cochlear responses is expected in the region where the BM and TM wave velocities are similar [4, 10]. It is unlikely that this region exceeds one wavelength of the TM travelling wave due to the longitudinal gradient in the TM mechanical properties [11]. Hence, effective interaction between the TM and BM, which results in the amplification of the cochlear responses should be observed over a shorter distance in the cochleae of *Tectb*$^{-/-}$ mice where the wavelength of the TM travelling wave is predicted to be shorter due to an increase in the TM elasticity.

5 Conclusion

Measurements from the Tectb-/- mutant mice reveal the counter demands of cochlear tuning and sensitivity and the need for a compromise between these requirements in the mammalian cochlea.

Acknowledgments

This work was supported by the Medical Research Council. We thank James Hartley for technical assistance. The main results of this work have been published before in [6].

References

1. de Boer, E., 1996. Mechanics of the cochlea: Modeling efforts. In: Dallos, P., Popper, A.N., Fay, R.R. (Eds.), The Cochlea. Springer, New York, pp. 258–317.
2. Abnet, C.C., Freeman, D.M., 2000. Deformations of the isolated mouse tectorial membrane produced by oscillatory forces. Hear. Res. 144, 29–46.
3. Freeman, D.M., Abnet, C.C., Hemmert, W., Tsai, B.S., Weiss, T.F., 2003. Dynamic material properties of the tectorial membrane: a summary. Hear. Res. 180, 1–10.
4. Ghaffari, R., Aranyosi, A.J., Freeman, D.M., 2007. Longitudinally propagating travelling waves of the mammalian tectorial membrane. Proc. Natl. Acad. Sci. USA 104, 16510-16515.
5. Zwislocki, J.J., Kletsky, E.J., 1979. Tectorial membrane: A possible effect on frequency analysis in the cochlea. Science 204, 639–641.
6. Russell, I.J., Legan, P.K., Lukashkina, V.A., Lukashkin, A.N., Goodyear, R.J., Richardson, G.P., 2007. Sharpened cochlear tuning in a mouse with a genetically modified tectorial membrane. Nat. Neurosci. 10, 215 - 223.
7. Dallos, P., Cheatham, M.A., 1976. Compound action potential (AP) tuning curves. J. Acoust. Soc. Am. 59, 591–597.
8. Allen, J.B., Fahey, P.F., 1993. A second cochlear-frequency map that correlates distortion product and neural tuning measurements. Hear. Res. 94, 809-816.
9. Zwislocki J. J., 1980. Theory of cochlear mechanics. Hear. Res. 2, 171-182.
10. Hubbard, A., 1993. A traveling-wave amplifier model of the cochlea. Science 259, 68-71.
11. Richter, C.P., Emadi, G., Getnick, G., Quesnel, A., Dallos, P., 2007. Tectorial membrane stiffness gradients. Biophys. J. 93, 2265-2276

Comments and Discussion

Braun: 1) Fig. 2 shows a total absence of a tail in the neural tuning curve for Tectb-/- mice. Is this finding not a further piece of evidence for local tuning within the organ of Corti?

2) In Tectb-/- mice neural tuning is clearly sharper than BM tuning (25.7 vs. 18.9 in Q10dB at the CF place of ca 53 kHz). Is this finding not a further piece of evidence for local tuning within the organ of Corti?

3) Fig. 2B shows that at 20 kHz the peaks of the neural tuning curves are strictly symmetrical for Tectb (-/-) mice, but not for Tectb (+/+) mice. The same figure also shows that at 35 kHz the peaks of the neural tuning curves are asymmetrical for both types of mice. These findings constitute a further indication that [even] in mammals auditory frequency selectivity appears to be based on local resonators. The frequency dependent differences between the two types of mice that are evident in Fig. 2B suggest that the longitudinal elastic coupling of the TM can influence the longitudinal cochlear spreading of motion of local resonators.

NOVEL ROLES FOR PRESTIN IN FREQUENCY TUNING AND NEURAL EXCITATION IN THE MOUSE COCHLEA

MARCIA M. MELLADO LAGARDE[1], MARKUS DREXL[1], ANDREI N. LUKASHKIN[1], JIAN ZUO[2], IAN J. RUSSELL[1]

[1]*School of Life Sciences, University of Sussex, Falmer, Brighton, BN1 9QG, UK.*
[2]*St. Jude Children's Research Hospital, Department of Developmental Neurobiology, Memphis, Tennessee, USA.*

The motor protein prestin in the outer hair cells is a prime candidate for the molecular amplifier that ensures the sensitivity, frequency tuning and dynamic range of the mammalian cochlea. Absence of prestin results in a 40-60 dB reduction in cochlear neural sensitivity. Here we show that sound-evoked basilar membrane (BM) vibrations in the basal cochleae of *prestin*$^{-/-}$ mice are as sensitive as those of their *prestin*$^{+/+}$ siblings. BM vibrations in *prestin*$^{-/-}$ mice are, however, broadly tuned to a frequency ~ a half octave below the characteristic frequency (CF) of similar BM locations in *prestin*$^{+/+}$ mice. The peak sensitivity of *prestin*$^{+/+}$ BM tuning curves matches the neural thresholds, while *prestin*$^{-/-}$ BM tuning curves at the best frequency are > 50 dB more sensitive than the neural responses. We conclude that prestin influences properties of the cochlear partition that are crucial for BM frequency tuning and for converting its vibrations into neural excitation.

1 Introduction

The remarkable acuity of hearing in mammals has its initial basis in the sharp tuning, high sensitivity and large dynamic range of their auditory sensory organ, the cochlea. These properties rely on active processes that take place in the outer hair cells (OHCs)[1] and have been united under the term "cochlear amplifier" [2]. Cochlear amplification implies the introduction of energy into the sound-induced vibrations of the cochlear partition [3] and generates non-linear mechanical responses [4]. Two mechanisms of electromechanical transduction that take place in the OHCs have been proposed as amplifiers: hair bundle motility and somatic motility. OHC somatic motility depends on the function of the protein prestin present in their basolateral membranes [5]. The properties of prestin and the cylindrical shape of the OHCs provide these cells with the mechanical capability to shorten and elongate their length following transmembrane voltage changes [6].

Prestin knockout mice (*prestin*$^{-/-}$) have been produced with a deletion in the *prestin* gene that removed about one third of the protein at the amino terminal [7]. In homozygous mutant mice the length of the OHCs is about 30% shorter than normal and there is a slightly more pronounced age-related hair cell loss, mainly in the basal region of the cochlea. However, it has been shown that up to 28 days after birth there is no significant hair cell loss in *prestin*$^{-/-}$ mice [8]. OHC hair bundles are normal in *prestin*$^{-/-}$ mice and so are the mechanoelectrical transducer currents, as revealed by cochlear microphonics (CM) [7, 9]. *Prestin*$^{-/-}$ mice have much higher thresholds (up to 60 dB) for auditory brainstem responses (ABR), compound action potentials (CAP) and distortion product otoacoustic emissions (DPOAEs) than their wild type siblings (*prestin*$^{+/+}$ mice), suggesting the absence of cochlear amplification [7, 9]. Without prestin the loss in sensitivity is larger at higher frequencies, supporting the idea that cochlear amplification is more relevant at those frequencies [4]. Prestin blockage by salicylate [10] also reduces the sensitivity and tuning of cochlear mechanical and neural responses [11].

Afferent cochlear neural responses, the information that is sent to the next stage in the auditory pathway, depend on the mechanical processes taking place in the organ of Corti during the analysis of incoming sounds. Both, the passive and active properties of the cochlear partition contribute to generate tuned, sensitive neural outcome in the mammalian cochlea. The aims of this study are to describe the cochlear mechanical responses in *prestin*$^{-/-}$ mice and to correlate them with the hearing loss revealed by their neural audiograms.

2 Materials and Methods

Mice (*prestin*$^{-/-}$, n = 5; *prestin*$^{+/+}$, n = 11) 3 to 4 weeks old were used in this study. The animals were anaesthetized with Urethane (2 mg/g bodyweigth). The level of anesthesia was monitored throughout the experiments by pedal reflex and audiovisual check of a stable heartbeat. Cochlear surgery began only when a deep level of anaesthesia was reached. The animals were tracheotomized and carbogen (95% O_2, 5% CO_2; flow rate ~ 0.3 L/min) was provided during the whole experiment. The heartbeat was monitored with skin electrodes inserted on both sides of the thorax. The core temperature was maintained at 38°C using a heating blanket and a heated head holder. All procedures involving animals were performed in accordance with the UK Home Office regulations with approval from the local ethics committee.

The right pinna and about 1/3 of the meatus were removed and the tip of a two-channel plastic probe housing the sound system was located at ~ 1mm from the tympanic membrane. The right bulla was opened using a ventrolateral approach to expose the round window. Cochlear stimulation and measurements were performed via the ear canal and the round window respectively, so that the cochlea remained intact. The sound system was calibrated in dBSPL re 20 μPa following a procedure described elsewhere [12, 13]. The method used to generate acoustic stimuli has also been described before [12], although in these experiments we used a custom-built condenser loudspeaker [14]. BM measurements were performed in the basal (60-65 kHz) region of the mouse cochlea by focusing the beam of a self-mixing laser-diode interferometer [13] through the round window. A lock-in amplifier was used to process the signal before acquisition by the data translation board (DT3010). Data acquisition and analysis were performed using programs written in Test Point (CEC) and Origin 7.0.

3 Results

OHC mechanoelectrical transduction is functional in *prestin*$^{-/-}$ mice, as seen from cochlear microphonics (CM) measured at the round window using low frequency (3kHz) acoustic tones (Fig. 1A). The CM is the result of the summed receptor currents from the

OHCs in the basal region of the cochlea [16]. CM level functions, although shifted in sensitivity, depend on sound pressure level (SPL) in a similar fashion in both *prestin*$^{+/+}$ and *prestin*$^{-/-}$ mice (Fig. 1A). CM waveforms in *prestin*$^{-/-}$ mice also resemble those of their wild type siblings (inset Fig. 1A). The higher levels necessary to obtain threshold CM in *prestin*$^{-/-}$ mice may be related to changes in the mechanical properties of the cochlear partition in the absence of prestin.

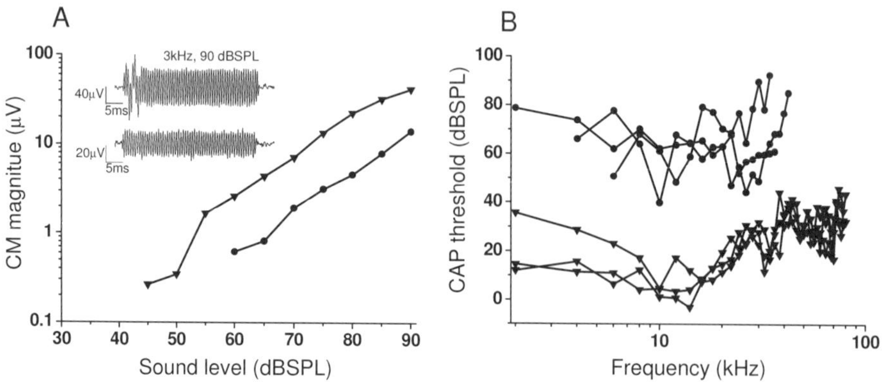

Figure 1: Cochlear electrical responses to acoustic stimulation measured in prestin+/+ and prestin-/- mice. A. Cochlear microphonic (CM) level functions in a prestin+/+ (triangles) and a prestin-/- (circles) mouse evoked by a 3 kHz tone. Inset: corresponding CM waveforms for 90 dBSPL (top: prestin+/+ mouse, bottom: prestin-/- mouse). B. CAP audiograms measured in 3 prestin+/+ (triangles) and 3 prestin-/- (circles) mice. Modified from [15], with permission.

CAP audiograms from *prestin*$^{+/+}$ and *prestin*$^{-/-}$ mice in response to tones between 2 and 80 kHz (Fig. 1B), reveal that the CAP thresholds of the mutant mice are more than 30 dB less sensitive that those of their wild type siblings. CAPs could not be measured for frequencies above 45 kHz at SPLs in the upper limit of the acoustic system (85 dB SPL).

BM displacements evoked by acoustic stimulation were measured *in vivo* in *prestin*$^{+/+}$ and *prestin*$^{-/-}$ mice. Threshold frequency tuning curves obtained from the basal region of the BM (CF 60-65 kHz) in the cochleae of *prestin*$^{+/+}$ mice show very sensitive responses at the CF (Fig. 2A, CF: 62 kHz). The tips of the tuning curves are sharply tuned, with $Q_{10\ dB} = 10 \pm 2$ (mean ± standard deviation, n = 5). The total phase lag between 10 and 80 kHz in figure 2C is about 3 cycles and includes a steeper phase accumulation associated with tuning at the CF (Fig. 2A). Cochlear mechanical sensitivity in these mice is largely reduced *post-mortem* and BM tuning is shifted to lower frequencies (about 0.25 octaves below the CF)(Fig. 2A). BM thresholds at the CF in *prestin*$^{+/+}$ mice *in vivo* (Fig. 2A) show excellent agreement with CAP thresholds recorded from the basal region of the cochlea (compare with Fig. 1B).

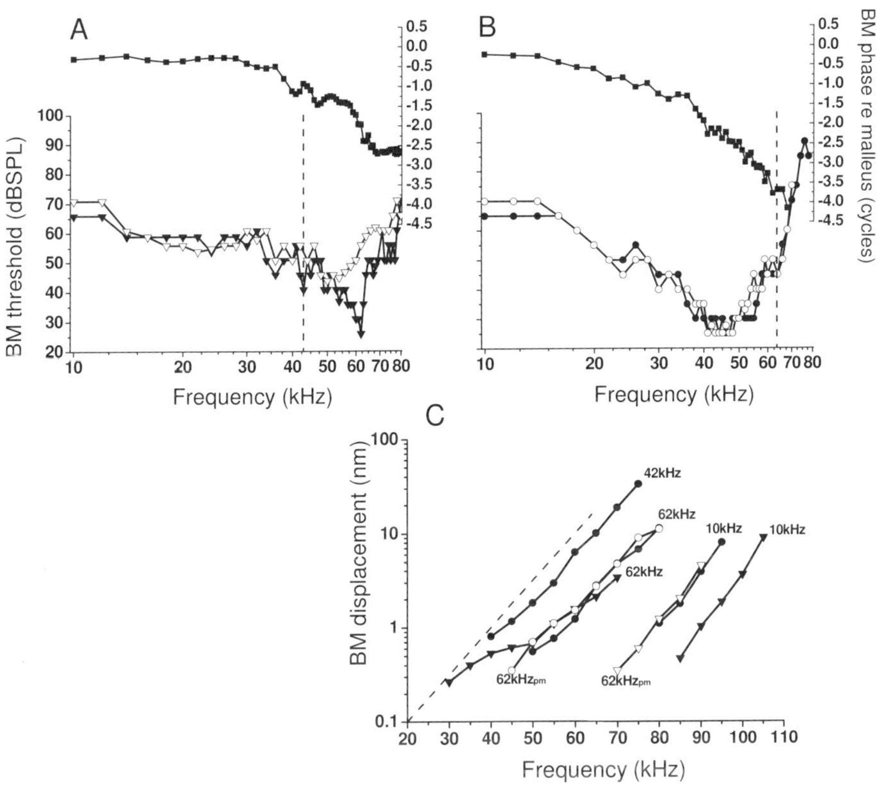

Figure 2: BM displacement evoked by acoustic stimulation in prestin+/+ and prestin-/- mice. A, B. BM threshold tuning curves (bottom, closed symbols) and phase (top, squares) in a typical prestin+/+ (A) and a typical prestin-/- (B) mouse. Open symbols represent post-mortem BM tuning curves. Vertical dashed lines represent half an octave below the CF for the prestin+/+ mouse and the estimated CF at half an octave above the BF for the prestin-/- mouse. C. BM displacement level functions from the prestin+/+ (closed triangles) and the prestin-/- (closed circles) mouse showed in A and B. Labels stand for the frequency of stimulation. The level functions were generated at the CF (62 kHz) and a frequency well below this (10 kHz) in the prestin+/+ mouse and at the CF (62 kHz), the BF (42 kHz) and a frequency well below those (10 kHz) in the prestin-/- mouse. Open symbols correspond to post-mortem level functions at 62 kHz for the prestin+/+ (triangles) and the prestin-/- (circles) mouse. Modified from [15], with permission.

BM tuning curves in *prestin*$^{-/-}$ mice (Fig. 2B) show similar sensitivity to that of *prestin*$^{+/+}$ mice but they are broadly tuned to a best frequency (BF) of about half an octave below the CF of the same BM location in their wild type siblings (Fig. 2B, BF: 44.6 ± 5.3 kHz, n = 8). The sensitivity of the BM at the BF (31 ± 7 dB SPL, n = 11) in *prestin*$^{-/-}$ mice is more than 50 dB higher than their neural sensitivity at the same frequency (compare Fig. 2B with Fig. 1B). At the estimated CF of the measurement location on the BM (Fig. 2B, dashed line), the difference between mechanical and neural sensitivity in *prestin*$^{-/-}$ is at least 30 dB. This matches the difference in sensitivity of BM responses between *prestin*$^{+/+}$ and *prestin*$^{-/-}$ mice at the CF (compare Fig. 2A and B). It is possible to predict the CF in the *prestin*$^{-/-}$ mice responses because the BM measurements in all the mice were performed in the same region of the cochlea following landmarks. The high BM sensitivity in *prestin*$^{-/-}$ mice does not change *post-mortem* (Fig. 2B), showing that it is not associated with active cochlear processing. The total phase lag of the BM response in *prestin*$^{-/-}$ mice between 10 and 70 kHz is about 3.5 cycles, which is similar to that of *prestin*$^{+/+}$ mice, but the phase accumulation at the BF or the CF is not as steep as for *prestin*$^{+/+}$ mice.

BM level functions at the CF measured *in vivo* in *prestin*$^{+/+}$ mice are sensitive and show compressive non-linearity, while the level functions at frequencies well below the CF (10 kHz) are linear and 56.5 ± 6.5 dB (n = 5) less sensitive (Fig. 2C). BM level functions measured at the CF *post-mortem* in the same *prestin*$^{+/+}$ mouse are also linear and insensitive (Fig. 2C). BM level functions measured *in vivo* in *prestin*$^{-/-}$ mice are linear at all frequencies including the CF (Fig. 2C) and are similar to those measured *post-mortem* (Fig. 2C). These results further support that the mechanical sensitivity in *prestin*$^{-/-}$ mice is not associated with cochlear amplification.

4 Discussion

The high sensitivity of cochlear mechanical responses to acoustic stimulation in *prestin*$^{-/-}$ mice is an unexpected finding. From the functional role of prestin as the molecular motor responsible for OHC somatic motility [5], and from the suppressive effect of salicylate, a prestin blocker [10], in cochlear mechanical responses [11], an overall reduction in mechanical sensitivity was expected in the cochleae of *prestin*$^{-/-}$ mice. Previous measurements of BM threshold tuning curves after salicylate application have shown a loss in sensitivity of about 30 dB and a shift of the tip of the tuning curve to frequencies about 0.2 octaves below the CF [11]. Similar changes in the BM responses were obtained in the present study in *prestin*$^{+/+}$ mice *post-mortem*. Indeed, *prestin*$^{-/-}$ mice showed a reduction in BM sensitivity at the corresponding CF of the wild type cochleae, which is in agreement with the hearing loss revealed by the CAP audiograms. However, thresholds of the mechanical responses in *prestin*$^{-/-}$ mice at frequencies below the CF could be up to 20 dB lower than in *prestin*$^{+/+}$ mice, a feature that is not apparent in the auditory neural responses. The structure of the IHCs and the afferent synaptic transmission in *prestin*$^{-/-}$ mice are preserved [8, 9], therefore, the mismatch found between mechanical and neural responses in these mice reveals disruption of the coupling between the BM vibrations and IHC excitation.

The CM measured in this study in *prestin*$^{-/-}$ mice is similar in waveform and level dependence to that measured in *prestin*$^{+/+}$ mice. This result agrees with previous measurements of CM in *prestin* mice and provides evidence that OHC mechanotransduction is functional in the basal cochlea of *prestin*$^{-/-}$ mice. CAP thresholds measured here in *prestin*$^{-/-}$ mice are also in agreement with earlier measurements of CAPs [9] and with ABR audiograms obtained in these mice [7].

Close similarity between the BM tuning curves measured *in vivo* and *post-mortem* and the absence of compressive non-linearity in the BM level functions reveal that the sensitive BM responses of *prestin*$^{-/-}$ mice are not amplified. This result suggests that prestin has a significant structural role in addition to its role in cochlear amplification in mammals. We propose that the presence of prestin increases the impedance of the cochlear partition, shifting its resonance to higher frequencies; while through amplification it may also increase cochlear mechanical sensitivity and tuning at a particular frequency (the CF). Additionally, prestin apparently guaranties appropriate coupling between the vibrations of the BM and excitation of the IHCs, thereby ensuring reliable transmission of mechanical energy into neural excitation. Indeed, the absence of prestin shifts the resonance of the cochlear partition to about half an octave below the corresponding CF, which matches the resonance attributed to the tectorial membrane (TM) [12, 17]. If the BM and the TM are tuned to a similar frequency in *prestin*$^{-/-}$ mice, no excitation of the IHCs should be expected. Prestin appears to contribute towards the creation of a mismatch in resonance between the TM and the BM that is required for neural excitation in the mammalian cochlea.

Acknowledgments

This work was supported by the Medical Research Council. M.M.L. was supported by a FENS – IBRO Fellowship. M.D. was supported by a DFG Fellowship. We thank James Hartley for technical assistance.

References

1. Dallos P., 1992. The active cochlea. J. Neurosci. 12, 4575-4585.
2. Davis H., 1983. An active process in cochlea mechanics. Hear. Res. 9, 79-90.
3. Lukashkin A.N., Walling M.N., Russell I.J., 2007. Power amplification in the mammalian cochlea. Curr. Biol. 17, 1340-1344.
4. Robles L., Ruggero M.A., 2001. Mechanics of the mammalian cochlea. Physiol. Rev. 81, 1305-1352.
5. Zheng J., Shen W., He D.Z., Long K.B., Madison L.D., Dallos P., 2000. Prestin is the motor protein of cochlea outer hair cells. Nature 405, 149-155.
6. Brownell W.E., Bader C.R., Bertrand D., de Ribaupierre Y., 1985. Evoked mechanical responses of isolated cochlear outer hair cells. Science 227, 194-196.
7. Liberman M.C., Gao J., He H.Z., Wu X., Jia S., Zuo J., 2002. Prestin is required for electromotility of the outer hair cell and for the cochlear amplifier. Nature 419, 300-304.
8. Wu X., Gao J., Guo Y., Zuo J., 2004. Hearing threshold elevation precedes hair-cell loss in prestin knockout mice. Brain Res. Mol. Brain Res. 126, 30-37.
9. Cheatham M.A., Huynh K.H., Gao J., Zuo J., Dallos P., 2004. Cochlear function in prestin knockout mice. J. Physiol. 560, 821-830.
10. Oliver D., He D.Z., Klocker N., Ludwig J., Schulte U., Waldegger S., Ruppersberg J.P., Dallos P., Fakler B., 2001. Intracellular anions as the voltage sensor of prestin, the outer hair cell motor protein. Science 292, 2340-2343.

11. Murugasu E., Russell I.J., 1995. Salicylate ototoxicity: the effects on basilar membrane displacement, cochlear microphonics, and neural responses in the basal turn of the guinea pig cochlea. Aud. Neurosci. 1, 139-150.
12. Legan P.K., Lukashkina V.A., Goodyear R.J., Kossl M., Russell I.J., Richardson G.P., 2000. A targeted deletion in alpha-tectorin reveals that the tectorial membrane is required for the gain and timing of cochlear feedback. Neuron 28, 273-285.
13. Lukashkin A.N., Bashtanov M.E., Russell I.J., 2005. A self-mixing laser-diode interferometer for measuring basilar membrane vibrations without opening the cochlea. J. Neurosci. Methods 148, 122-129.
14. Schuller G., 1997. A cheap earphone for small animals with good frequency response in the ultrasonic frequency range. J. Neurosci. Methods 71, 187-190.
15. Mellado Lagarde M.M., Drexl M., Lukashkin A.N., Zuo J., Russell I.J., 2008. Prestin's Role in Cochlear Frequency Tuning and Transmission of Mechanical Responses to Neural Excitation. Current Biol. 18, 200-202.
16. Patuzzi R.B., Yates G.K., Johnstone B.M., 1989. The origin of the low-frequency microphonic in the first cochlear turn of guinea-pig. Hear. Res. 39, 177-188.
17. Gummer A.W., Hemmert W., Zenner H.P., 1996. Resonant tectorial membrane motion in the inner ear: its crucial role in frequency tuning. PNAS 93, 8727-8732.

Comments and Discussion

Braun: Is the proposal "that the presence of prestin increases the impedance of the cochlear partition, shifting its resonance to higher frequencies" not unrealistic considering the recent findings of Wu et al. (2008) and Cheatham et al. (2008)? These authors developed a new prestin knockin mouse. Here prestin is still present in the OHCs, but its motor function is disabled. Thus the passive mechanical properties of OHCs, organ of Corti, and cochlear partition are unchanged in these animals. However, Cheatham et al. (2008) found that the new prestin knockin mouse had exactly the same functional deficits as the known prestin knockout mouse. Do these new findings not suggest that the function of prestin is restricted to active mechanics?

Wu, X. et al. (2008) Creation of a Novel Prestin Knockin Mouse Model with Normal Cochlear and Outer Hair Cell Structure and Mechanical Properties. ARO Midwinter meeting 2008, Abstract 141.

Cheatham, M. A. et al. (2008) Prestin-Based Somatic Electromotility is the Cochlear Amplifier. ARO Midwinter meeting 2008, Abstract 142.

BIAS-TONE EFFECTS ON THE FIRST-PEAK VERSUS LATER PEAKS OF AUDITORY-NERVE RESPONSES

J. J. GUINAN JR.

Eaton-Peabody Lab, Massachusetts Eye & Ear Infirmary, and Harvard Medical School, 243 Charles St. Boston MA 02114, USA

Preliminary experiments show that low-frequency "bias" tones at sound levels that produce twice-a-cycle suppression of low-level, cochlear-amplified auditory-nerve (AN) responses, suppress the first peak of AN click responses in different ways depending on click level. At moderate click levels, the most common bias-tone effect was twice-a-cycle suppression, but at high click levels where the AN first peak switches from rarefaction to condensation clicks, bias tones sinusoidally modulated the click response. The results are consistent with the hypothesis that AN first-peak responses to high-level clicks originate from a different cochlear vibrational mode than other AN responses.

1 Introduction

The first peak of responses to moderate-level clicks in cat auditory-nerve (AN) fibers with characteristic frequencies (CFs) less than 6 kHz is strongly inhibited by stimulation of medial olivocochlear (MOC) efferents [1]. In contrast, the first peak of click responses in guinea-pig basilar-membrane (BM) motion in the cochlear base is not affected by MOC stimulation [2]. The lack of MOC inhibition of the BM first peak is consistent with a variety of measurements that indicate the first peak of the BM response to clicks is passive, i.e. it grows linearly with level and is little changed by death [3]. A passive click-response first peak is also consistent with many cochlear models and measurements of basal-turn click-response "glides" which indicate that the first peak in the classic traveling wave is due to below-CF energy and is not significantly amplified by the cochlear traveling-wave amplifier [4]. Considering the passive nature of the first peak of the BM traveling wave, it is not surprising that it is not changed by MOC stimulation. However, this leaves the MOC inhibition of the first peak of the AN click response unexplained. Indeed, it is difficult to see how any simple change in cochlear traveling-wave models made to mimic BM motion in the base of the cochlea can account for the MOC inhibition of the AN first peak in the apical half of the cochlea.

We have hypothesized that the auditory-nerve initial peak (ANIP) response to clicks at moderate sound levels is due to a motion that bends inner-hair-cell (IHC) stereocilia but is separate from the motion of the classic traveling wave [1, 2]. Since MOC fibers synapse on outer hair cells (OHCs) and inhibit the ANIP response, we also hypothesize that the ANIP motion is strongly influenced, perhaps produced, by active processes in OHCs. Although recent work has revealed many ways in which cochlear structures can have vibration patterns that are not locked to BM motion but lead to the bending of IHC stereocilia [5-7], it is not clear, which, if any, of these might be responsible for the ANIP response and its MOC inhibition.

A further complication is that at high levels, the ANIP response "reverses". At moderate click levels (below ~100 dB SPL) rarefaction clicks evoke the ANIP response (termed the "ANIPr" response) but, at in many fibers at higher click levels, condensation clicks evoke the ANIP response (termed the "ANIPc" response) [8]. We have suggested that the ANIPr and ANIPc responses are due to separate mechanical drives that bend IHC stereocilia and excite AN fibers, presumably from different vibrational motions of the organ of Corti [8].

Insight into the processes involved in producing ANIP responses might be obtained from the effects of low-frequency "bias" tones. Bias tones move the organ of Corti up and down thereby bending OHC and IHC stereocilia and changing their operating points. This reduces the cochlear traveling-wave amplifier when the bias tone pushes the OHC stereocilia away from the high-slope region of the stereocilia receptor-current-versus-angle function [9, 10]. By placing clicks at various phases of the bias tone, we can determine whether ANIP responses depend on OHC stereocilia position in the same way as the later peaks that, at low levels, are amplified by the traveling-wave amplifier.

2 Methods

Recordings were made from AN fibers in anesthetized cats using methods described previously and approved by our institutional animal review committee [1, 8]. To minimize ringing, clicks were produced by a reverse-driven condenser earphone. Bias tones (39-100 Hz) were produced by a DT48 coupled through the hollow ear bars. AN fiber responses were measured while clicks were presented at 8-10 different phases of the bias-tone response. Two paradigms which produced different patterns of AN adaptation were used: (1) Clicks were presented at a fixed phase of the bias tone and runs were done with clicks at 8-10 different phases, and (2) Clicks were presented at intervals that were slightly shorter than the bias-tone period so that over many bias-tone cycles the clicks occurred at 8-10 different bias-tone phases.

Our strategy was to choose a bias tone level that (1) produced moderate, twice a bias-tone-cycle suppression of responses to clicks that were 10-15 dB above the fiber's click threshold, and (2) did not produce significant bias-tone modulation of the fiber's "spontaneous" firing rate. Responses to clicks at 10-15 dB above threshold depend heavily on cochlear amplification and condition 1 was intended to "calibrate" the bias tone so that it was at a level that bent OHC stereocilia twice a cycle into low-slope regions that reduced the gain of the traveling-wave cochlear amplifier. Condition 2 was to insure that the bias tone did not significantly bend IHC stereocilia. Presumably, since IHC stereocilia are not imbedded in the tectorial membrane their input is high-pass filtered and is less affected by bias tones than OHC stereocilia.

3 Results

Our preliminary results show that when the bias tone was at a low enough level to produce no significant modulation of spontaneous rate, the effect on the click responses was small. Unfortunately we did not usually record enough spikes to get clear bias-tone patterns under these conditions. Fortunately, the bias-tone effect on "spontaneous" activity had a different phase than the effect on click responses, which allowed the effect of bias tones on click responses to be unambiguously determined, even though these bias tones may have modulated the "spontaneous" activity.

The bias-tone effect on ANIPr responses from moderate-level clicks normally was suppression twice each bias-tone cycle similar to that seen in responses to low-level

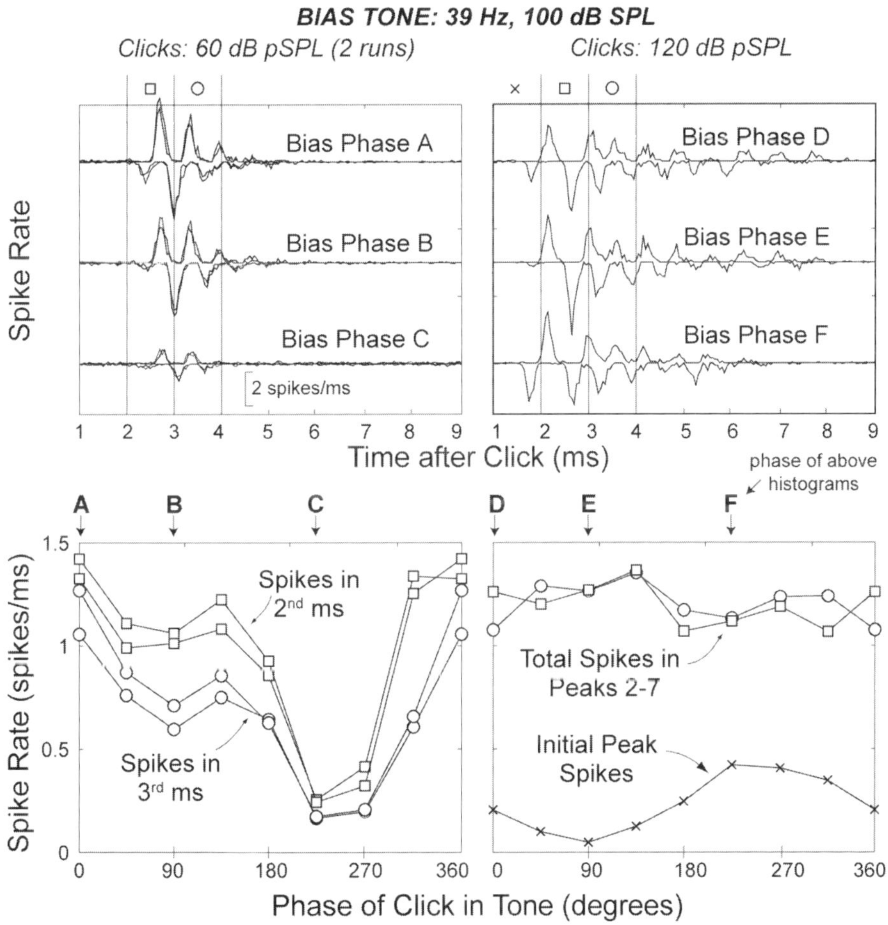

Figure 1. Low-frequency bias tones have a different effect on the first peak of cat AN responses to high-level vs. low-level clicks. Left: 60 dB pSPL clicks. Right: 120 dB pSPL clicks. Top panels are compound PSTs (PSTs from rarefaction clicks plotted upward and from condensation clicks downward), each at one phase of the bias tone (letters indicate tone phases at bottom). The amplitudes of different click-response peaks were quantified by combining spikes (from condensation and rarefaction clicks) in 1 ms bins demarcated by the vertical lines at top. For low-level clicks (left) the response was suppressed at 90 and at 225-270 degrees. In contrast, the earliest peak in the response to high-level clicks (the ANIPc response) was largest at the bias phase that most suppressed the low-level response. At this high level, the later peaks were barely suppressed at any bias-tone phase, consistent with the cochlear amplifier no longer affecting them. Click rate=80/s. CF=1.56 kHz.

sounds. However, the bias-tone effect depended somewhat on fiber CF. More work needs to be done to determine the full pattern of bias-tone effects on ANIPr responses.

The most dramatic effect of the bias tone was on ANIPc responses. Bias-tone levels that produced twice-a-bias-tone-cycle suppression at low click levels produced a sinusoidal modulation of ANIPc responses. An example is given in Figure 1 which shows cat AN single-fiber responses to clicks at 8 different phases of a 39 Hz bias tone. For low-level clicks, where the AN response was ringing at CF, AN responses were reduced at two phases of the bias tone, approximately 180° apart, and one reduction was more than the other (Fig. 1, left). This pattern is typical for effects of near-threshold bias tones on BM, IHC and AN responses to low-level CF tones. Such a pattern is thought to be produced by the bias tone bending OHC stereocilia toward their saturation region, with more nonlinearity in one direction than the other, thereby decreasing the gain of the cochlear amplifier for CF tones twice each bias-tone cycle [10]. In contrast, at high click levels, the bias tone modulated the ANIPc response in a near-sinusoidal, once-a-cycle pattern with the maximum reduction of the response at the opposite phase of the bias tone compared to the phase that produced the largest reduction in the low-level-click response (Fig. 1).

4 Discussion

Our data provide evidence for three different drives that excite AN fibers. The traveling wave and ANIPr drives are similar in that both are inhibited by MOC efferents and both show twice-a-cycle suppression by the bias tone. These properties indicate that the two drives depend on both OHC motility and the slope of the OHC stereocilia current-vs.-angle function. In contrast, the ANIPc drive is modulated sinusoidally by the bias tone and, although not studied extensively, our data indicate that the ANIPc response is not inhibited by MOC stimulation [1]. These features are consistent with the ANIPc drive being passive, i.e. not affected by active cochlear processes. The sinusoidal modulation of the ANIPc response by the bias tone is presumably due the effects of changes in the shape of the organ of Corti produced by the bias tone.

The reversal of the ANIP response at high sound levels is reminiscent of the reversal of AN responses to tones [11]. MOC stimulation does not inhibit the response to tones above this reversal [12] which is similar to the lack of MOC effect on the ANIPc response. A response maximum at the bias-tone phase that produces a minimum response on low-level tones has also been seen before in IHC receptor potentials for certain below-CF tones [13, 14]. The relationship of these to the bias-tone phase effects on the ANIPc response is unclear, but all might be manifestations of the same underlying phenomena.

Our working hypothesis is that the three drives seen in the AN click responses are due to three different vibrational modes, or three different combinations of vibrational modes of the organ of Corti. The question is: What are these motions? The bulk of the response (i.e. everything but the first peak) is presumably due to the traditional traveling wave which elicits up-and-down motion of the organ of Corti, and perhaps also other motions such as tilting of the reticular lamina [5]. Of the ANIPc and ANIPr motions, we consider the ANIPc first. It seems unlikely that the ANIPc motion is due to a direct effect of a cochlear fluid-compression wave because the ANIP latency is much too long (e.g. the first peak to high-level clicks is ~2 ms for CFs near 1 kHz [8]). Considering that the ANIPc response occurs just before, and in step with, the traveling wave response, we hypothesize that the ANIPc response is a passive response to the first peak of the traveling wave which itself is passive [2, 3]. Then, since the ANIPr response has approximately the same latency but is strongly influenced by OHCs, we hypothesize that the ANIPr response is due to an active response elicited by the first peak of the traditional traveling wave, an active response that we presume is overwhelmed at high levels by an opposite-polarity passive response motion. One suitable active response would be tilting of the reticular lamina [5]. With this hypothesis the outer hair cells contract (or elongate), which tilts the reticular lamina and squeezes (or enlarges) the space between the reticular lamina at the IHC and the tectorial membrane above it [5]. We cannot now specify the direction of the reticular lamina motion but the result must bend IHC stereocilia in the excitatory direction both at the first peak of very high-level condensation clicks and at lower levels for rarefaction clicks. Reticular lamina tilting may also occur in response to the later peaks of the traveling wave. Considering the great difference seen in the MOC inhibition of the AN first peak compared to the second and later peaks [1], there must be some difference between the ANIPr and traveling-wave motions. One possibility is that the traveling wave amplification is due to OHC squeezing/elongating forces being coupled to the BM by the fluids inside of the organ of Corti and that the additional tilting of the reticular lamina is a secondary phenomenon that is not important in traveling wave cochlear amplification. Micromechanical measurements indicate that friction in the sub-tectorial space restricts tilting of the reticular lamina to frequencies below ~4 kHz [5] which is near the upper frequency at which MOC inhibition of ANIPr responses have been seen [1]. Thus, reticular-lamina tilting may be the motion that produces the ANIPr response but has little or no influence on basilar-membrane motion or the traveling wave. It is interesting to note that if this hypothesis is correct, then the ANIPr response is due to active OHC processes and might be called a "cochlear amplifier" that is separate from the traveling-wave cochlear amplifier.

The above hypothesis fits many aspects of the data but it is far from proven. It is also possible that the ANIPr response is due to other things such as OHC-induced waves in the tunnel of Corti or in the tectorial membrane [6, 7] and that these might arrive slightly ahead of the traveling wave [1]. All of these phenomena are consistent with the hypothesis that AN responses are driven by multiple vibrational modes of the organ of Corti, modes that do not necessarily have a simple relationship to the traveling-wave vibrational mode.

Acknowledgments

Supported by NIDCD RO1 DC00235 and P30 DC005209.

References

1. Guinan, J.J., Jr., Lin, T., Cheng, H. 2005. Medial-olivocochlear-efferent inhibition of the first peak of auditory-nerve responses: Evidence for a new motion within the cochlea. J Acoust Soc Am 118, 2421-2433.
2. Guinan, J.J., Jr., Cooper, N. 2008. Medial olivocochlear inhibition of basilar-membrane responses to clicks: Evidence for two modes of cochlear mechanical excitation. J Acoust Soc Am (in press).
3. Recio, A., Rich, N.C., Narayan, S.S., Ruggero, M.A. 1998. Basilar-membrane responses to clicks at the base of the chinchilla cochlea. J Acoust Soc Am 103, 1972-1989.
4. Shera, C.A. 2001. Frequency Glides in Click Responses of the Basilar Membrane and Auditory Nerve: Their scaling behavior and origin in traveling-wave dispersion. J Acoust Soc Am 109, 2023-2034.
5. Nowotny, M., Gummer, A.W. 2006. Nanomechanics of the subtectorial space caused by electromechanics of cochlear outer hair cells. Proc Natl Acad Sci U S A.
6. Karavitaki, K.D., Mountain, D.C. 2007. Evidence for outer hair cell driven oscillatory fluid flow in the tunnel of corti. Biophys J 92, 3284-93.
7. Ghaffari, R., Aranyosi, A.J., Freeman, D.M. 2007. Longitudinally propagating traveling waves of the mammalian tectorial membrane. Proc Natl Acad Sci U S A 104, 16510-5.
8. Lin, T., Guinan, J.J., Jr. 2000. Auditory-nerve-fiber responses to high-level clicks: interference patterns indicate that excitation is due to the combination of multiple drives. J Acoust Soc Am 107, 2615-30.
9. Ruggero, M.A., Robles, L., Rich, N.C. 1992. Two-Tone suppression in the basilar membrane of the cochlea: Mechanical basis of auditory-nerve rate suppression. J. Neurophysiol. 68, 1087-1099.
10. Cai, Y., Geisler, C.D. 1996. Suppression in auditory-nerve fibers of cats using low-side suppressors. III. Model results. Hearing Res 96, 126-140.
11. Liberman, M.C., Kiang, N.Y.S. 1984. Single-neuron labeling and chronic cochlear pathology. IV. Stereocilia damage and alterations in rate- and phase-level functions. Hear Res 16, 75-90.
12. Gifford, M.L., Guinan, J.J., Jr. 1983. Effects of crossed-olivocochlear-bundle stimulation on cat auditory nerve fiber responses to tones. J. Acoust. Soc. Am. 74, 115-123.
13. Patuzzi, R., Sellick, P.M. 1984. The modulation of the sensitivity of the mammalian cochlea by low frequency tones. II.Inner hair cell receptor potentials. Hearing Res. 13, 9-18.
14. Cheatham, M.A., Dallos, P. 1997. Low-frequency modulation of inner hair cell and organ of Corti responses in the guinea pig cochlea. Hearing Res. 108, 191-212.

Comments and Discussion

Gummer: With respect to your suggestion that results of the bias-tone experiments can be explained by "changes in the shape of the organ of Corti produced by the bias tone, e.g. a change in the distance between the reticular lamina and the tectorial membrane [5]", it is worth emphasizing that the counter-phasic motion between reticular lamina and tectorial membrane at the inner hair cell, which is elicited by somatic electromotile force from the outer hair cells, is also limited to frequencies of about 3 to 4 kHz (in all cochlear turns) (see Nowotny and Gummer, 2006). This is consistent with your observation that first-peak inhibition is also limited to neurones with characteristic frequency up to about 4 kHz.

Guinan: Yes, I agree. We cannot say for certain that tilting of the rcticular lamina is the motion that produces the ANIPr response and is inhibited by medial efferents. However, the fact that viscous drag in the space between the reticular lamina and the tectorial membrane limits this motion to driving frequencies of 4 kHz and below [5], and this is very similar to the highest frequency at which we found efferent inhibition of the first peak of click responses [1], provides support for this hypothesis.

DUAL TUNING IN THE MAMMALIAN COCHLEA: DISSOCIATION OF NEURAL AND BASILAR MEMBRANE RESPONSES AT SUPRA-THRESHOLD SOUND LEVELS – A META-ANALYSIS

MARTIN BRAUN

Neuroscience of Music, Gansbyn 14, S-67195 Klässbol, Sweden

In the basal turn of the cochlea, tuning of auditory nerve fiber (ANF) and basilar membrane (BM) responses are closely correlated at threshold levels. Their correlation also at supra-threshold levels has always been an implicit postulate of BM transmission line theory. A test of this postulate by empirical data, however, has never been published. In order to elucidate this discrepancy, a large number of ANF recordings was analyzed and compared to equivalent BM recordings. The results show that, at least in the basal turn of the cochlea, the best frequency (BF) of BM responses drops by about half an octave between 30 and 100 dB SPL. The comparable neural data never reflect anything similar. 1) In the majority of cases (52%) BF is totally unaffected by sound level. 2) In a minority of cases (36%) BF drops slightly at the highest sound levels, but only by about half as much as for the BM. 3) In a relevant number of cases BF is bimodal at the highest sound levels, such that the antagonistic majority and minority versions appear concurrently. These results indicate a dual tuning, such that BM and organ of Corti (OC) are tuned separately, and they are consistent with the concept of overload protection of OC sensitivity as the genuine function of BM tuning.

1 Introduction

For ten years now, we have known that in the mammalian cochlea responses of the basilar membrane (BM) and of auditory nerve fibers (ANFs) can be closely related at the threshold of hearing [1, 2]. As usual, however, correlation says nothing about cause and effect. Currently, there are two competing and mutually exclusive schools of thought. The first one suggests that at threshold of hearing the chain of activation is BM-OHC-BM-IHC-ANF [3]. The second one suggests that it is OHC-IHC-ANF [4, 5], with the simultaneous OHC-BM branch as an epiphenomenon [6, 7].

Important information on the matter can be expected from data on BM and ANF responses at supra-threshold sound levels. If these are correlated, like they are at threshold (outcome A), the second hypothesis ("OHC first") will be unrealistic, because at high sound levels the relative contribution of OHC motility to mechanical forces in the cochlea is too weak to dominate BM behavior. If BM and ANF responses are not correlated at supra-threshold sound levels (outcome B), the first hypothesis ("BM first") will be unrealistic, because indications for a chain of activation BM-IHC-ANF will then be missing.

2 Methods

From the published literature two bodies of evidence were scrutinized: (a) all data on level dependence of the best frequency (BF) of single ANFs that appeared after the introduction of single fiber recording in this field in the mid-1960s, and (b) all data on level dependence of BF of BM responses that appeared after the introduction of laser interferometry in this field in 1990. The major trends within each of the two data bodies were determined and then compared across data bodies.

3 Results

3.1 Basilar membrane

For the basal turn of non-specialized mammalian cochleae sufficient data are available, and the results are uniform. A typical example is shown in Fig. 1. Here BF is stable at 9.5 kHz from 0 to 20 dB, but already at 30 dB it has migrated to 9.25 kHz. The migration toward lower BFs then continues monotonously, reaching 9.0 kHz at 50 dB, 8.5 kHz at 70 dB, 7.5 at 80 dB, and 6.5 at 100 dB. In octaves (oct) the cumulated shifts are 0.04, 0.08, 0.16, 0.34, and 0.55, respectively.

3.2 Auditory nerve fibers

There are two large-scale investigations of the sound-level dependence of ANF firing rates in the squirrel monkey, by Rose *et al.* [8] (R71) and by Geisler *et al.* [9] (G74). Both studies can be considered as normative, because of (a) the very large quantity of the collected and published data and (b) the absence of deviating results elsewhere in the literature.

R71 recorded data from 48 animals and published 14 figures with iso-intensity curves. In 13 of them BF did not vary with sound level. In one (Fig. 2B) BF shifted slightly toward lower frequencies, from 2.1 to 1.7 kHz (0.30 oct) in the range from 50 to 90 dB.

G74 recorded data from 31 animals and published 9 figures with iso-intensity curves. In six of them BF did not vary with sound level. In two (Figs. 2A , 2B) BF shifted slightly toward lower frequencies, from 4.6 to 4.3 kHz (0.10 oct) in the range from 50 to 80 dB and from 7.3 to 6.9 kHz (0.08 oct) in the range from 50 to 80 dB. In one (Fig. 8) BF shifted slightly toward higher frequencies, from 3.5 to 3.7 kHz (0.08 oct) in the range from 65 to 95 dB. Further, G74 stated the proportional distribution of types of BF shifts as present in their total data base: "52 % of the fibers sampled showed no change in best frequency with intensity changes" (p. 1161); in 36 % BF dropped "slightly", and in 12 % it increased with level.

In five of the nine iso-intensity plots of G74 BF behavior is clearly visible even at the highest sound levels, i.e. not hidden by high-level saturation. In four of these five, BF shows a bimodal behavior at highest levels. Three examples are displayed in Fig. 1, where the broken lines in the ANF plots indicate secondary BFs on the low-frequency side, with deviations from the normal BFs of 0.21, 0.21, and 0.26 oct, respectively, at 100 dB.

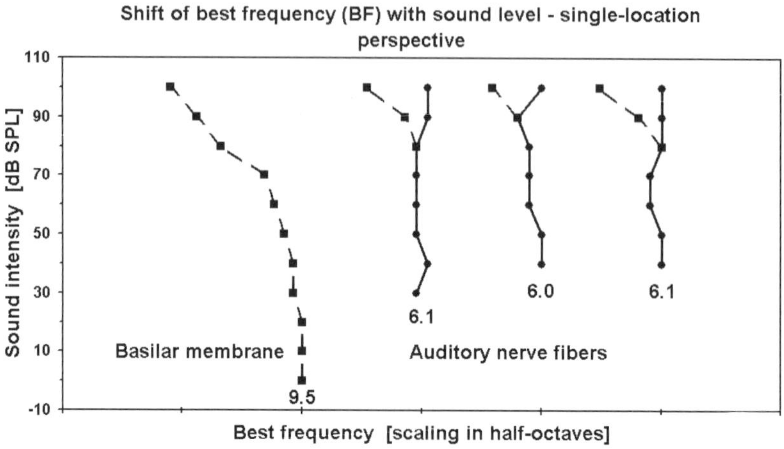

Figure 1. BF versus sound level. X-axis: BF as defined by the peaks in the iso-intensity curves of BM and ANF responses. Y-axis: sound intensity of the stimulus. The four curves are spaced in half-octaves re threshold BF, which is stated in kHz below each curve. Leftmost curve: plotted from the BM velocity data in Fig. 1A of Ruggero *et al.* [2]. Other three curves: plotted from the ANF firing-rate data (3 fibers of 2 animals) in Figs. 7A-C of Geisler *et al.* [9]. Stimuli were 128 to 512 tone bursts of 5 to 100 ms per stimulus condition for the BM recordings and 4 to 8 tones of 500 ms per stimulus condition for the ANF recordings.

3.3 *Comparison BM versus ANF*

1) The majority of ANFs has a stable BF across sound levels and shows no relation at all to the BM's shift of BF.
2) A minority of ANFs shows a BF shift at the highest sound levels that goes into the same direction as observed at the BM, but is only about half as large.
3) Some ANFs show a bimodal BF at the highest sound levels, with one mode at the fiber's normal BF and the other mode reflecting a shift that goes into the same direction as observed at the BM, but is only about half as large.

3.4 *Stability of BF of ANF responses in cross-correlation data*

Bimodal BFs, similar to those in Fig. 1, have also been documented in cross-correlation analyses of ANF responses to broadband noise at high sound levels [10; Fig. 3]. Further, since 2005 we have data from Wiener-kernel analysis of ANF responses to broadband noise even from the basal cochlear turn [11; Fig. 14]. Because these data are from the same species and from a similar frequency place as the BM data in Fig. 1, a direct comparison is possible. For sound levels from 33 to 68 dB SPL in ERB the BF shift of the ANF is very small: from 12.1 to 11.8 kHz (0.036 oct). The comparable BF shift of the BM (Fig. 1), however, is more than three times as large: from 9.25 to 8.5 kHz (0.12 oct).

3.5 *Level independence of BF in further neural pathway*

Concerning the further neural pathway, a BF shift that is similar to the one in BM responses has never been reported. To the contrary, in the cochlear nucleus it is common

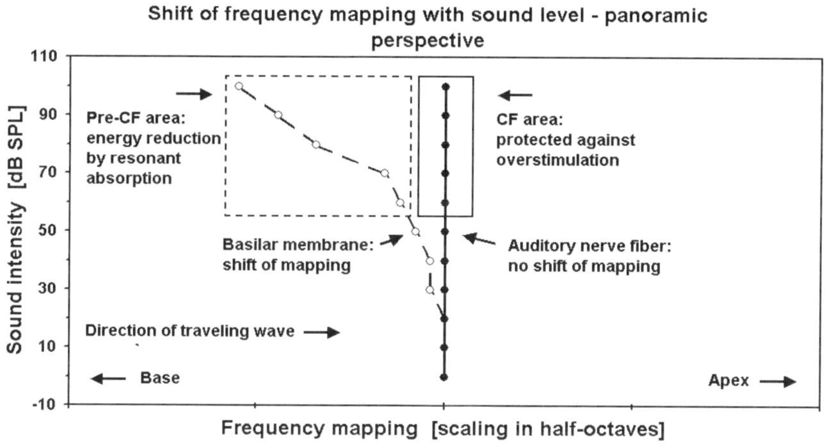

Figure 2. Cochlear frequency mapping versus sound level. X-axis: BF place position along the length of the cochlea. Y-axis: sound intensity of the stimulus. BM data according to sample in Fig. 1. ANF data according to the summarized results of the present study.

that neurons which apparently receive their major excitatory input from the auditory nerve have a strictly constant BF, even if their dynamic range is >80 dB [12; Fig. 3C]. Also for the inferior colliculus the level independence of BF of sharply tuned neurons is well documented [13; Figs. 2 & 11].

4 Discussion

4.1 Dual tuning

The finding of a clear dissociation between BM tuning and ANF tuning at medium and high sound levels indicates a dual mechanism. Besides the well-known BM tuning, which also occurs post mortem at levels above ca 60 dB, there appears to be an independent tuning mechanism in the organ of Corti (OC) that underlies ANF behavior.

The fact that a minority of ANFs partially, but never fully, follows the BF shift of the BM further confirms a dual mechanism. It can be understood as an effect of the well-known phenomenon of frequency pulling in the cochlea [14]. Limit cycle oscillators in the OC, as a possible cause of the BF of ANFs, could under certain conditions be partially entrained by strong nearby BM oscillations.

4.2 Function of dual tuning

The dissociation between BM and ANF tuning at supra-threshold sound level is also reflected in the well-documented phenomenon of the "half-octave shift in temporary threshold shift (TTS)" [15]. Here, exposure to a high-level tone typically reduces hearing sensitivity at a frequency half an octave above the exposure frequency, while leaving sensitivity fully intact at the exposure frequency itself. This dissociation of exposure frequency and frequency range of damage precisely mirrors the dissociation of BM and neural tuning. Most importantly, the "half-octave shift in TTS" protects sensitivity at the exposure frequency. As shown in Fig. 2, a section of the BM that lies basalward of a given characteristic frequency (CF) absorbs vibrational energy before this energy reaches the place of CF. Thus, the present results further substantiate earlier observations that overload protection appears to be the genuine function of the BM traveling wave in the mammalian inner ear [4, 16, 17].

5 Conclusions

1) The "BM-first" hypothesis of cochlear activation is not compatible with the present findings, because BM tuning dissociates from OC tuning at sound levels above 20 dB SPL.
2) The "OHC-first" hypothesis of cochlear activation is compatible with the present findings, because in the sound level range that is dominated by OHC motility the apparent BM tuning is almost congruent with OC tuning.
3) The BM's passive tuning at medium and high sound levels is strategically placed for a protection of the sensitivity of the OC's active tuning, because it absorbs energy basalward of any given CF in the OC.

6 Addendum

After the completion of this study, Mellado Lagarde *et al.* [18] reported data on the dissociation between BM and neural responses that are in remarkable agreement with the present results and conclusions. In mice where the gene for the OHC motor protein prestin was deleted CF of BM responses was reduced by half an octave. Further, neural responses were less sensitive by >50 dB SPL than BM responses. These new data appear to reflect the same co-occurrence of two separate tuned mechanisms as described in the present study, because both the half-octave disparity and a striking independence of ANF and BM responses were observed again. Also here indications for a chain of activation BM-IHC-ANF were missing, and thus the "BM-first" hypothesis again appeared to be unrealistic.

References

1. Narayan, S.S., Temchin, A.N., Recio, A., Ruggero, M.A., 1998. Frequency tuning of basilar membrane and auditory nerve fibers in the same cochleae. Science 282, 1882-1884.
2. Ruggero, M.A., Narayan, S.S., Temchin, A.N., Recio, A., 2000. Mechanical bases of frequency tuning and neural excitation at the base of the cochlea: comparison of basilar-membrane vibrations and auditory-nerve fiber responses in chinchilla. Proc. Natl. Acad. Sci. USA 97, 11744-11750.
3. Robles, L., Ruggero, M.A., 2001. Mechanics of the mammalian cochlea. Physiol. Rev. 81, 1305-1352.

4. Braun, M., 1994. Tuned hair cells for hearing, but tuned basilar membrane for overload protection: evidence from dolphins, bats, and desert rodents. Hear. Res. 78, 98-114.
5. Nowotny, M., Gummer, A.W., 2006. Nanomechanics of the subtectorial space caused by electromechanics of cochlear outer hair cells. Proc. Natl. Acad. Sci. USA 103, 2120-2125.
6. Dancer, A., Avan, P., Magnan, P., 1997. Can the traveling wave be challenged by direct intra-cochlear pressure measurements? In: Lewis, E.R., Long, G.R., Lyon, R.F., Narins, P.M., Steele, C.R., Hecht-Poinar, E. (Eds.), Diversity in Auditory Mechanics. World Scientific, Singapore, pp. 340-346.
7. Pujol, R., Eybalin, M., Lavigne-Rebillard, M., Lenoir, M., Puel, J.L., Rebillard, G., Ruel, J., Trigueiros-Cunha, N., Wang, J., 1999-2007. Promenade around the Cochlea. Section "Cochlea", subsection "physics". http://www.cochlea.org/
8. Rose, J.E., Hind, J.E., Anderson, D.J., Brugge, J.F., 1971. Some effects of stimulus intensity on response of auditory nerve fibers in the squirrel monkey. J. Neurophysiol. 34, 685-699.
9. Geisler, C.D., Rhode, W.S., Kennedy, D.T., 1974. Responses to tonal stimuli of single auditory nerve fibers and their relationship to basilar membrane motion in the squirrel monkey. J. Neurophysiol. 37, 1156-1172.
10. Møller, A.R., 1978. Frequency selectivity of the peripheral auditory analyzer studied using broad band noise. Acta. Physiol. Scand. 104, 24-32.
11. Recio-Spinoso, A., Temchin, A.N., van Dijk, P., Fan, Y.H., Ruggero, M.A., 2005. Wiener-kernel analysis of responses to noise of chinchilla auditory-nerve fibers. J. Neurophysiol. 93, 3615-3634.
12. Spirou, G.A., Davis, K.A., Nelken, I., Young, E.D., 1999. Spectral integration by type II interneurons in dorsal cochlear nucleus. J. Neurophysiol. 82, 648-663.
13. Ramachandran, R., Davis, K.A., May, B.J., 1999. Single-unit responses in the inferior colliculus of decerebrate cats. I. Classification based on frequency response maps. J. Neurophysiol. 82, 152-163.
14. Long, G.R., Tubis, A., Jones, K.L., 1991. Modeling synchronization and suppression of spontaneous otoacoustic emissions using Van der Pol oscillators: Effects of aspirin administration. J. Acoust. Soc. Am. 89, 1201-1212.
15. McFadden, D., 1986. The curious half-octave shift: evidence for a basalward migration of the traveling-wave envelope with increasing intensity. In: Salvi, R.J., Henderson, D., Hamernik, R.P., Coletti, V. (Eds.), Basic and Applied Aspects of Noise-induced Hearing Loss. Plenum Publishing, New York, pp. 295-312.
16. Braun, M., 1993. Basilar membrane tuning re-examined: frequency selective damping of high level input may be its genuine function. In: Duifhuis, H., Horst, J.W., van Dijk, P., and van Netten, S.M. (Eds.), Biophysics of Hair Cell Sensory Systems. World Scientific, Singapore, p. 406.
17. Braun, M., 1996. Impediment of basilar membrane motion reduces overload protection but not threshold sensitivity: evidence from clinical and experimental hydrops. Hear. Res. 97, 1-10.
18. Mellado Lagarde, M.M., Drexl, M., Lukashkin, A.N., Zuo, J., Russell, I.J., 2008. Prestin's role in cochlear frequency tuning and transmission of mechanical responses to neural excitation. Curr. Biol. 18, 200-202.

SLOW OSCILLATORY COCHLEAR ADAPTATION TO BRIEF OVER STIMULATION: COCHLEAR HOMEOSTASIS DYNAMICS

DAVID T. KEMP, OLIVER J. BRILL*

Centre for Auditory Research, UCL Ear Institute, 332 Gray's Inn Road, London WC1X 8EE

Oscillatory changes in the intensity of OAEs are seen in human ears after brief loud sound exposures. The oscillations have a period of about 200s (frequency, 5mHz) and they parallel oscillations seen in hearing threshold. We propose that these fluctuations are a byproduct of a control system affecting the organ of Corti which governs cochlear amplifier gain and contributes to cochlear homeostasis. We have measured OAEs in 22 normally hearing ears after 7-120s exposure to 105dBA low frequency tones. By fitting individual experimental data to a simple a under-damped negative feedback model regulating a notional 'cochlear gain potential' we can extract parameters which allow us to accurately predict the dynamic response of individual human cochleae to a wide range of exposure durations. We propose that the negative feedback loop serves to maintain the potential for high cochlear amplifier gain during widely varying acoustic load conditions ensuring rapid recovery of threshold after acoustic overloads. We find the feedback system fatigues slightly with exposure time and this may be related to the onset of temporary threshold shift.

1 Introduction

The cochlea has to accommodate an enormous range of stimulation intensities yet its sensitivity in quiet is normally remarkably stable. This is despite the fact that the sensory mechanism itself depends on a delicate physiological balance that is so easily damaged. It is reasonable to assume that the functional stability of cochlear sensitivity is under the control of powerful homeostatic mechanisms. This stability breaks down when high enough levels of exposure cause temporary threshold shift (TTS). TTS has been repeatedly investigated but the functional dynamics of stability control prior to TTS have not. The control dynamics may point to specific cellular processes and may reveal differences between individuals not evident from threshold sensitivity measurements.

Human hearing sensitivity shows oscillatory fluctuations at about 5mHz (200s period) following acoustic overload. Increases as well as decreases in hearing sensitivity occur. The phenomenon is known as the 'bounce' after Hirsh and Ward 1952 [1]. In 1986 Kemp [2] reported a corresponding oscillation in OAE level and suggested it was the response of a second order control system regulating OHC 'energy levels'. In 1997 Kirk and Patuzzi [3] and Kirk et al [4] performed intra cochlear investigations in guinea pigs and found a similar 'bounce' effect. Post exposure instability in CM persisted even with synapses blocked by tetrodotoxin. They considered an origin in sound induced changes in ionic concentrations affecting cochlear amplifier gain and proposed a 'double exponential' model in which two opposing processes with different rates of exponential decay combined to give a biphasic time course. Later O'Beirne and Patuzzi [5]

*Now at OAE Research Laboratory, Otodynamics Ltd, Hatfield UK, AL108BB suggested the involvement of a homeostatic mechanism mediated by the intracellular concentration of Ca++. Kevanishvilli et al's 2007 study of post exposure OAEs changes in human ears [6] appeared to support Kirk and Patuzzi's double exponential model but their experiments had a much lower time resolution than our study.

1.1 Rationale of this study

The aim of this study is to describe the functional dynamics of cochlear amplifier homeostasis during and after mild over-stimulation. We start with the reasonable assumption that cochlear sensitivity, which is critically determined by hair cell status and operating point, is stabilized by homeostatic processes. We hypothesize that strong acoustic stimulation depletes the resources of the organ of Corti (or otherwise changes its status) and that this decreases cochlear sensitivity through a reduction in cochlear amplifier gain in the quiet. We introduce the term *"cochlear gain potential"* or CGP to describe the underlying (composite) physiological parameter on which cochlear amplifier gain ultimately depends. We propose that strong stimulation depresses the CGP but that this is actively compensated for by a strong negative feedback loop [2]. Compensation will be less than 100% and the loop may have oscillatory characteristics. The residual changes in CGP after exposure will change the amplitude of place-fixed (i.e. reflection) OAEs. We make the reasonable assumption that for small enough perturbations of cochlear status any changes in CGP will be faithfully translated into OAE intensity changes providing a tool to explore the dynamics of one aspect of cochlear homeostasis.

2 Method

We observed post exposure OAE fluctuations in 22 volunteer human subjects. Transient evoked OAEs were used so as to maximize simultaneous data collection across frequency and so minimize the number of exposures needed. TEOAEs were obtained from consecutive 10s duration samples taken every 12s for several minutes after sound exposures. These were compared to a pre-exposure recording. Uniform click stimulation at 50/s was employed without 'nonlinear' correction for stimulus artifacts. The later were minimized by careful probe placement and the use of moderate click levels (74dBSPL p.e.). TEOAEs can be retrospectively frequency analyzed. Changes in the wideband (800-5kHz) and half octave OAE levels were determined for each 10s sample. We used exposure tones from 150Hz to 2kHz. Our longest and most intense exposure was a 150Hz tone at 105dB(A) lasting 2m. This sound is loud but easily tolerated causing no significant temporary threshold shift. Accumulated subject sound exposures were carefully monitored and controlled to be well within the E.U. daily exposure limit for normal working environments. Experiments were conducted with informed consent and as approved by the UCL Ethics Committee.

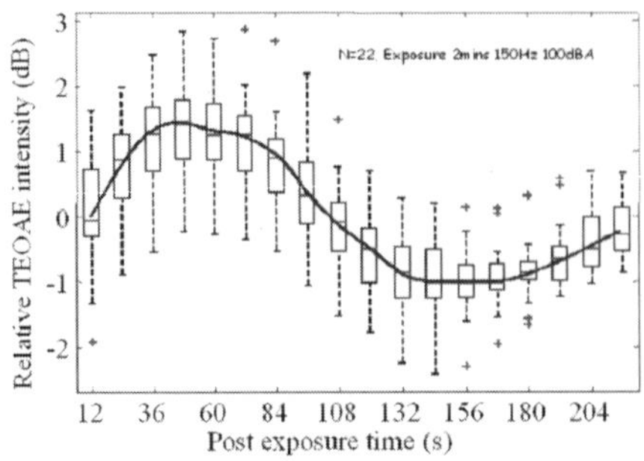

Figure 1. Changes in TEOAE intensity following exposure to a 150Hz, 105dB(A) tone for 2 minutes. Boxplot is of responses from 22 ears with excluded data sets marked singlely. Key points of reference are the initial post exposure level, the maximum enhancement and maximum depression. Note that for this exposure the initial level equals the pre-exposured level indicated by '0dB'.

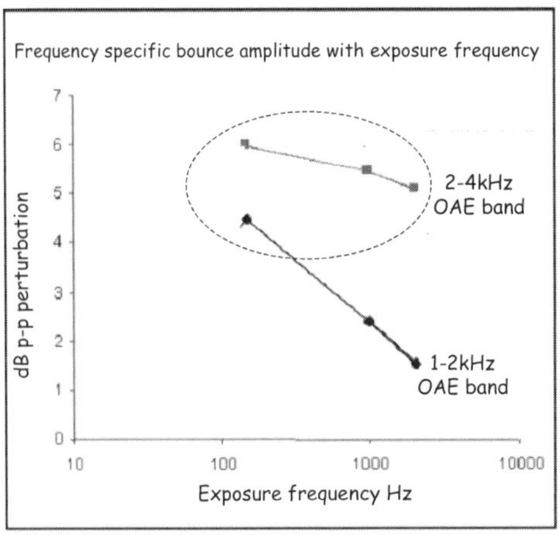

Figure 2. Evidence that the effects seen are local and not cochlea wide. The amplitude of post exposure OAE oscillations in the 1-2 and 2-4kHz OAE bands with exposure frequencies indicated by the abscissa. The effect of the 105dBSPL exposure is frequency dependent. OAE frequencies below the exposure frequency are much less affected than those above. Ellipse encloses data $F_{OAE}>F_{exp}$

3 Results

Figure 1 shows the wideband results from 22 ears for a 2m 150Hz exposure. Initial post exposure OAE is depressed by an insignificant amount, and is then enhanced by more than 1dB at 50s before becoming depressed by about 1dB around 160s. Half octave analysis showed the peak enhancement was consistent across OAE frequency but the subsequent suppression exhibited a broad maximum from 1.5k to 2.5kHz.

We found that the bounce effect is not limited to very low frequency exposure tones as implied by [2]. Using different exposure frequencies we found the effect to be strongest at OAE frequencies at or above the exposure frequency indicating a localized phenomenon. (figure 2). We found no change in the OAE bounce pattern when an experiment was repeated with the cochlear reflex activated by contralateral suppression. This argues against neural involvement [3].

We observed the OAE bounce with exposures times from 7.5 to 120s. Figure 3a shows that the overall magnitude of OAE modulation increased systematically with lengthening exposure duration. The time course of key bounce features are shown in figure 3b (center) as a function of exposure duration. Peak enhancement increases and peak depression deepens with increasing exposure duration. The initial depression deepens at first but then recovers.

4 Model

We tried to fit our data to Kirk et al's (1997) model of opposing exponentially decaying factors but this did not fit the second suppression. More importantly model parameters had to be varied greatly with exposure duration in order to fit the experimental data. A simple analogue model in which the OAE amplitude is modulated by a damped sine wave triggered at both the start and end of exposure fitted our data much better. Using a MATLAB program we allowed the initial velocity, decay time constant and frequency of the oscillation to be varied to obtain best fit. Figure 4 shows how well the analytical model could fit data from different exposures with only minimal parameter exposure dependencies.

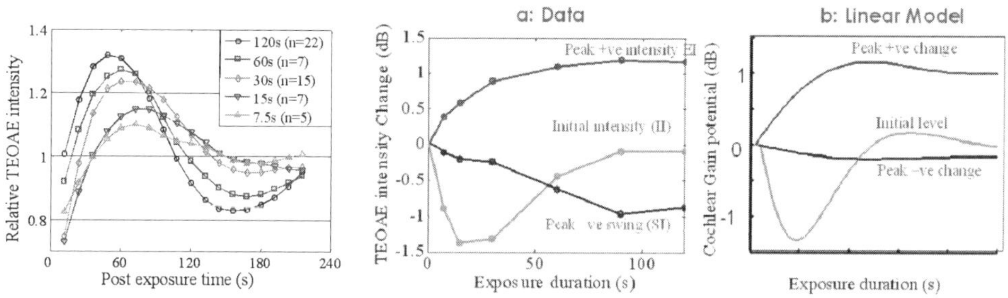

Figure 3. Left Post exposure OAE time course with exposures from 7.5 to 120s of a 105dBSPL 150Hz tone. Centre: The initial, peak enhancement (Peak +ve) and peak depressions (Peak –ve) as a function of exposure duration. Right: comparable data derived from the model.

In this simple model the changes in post exposure oscillation with exposure duration are not due to fatigue related parameter changes but are the result of interference between onset and offset oscillations. The relevance of the model is underlined by figure 3(right) in which model predictions of initial depression, peak enhancement and peak depression reproduce experimental data (3b) reasonably well. This gave us confidence in the analytical model and we designed an analogue circuit model of the negative feedback system based on our observations. The model is described in figure 5. Figure 6 illustrates how this model reacts dynamically to a fixed duration exposure 'load' with fixed components values but a range of feedback gain FB. As FB is increased the sustained per-exposure drop in CGP is greatly reduced. Instead there is a transient oscillation after onset end and offset of the load. In this second order system a feedback gain of about 8 matches the oscillations in our data best. The exposure duration in figure 6 is much longer than in our experiments so the transient disturbances do not interfere.

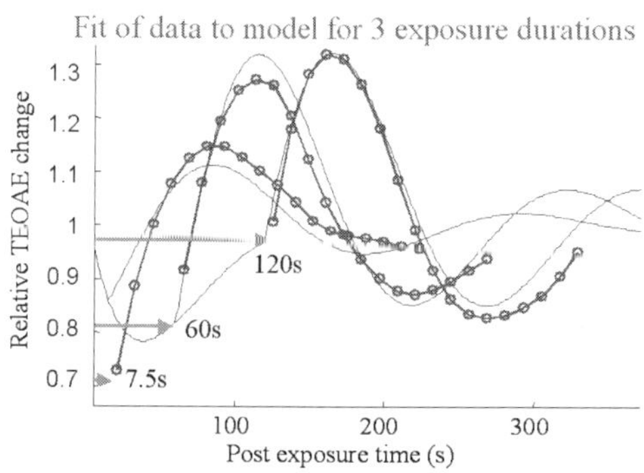

Figure 4. Thin lines show model predictions of the response to three exposure durations. Model parameters were optimized for overall best fit with experimental data (thick lines with circles). Arrow indicates the end of exposure and the start of experimental OAE data. Model parameter sets at start and after exposure were allowed to differ but remained similar. Worst fit is for the shortest duration (7.5s). Note that in the model oscillation starts during the exposure period (ie the thin line during the exposure marked by rhe the arrow).

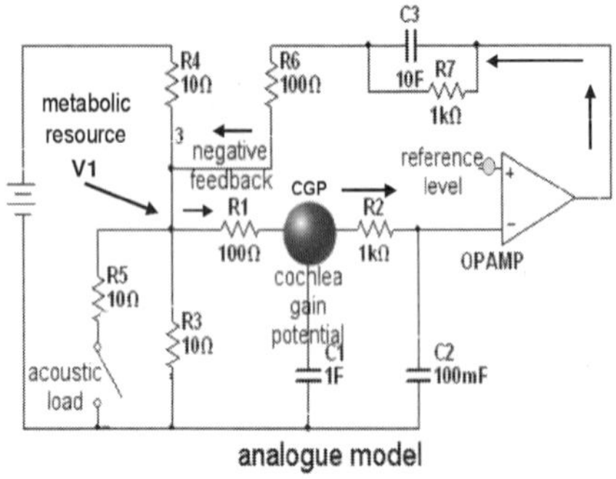

Figure 5. (right) This circuit controls the cochlear gain 'potential' CGP (ball) which is locally stabilized by capacitor C1. C1 is charged from a metabolic resource potential V1 through R1. Acoustic loading (switch) immediately lowers the potential of this resource causing a downward drift in the CGP. Any drop is detected by the comparator OpAmp via the integrative components R2 and C2 which introduce a time delay. Change is amplified and strong negative feedback is delivered back to the resource V1 via R7 and R6. R7 bypassed by C3 which has a time constant longer than all others in the system. The feedback acts to partially restore V1 but not quickly enough to prevent some oscillation of CGP if the gain is high enough. The feedback strength diminishes long term as C3 becomes fully charged allowing CPG to slowly drift down during sustained loading.

Our experimental exposures were short enough for onset and offset oscillations to interfere linearly causing the shape changes with duration seen in figure 4. In figure 6 the capacity of C3 was large enough to ensure no loss of feedback during the exposure. In figure 7a, C3 is reduced to create a slow decrease in sustained feedback during the exposure. We found the best fits of individual data to the model could account for about 60% of variance if we constrained onset and offset parameters to be identical and allowed no drift. This rose to 94% of variance when drift was allowed. Figure 7 b and c illustrates this. Our negative feedback model was able to predict the behavior of the OAE bounce phenomenon to different load durations very accurately. The parameters obtained by fitting data to the model must each have physiological significance - but what? We looked at what determined the observable features. In the model oscillation amplitude is governed by both sensitivity to loading and the feedback gain so is ambiguous. Oscillation frequency depends on both the system time constant and feedback gain so differences in frequency are also ambiguous. But oscillation decay time constant is relatively independent of feedback gain. This decay time could be a physiologically important factor because it originates in the system's internal time constants and is likely to relate to specific chemical processes. In contrast, increases or decreases in oscillation frequency without decay time constant changes would indicate a change in feedback gain and this could relate

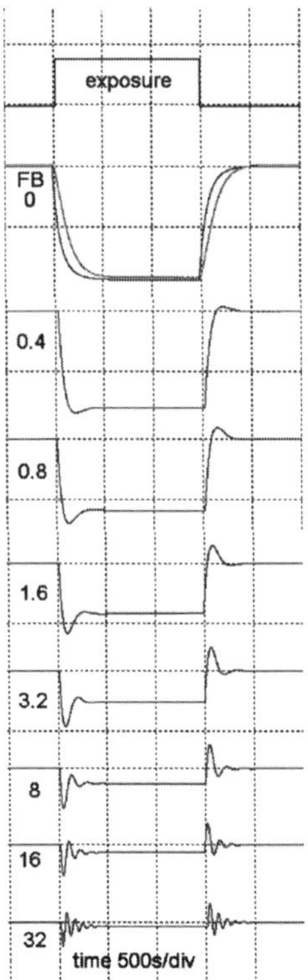

Figure 6 (left): From the model in figure 5 a series of responses of the cochlear gain potential to the load imposed by a fixed exposure for various amounts of negative feedback gain (FB). As feedback is increased the DC drop in cochlear gain potential during the exposure is reduced by 1/(1+FB). As FB gain is increased complementary oscillations appear at the start and finish of exposure. These increase in frequency and in the number of oscillations as FB increases, exhibiting an exponential decay. For feedback gain of 32 (bottom) the DC drop is reduced to only 3% of the no-feedback condition.

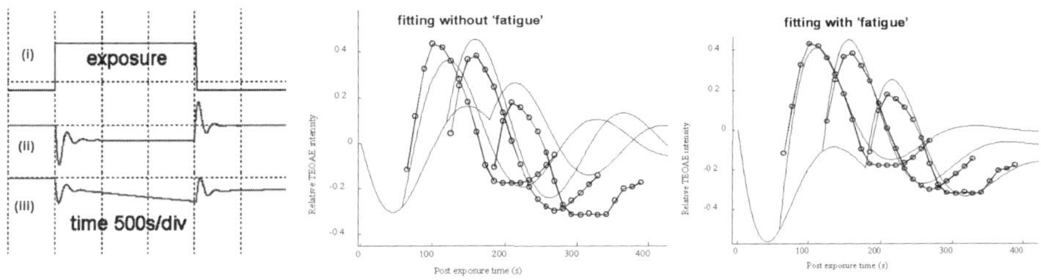

Figure 7. a) Left: (i) The exposure schematic as in figure 6. (ii) Response for gain 8. (iii) A reduction in the time constant of C3, R7 in Fig 5 to 1000s allows the negative feedback to fade causing a downward drift in the restored CGP. b) Center: Best '3 parameter' fit to an individual ear's data with no drift. c) Right: Best fit to same ear with drift and onset offset parameter differences allowed.

to electromotile or micro-mechanical factors. Individual parameters govern the response to over-stimulation and may relate to TTS susceptibility.

5 Conclusion

The post exposure oscillations seen in OAEs intensity are evidence of the action of a homeostatic system and have little to do with fatigue. They can be accurately modeled by a quasi-linear negative feedback system. In the model a cochlear parameter (CGP) important for sensitivity tends to be depressed by strong stimulation but the degree of depression is largely compensated for by a tight negative feedback mechanism with loop gain about 8 and a time constant of about 30 seconds. The feedback is under-damped and inherent delays result in oscillations at ~ 5mHz. Interference between oscillations in CGP initiated by the start and end of exposure explains the precise form of the OAE bounce [7]. The mechanism appears to act locally on exposed areas of the basilar membrane and not globally. We believe the decay time constant will relate to a chemical process associated with outer hair cell function (eg Ca^+ or K^+ recycling). We found that the feedback fatigues (or adapts) slowly (time constant ~1000s) causing an accumulation of sensitivity loss which may be related to TTS.

Acknowledgments

This work was partially supported by Otodynamics Ltd, UK.

References

1. Hirsh, I.J., Ward, W.D., 1952. Recovery of the auditory threshold after strong acoustic stimulation. J. Acoust. Soc. Am. 24 (2), 131-141.
2. Kemp, D.T., 1986. Otoacoustic emissions, travelling waves and cochlear mechanisms. Hearing Research 22, 95-104.
3. Kirk D.L., Patuzzi, R.B., 1997. Transient changes in cochlear potentials and DPOAEs after low-frequency tones: the 'two minute bounce' revisited. Hearing Research 112, 49-68.

4. Kirk D.L., Moleirinho A., Patuzzi, R.B., 1997. Microphonic and DPOAE measurements suggest a micromechanical mechanism for the 'bounce' phenomenon following low frequency tones. Hearing Research 112, 69-86.
5. O'Beirne, G.A., Patuzzi, R.B., 2002. Modelling the role of the outer hair cells in cochlear regulation and tinnitus. Proc 7th Intl. Tinnitus Seminar, pp. 62-67.
6. Kevanishvilli Z., et al., 2006. Behaviour of evoked otoacoustic emission under low-frequency tone exposure: Objective study of the bounce phenomenon in humans. Hearing Research 222, 62-69.
7. Kemp, D.T, Brill, O.J., 2007. Slow oscillatory cochlear adaptation to brief low frequency overstimulation: The human OAE 'bounce' effect revisited'. 30th ARO Midwinter Res Mtng., Abs. 412.

ADAPTIVE BEHAVIOUR SHOWN IN EAR-CANAL PRESSURES RELATED TO DISTORTION PRODUCT MAGNITUDES

E.L. LEPAGE, N.M. MURRAY

OAEricle Laboratory, P.O. Box 2564, Mt. Claremont, W.A. 6010, Australia.

A two-tone probe-masker experiment is conducted on human ears using a standard DPOAE probe sealed in the ear canal and the frequency-domain measures of response are correlated with time-domain measures. The onset of the probe tone burst of 25ms of fixed frequency (3kHz) and level (70dB SPL) was followed 8 ms later by a "masker" tone burst of 7 ms which varied over a grid from 0.5 octaves above and below the probe tone, and from 50 to 90 dB SPL. This grid of 13x9 tone complexes was repeated 10 times to test reproducibility. Ear canal pressure waveform displayed adaptive baseline shifts towards rarefaction and condensation. A previous presentation showed that, with onset and offset of the masker tone, the pressure waveform displayed two-tone-suppression-type contours for level versus frequency. In this presentation quadratic and cubic distortion products are compared with the variables derived from linear regression of baseline slope changes for the two onsets and two offsets. Scatter plots of the DPOAE levels with the associated baseline variables, for each of the 10 repetitions showed marked variability suggestive of history-dependent operating point change from one repetition to the next, suggesting an underlying Boltzmann characteristic. The frequency and time waveform variables showed a wide array of high correlations, suggesting that some aspects of distortion product generation and adaptive behaviour are directly related.

1 Introduction

1.1 Previous evidence supporting the notion of homeostatic regulation

In response to low level acoustic stimulation the activity of the outer hair cells (OHC) is postulated to boost high-frequency vibration of cochlear structures for ease of detection by the inner hair cells (IHC). More generally OHC motility may change the configuration of structures in the organ of Corti, but their role in doing so has received much less attention than fulfilling the need that they function as a cochlear amplifier. More generally still, the capabilities of the organ of Corti must be able to accommodate substantial percentage changes in the volume of key vessels, such as the OHC and IHC, and indeed scala media, so that hearing sensitivity is not lost due to the resulting biases. Indeed, not all measureable position shifts [1] are interpreted as outer hair cell (OHC) motile behavior: mechanical measurements by Flock have observed substantial "dc-shifts" in the motion of the Hensen cells signifying the presence of sound exposure-produced hydrops [2]. Quadratic distortion products (QDP) undergo baseline shifts and these have been interpreted as operating point shifts in OHC potentials, and treated as diagnostic of hydrops [3]. Transducer function characteristics have been derived on the basis of applying an external bias [4]. However, the question of the extent to which OHC adaptation is manifest in the ear canal pressure has only recently been addressed [5]. The question is relevant to the clinical application of transient-evoked otoacoustic emissions (EOAE) which typically contain considerable low frequency "noise" which is routinely removed by high-pass filtering around 700 Hz [6]. This noise is clearly not external to the ear but is origin is unknown.

1.2 Working hypothesis

Measurement noise in clinical otoacoustic emissions is due to adaptive OHC response, *hydropic* response, or both, with time courses not allowed for in the usual recording paradigm. If so, it may indicate that, far from improving the characterisation of active processes, the act of signal averaging of the OAE signal may smear and obscure the dynamic nature of the regulatory processes controlling emission levels. If the altogether-remarkable hypothesis that "most OAE noise is not noise" holds, then the significance of the approach should be clearly demonstrable from the data of any normally-hearing subject. Cochlear mechanical behaviour may appear nonlinear in a general sense whereas the underlying process may be primarily adaptive and vulnerable in nature with a time-course not allowed for in any experimental paradigm.

2 Methods

The experimental approach utilizes the same type of nonlinear residue extraction technique employed by Kemp [6] in transient (TEOAE) collection in comparing records obtained successively. A standard distortion product otoacoustic emission (DPOAE) probe is sealed in the human ear canal. A two-tone masking paradigm is used; a constant 25 ms probe tone of f_1=3 kHz at 70 dB SPL is repeated at 50 ms intervals. A masker tone of frequency f_2, 7 ms duration is added 9 ms after the start of the probe tone. Both bursts employ 1 ms rise/fall times. The masker is varied in frequency (probe ± 0.5 oct in 1/12 oct steps), and level (probe ± 20 dB in 5 dB steps). Each digitally-generated two-tone complex is immediately repeated, but with the repeat phase-reversed (see Fig.1a). The whole sequence-grid of these "phase-reversal pairs" of 9 levels and 13 frequencies is repeated ten times, taking 2 minutes, and the 10 response sequences was gathered with the intention of averaging them to eliminate external noise. Adult subjects (5 male, 4 female, whose audiometric and TEOAE characteristics

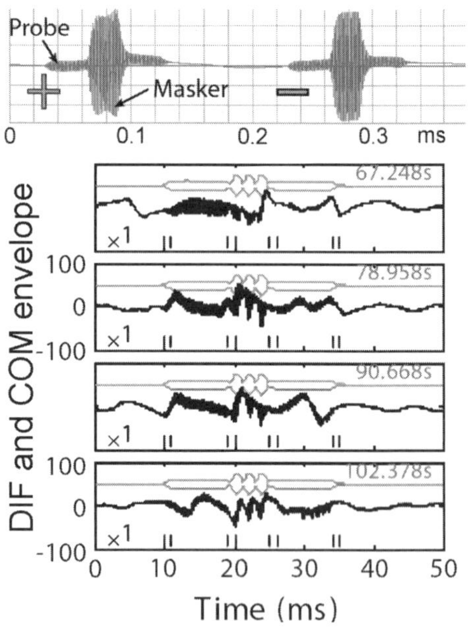

Figure 1 Protocol to examine time dependence of two tone interactions in human ear canal, using constant-probe/varying-masker paradigm, (a) basic stimulus immediately inverted, (b) 5 of 10 repetitions for one masker frequency and level.

are previously determined) are seated comfortably with head supported by a head rest. The signals are generated and recorded at 44.1 kHz by a NAL in-house-built signal conditioning and recording system incorporating a Lynx 4 channel, 22 bit sound card. Sound calibrations were carried out using a 2cc coupler and B&K 2604 amplifier. The sampled-waveform was high-pass filtered (corner frequency 110 Hz) to remove heart beat pressure pulsations.

The sum and difference of each phase-reversal pair is computed. The difference of these two pairs emphasizes what is common ("COM") to the pair which is primarily the delivered signal. This is represented in Fig. 1b by its envelope (Hilbert transform - grey line). The nonlinear residue ("DIF") is derived from summing the pair. This presentation is confined to analysis of the activity within segment during which both tones are present. Baseline pressure changes are determined from linear regressions on the time waveform to determine the dc-offset (intercept) plus rate of change (slope) of the DIF. Also computed are the integral and derivative of the smoothed DIF as well as the estimated asymmetry of the DIF about the smoothed value. The distortion products magnitudes are computed using a 2048-point Fast Fourier transform of the time waveform after first applying a Hanning window. Shown are the magnitudes of the spectral peaks at the expected frequencies of the various quadratic $|f_1 \pm f_2|$ and cubic products $|mf_1 \pm nf_2|$ (QDPs, CDPs). The results presented are from the DIF determination.

3 Results

Figure 1b shows for one subject, female normal hearing, strong transient emissions, a subset of the raw data obtained from the 2 minute sitting. The top traces in each panel show the "COM" signal (essentially the usually recorded DPOAE response) for the fixed probe clearly amplitude-modulated at the difference frequency f_2-f_1 (QDP). Shown are the 5 successive repetitions of the *unaveraged* responses of the "DIF" component which, by linear subtraction, emphasises the nonlinear residue. At first the records appear to be significantly affected by noise, previously considered external to the processes of interest. Closer examination shows a wide variety of behaviours which may be interesting in terms of OHC operating point control and cochlear homeostasis. The most important feature is the suggestion that, after each of the 117 tone complexes in the grid, returning to the same stimulus condition finds that the size of *ac* responses may vary significantly. Secondly the asymmetry of the waveforms varies markedly from one level to the next, from one repetition to the next (in between are all other frequencies and levels). Thirdly, some waveforms are modulated at the QDP frequency leading often to a rapid baseline rise, while others the frequency is at $2f_2$ [5] and highly asymmetric, mostly with an unchanging or falling dc value. All these factors suggest that the operating point is history dependent. To signal-average such data repetitions could smear across changes in active state.

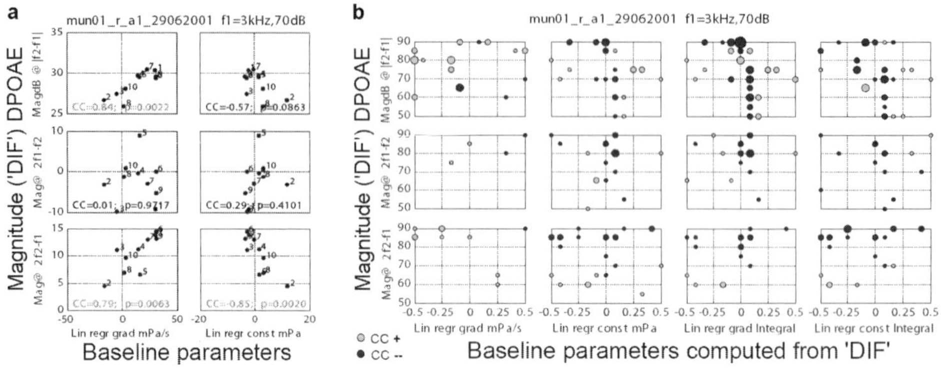

Figure 2(a) Cross-correlating quadratic and 1st order cubic distortion products against baseline adaptive behaviour, characterised by linear regression of the waveform during the segment when the masker is present, viz using slope and constant. (b) Each box represents a grid of all the masker frequencies (oct relative to probe) and levels (dB SPL). There are frequent incidences of high correlation between the magnitude of the DPOAEs and the adaptive behavior (here also included regression of the integral of the baseline variation).

Figure 2a shows the result of cross correlating the DIF parameters computed from the baseline (time) variables for just the two-tone segment versus the spectral for the same time segment. In each subplot are shown 10 points representing the values of the variables chosen for the 10 repetitions of the two-tone stimulus. The two columns represent, respectively, 1) the gradient of the DIF component, and 2) the linear regression constant (baseline starting value). Descending subplot rows represent 1) the DIF magnitude of the $|f_2\text{-}f_1|$ peak (for half the repetitions $f_2<f_1$), 2) the DIF magnitude of the $2f_1\text{-}f_2$, and 3) the DIF magnitude of the $2f_2\text{-}f_1$ peak. In each subplot, the correlation coefficient and its significance (p-val) is listed within the box. As commented above it should not be all that remarkable that many of the cross-comparisons for the same data are highly correlated. On the other hand, that which *is* remarkable is that these variables are correlated despite their wildly fluctuating behaviors as is indicated by the number of the repetition shown against each point. Considering that the baseline parameters contain information which does not occur at the DP frequencies, the fact that these correlations can be high, suggests that for a subset of stimulus conditions distortion product magnitudes may be related to outer hair cell operating point, although we are only able to observe the small deviations because of the middle ear band-pass characteristic and the filtering applied.

Each of the subplots of Figure 2b charts statistical outcomes for masker (f_2) level versus masker frequency (oct) relative to the probe (f_1) frequency. Every point in each box represents a significant cross-correlation between DP strength and a baseline parameter. For the complete set of 4212 comparisons, 664 are significant as listed in each panel shown. The significance level of each point is represented by its size - (small $p<0.05$; medium $p<0.01$ and large $p<0.001$). No correction is applied since each point represents an independent stimulus condition. Positive and negative correlations are

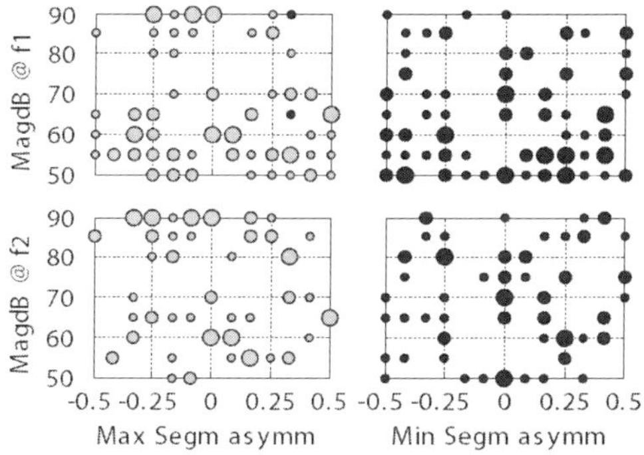

Figure 3 Estimates of asymmetry of the time waveforms require examining the responses to the primary tones. Correlating the DIF values for the magnitude of the primaries f_1 and f_2 plotted versus time waveform asymmetry during the nonlinear interaction. Again, the level of significance is represented by the size of the symbol, the direction of the correlation by the shading.

shaded differently. Clusters with correlations of one polarity are often flanked by others of opposite polarity, while high correlations still occur at the lowest levels. Also significant were indications of positive and negative waveform asymmetry of the DIF *ac* response (Fig. 3), again suggestive that these responses are displaying waveform distortion due to saturation. The tendency to clustering of the responses according to the predominance of one primary level or the other may suggest some form of level-dependent competition between the primaries.

4 Discussion

The sensitivity of the technique for revealing these correlations may be tied to the nonlinear ("DIF") extraction technique, which also is responsible for showing that the response to the probe signal is not always completely cancelled. Although ten repetitions of the signal sequence were collected to allow for signal averaging (as was previously used to show two-tone suppression contours on these same data [5]), in *none* of the OAE probe data presented here has signal averaging been used to improve the signal-to-noise ratio (SNR). After viewing the raw data it was evident that individual repetitions (interspersed by all other frequencies and levels in the masker grid) exhibited responses which appeared to represent the cochlea in different states, in which case averaging would merely obscure important phenomena. The various forms of high-pass filtering in this human measurement removed the possibility of actual operating point determination, yet the responses suggest that the resulting waveforms were nevertheless dominated by the current effective state, which was found to be different each time the stimulus conditions were repeated, i.e. strongly history-dependent.

This analysis has followed the realisation that these data may not be as strongly contaminated by external sources of noise as previously thought. This surprising outcome cannot simply be due to the expectation that spectral parameters should correlate to time-series estimates - if this were the case all 4212 comparisons would be significant. Instead, significance occurs in clusters which have meaning for 2TS and DP generation. These data test the hypothesis that clinical algorithms used to improve the SNR may actually obscure vital behavior necessary to fully characterise the active processes. The "noise" previously deemed to be external to the process being characterised (e.g. strength of the $2f_1$-f_2 DPOAE) is instead intimately related to it. While we have no complete description of what is causing these changes in state, at least some of the masker levels will be perturbing the ear generating fatigue which is strongly associated with operating point shift changing over the whole range of the transducer function. Very often it is in a linear central region, or alternately saturated above and below this mid-point. Subsequent baseline movement appears to be controlled by processes invoked when saturation result in strong asymmetry of the ac waveform. These data further support the notion of feedback-control of operating point.

5 Conclusions

1) These results suggest that it may be possible to directly access cochlear homeostatic processes remotely from ear-canal pressure.
2) Very high correlations have been found between baseline "noise" in the unaveraged signal and the magnitude of DPOAEs, particularly the QDP.
3) Lack of taking sufficient account of such a control mechanism could also account for our difficulties with comparing TEOAEs in longitudinal data.
4) Signal averaging may be inherently destructive of operating point information producing the statistical variability evident in clinical data.

Acknowledgments

This work was carried out at the National Acoustic Laboratories under project HLPR98.1, with technical assistance from J. D. Seymour. We thank E.S. Olsen and W.W. Jedrezejczak for comments on this manuscript.

References

1. LePage EL., 1987. Frequency-dependent self-induced bias of the basilar membrane and its potential for controlling sensitivity and tuning in the mammalian cochlea. J.Acoust.Soc.Am.; 82:139-54.
2. Flock A, Flock B., 2000. Hydrops in the cochlea can be induced by sound as well as by static pressure. Hear Res, Dec;150(1-2):175-88.
3. Sirjani DB, Salt AN, Gill RM, Hale SA., 2004. The influence of transducer operating point on distortion generation in the cochlea. J Acoust Soc Am Mar;115(3):1219-29.
4. Bian L, Chertoff ME, Miller E., 2002. Deriving a cochlear transducer function from low-frequency modulation of distortion product otoacoustic emissions. J. Acoust Soc Am., Jul;112(1):198-210.
5. LePage, E. L. ; Murray. N.M.; Seymour, JD., 2005. Novel otoacoustic baseline measurement of two-tone suppression behaviour from human ear-canal pressure. In Proceedings of the ninth "Mechanics of Hearing" workshop; OHSU, Portland, OR.
6. Kemp DT., 1979. Evidence of mechanical nonlinearity and frequency selective wave amplification in the cochlea. Arch Otorhinolaryngol., 224(1-2):37-45.

THE INFLUENCE OF LANGUAGE EXPERIENCE ON CONTRALATERAL SUPPRESSION OF CLICK-EVOKED OTOACOUSTIC EMISSIONS IN YOUNG ADULTS

SHAUM P. BHAGAT, JINGJING XU

School of Audiology and Speech-Language Pathology, The University of Memphis, 807 Jefferson Avenue, Memphis, Tennessee 38105, U.S.A.

Click-evoked otoacoustic emissions (CEOAEs) were recorded from the right ear of eight native Mandarin speakers and eight native English speakers in conditions with and without a contralateral suppressor. The CEOAE stimuli consisted of linear clicks presented at 65 dB peSPL. Two types of 70 dB SPL contralateral suppressors (white noise and synthesized Mandarin tone) were utilized. The amount of contralateral suppression was quantified in nine 2-ms time bands. No significant differences in contralateral suppression occurred between language groups for the white noise suppressor. Notably, significantly more contralateral suppression was evoked by the synthesized Mandarin tone in native Mandarin speakers compared to native English speakers.

1 Summary

In tone languages, differences in lexical meaning are conveyed by a change in pitch at the syllable level. Mandarin Chinese is a language that utilizes contrasting lexical tones with different pitch contours. Brain imaging [1] and evoked potential [2] studies have demonstrated differences in cortical activation between native Mandarin and native English speakers in the auditory processing of Mandarin tones. Krishnan *et al.* [3] reported that these language group differences in the neural encoding of lexical tones are also evident at the level of the rostral brainstem. The aim of this study was to examine if language experience modulates auditory efferent activity as measured via the contralateral suppression of click-evoked otoacoustic emissions (CEOAEs).

The subjects were sixteen female young adults (m=26.8 years, sd = 4.2). Eight of the subjects were native English speakers with no knowledge of Mandarin Chinese or any other tone language. The remaining eight subjects were native Mandarin speakers. All of the subjects indicated a right-hand preference. Each of the subjects had normal hearing, defined as pure-tone thresholds at or better than 15 dB HL from 250-8000 Hz, and exhibited normal tympanometry and normal CEOAEs. Normal CEOAEs were defined as having a stimulus stability of at least 70% and overall waveform reproducibility exceeding 80%. During the contralateral suppression (CS) paradigm, CEOAEs were evoked from the right ear by linear click trains, calibrated in the ear canal at 65 dB peSPL. The CS stimuli had equal total durations, and consisted of a white noise burst and a synthesized high-falling Mandarin monosyllable, each presented to the left ear at 70 dB SPL. CEOAE recordings were interleaved with and without the CS stimuli, and the order of presentation was randomized. Each CEOAE waveform consisted of 260 low-noise samples.

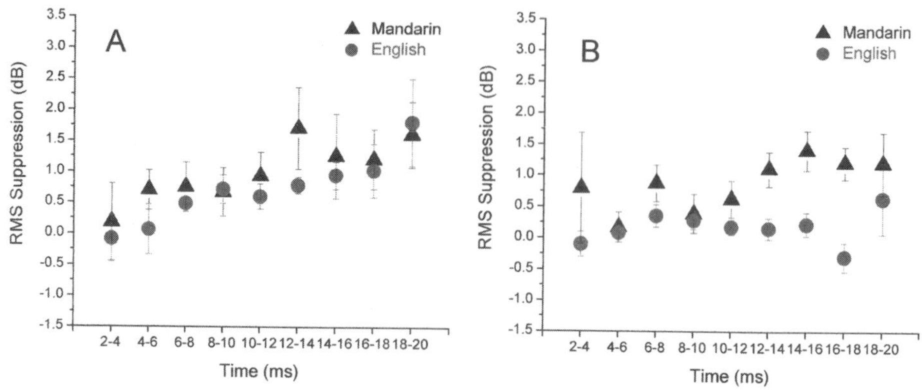

Figure 1. Mean RMS suppression across nine 2-ms time bands for the white noise suppressor (panel A) and Mandarin tone suppressor (panel B). Error bars= 1 SE of the mean.

The amount of RMS suppression for each type of CS stimulus was calculated in each subject in nine 2-ms time bands using the Kresge Echomaster 4.0 software program. Mean amounts of RMS suppression for each language group are plotted in Figure 1.

No significant (p=0.453) language group difference in the amount of RMS suppression of CEOAEs occurred for the white noise condition, but the synthesized Mandarin tone evoked significantly more (p=0.01) RMS suppression of CEOAEs in native Mandarin speakers compared to native English speakers. These language group differences were most prominent in the 10-20 ms range. The results of this study support previous investigations that have found physiological differences between native Mandarin and English speakers in the auditory processing of Mandarin lexical tones. These preliminary findings suggest that the neural mechanisms shaped by linguistic experience may extend to the caudal brainstem, and may influence control of cochlear micromechanics. This interpretation is consistent with the accumulating evidence of experience-dependent plasticity that is manifested in the human auditory system.

Acknowledgments

We wish to thank Dr. Ravi Krishnan, of Purdue University for providing us with the synthesized Mandarin tone stimuli that were used in this investigation.

References

1. Klein, D., Zatorre, R., Milner, B., Zhao, V., 2001. A cross-linguistic PET study of tone perception in Mandarin Chinese and English speakers. Neuroimage, 13 (4) 646-653.
2. Chandrasekaran, B., Krishnan, A., Gandour, J., 2007. Mismatch negativity to pitch contours is influenced by linguistic experience. Brain Res., 1128 (1) 148-156.
3. Krishnan, A., Xu, Y., Gandour, J., Cariani, P., 2005. Encoding of pitch in the human brainstem is sensitive to language experience. Cogn. Brain Res., 25 (1) 161-168.

AMPLITUDE AND FREQUENCY MODULATIONS OF SPONTANEOUS OTOACOUSTIC EMISSIONS

LIN BIAN

Auditory Physiology Lab, Dept. of Speech and Hearing Science, Arizona State University Tempe, AZ 85287-0102, USA

It has been speculated that the spontaneous otoacoustic emissions (SOAEs) are associated with the mechanical feedback from the cochlear outer hair cells. In humans, the amplitudes and frequencies of SOAEs could be modulated by a low-frequency bias tone. The effects on the SOAE magnitudes were an amplitude modulation and a suppression. In the spectral domain, there was an upward shift of the SOAE frequencies with the bias tone level. In the time domain, variations of the SOAE amplitudes and frequencies followed the bias tone phase. Increasing the biasing pressure in either direction reduced the SOAE amplitudes and elevated the frequencies. The amplitude modulation pattern was consistent with the first derivative of a sigmoid-shaped nonlinear function representing hair cell transduction. Both amplitude and frequency modulations of SOAEs indicate that the nonlinear transducer characteristics and mechanical properties of the cochlear hair cells can influence the SOAE generation.

1 Introduction

Our inner ears can produce soft sounds even without any acoustic input. These self-generated tonal sounds or spontaneous otoacoustic emissions (SOAEs) are related to normal hearing and diminished with age [1]. The SOAEs are distributed in low- to mid-frequency range where the arrangement of cochlear outer hair cells (OHCs) or the efferent innervations to these cells are more or less irregular [2,3]. These emissions are possibly produced by the mechanical feedback of these OHCs' activities that are thought to enhance hearing sensitivity [1,4]. A standing wave could also form between the stapes and the generating sites when the OHC vibrations propagate towards the ear canal [5]. In any case, the OHC is an essential element in the generation of SOAEs.

The major function of the OHC is a bi-directional transduction, i.e., it senses the acoustical input and controls the cochlear partition vibration to optimize the neural transmission in the inner hair cells. A key feature of these reciprocal transducers is nonlinearity, because stimulated by two tones the inner ear can generate distortion products (DPs) and create mutual suppression [6]. These nonlinear effects are also observed with SOAEs when an external tone is presented to the ear [7]. In addition to suppression and distortion, the SOAE frequencies can be shifted with changes in cochlear conditions, such as inner ear fluid pressure [8]. Recently, a low-frequency biasing technique has been used to assess the dynamical nonlinearity of the inner ear [9,10], because the bias tone can shift the cochlear partition and produce an amplitude modulation (AM) of the DPs. To probe the SOAE generation, effects of low-frequency biasing on SOAEs were measured in humans. Here, various dynamic effects are reported and possible mechanisms explored.

2 Methods

Eleven normal-hearing subjects with SOAE magnitudes more than 20 dB above noise floor (ILO-92, Otodynamics) participated in the study. Their middle ear and outer ear functions were normal based on an otoscopic examination and a middle ear immittance measure. Their SOAEs were recorded with a probe microphone (Etymōtic Research, ER-10B+) while presenting a bias tone with various signal conditions.

2.1 Bias Tones

The bias tone was generated and controlled by a software written in LabVIEW (National Instruments, NI), then presented with an earphone (JVC HA-FX55) connected to the ER-10B+. The bias tone frequencies (f_{bias}) were 25, 32, 50, 75, and 100 Hz. The bias tone level (L_{bias}) was automatically attenuated from its maximal value to 0 in 41 steps. The maximal level ranging from 3 to 4 Pa was determined by increasing the L_{bias} until a significant reduction of the largest SOAE was observed. The bias tone was 1 s long to result in a 1-Hz frequency resolution in the frequency analysis.

2.2 Signal Processing and Data Analysis

To analyze the behaviors of SOAEs under low-frequency biasing, the ear-canal acoustical signal was processed in three steps using Matlab, namely 1) spectral analysis, 2) linked-window analysis, and 3) temporal feature extraction.

1. General spectral analysis: Features of the SOAEs and their sidebands were obtained from a fast Fourier Transform (FFT) of the entire signal.
2. Linked-window analysis: Windows fixed at the peaks and troughs of the bias tones with various amplitudes were linked and submitted to an FFT to obtain the amplitude and frequency of the SOAEs. Linking these windows avoided significant reductions in frequency resolution of the FFT. Each fixed-window length was adjusted between ¼ ~ ½ biasing period to yield a maximal peak at an SOAE frequency.
3. Temporal feature extraction: The spectral contents around an SOAE were windowed and inverse fast Fourier Transformed (IFFT) to obtain an analytic signal. The absolute value of the signal yielded the SOAE amplitude envelope. The phase derivative with respect to time gave rise to the instantaneous SOAE frequency. Both instantaneous amplitude and frequency were averaged over multiple biasing cycles to provide a period modulation pattern for each measure.

3 Results

3.1 Spectral Fine-Structures

The general characteristics of the SOAEs in the subjects were described elsewhere [11]. The present paper focuses on the spectral fine-structures around the SOAE frequencies (Fig. 1). Three effects of the bias tone were observed: 1) suppression of SOAE amplitude, 2) generation of sidebands, and 3) upward shift of SOAE frequency.

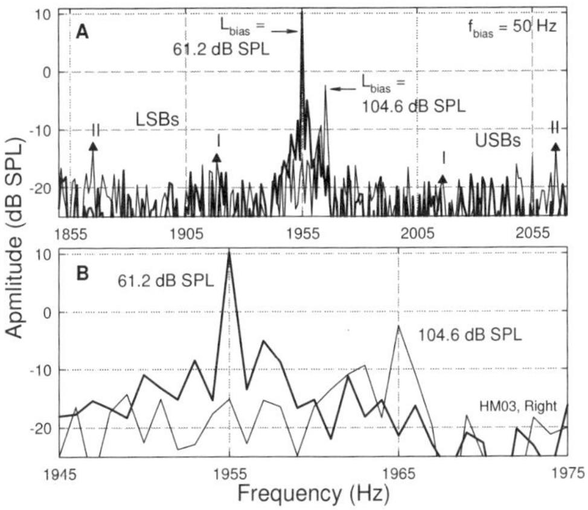

Figure 1. Effects of bias tone on an SOAE. A: the thick line is the SOAE spectrum when the bias tone is ineffective. The thin line shows three biasing effects: reduction in SOAE amplitude, increase in SOAE frequency, and presence of multiple sidebands (▲). The frequency intervals between the sidebands and the SOAE equal to integer multiples of the f_{bias} (50 Hz). The sidebands below and above the SOAE frequency are termed lower and upper sidebands (LSBs, USBs), respectively. B: a zoomed-in view of panel A showing an amplitude suppression and a frequency increase of the SOAE.

As the L_{bias} increased, the SOAE magnitude and frequency showed progressive changes (Fig. 2). The first noticeable change was a frequency increase that occurred at L_{bias} about 75 dB SPL. Then the lower and upper sidebands (LSBs and USBs) became measurable around 85 dB SPL. Finally, the SOAE amplitude showed a significant reduction when the L_{bias} was above 100 dB SPL. The presence of sidebands at integer multiples of f_{bias} indicated temporal modulations of both SOAE amplitude and frequency. Usually, the sidebands closer to the SOAE were larger. Often, the magnitudes of LSBs were greater than the corresponding USBs.

3.2 Quasi-static Modulation Patterns

The SOAE magnitudes and frequencies measured with the linked-window FFT gave rise to the quasi-static modulation patterns. It reflects a "static" change in SOAE when the cochlear partition is biased in either positive or negative sound pressure direction. The quasi-static modulation patterns of SOAE amplitude and frequency showed opposite behaviors (Fig. 3). The SOAE magnitude demonstrated a "bell-shaped" pattern, whereas the frequency displayed a "U" shape. In other words, the SOAE amplitude was reduced at high biasing pressures in both positive and negative directions, while its frequency

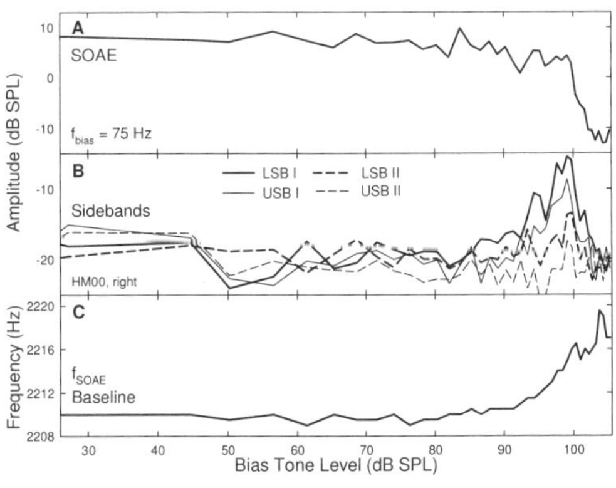

Figure 2. Changes of SOAE amplitude, sidebands, and frequency as functions of L_{bias}. A: suppression of SOAE amplitude at high L_{bias} (> 100 dB SPL). The amplitude fluctuates in the L_{bias} range between 50 and 100 dB SPL. B: rises of the sidebands. Sidebands become measurable and increase between 85 and 100 dB SPL L_{bias}. The sidebands roll over when SOAE magnitude is significantly suppressed. C: frequency shift. The SOAE frequency fluctuates at 50 dB SPL and continuously increases above 75 dB SPL L_{bias}. The SOAE frequency reaches the maximum when its amplitude is diminished.

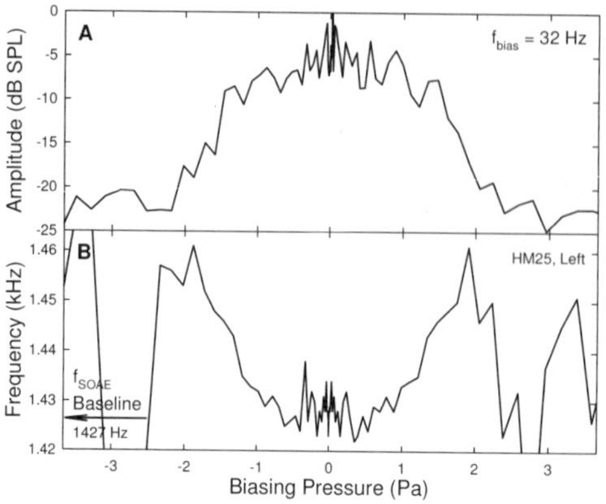

Figure 3. Quasi-static modulation patterns. A: the SOAE amplitude is suppressed while the cochlear partition is biased in either positive or negative sound pressure direction. The SOAE amplitude fluctuates around its maximal value near zero biasing pressure. B: the SOAE frequency is elevated when the biasing pressure is increased in both directions. The SOAE frequency fluctuates at its baseline value around zero biasing pressure. Note: further increase in biasing pressure (> ±2.5 Pa) diminishes the SOAE amplitude causing large fluctuations in its frequency.

peaked. When the SOAE was diminished at the extreme biasing pressures, its frequency became immeasurable marked by large fluctuations. The maximal reduction of SOAE magnitude was up to 20 dB and the elevation of frequency could be as high as about 30 ~ 40 Hz. It was noted that both the amplitude and frequency were unstable when the biasing pressure was below ±0.5 Pa.

3.3 Period Modulation Patterns

The temporal modulation patterns of SOAE amplitude were obtained from conditions with relatively high bias tone levels where the AM was the dominant effect. For SOAE frequency, the modulation pattern was derived at moderate biasing levels, because the magnitude was relatively stable and the frequency estimate was more reliable.

Both the SOAE amplitude and frequency were modulated depending on the bias tone phase (Fig. 4). The amplitude pattern contained two "bell-shaped" peaks each corresponding to a zero-crossing of the bias tone. At the biasing peak and troughs, the

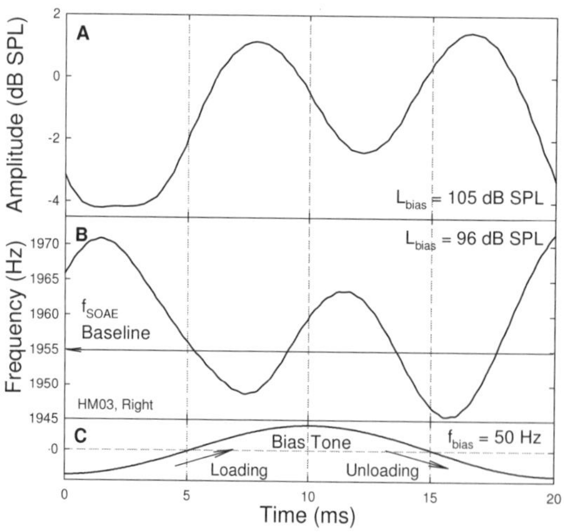

Figure 4. Period modulation patterns. A: SOAE amplitude shows a double-peaked pattern. Each peak corresponds to a zero-crossing of the bias tone. Both positive and negative biasing extremes suppress the SOAE. B: instantaneous SOAE frequency demonstrates a "W" pattern with elevations at the biasing peaks or troughs. The frequency drops below its baseline near the zero-crossings of the bias tone. The phase-dependent effects occur with a short time delay with reference to the bias tone in panel C. Note: reduced effects at the bias tone peak. C: the corresponding bias tone. Loading and unloading: increase and decrease in biasing pressure.

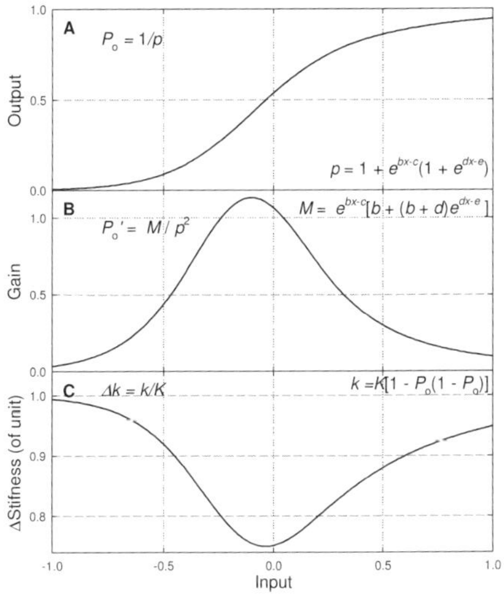

Figure 5. A nonlinear transducer model. A: a transducer function in the form of a second-order Boltzmann function represents the open probability (P_o) of OHC transduction channels. The operating feature is a saturating nonlinearity. Parameters used: b = -1.7, c = -1.3, d = -4, and e = 0.15. B: the first derivative of the Boltzmann function or P_o' showing a "bell" shape. The derivative reflects the slope or the gain of the OHC transducer nonlinearity P_o. C: calculated transduction channel gating stiffness k expressed as a change from the unit value K. The stiffness reaches a minimum when 50% of the channels are in open state. The transducer apparatus becomes more stiffened at the two displacement extremes. Note: the asymmetry in these features of the cochlear transducers. This nonlinear model reflects the averaged transfer characteristics of a group of OHCs contributing to the production of an SOAE.

SOAE was reduced. The suppression was asymmetrical, because the effect was smaller at the biasing peak, i.e., the depth of the AM was larger at the biasing trough. However, the SOAE frequency showed a completely opposite modulation pattern. The frequency was increased at the biasing extremes and reduced below its baseline at the zero-crossings of the bias tone. The frequency excursion was higher than the frequency shift measured from the overall FFT (Fig. 2C). It should be noted that these effects occurred with a short time delay (< 2 ~ 3 ms) with respect to the bias tone phase.

4 Discussion

Low frequency biasing of SOAE produces several fast-acting effects that occur within a few milliseconds. Thus, any efferent-mediated responses in the inner ear and the middle ear could be ruled out [12]. The suppression and modulation of SOAE amplitudes are consistent with the well-observed two-tone interactions in the inner ear, namely two-tone suppression and intermodulation distortion [6]. Indeed, the DPs generated from two-tone stimulation could be viewed as a spectral representation of a nonlinear AM with the beat frequency [13]. It is most likely that the low-frequency biasing effects on SOAEs initiate from structures within the cochlea.

The structural and functional basis for two-tone interaction is the nonlinearity in hair cell transduction [14]. Mechanical two-tone responses observed in cochlear partition vibrations and in ear-canal acoustics are possible consequences of the reciprocal transduction of the OHCs. The OHC transduction is highly nonlinear, and has been described by a three-state model [15]. The open probability (P_o) of hair cell transduction channels is expressed by a Boltzmann function with a saturating nonlinearity (Fig. 5A): $P_o = 1/p$, and $p = 1 + e^{bx-c}\cdot(1 + e^{dx-e})$, where x is the hair bundle displacement, (b, d) are sensitivity parameters, and (c, e) are constants relating to the resting position. The gain of the cochlear transducer or the first derivative of the Boltzmann function shows a "bell" shape (Fig. 5B) similar to the AM pattern of SOAEs. Thus, the modulation of SOAE amplitude reflects the gain control in OHC transducers, since a surge in biasing pressure would saturate the transduction channel and reduce the gain.

An SOAE may result from the mechanical activities of a group of OHCs that operate with the highest gain near hearing threshold. Its frequency is determined by the resonant frequency of the cochlear partition which is proportional to the stiffness ($\sqrt{k}$) of the OHCs [4]. Based on the gating compliance model [15], the stiffness is nonlinear and can be estimated as a change from a unit value K: $k = K\,[1 - P_o(1 - P_o)]$. The stiffness change (Fig. 5C) demonstrates a "U" shape that is similar to the SOAE frequency modulation pattern. Plus, adjusting the transducer gain and adaptation of transduction channel gating require a limited time [16]. Therefore, a time-dependent nonlinear transducer mechanism may play an essential role in the generation of SOAEs.

Acknowledgments

This is work was supported by NIH/NIDCD grant DC006165.

References

1. Probst, R., Lonsbury-Martin, B. L., Martin, G. K., 1991. A review of otoacoustic emissions. J. Acoust. Soc. Am. 89, 2027-2067.
2. Lonsbury-Martin, B. L., Martin, G. K., Probst, R.,Coats, A., 1988. Spontaneous otoacoustic emissions in a nonhuman primate. II. Cochlear anatomy. Hear. Res. 33, 69-94.
3. Thiers, F. A., Burgess, B. J., Nadol, Jr. J. B., 2002. Reciprocal innervation of outer hair cells in a human infant. J. Assoc. Res. Otolaryngol. 3, 269-278.
4. Dallos, P., 1996. Overview: Cochlear neurobiology. In: Dallos, P., Popper, A. N., Fay, R. R., (Eds.), *The Cochlea.* Springer, NY, pp. 1-43.
5. Shera, C. A., 2003. Mammalian spontaneous otoacoustic emissions are amplitude-stabilized cochlear standing waves. J. Acoust. Soc. Am. 114, 244-262.
6. Robles, L., Ruggero, M. A., 2001. Mechanics of the mammalian cochlea. Physiol. Rev. 81, 1305-1352.
7. Long, G. R., Sun, C., Talmadge, C. L., 1993. Interaction between spontaneous otoacoustic emissions and external tones: suppression, frequency shifting and distortion product generation. In: Duifhuis, H., Horst, J. W., van Dijk, P., van Netten, S. M., (Eds.), *Biophysics of Hair Cell Sensory Systems.* World Scientific, Singapore, pp. 40-46.
8. de Kleine, E., Wit, H. P., van Dijk, P., Avan, P., 2000. The behavior of spontaneous otoacoustic emissions during and after postural changes. J. Acoust. Soc. Am. 107, 3308-3316.
9. Bian, L., Chertoff, M. E., Miller, E., 2002. Deriving a cochlear transducer function from low-frequency modulation of distortion product otoacoustic emissions. J. Acoust. Soc. Am. 112, 198-210
10. Bian, L, and Scherrer, N. M., 2007. Low-frequency modulation of distortion product otoacoustic emissions in humans. J. Acoust. Soc. Am. 122, 1681-1692

11. Bian, L, and Watts, K. L., 2008. Effects of low-frequency biasing on spontaneous otoacoustic emissions: Amplitude modulation. J. Acoust. Soc. Am. 123, 887-898.
12. Cooper, N. P., and Guinan, Jr., J. J., 2003. Separate mechanical processes underlie fast and slow effects of medial olivocochlear efferent activity. J. Physiol. 548, 307-312.
13. Brown, A. M., 1994. Modulation of the hair cell motor: A possible source of odd-order distortion. J. Acoust. Soc. Am. 96, 2210-2215.
14. Geisler, C. D., Yates, G. K., Patuzzi, R. B., Johnstone, B. M., 1990. Saturation of outer hair cell receptor currents causes two-tone suppression. Hear. Res. 44, 241-256.
15. Markin, V. S., Hudspeth, A. J., 1995. Gating-spring models of mechanoelectrical transduction by hair cells of the internal ear. Annu. Rev. Biophys. Biomol. Struct. 24, 59-83.
16. van der Heijden, M., Joris, P. X., 2005. The speed of auditory low-side suppression. J. Neurophysiol. 93, 201-209.

Comments and Discussion

Shera: Very interesting. If I understand your model, though, doesn't it predict SOAE frequency fluctuations much larger than those you observe? The model predicts frequency fluctuations df/f = 0.5 dk/k, where k is the stiffness. From Fig 5, dk/k is about 0.25 so the predicted df/f is about 12.5%. The SOAE data in Fig. 4, though, suggest that df/f is less than 1%. Any idea what might account for the discrepancy?

Bian: Your interpretation is correct. The gating stiffness model only attempted to provide a qualitative description of the underlying mechanism for the SOAE frequency behavior under the acoustic biasing. However, the prediction could be made more precise if the macroscopic nature of the SOAEs is considered. The SOAE frequency as a measure in the ear canal reflects a mechanical property of the cochlear partition at the generating site which includes not only the hair bundles but also the cell bodies, the supporting cells, the tectorial membrane, and the basilar membrane. In fact, the gating stiffness contributes to only about 30~40% of the hair bundle stiffness (Kros, 1996). Assuming that a relatively large constitute of 50% of the cochlear partition stiffness is from the hair cell bundles, this could bring the contribution of the gating springs down to 15~20%. This estimate is consistent with a 20% reduction in cochlear partition stiffness by disrupting the hair bundle tip-links in an in vitro preparation of gerbil cochlea (Chan and Hudpeth, 2005). Therefore, the reduction of gating stiffness by about 12.5% corresponds to roughly 2% frequency change in the SOAEs which is in close agreement with the present observation.

C. J. Kros (1996) In: The Cochlea, P. Dallos, A. N. Propper, and R. R. Fay eds., Springer, NY, pp.318.
D. K. Chan and A. J. Hudspeth (2005) Biophys. J. 89, 4382.

SHIFTING THE OPERATING POINT OF COCHLEAR AMPLIFICATION? IMPACT OF LOW FREQUENCY BIASING AND CONTRALATERAL SOUND STIMULATION ON DPOAE

ANNA WITTEKINDT, CORNELIUS ABEL, MANFRED KÖSSL

Institute for Cellbiology and Neuroscience, Goethe University Frankfurt
Siesmayerstr. 70, 60323 Frankfurt, Germany
e-mail: Anna.Wittekindt@bio.uni-frankfurt.de

The mammalian efferent medial olivo-cochlear system is known to modulate active amplification of low-level sound in the cochlea. We investigated the effect of contralateral acoustic stimulation (CAS), known to elicit efferent activity, on distortion product otoacoustic emissions (DPOAEs) in the gerbil and, in second approach, biased the position of the cochlear partition and hence the operating point of the cochlear amplifier periodically by a low frequency tone (5 Hz). The study focussed on the quadratic distortion product f2-f1 that is sensitive to changes in the operating point of the amplifier transfer function. During CAS, a significant increase of the amplitude of f2-f1 was found while 2f1-f2 was less affected. Biasing by the low frequency tone resulted in a phase related amplitude modulation of f2-f1. This modulation pattern was changed pronouncedly during CAS, in dependence on the CAS-level. The current results suggest that efferent effects on DPOAEs might be produced by changes in the operating point of the cochlear amplifier and were in good agreement with a simple model based on a Boltzman function.

1 Introduction

The mammalian organ of Corti is densely innervated by efferent fibers, which originate in the brain stem and, through synapses on outer hair cells, can modulate the cochlear mechanics (reviews: [1] and [2]). Activation of the olivo-cochlear bundle has a substantial effect on DPOAEs that are generated as a byproduct of nonlinear mechanical amplification of low level sound by outer hair cells [3]. The cubic DPOAE 2f1-f2 (CDT) depends on odd components, whereas the quadratic f2-f1 emission (QDT) relies on even components of the cochlear amplifier transfer function and is sensitive to changes in the operating point (OP) of the transfer function if the OP is situated near the point of inflection [4-6].

For a better understanding of the impact of the efferent system on the cochlear amplifier transfer characteristics, the present study investigates the effect of contralateral broadband noise stimulation, known to elicit efferent activity [7], focusing on the quadratic f2-f1 distortion product. To test the hypothesis, that efferent activity can shift the OP of the cochlear amplifier we additionally biased the position of the cochlear partition and hence the OP of the cochlear amplifier periodically by a low frequency tone (5 Hz) under different conditions of contralateral acoustic stimulation (CAS). The results are compared qualitatively with a simple model using a Boltzman function to simulate the consequences of OP shifts on the distortions.

2 Methods

For DPOAE recordings, 9 mongolian gerbils of both sexes were lightly anesthetized with ketamine and xylazine and placed on a heating pad to keep body temperature constant. The acoustic system incorporated two reverse driven condenser microphones (B&K 4133, ½"), a headphone loudspeaker (DT 880, Beyerdynamic) for presentation of the low-frequency biasing tone, and a microphone (B&K 4190, ½") for measuring the acoustic signal at the tympanum (for details see [4]). For contralateral acoustic stimulation (CAS) a calibrated reverse driven microphone (B&K 4133) was introduced into the contralateral meatus. The microphone signal was amplified and sampled at 100 kHz. A two-tone stimulus (f1 = 5.47 kHz; f2 = 7 kHz; levels L1 = L2 = 20-60 dB SPL) was used to evoke DPOAEs. In the biasing-experiments, a 5 Hz bias tone was additionally presented at various sound pressure levels (90-110 dB SPL). For CAS, broad band noise (BBN, levels between 10-70 dB SPL) was used. DPOAE magnitudes were calculated with a moving window method for FFT extraction.

To assess the dependence of cochlear distortions on the OP of cochlear transducer functions, the output of a Boltzman function (Eq. (1) with $x_1 = x_2 = -0.06$; $a_1 = 3a_2 = 12.8$, s = variable OP shifts, see [4]) for a double sinusoidal input was calculated.

$$y = \frac{1}{1 + e^{a_2(x_2 - x - s)}\left(1 + e^{a_1(x_1 - x - s)}\right)} \quad (1)$$

The input to this nonlinear function was additionally modulated by a low frequency sinus (5 Hz), resembling the low frequency biasing in the experiments. The magnitudes of the f2-f1 distortion product was extracted from the output of the Boltzman function using the same analysis procedure as for the experimental data.

3 Results

3.1 Basic effects of contralateral noise on f2-f1 and 2f1-f2

DPOAEs recorded during continuous two-tone stimulation and in the absence or presence of contralateral broad band noise are shown in the example in Fig. 1A. During the period of CAS, an increase of the level of the quadratic f2-f1 distortion was observed in all animals at ipsilateral stimulus levels between 30-55 dB SPL. In two animals an additional f2-f1 level decrease was observed for certain stimulus levels. The f2-f1 enhancement was dependent on the contralateral noise level. The cubic 2f1-f2 distortion was not or much less affected by CAS. Analysis of the pooled data (Fig. 1B) of the CAS induced DPOAE level increase (compared to the period before CAS) yielded a significant QDT level shift for CAS levels of 30 dB SPL and above with maximum changes of about 4 dB and only a very slight suppression of 2f1-f2.

The observation of a pronounced influence of CAS on the quadratic f2-f1 DPOAE, which is known to be a sensitive parameter for the operating state of the cochlear

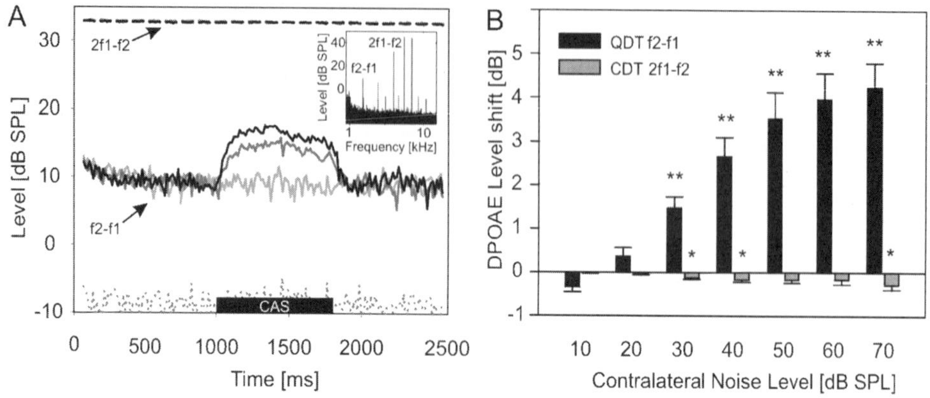

Figure 1. Effects of contralateral acoustic stimulation on DPOAEs. A: Example of DPOAE measurements (L1 = L2 = 44 dB SPL) under three different CAS-conditions: 20 (grey), 40 (dark grey) and 60 (black) dB SPL; f2-f1 is indicated by solid lines, 2f1-f2 by dashed lines, mean background noise level by the grey dotted line, CAS-period by the black bar. The inset gives an exemplary spectrum of the ear canal recording. B. Averaged DPOAE level-shift (mean +/- SE) during CAS (n = 8 animals for CAS = 30-70 dB SPL, n = 7 for CAS = 20 dB SPL, n = 5 for CAS = 10 dB SPL); asterisks indicate significant difference from zero (Wilcoxon signed rank test, * = p < 0.05, ** = p < 0.01).

amplifier, opens the possibility that the operating point of the cochlear amplifier could have been shifted due to CAS induced activity of the olivocochlear efferents. Of course, the amplitudes of both types of DPOAEs could also change if the overall gain of the amplifier is changed by efferent modulation.

3.2 Low frequency biasing induced DPOAE modulation – effect of CAS

To induce a shift of the OP, we applied an ipsilateral periodic low frequency tone for biasing the cochlear partition and, in a second step, used CAS for efferent activation. The effects of periodical shifts of the OP due to low frequency biasing were simulated by using a Boltzman function (Fig. 2).

Biasing of the cochlear partition by the 5 Hz tone resulted in a specific, phase related f2-f1-level modulation pattern, whereas 2f1-f2 was nearly not affected (Fig. 3A). The shape of the f2-f1 modulation pattern depended on the level of the bias tone: at lower

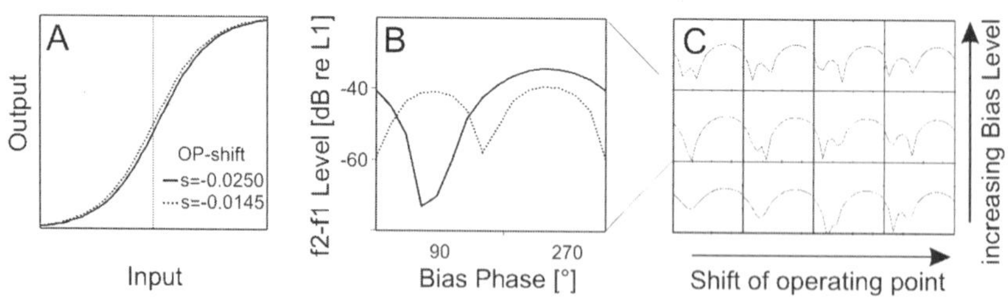

Figure 2. Simulation of the f2-f1 distortion generation during periodically biasing the input signal. A: Two nonlinear transfer functions with different OP positions. B: Amplitude of f2-f1 distortion from the output of A as a function of bias phase. C: f2-f1 modulation patterns for different combinations of bias tone magnitudes and OP shifts.

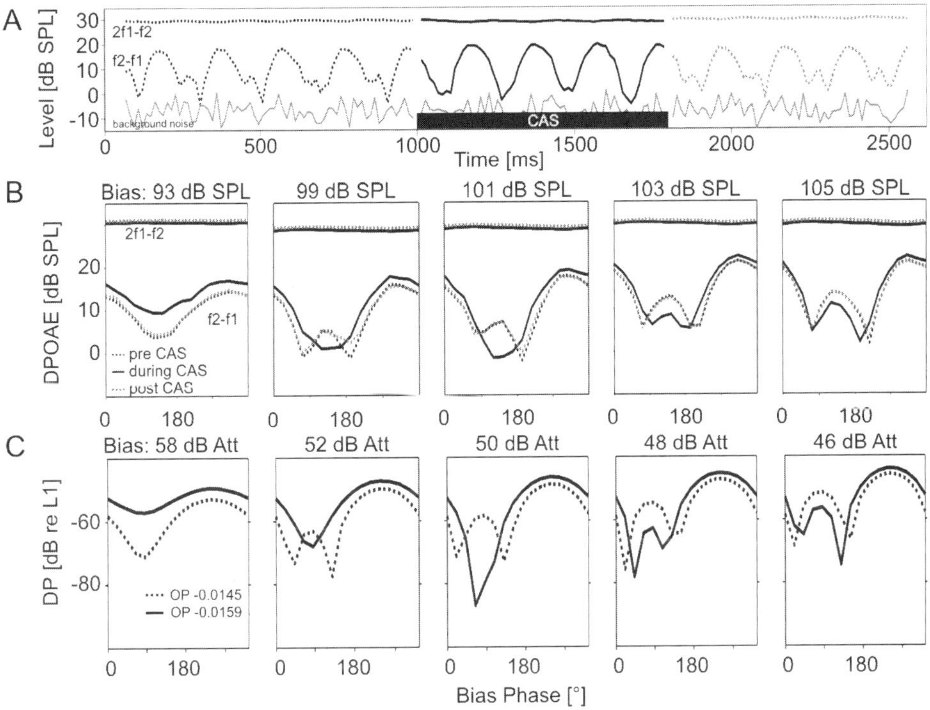

Figure 3. DPOAE modulation during low frequency biasing and intermitted CAS (50 dB SPL). A: Time course of DPOAE-level (L1 = L2 = 34 dB SPL) during low frequency biasing (5 Hz, 101 dB SPL). B: DPOAE modulation patterns in the presence or absence of simultaneous CAS under five conditions of bias tone level (93-105 dB SPL). DPOAE-level is averaged across and plotted against the corresponding bias tone phase for timeframes before (dotted black), during (solid black) and after (dotted grey) CAS. C: Simulation of f2-f1 amplitude during low frequency biasing for two resting positions of the OP (solid or dotted line) and increasing bias tone amplitudes.

bias tone levels, a distinct minimum and maximum of f2-f1 level occurred during one cycle of the bias tone, which changed to a "double-modulation" pattern with a second maximum when the bias level was increased. Comparison of the low frequency induced DPOAE modulation pattern between the different conditions with or without contralateral noise, yielded a pronounced effect of CAS. The resulting DPOAE modulation during low frequency biasing was in good agreement to the predictions of the model. Fig. 3B illustrates the effect of increasing bias tone level on the f2-f1 distortion by exemplary simulations for increasing bias amplitudes, each for two different assumed OP to provide qualitative equivalents to the measured data.

Both low frequency tone induced phase related f2-f1 modulation pattern and CAS induced changes of this pattern were observed in all tested animals. Strength of pattern change was correlated with the level of contralateral stimulation and increased to higher CAS levels (Fig. 4).

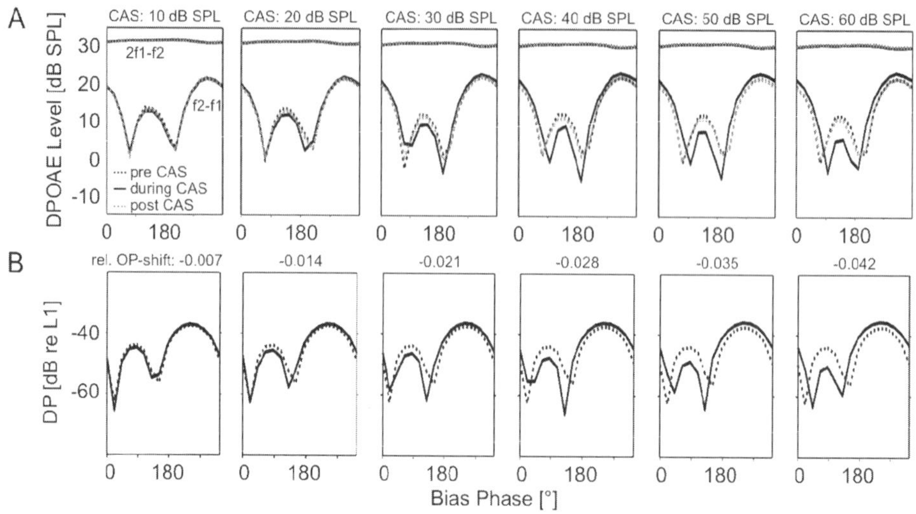

Figure 4. A: Examples of DPOAE modulation patterns during low frequency biasing (5 Hz, 107 dB SPL; L1 = L2 = 36 dB SPL) in the presence or absence of CAS of increasing level (10-60 dB SPL). Diagrams equivalent to Fig. 3. B: Simulation of f2-f1 distortion amplitude during low frequency biasing. DP modulation patterns for different OP shifts (solid lines); dotted lines give the pattern for the OP serving as reference (s = -0.0166).

4 Discussion

The strong influence of CAS on the quadratic DPOAE at f2-f1, which increased to higher CAS levels and was much bigger than the effect on the cubic DPOAE at 2f1-f2, can be seen analogous to previous studies that report an effect of CAS on f2-f1 [8-10]. A role of middle ear muscle reflex for the CAS induced effects observed in the present study is rather unlikely, because the effect already occurred at low contralateral levels (~30 dB SPL) and mainly consists of a level enhancement of the DPOAE. Most previous studies report a level reduction of f2-f1 due to efferent manipulations or bipolar effects, but level decrease versus enhancement may vary well depend on the starting state of the operating point and on the primary tone level which could be different for different preparations. In addition, at higher primary tone levels a modulation of 2f1-f2 may also be expected. Furthermore, in some of these studies, CAS duration was much longer than in the present study which may induce additional slow efferent effects [11].

The difference tone f2-f1 is sensitive to changes in the OP of the cochlear amplifier transfer function during low frequency biasing. Our biasing results are comparable to previous studies using that method [4-6]. The f2-f1 modulation pattern changed in our experiments systematically with the level of the bias tone and the additional application of contralateral noise produced pronounced changes of the DPOAE modulation pattern. According to our model, the CAS induced change of the modulation pattern is consistent with an additional shift of the operating point of the transfer function. Thus the observed change of the modulation pattern during CAS might be evoked by a shift of cochlear amplifier OP during CAS induced efferent activity. Of course, the comparison between DPOAE modulation patterns in the experimental data and the simulation only provides a qualitative assessment and real values for efferent shifts of the OP have to be obtained by other methods. However, the observed effects on f2-f1 could also occur due to a change of gain or a combination of both amplification and OP changes [2].

The concept that efferents are acting through a shift of the operating state of the transfer characteristics of the cochlear amplifier has been suggested by other authors [2] and could be based on changes of the mechano-electrical properties of the outer hair cells and calcium-mediated structural changes [12].

Acknowledgments

We thank A. Lukashkin for the helpful comments. This study was supported by DFG (Ko-987/8-2), Jürgen-Manchot-Stiftung, and FAZIT-Stiftung.

References

1. Guinan, J.J., 1996. Physiology of olivocochlear efferents. In: Dallos, P., Popper, A.N., Fay, R.R. (Eds.), The Cochlea. Springer, New York, pp. 435-502.
2. Russell, I.R., Lukashkin, A.N., 2008. Cellular and molecular mechanisms in the efferent control of cochlear nonlinearities. In: Manley, G.A., Fay, R.R., Popper, A.N. (Eds.), Active Processes and Otoacoustic Emissions. Springer, New York, pp. 343-379.
3. Kemp, D.T., 2002. Otoacoustic emissions, their origin in cochlear function, and use. Br Med Bull 63, 223-241.
4. Frank, G., Kössl, M., 1996. The acoustic two-tone distortions 2f1-f2 and f2-f1 and their possible relation to changes in the operating point of the cochlear amplifier. Hear Res 98, 104-115.
5. Bian, L. 2004., Cochlear compression: effects of low-frequency biasing on quadratic distortion product otoacoustic emission. J Acoust Soc Am 116, 3559-3571.
6. Lukashkin, A.N., Russell, I.J., 2005. Dependence of the DPOAE amplitude pattern on acoustical biasing of the cochlear partition. Hear Res 203, 45-53.
7. Puel, J.L., Rebillard, G., 1990. Effect of contralateral sound stimulation on the distortion product 2F1-F2: evidence that the medial efferent system is involved. J Acoust Soc Am, 87 1630-1635.
8. Kirk, D.L., Johnstone, B.M., 1993. Modulation of f2-f1: evidence for a GABA-ergic efferent system in apical cochlea of the guinea pig. Hear Res 67, 20-34.
9. Chang, K.W., Norton, S.J., 1997. Efferently mediated changes in the quadratic distortion product (f2-f1) J Acoust Soc Am 102, 1719-1733.
10. Kujawa, S.G., Fallon, M., Bobbin, R.P., 1995. Time-varying alterations in the f2-f1 DPOAE response to continuous primary stimulation. I: Response characterization and contribution of the olivocochlear efferents. Hear Res 85, 142-154.
11. Cooper, N.P., Guinan, J.J., 2003. Separate mechanical processes underlie fast and slow effects of medial olivocochlear efferent activity. J Physiol 548, 307-312.
12. Frolenkov, G.I., Mammano, F., Belyantseva, I.A., Kachar, D.C.B., 2000. Two distinct Ca(2+)-dependent signaling pathways regulate the motor output of cochlear outer hair cells. J Neurosci, 20, 5940-5948.

THE EFFECT OF EAR CANAL PRESSURE ON SPONTANEOUS OTOACOUSTIC EMISSIONS: COMPARISON BETWEEN HUMAN AND LIZARD EARS

P. VAN DIJK

University Medical Center Groningen
P.O.Box 30001, 9700 RB Groningen, The Netherlands

G. A. MANLEY

Lehrstuhl für Zoologie, Technische Universität München
Hochfeldweg 2, 85350 Freising, Germany

The center frequency, height and width of peaks in SOAE spectra depend on ear canal pressure. The width is interpreted as a measure of the inner ear source-signal-to-(e.g. thermal)-noise ratio. In humans, width increases with decreasing height. Apparently, ear canal pressure modifies the amplitude of the inner ear emission source signal. In lizards, the relation between peak width and height is not consistent. Here, middle ear transmission changes may account for many the observed amplitude effects.

1. Introduction

Spontaneous otoacoustic emissions (SOAEs) are weak acoustic signals emitted from the inner ear. They have been recorded from amphibians, lizards, birds and mammals[1]. They are believed to be caused by instabilities in the hair cell epithelium in the inner ear. By active mechanical feedback, hair cells improve the detection threshold and frequency selectivity of the ear. SOAEs occur as a result of instabilities in the active mechanics[2].

SOAEs in human are influenced by static ear canal pressure. For example, Schloth and Zwicker[3] showed that with increasing or decreasing static pressure, the amplitude of SOAE was reduced and the frequency slightly increased. It was, however, unclear whether these changes resulted from changes in the middle ear transmission, or whether the inner ear emission source signal was modified.

In this paper, we investigated whether ear canal SOAE amplitude changes with static pressure due to a changed middle ear transmission, or whether the amplitude of the source signal is changed. We studied this for individual emission peaks in the spectrum by determining the relationship between peak height and width. The finite spectral width of SOAEs must be caused by inner ear noise to which the emission generator is exposed[4]. For a simple oscillator (such as the Van der Pol oscillator[5]) exposed to (e.g. thermal) broad-band noise, the power spectrum $S(\omega)$ of the oscillator signal is given by a Lorentz curve, where the peak width (FWHM) $\Delta\omega$ is related to the peak height $S(\omega_0)$ and the noise spectral density $N(\omega_0)$ by

$$\frac{\Delta\omega}{\omega_0} = \sqrt{\frac{N(\omega_0)}{S(\omega_0)}}. \qquad (1)$$

Thus, if the oscillation amplitude decreases, the peak spectral width will decrease and, consequently, the peak width $\Delta\omega$ increases. We tested whether this model applies to SOAE from humans and lizards when the emission amplitude is modified by the application of static ear canal pressure. The primary noise source in the inner ear may be thermal noise and is not expected to change with static pressure. Consequently, if static pressure decreases the inner ear source amplitude, the model predicts that the peak width would increase. Specifically, then $\Delta\omega \propto 1/\sqrt{S(\omega_0)}$. On the other hand, if static pressure only modifies the middle ear transmission and leaves the inner ear source amplitude unchanged, the width would be constant.

2. Material and Methods

SOAEs were recorded in 7 human subjects and 14 lizards of 5 species belonging to 5 families (*Riopa fernandi, Gerrhosaurus major, Cordylus tropidosternum, Tupinambis nigropunctatus, Varanus exanthematicus*).

In humans, emissions were recorded by connecting an Etymoic Research ER10C microphone to the ear. The microphone signal was amplified by 40 dB and digitized with the built-in A/D converter of an Apple G5 computer. The static ear canal pressure was varied using a GSI Tympstat clinical tympanometer in manual pressure control mode. Emissions recordings were made at pressures from -200 to +200 daPa in 50 daPa steps. For each pressure, the record length was 1 minute. During the measurements, the subject sat in an upright position. Emission spectra were viewed online using a SR560 spectrum analyzer and were recomputed offline from the digitized signals.

To measure SOAEs in lizards, the animals were anesthetized using isoflourane with an initial concentration between 1.0 and 1.4 Vol% added to the air. After a few minutes, the lightly anesthetized animal was place on a heating blanket on a gas-damped table top inside a sound-attenuating chamber, with the animal anesthetised using 0.7 Vol %. The animal's temperature was held as constant as possible (usually $\pm$ 0.1°C). The opening of a plastic tube that contained the measurement microphone (Etymotic, ER-10B) was sealed over the ear canal and connected to a 50ml plastic syringe containing air. This syringe was used to change the air pressure outside the eardrum in increments of 1 daPa. The pressure was measured using a pressure-sensor system (Greisinger electronic, GMH 3150). The microphone signal was amplified 40 dB and analyzed using a spectrum analyzer (SR 760 FFT).

All emission peaks in the spectra obtained from humans and lizards were fitted to a Lorentzian curve in order to obtain the peak height, width and center frequency. Thus we obtained the dependence of these peak parameters on static pressure in the ear canal.

3. Results

The frequency, amplitude and width of 41 peaks in human emission spectra were analyzed. Typically, a Lorentz curve fitted well to spectral SOAE peaks. On average, emission frequency increased 4.4 (s.d. 7.5) Hz at -200 daPa and 1.9 (s.d. 9.1) Hz at

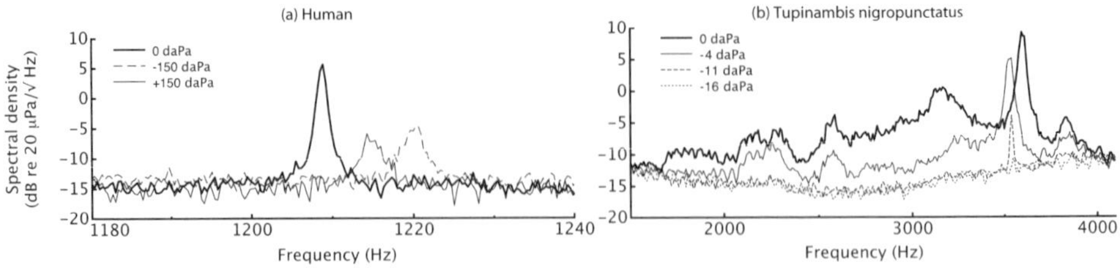

Fig. 1. Dependence of SOAE spectra on ear canal pressure. (a) A human subject. Both with increased and decreased ear canal pressure the frequency of the emission peak increased, the amplitude decreased and the width of the peak increased. (b) A lizard *Tupinambis nigropunctatus*. With decreasing pressure, the emission peaks decrease in amplitude. The peak frequencies near 2600 Hz, 3100 Hz, and 3600 Hz stayed constant, increased and decreased, respectively, as the static pressure became more negative. In addition to the peaks, a broad emission plateau is present between 1500 and 4000 Hz. The level of this plateau decreases as the pressure became more negative.

+200 daPa. Average amplitude was changed by -3.4 (s.d. 3.5) dB at -200 daPa and -4.0 (s.d. 3.6) dB at +200 daPa (see Fig. 1a and 2 for an example).

These standard deviations reflect the large range of behaviours observed. Spectral peak widths (FWHM) ranged from 0.24 Hz to 46.2 Hz. In the majority of emission peaks, the width varied with ear canal pressure and was inversely proportional to the square root of the peak spectral density of the SOAE. Eq. 1 predicts this relationship; see Fig. 2d for an example and Fig. 3 for results across subjects and SOAE peaks.

Comparing the effects of pressure on SOAE frequency in all lizard species revealed no consistent picture, even within one group of lizards with one type of papillar structure. The SOAEs differed greatly in amplitude and the magnitudes of observed effects also differed. Thus in species with sallets (*Riopa*, *Cordylus* and *Gerrhosaurus*), which tend to have smaller SOAE amplitudes, the changes observed are much smaller than, for example, in *Varanus*. As a result of a fall in pressure outside the eardrum, frequency increased or decreased, with a maximal change of less than 7% (in most cases less than 3%). Similarly small changes were seen due to increasing pressures outside the eardrum. In *Tupinambis* there was an equal probability of frequency rising or falling. The other species demonstrate no convincing deviations from this pattern, although in *Gerrhosaurus*, the largest shifts were in a positive direction. Nonetheless in most cases, frequency did not remain stable, but was shifted a few percent. In general, the frequency changed steadily as pressure outside the eardrum fell, until the SOAE was so small as to be unmeasurable. SOAE amplitudes more consistently fell as a result of negative air pressure, independently of whether the frequency was shifting up or down. In the majority of cases, an increase in air pressure outside the eardrum caused a small increase in SOAE amplitude.

The relationship between peak width and height was inconsistent across lizard species and across the peaks of individual emissions. None of the lizards complied to the relation predicted for a simple oscillator (Eq. 1). However, in *Varanus*, the width

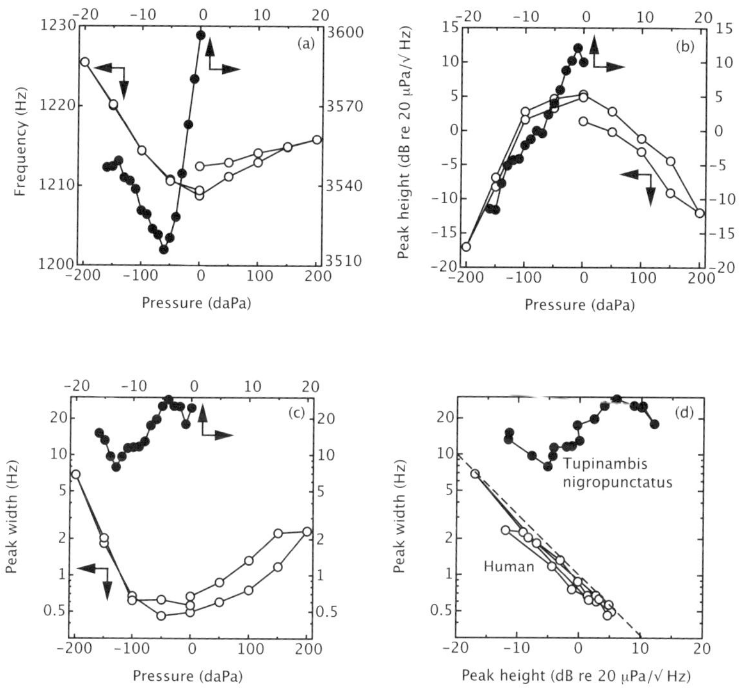

Fig. 2. Examples of the relationship between static ear canal pressure and SOAE peak frequency, height, and width. Open symbols: a human. Closed symbols: *Tupinambis nigropunctatus*. (a) SOAE frequency vs. static pressure. (b) Height vs. pressure. (c) Width vs. pressure. (d) Width vs. height. In panel (a), (b) and (c) the arrows point to the axes relevant to each data set. Panel (d) shows that the human SOAE peak closely follows the predicted relation between width and height (Eq. 1). The lizard SOAE deviates strongly from this behaviour.

and height were often negatively correlated, similar to humans. In *Tupinambis*, *Cordylus*, and *Gerrhosaurus*, the width was not clearly related to peak height, and for several emission peaks width was approximately constant. In *Riopa*, width was often positively related to height.

4. Discussion

The SOAEs in humans were generally consistent with earlier reports[3, 6]. In addition, we show here that the inner ear source signal amplitude is affected. This follows from the relationship of width being proportional to 1/sqrt(height).

In the lizards, similar effects occur over a much smaller range of pressures (± 20 daPa). No consistent changes of the peak width were observed in most species. This suggests that for most lizard data, the results are best explained by middle ear transmission changes. This transmission is expected to change rapidly with pressure, since normally the tympanic membrane of lizards is not tightly stretched.

Two views on the mechanics of SOAEs have been proposed. In humans, SOAEs may correspond to mechanical standing waves on the basilar membrane: a standing

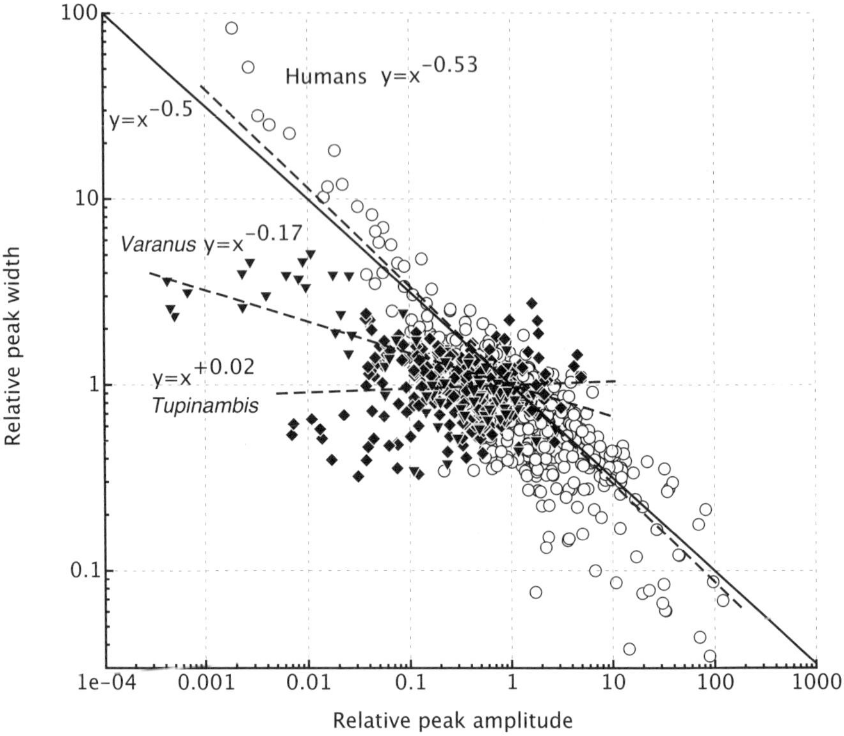

Fig. 3. Relative SOAE peak width vs. relative peak height. For each SOAE peak, the width and height was normalized to 1 for the first recording at 0 daPa static pressure. Open circles: humans. Closed triangles: *Varanus exanthematicus*. Closed diamonds: *Tupinambis nigropunctatus*. The solid line corresponds to the model equation 1. The dashed lines are curves fitted to the three data sets, respectively.

wave may exist between the stapes at the basal end of the cochlea and the characteristic place of the emission center frequency on the basilar membrane. SOAEs occur at those frequencies for which a standing wave can built up. In a cochlear model where the reflection at the characteristic place occurs due to slight irregularities in the basilar membrane impedance, the standing wave can only build up at frequencies that are spaced at specific distances[7]. This specific spacing of SOAE frequencies is consistent with observations on most SOAEs in humans[8].

An alternative view of SOAE generation is that SOAEs occur as a result of locally strongly active "point source" along the sensory epithelium. In this view, a group of hair cells is coupled and produces coherent acoustic energy, which is emitted without substantial reflection at the stapes. In contrast to the standing wave hypothesis, the SOAE frequencies are primarily determined by local resonance properties of the sensory epithelium and inner ear geometry, rather than being "dictated" by the standing wave.

Our results in humans are consistent with the standing wave hypothesis, since in that case a change of stapes impedance is expected to change the inner ear signal, apparently including its amplitude. However, point sources are not excluded by this result: for a point source, a change of impedance load would also change the amplitude of the signal.

In the lizards, we found no evidence for a consistent change of SOAE width. Here, middle ear transmission is probably the determining factor in modifying the emission signal. Some of the data from *Varanus* showed behaviour similar to that in humans over part of the pressure range. All the peaks that showed this behaviour were large ($>$ 15 dB above the noise). Thus the behaviour of lizard SOAE peaks may be due to both factors, perhaps depending on their amplitudes.

SOAE plateaus found in some lizard data (see this paper and previous reports[9]) are obviously reduced in amplitude when the air pressure changes. These plateau emissions appear to simply contradict the standing wave hypothesis: here, there appears to be no mechanism that promotes specific frequencies, such as the standing wave condition.

In general, the two hypotheses of the standing wave and point source SOAE generators are not mutually incompatible. A signal emitted by a point source will always be partially reflected at the stapes. This will again be partially reflected at the more apical characteristic place. There may thus be complementarity between the two mechanisms: local resonances and couplings may cause a group of hair cells to oscillate, while wave reflections may promote and thus amplify certain emission frequencies.

References

1. Manley, G. A., van Dijk, P., 2008. Otoacoustic emissions in amphibians, lepidosaurs and archosaurs. In: Active Processes and Otoacoustic Emissions. Manley, G. A., Fay, R. R., Popper, A. N. (eds.). Springer–Verlag, pp. 211–260.
2. Hudspeth, A. J., 2005. How the ears works work: mechanoelectrical transduction and amplification by hair cells. C. R. Biol. 328, 155-162.
3. Schloth, E., Zwicker, E., 1983. Mechanical and acoustical influences on otoacoustic emissions. Hear. Res. 11, 285-293.
4. van Dijk, P., Wit, H. P., 1990. Synchronization of spontaneous otoacoustic emissions to a 2f1-f2 distortion product. J. Acoust. Soc. Am. 88, 850-856.
5. Stratonovich, R. L., 1963. Topics in the Theory of Random Noise, Volume 1. Gordon and Breach, New York.
6. Hauser, R., Probst, R., Harris, F. P., 1993. Effects of atmospheric pres-sure variation on spontaneous, transiently evoked, and distortion product otoacoustic emissions in normal human ears. Hear. Res. 69, 133-145.
7. Shera, C. A. 2003. Mammalian spontaneous otoacoustic emissions are amplitude-stabilized cochlear standing waves. J. Acoust. Soc. Am. 114, 244-262.
8. Talmadge, C., Long, G., Murphy, W., Tubis, A., 1993. New off-line method for detecting spontaneous otoacoustic emissions in human sub jects. Hear. Res. 71, 170-182.
9. Manley, G. A., Gallo, L., Köppl, C., 1996. Spontaneous otoacoustic emissions in two gecko species, Gekko gecko and Eublepharis macularius. J. Acoust. Soc. Am. 99, 1588-1603.

Comments and Discussion

Braun: In their discussion section, the authors wrote: "This specific spacing of SOAE frequencies is consistent with observations on most SOAEs in humans [8]." Do the data shown in Fig. 8 of the referenced paper [8] not clearly contradict this interpretation? There are no data points for the double of the so-called "minimum separation" of SOAEs. A later analysis by Braun (1997), which is based on five times as many human SOAE data, fully confirmed the findings of [8] in this respect: Human SOAEs show a preferred minimum frequency separation, but definitely no equidistant spacing. Are these findings not much more compatible with the suggestion by van Hengel et al. (1996) that the "minimum separation" may be caused by mutual suppression of neighboring SOAE sources?

Braun, M., 1997. Frequency spacing of multiple spontaneous otoacoustic emissions shows relation to critical bands: A large-scale cumulative study. Hear. Res. 114, 197–203.

van Hengel, P. W., Duifhuis, H., van den Raadt, M. P., 1996. Spatial periodicity in the cochlea: the result of interaction of spontaneous emissions? J. Acoust. Soc. Am. 99, 3566–3571.

DEPENDENCE OF DISTORTION-PRODUCT OTOACOUSTIC EMISSION COMPONENTS ON PRIMARY-LEVEL RATIO

GLENIS R. LONG, CHANGMO JEUNG

Speech-Language-Hearing Sciences, Graduate Center City University of New York New York, NY 10016, U.S.A.

CARRICK L. TALMADGE

National Center for Physical Acoustics, University of Mississippi, Oxford, MS 38677, U.S.A.

Distortion product otoacoustic emissions (DPOAE) were evaluated in normal hearing subjects using logarithmically sweeping tones (2s/octave), which maintained a constant frequency ratio (f2/f1= 1.22). The levels of the two major components were extracted using a least-squares fit procedure, which permits dynamic fitting of the delay of a given component. While the shape of the extracted generator component input/output function depends on the paradigm, the magnitude of the reflection component tends to saturate at low and moderate levels (which does not depend on primary level ratio). Changes in fine-structure frequency also depends on the primary ratio and appears to be related to the growth of the generator component.

1 Introduction

1.1 Extracting Distortion Product Otoacoustic Emissions (DPOAE) components

DPOAE are signals generated in the cochlea in response to two external tones (primaries) f_1 and f_2 (when f2 > f1). The most commonly investigated DPOAE is $2f_1$-f_2. DPOAE can be detected in the ear canal and potentially provide a noninvasive tool to evaluate cochlear function. Examination of the properties of DPOAE measured with adequate frequency resolution, have revealed that the ear canal signal is made up of at least two components coming from different cochlea regions which interact to generate pseudo periodic fluctuations in amplitude and phase [1]. The distortion is generated at the region of maximal overlap of the primaries. Some energy travels basal-ward and out through the middle ear. However, some of the energy travels apically to the characteristic region for that frequency and is reflected by the same processes responsible for other otoacoustic emissions [2]. The two components have the same frequency but different phase properties so that they can interact either constructively or destructively. Consequently, changes in the ear canal signal can be difficult to interpret. Separating the two components can be very time consuming, but we exploit the latency delays of the OAE to permit the extraction of the two components from DPOAE obtained with continuously sweeping primaries[3]. We have now modified the analysis to permit the specification of the expected frequency dispersion of the different OAE components to permit extraction of both major components from the sound wave file. We have compared this new analysis to a version of the time windowing procedure based on the use of Inverse Fast

Fourier Transforms [4] and have shown that while the results are very similar at high signal to noise ratios, the new procedure is better at extracting the two components when the signal to noise ratio is poorer and when the components are not well separated in time (at high frequencies).

1.2 Choosing the Primary Levels

Whenever one is collecting DPOAE, decisions have to be made as to the optimal level ratio of the primaries. When DPOAE are obtained with equal-level primaries, they grow with a slope of a little less than 1dB/db [5]. Brown and Gaskill [5] demonstrated that optimal DPOAE levels are obtained when the level of the lower frequency primary (L_1) is greater than the level of the lower frequency primary (L_2). Whitehead et al.[6] and others have provided additional support for this claim (reviewed in [7]). In 1998 Kummer et al. [8] developed a formula based on Brown and Gaskill's data which they called the Scissor procedure. Recent research by Neely et al. [7] has lead to what the authors call the "optimal" level ratio which is very similar to that developed by Whitehead et al. [6]. All of these formula have been based on the total DPOAE, we need a better understanding of the impact of these level ratios on the different DPOAE components [9] if we are to better understand the mechanisms responsible for DPOAE generation and optimize primary levels for clinical and research uses.

2 Methods

DPOAE data were collected from 5 subjects (4 female and one male) seated in a reclining chair in a double walled IAC booth. We measured DPOAE using equal-level primaries (L_1=L_2 from 40 to 75 dB in 5 dB steps) and the L_1/L_2 recommended for the scissors (L_1=0.4L_2+39 dB) and Boys-Town (L_1=0.45L_2+44 dB) paradigms (L_2 was varied from 25 to 75 dB in 5 dB steps) and f_2/f_1 was always 1.22. The primaries were swept over three octaves in 6 s (2s/octave). Up and down sweeps (which provide independent estimates of the DPOAE properties) were alternated and the resulting sound files streamed to disk. All sweeps at each level were collected in one block, and two paradigms were collected in one two hour session. The number of sweeps depended on the expected signal to noise ratio. After elimination of noisy files, the remaining files were averaged and analyzed. By varying the analysis window [3] it was possible to extract the combined DPOAE or separate the two major components.

When a short analysis window was used (8Hz bandwidth) both components fell within the analysis window so both components were detected and the fine structure could be examined. The data from the individual levels were fitted with a model of DPOAE fine structure [10] to extract the frequency of the fine structure maxima. In order to evaluate changes in the frequency of the fine structure maxima with level, the frequency of each maximum was subtracted from the frequency of the nearest maximum at the next highest level (for L_2 below 65 dB SPL). The amount of shift was normalized to the frequency of the maxima.

A longer analysis window (1-Hz Bandwidth) permitted the separation of the two components. The resulting input/outputs (I/O) functions could be plotted as a function of frequency. Since the reflection component has similar origins to the TEOAE and SFOAE the level of this component varied significantly as a function of frequency. In an attempt to smooth this data the rms amplitude of each component within a 1/3 octave bandwidth was calculated. Once the data was smoothed, we estimated the I/O functions every 500 Hz from 1000-2500 and calculated the average and standard deviations of these I/0 in order to evaluate the overall growth of the two components as a function of primary level.

3 Results

3.1 Changes in Fine-Structure Frequency with Level

As shown in Fig. 1, when $L_1 = L_2$ the frequency of the maxima decrease systematically as the primary level increases, but there is little change in the frequency of the maxima when the scissors paradigm is used. When this paradigm is used, the DPOAE levels are highly compressed so there is little difference in the level of the maxima for adjacent primary levels. The Boys-Town paradigm produces a rapid decrease in frequency with increasing primary level at lower primary levels, but this becomes less steep at higher primary levels when the DPOAE begins to saturate.

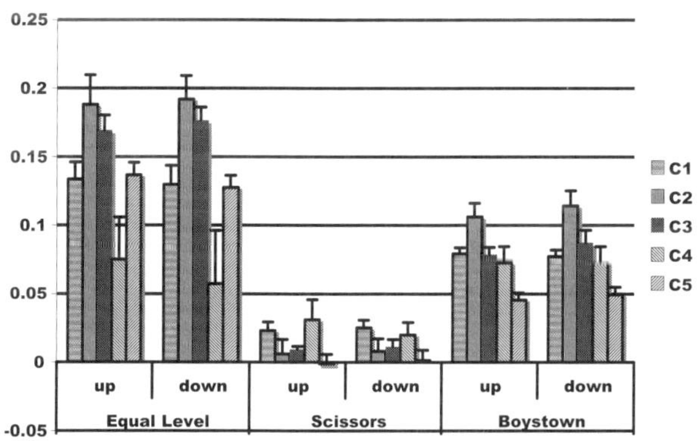

Figure 1. Decrease in the frequency of the fine structure maximum (% frequency change per dB change in the primaries) for all 5 subjects and each paradigm. Separate estimates were obtained for data obtained with up and down sweeps.

3.2 I/O for the Two Major Components as a Function of Primary Ratio

When both primaries are changing in level, one can plot the I/O function either as a function of L_2 (Figure 2) or as a function of L_1 (Figure 3). The data from $L_1=L_2$ are the same in both Figures. Data from the scissors paradigm are only plotted for L_2 below 65 dB (because this paradigm uses equal-level primaries above this level). When $L_1=L_2$, the I/O function for the generator component (left column) grows at a little less than 1dB/dB (indicated by the solid line). In contrast, the reflection component saturates at moderate levels. When the scissors paradigm is used, the generator component is larger than that

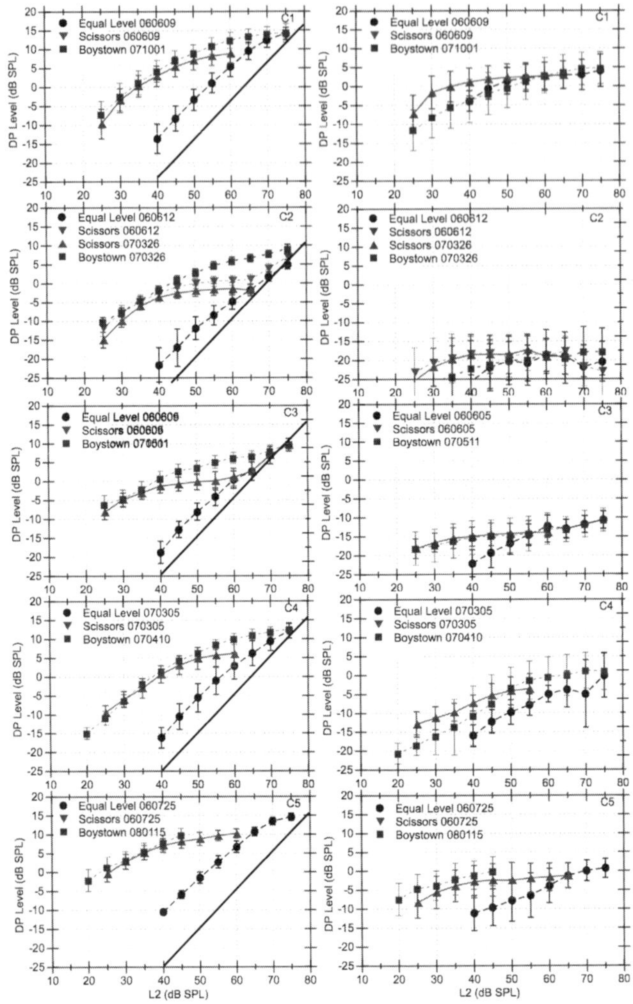

Figure 2. Input/Output functions from 5 subjects as a function of L_2 level for the generator component (left column) and reflection component (right column) The line indicates a slope of 1dB/dB for comparison.

generated by the equal level paradigm at low stimulus levels, and the level growth is compressed at the higher levels used. The Boys-Town paradigm produces similar levels for the generator component as the scissors procedure at lower stimulus levels, but produces the highest DPOAE level at the higher primary levels before the I/O function saturates. Reflection components from individual subjects tend to saturate at the same level for all paradigms. The level of this saturation is associated with the number and magnitude of the spontaneous otoacoustic emissions (SOAE) in each subject (C1, C4 and C5 all have several large SOAE – C2 has no detectable SOAE).

Plotting the data as a function of L_1 (Figure 3) makes it very clear that the reflection component saturates at a similar level for all the different level paradigms. The enhancement of the generator component with the non-equal-level paradigms can be seen in the elevation of the scissor and Boys-Town paradigms above the equal level paradigms (L1 between 45 and 60 dB for Scissors, and L1 above 65 for the Boys-Town paradigm) before the component becomes compressed. We kept the same level ratio across all frequencies. The advantages of the Boys-Town paradigm, might be more obvious if we had varied the level ratio with frequency as recommended [7].

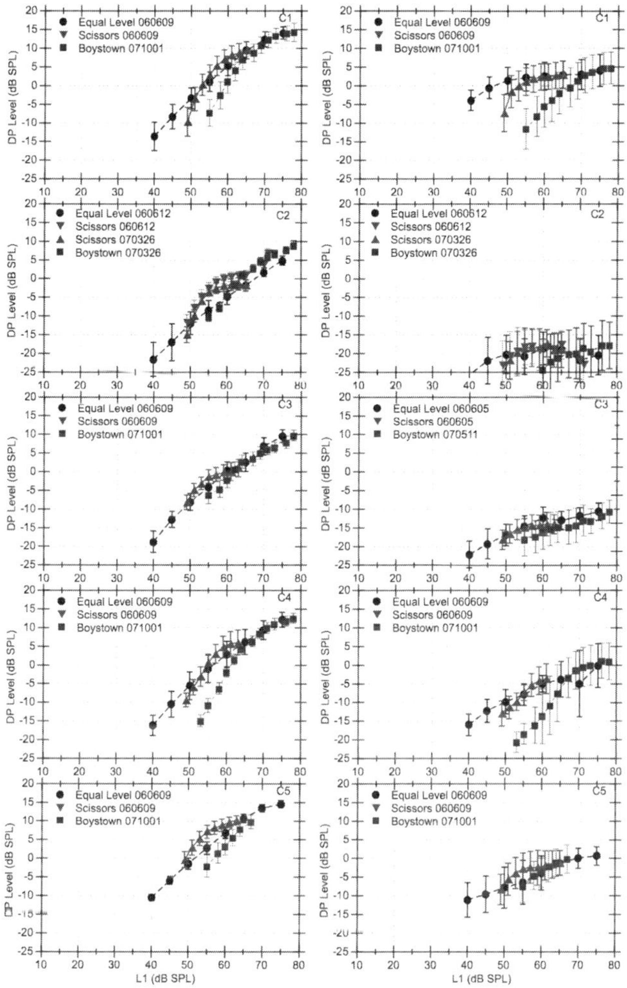

Figure 3. Input/Output functions as a function of L_1 level for the generator component (left column) and reflection component (right column).

4 Discussion

Whenever the Total DPOAE is measured without sufficient detail to know whether the isolated data points fall near a maxima or minima of the fine structure, one cannot know whether changes in the DPOAE levels are indicative of reduced/enhanced DPOAE generation, cancellation or enhancement of the two components. The change in frequency of level maxima and minima, means that the position of an individual frequency in the fine structure depends on primary level. This can produce an I/O function that does not represent what is happening at any cochlear region. As long as one does not want to measure above 65 dB SPL, the lack of change in the frequency of the fine structure in the scissors procedure may mean that this procedure is less likely to be influenced by changes in the fine structure. The large differences in the shape of the I/O functions across paradigms means that care must be taken when using DPOAE I/O functions to predict compression on the basilar membrane. We need a much more thorough understanding of the sources of DPOAE before making any such claims.

The similarity of the reflection component saturation level across paradigms is not unexpected, as we expect that the saturation level to be related to the properties of cochlear processes and not the properties of the primaries. This means that this component has the potential to provide useful information about the state of the cochlea.

The pattern of change in fine-structure frequency appears to depend on the amount of saturation of the generator component. In this paper, we have chosen to concentrate on the levels of the different components. Changes in fine-structure frequency are determined by changes in the relative phase of the two components. A change in the level of the generator component could result either a level dependent change in the phase of either component. No conclusion can be reached until the planned examination of phase changes of each component is completed.

Acknowledgments

Supported by the National Institute on Disability and Rehabilitation Research, U.S. Department of Education, Rehabilitation Engineering Research Center — Hearing Enhancement and the National Organization for Hearing Research. The authors wish to thank Simon Henin for help with the analysis.

References

1. Talmadge, C.L., Long, G.R., Tubis, A and Dhar, S., 1999. Experimental confirmation of the two-source interference model for the fine structure of distortion product otoacoustic emissions. J. Acoust. Soc. Am. 105, 275-92.
2. Shera, C.A. and J.J. Guinan, Jr., 1999. Evoked otoacoustic emissions arise by two fundamentally different mechanisms: a taxonomy for mammalian OAEs. J. Acoust. Soc. Am. 105, 782-98.
3. Long, G.R., Talmadge, C.L., and Lee, J., 2008. Measuring Distortion Product Otoacoustic Emissions using Continuously Sweeping Primaries J. Acoust. Soc. Am., in press.
4. Withnell, R.H., L.A. Shaffer, and C.L. Talmadge, 2003. Generation of DPOAEs in the guinea pig. Hear. Res., 178, 106-17.
5. Brown, A.M. and S.A. Gaskill, 1990. Measurement of acoustic distortion reveals underlying similarities between human and rodent mechanical responses. J. Acoust. Soc. Am. 88, 840-9.
6. Whitehead, M.L. et al., 1995. Dependence of distortion-product otoacoustic emissions on primary levels in normal and impaired ears. I. Effects of decreasing L2 below L1. J. Acoust. Soc. Am. 97, 2346-58.
7. Neely, S.T., T.A. Johnson, and M.P. Gorga, 2005. Distortion-product otoacoustic emission measured with continuously varying stimulus level. J. Acoust. Soc. Am. 117, 1248-59.
8. Kummer, P., T. Janssen, and W. Arnold, 1998. The level and growth behavior of the 2 f1-f2 distortion product otoacoustic emission and its relationship to auditory sensitivity in normal hearing and cochlear hearing loss. J. Acoust. Soc. Am. 103, 3431-44.
9. Dhar, S., et al., 2005. The effect of stimulus-frequency ratio on distortion product otoacoustic emission components. J. Acoust. Soc. Am. 117, 3766-76.
10. Dhar, S., et al., 2002. Multiple internal reflections in the cochlea and their effect on DPOAE fine structure. J. Acoust. Soc. Am. 112, 2882-97.

OTOACOUSTIC EMISSIONS EVOKED BY TWO-TONE BURSTS USING LINEAR AND NON-LINEAR PROTOCOL

W.WIKTOR JEDRZEJCZAK

Institute of Physiology and Pathology of Hearing, ul. Zgrupowania AK 'Kampinos' 1, 01-943 Warszawa, Poland

JACEK SMURZYNSKI

Department of Communicative Disorders, East Tennessee State University, Box 70643, Johnson City, TN 37614, USA

KATARZYNA J. BLINOWSKA

Department of Biomedical Physics, Institute of Experimental Physics, Warsaw University, ul. Hoza 69, 00-681 Warszawa, Poland

KRZYSZTOF KOCHANEK, HENRYK SKARZYNSKI

Institute of Physiology and Pathology of Hearing, ul. Zgrupowania AK 'Kampinos' 1, 01-943 Warszawa, Poland

Otoacoustic emissions (OAEs) were recorded for tone bursts presented separately and as a two-tone burst complex. Signals were delivered at 70 dB SPL using a non-linear processing method and at 60 dB SPL using a linear method. Two types of stimuli were applied to record tone burst OAEs (TBOAEs): (a) cosine-windowed tone bursts of 5-ms duration with center frequencies of 1, 1.5, 2 and 3 kHz, (b) complex stimuli consisting of a digital addition of the 1-kHz tone burst together with either the 1.5-, 2- or 3-kHz tone burst. Recorded signals were processed using the method of adaptive approximations by matching pursuit (MP). The MP method allowed decomposition of signals into waveforms of defined frequency, latency, time span, and amplitude. This approach provided a high time–frequency (t–f) resolution and identified patterns of resonance modes that were characteristic for TBOAEs recorded in each ear. The effect of suppression was statistically significant only for the 1, 1.5-kHz condition and slightly greater for the non-linear mode than for the linear mode. For the dual stimulation by 1 and 1.5-kHz, the MP revealed the existence of closely positioned resonance modes associated with responses recorded individually with the stimuli differing in frequency by 500 Hz. The comparison of t–f distributions calculated for dual (two-tone bursts) and sum-of-singles conditions exhibited mutual suppression of resonance modes common to both stimuli.

1 Introduction

Animal physiological studies and psychoacoustical experiments in humans provided supporting evidence of a compressive nonlinearity in cochlear processes [1, 2]. It is postulated that the compressive region resulting from a tone at characteristic frequency extends over an octave. The responses evoked by a tone from a location on the basilar membrane (BM) lying within that tone's amplification region can be reduced in amplitude by the simultaneous presentation of a second tone. Two-tone suppression was reported in BM vibrations and at the level of the cochlear nerve. For multicomponent stimuli, mutual suppression occurs when stimuli components are within the compressive region of the BM [3]. Studies of suppression of OAEs have shown that the shapes of suppression tuning curves measured for all classes of OAEs were similar to neural and psychoacoustic tuning curves [4]. The issue of the independence of 'emission channels' has been investigated by comparing OAEs elicited by broad-band stimuli with those recorded using energy concentrated in narrow bandwidths [5, 6].

The purpose of the present study was to analyze OAEs recorded with tone bursts presented separately and as a two-tone burst complex for different frequency separations of the stimuli using the matching pursuit (MP) method of adaptive approximation. Therefore, more detailed data of cochlear suppression effects than those based on the FFT method could be obtained.

2 Materials and method

2.1 Instrumentation and stimuli

Tone burst evoked otoacoustic emissions (TBOAEs) were collected for 21 normally hearing young adults using the ILO88 system. One ear of each subject was selected for testing. Two types of stimuli were applied: (a) cosine-windowed tone bursts of 5-ms duration with center frequencies of 1, 1.5, 2 and 3 kHz, (b) complex stimuli consisting of a digital addition of the 1-kHz tone burst together with either the 1.5-, 2- or 3-kHz tone burst. All bursts had a zero-starting phase. Stimuli were presented at levels of either 60 or 70 dB SPL using 'a linear presentation mode' or 'a non-linear presentation mode', respectively. Responses were analyzed in the window of 5.5-20.5 ms after the onset of the stimuli. TBOAEs obtained with single-tone bursts were superimposed off-line (creating 'sum of singles' responses) and were compared to those of the two-tone burst complex ('dual responses').

2.2 Method of data analysis

The signals were analyzed using the matching pursuit (MP) method (introduced by Mallat and Zhang [7]). The MP algorithm relies on adaptive decomposition of the signal into waveforms (also called atoms) from a large and redundant set of functions (called dictionary). Finding an optimal approximation of a signal by selecting functions from a very large and redundant set is a computationally intractable problem, therefore sub-optimal solutions were applied. The waveforms were fitted in an iterative procedure, starting with the atom giving the highest product with the signal, which means that it accounted for the largest part of the signal energy. Then, the next atoms were fitted to the residues. This procedure was continued until 99.5 % of the original signal energy was expressed. A dictionary consisting of 10^6 Gabor functions (sine modulated Gaussians) was used, with these functions being described by the parameters such as frequency, latency, time span, amplitude, and phase.

It has been shown that 7-10 resonance modes are enough to describe click evoked OAE accurately [8]. For OAEs evoked by tone burst stimuli, which have narrower spectra than click stimuli, lower number of waveforms would be needed for accurate signal decomposition. Therefore, in the current analyses, seven resonant modes corresponding to seven highest energy atoms found by MP procedure were considered. Those modes accounted for at least 95% of the energy of the signals and consequently characterized the main features of TBOAE signals. The t-f distributions of signals' energy were constructed by summation of Wigner transforms of atoms, according to the procedure described previously [8]. The Wilcoxon rank sum test was used for statistical analyses of the data with the criterion of significance set at $p<0.05$.

3 Results

The t-f distributions of energy, spectra and time profiles of energy for dual and sum of singles stimulations averaged over 21 ears are shown in Fig. 1, together with the resulting differences from the subtraction for each of those two averaged measures (sum of singles minus dual). The t-f maps are calculated only for OAEs recorded in linear protocol, while spectra and time evolutions are shown also for non-linear data. To calculate t-f maps, short time Fourier transform (STFT) was applied to individual results. For each subject, three t-f maps were created: dual, sum of maps of singles, and difference (sum of singles minus dual). Then, each type of maps was averaged for all subjects resulting in group data. The largest difference between the t-f distributions and the spectra obtained for dual and sum of singles conditions was found for the stimulus frequencies closest to each other (1 and 1.5 kHz; Fig. 1A). The difference peaked at around 1250 Hz, which corresponded to the suppression of components in the frequency region in the middle between the two burst stimuli. The suppression effect was smaller for linear than for non-linear mode but the difference was not statistically significant.

Figure 2 depicts an example of data analyses of TBOAEs recorded in an ear for combination of frequencies, 1 and 1.5 kHz for linear protocol (top four panels, marked as Fig. 2A), and non-linear protocol (bottom four panels, Fig. 2B). Results obtained for the complex (dual) stimuli and for the off-line summations of data collected with single-tone bursts (sum of singles) are shown in the left and in the right column, respectively. For each frequency combination and for each type of stimulation (sum of singles and dual), two panels are depicted: the upper panel corresponding to the spectrum and the lower one to a t-f distribution of TBOAE energy. In panels representing spectra, the envelopes do not allow to determine contributions of individual components, i.e., weaker components are masked by the envelope slopes of the stronger ones. However, the MP method provided detailed information about individual resonance modes.

In Fig. 2, the modes are marked by lines of length proportional to their amplitudes and placed at corresponding frequencies. The examination of the 'sum of singles' spectrum revealed the existence of several closely positioned resonance modes associated with responses recorded individually with stimuli differing in frequency by 500 Hz. The resonance modes extracted from the responses evoked by 1-kHz and 1.5-kHz bursts overlap in the regions around 1 and 1.2 kHz.

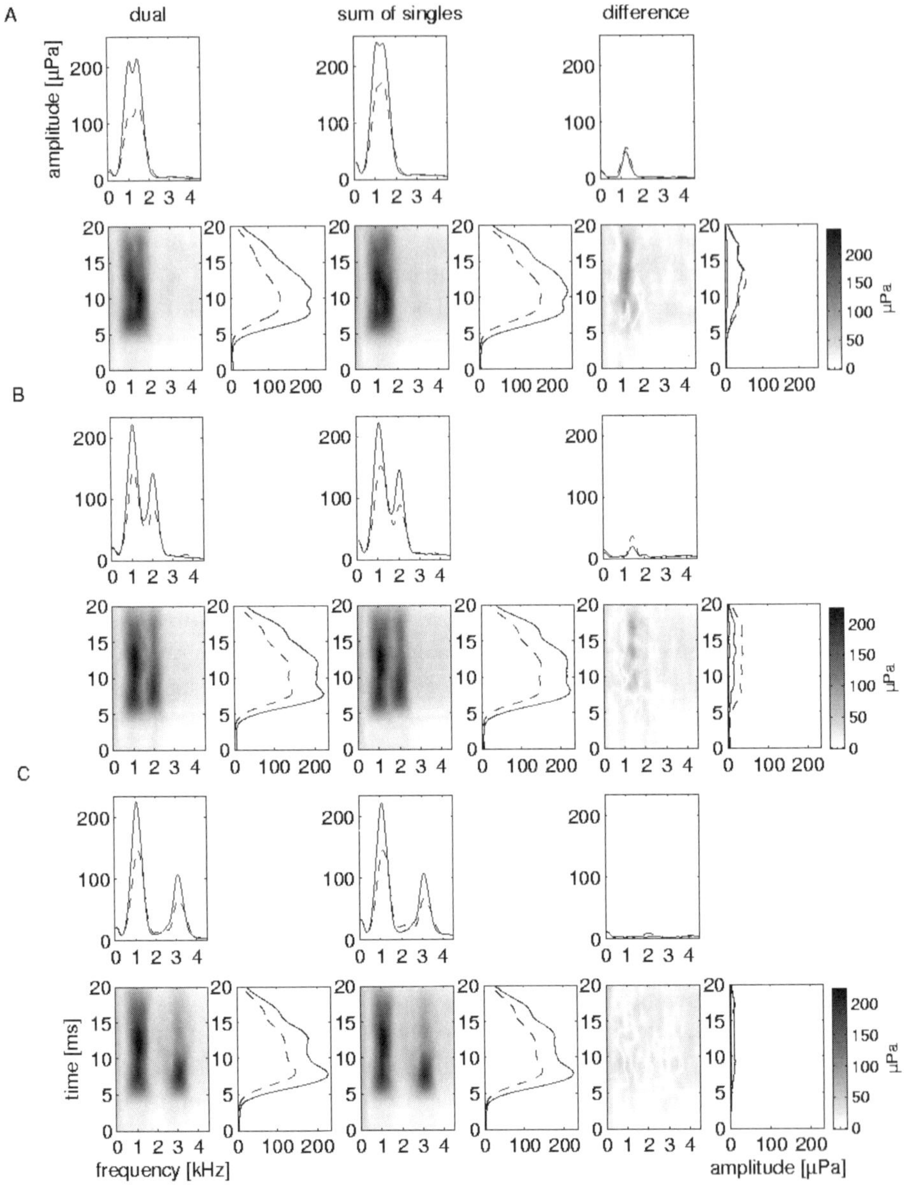

Figure 1. Time–frequency distributions of TBOAE energy and corresponding spectra averaged over 21 ears for linear stimulation. Corresponding spectra are plotted on top and the time evolutions of total energy are plotted to the right of the t–f maps (solid lines). The t–f distributions, spectra, and the energy evolutions in time are plotted resulting from calculating the difference between responses obtained for sum of singles and dual stimulations (sum of singles minus dual). A: stimulation by bursts of 1- and 1.5-kHz. B: 1- and 2-kHz. C: 1- and 3-kHz. For comparison, the spectra and time evolutions of responses recorded by non-linear paradigm are shown (dashed lines).

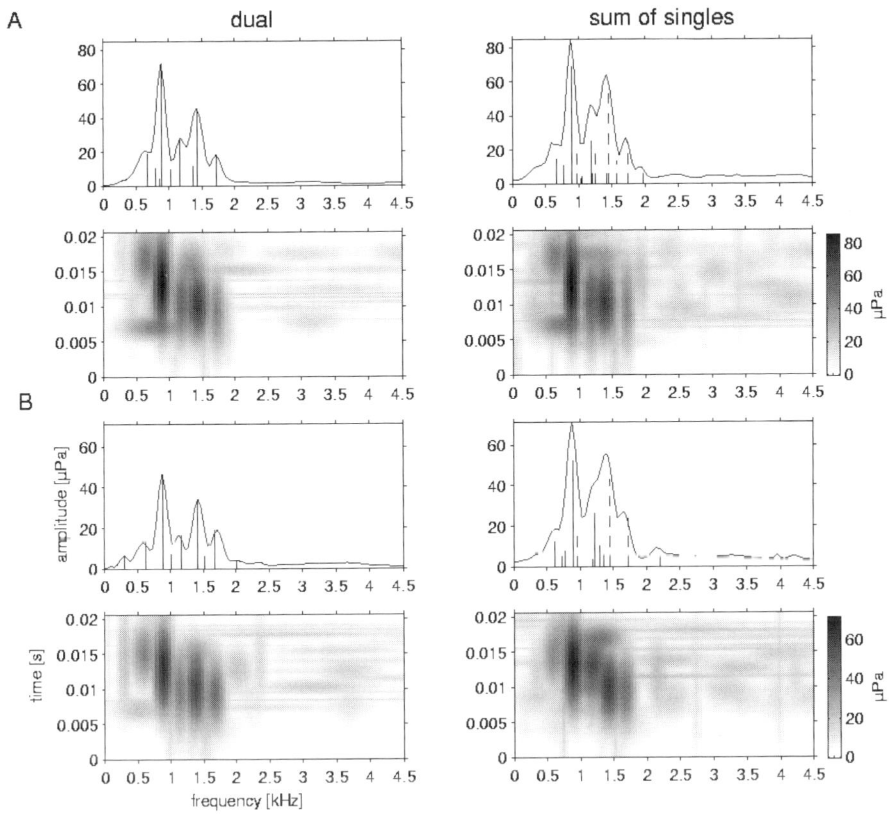

Figure 2. An example of the t-f distributions of TBOAE energy and corresponding spectra recorded in one ear for bursts of 1- and 1.5-kHz. Results for dual stimuli and sum of singles are shown in the left and the right columns, respectively. A: data recorded with the linear protocol. B: data recorded with the non-linear protocol. The upper panels correspond to the spectra. Resonance modes are marked as the bars of height proportional to their amplitudes. For sum of singles analyses (right columns), solid lines correspond to resonance modes excited by 1-kHz tones and dashed lines to those excited by 1.5-kHz tones. The lower panels correspond to the t-f distributions of TBOAE energy.

The effect of suppression of modes for closely-spaced dual stimuli was observed for all subjects but the degree of suppression depended on the individual patterns of resonant modes. It was hypothesized that the effect is due to the suppression of resonant modes when those are evoked by both frequencies and they overlap. Such a hypothesis was tested statistically by doing the following data processing. First, for TBOAEs recorded with the single stimuli of 1 and 1.5 kHz, seven strongest modes were evaluated (accounting in average for 95% of signals' energy). Those modes that were present in both recordings and their frequencies differed by less than 50 Hz were identified. The value of the accuracy of the determination of frequencies was set at 50 Hz based on an estimation derived from previous studies [6, 8]. Next, the amplitudes of the selected modes (differing by less than 50 Hz) recorded using a dual stimulus and obtained as sum of singles were compared. The differences in amplitudes were statistically significant for both recording protocols ($p<0.01$). When a similar analysis was performed for modes that were not overlapping (separated by more than 50 Hz) no significant difference was found.

4 Conclusions

The application of the MP method which makes identification of the resonance modes possible may be helpful in testing models of OAE generation, especially for evoking stimuli with complex spectra, e.g., those with multifrequency components. The results of the present study support the idea that the suppression observed in case of dual stimuli with closely-spaced components is due to mutual attenuation of the overlapping resonance modes. When the stimulation frequencies are far apart, there is no suppression due to the absence of overlapping modes. Generation mechanisms may possibly involve mutual suppression and/or enhancement of resonance modes with neighboring frequencies and diverse phase relationships. The distribution of resonance modes was similar for both protocols whereas the suppression effect was slightly greater for the non-linear mode than that for the linear mode, in agreement with the previous study [5].

Acknowledgment

This research was partially founded by the Polish Ministry of Science and Higher Education within the framework of the 2007-2010 scientific support program (Grant No. N N518 0924 33).

References

1. Robles, L., Ruggero, M.A., 2001. Mechanics of the mammalian cochlea. Physiol. Rev. 81, 1305-1352.
2. Moore, B.C.J., 2003. An Introduction to the Psychology of Hearing. London, UK: Academic Press.
3. Rhode, W.S., Recio, A., 2001. Basilar-membrane response to multicomponent stimuli in chinchilla. J. Acoust. Soc. Am. 110, 981-994.
4. Harris, F.P., Glattke, T.J., 1992. The use of suppression to determine the characteristics of otoacoustic emissions. Sem. Hear. 13, 67-80.
5. Yoshikawa, H., Smurzynski, J., Probst R., 2000. Suppression of tone burst evoked otoacoustic emissions in relation to frequency separation. Hear. Res. 148, 95-106.
6. Jedrzejczak, W.W., Smurzynski, J., Blinowska, K.J., 2008. Origin of suppression of otoacoustic emissions evoked by two-tone bursts. Hear Res. 235, 80-89.
7. Mallat, S.G., Zhang, Z., 1993. Matching pursuit with time-frequency dictionaries. IEEE Trans. Signal Process. 41, 3397–3415.
8. Jedrzejczak, W.W., Blinowska, K.J., Konopka W., 2006. Resonant modes in transiently evoked otoacoustic emissions and asymmetries between left and right ear. J. Acoust. Soc. Am. 119, 2226-2231.

SECTION III

NEW MEASUREMENT TECHNIQUES

T. Ren, M.A. Cheatham and C.-P. Richter

DISTORTION PRODUCT OTOACOUSTIC EMISSIONS EVOKED BY TONE COMPLEXES

SEBASTIAAN W. F. MEENDERINK, MARCEL VAN DER HEIJDEN

Department of Neuroscience, Erasmus MC, Dr. Molewaterplein 50
Rotterdam, 3015 GE, the Netherlands

Distortion product otoacoustic emissions are a manifestation of nonlinear interaction between two or more stimulus components within the cochlea. Most studies employ a two-tone stimulus to evoke them. In this study we used a more complex stimulus, in which one of the primaries of the customary two-tone stimulus was replaced by an irregularly spaced tone complex. We obtained data from Mongolian gerbils, and investigated whether the novel stimulus has added value in terms of measurement efficiency, and whether it allows the derivation of information on cochlear mechanics that cannot be derived from two-tone stimuli.

1 Introduction

To date, most studies on distortion product otoacoustic emissions (DPOAEs) have used two-tone stimuli [1]. In this study we explore the use of multi-tone stimuli. The main motivation was our intention to combine, within single experiments, single-unit recordings of the auditory nerve with DPOAE measurements. Particularly for high-frequency stimulation, which does not evoke phase-locked responses, the use of complex stimuli has many advantages over tone pairs. The present study is confined to emissions.

We investigated how the DPOAE spectrum is affected when the traditional tone pair is replaced by a tone complex, whether such a stimulus has added value in terms of measurement efficiency, and whether it allows the derivation of information on cochlear mechanics that cannot be derived from two-tone stimuli.

Our stimuli were based on the irregularly spaced tone complexes (termed zwuis stimuli) introduced in recent neurophysiological work [2]. The essence of this stimulus is a careful choice of the tone frequencies that allows the distinction of different types and orders of distortion products in the response. We adapted the zwuis stimulus to optimize DPOAE generation by replacing one of the primary tones of the customary two-tone stimulus by a zwuis-compliant tone complex. Our data show that this novel stimulus paradigm evokes otoacoustic emissions that are very similar to DPOAEs evoked by tone pairs. In addition, a new family of DPs is produced that provides a new window on the generation and propagation of OAEs.

2 Methods

We present data obtained from the Mongolian gerbil (*Meriones unguiculatus*) anaesthetized by *i.p.* injection of ketamine/xylazine mixture. The right pinna of the animal was removed and a custom-built probe connected to a B&K 4192 microphone

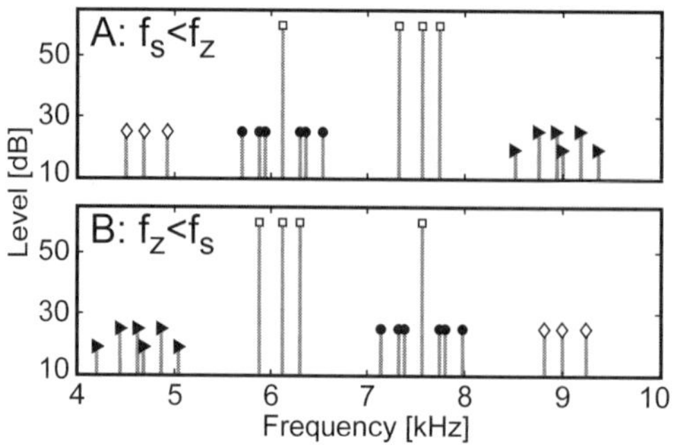

Figure 1. Cartoons of amplitude spectra illustrating the used stimuli and the resulting groups of OAEs. □: stimulus tones; ▶: near group; ◊: far group; •: suppression group. DPOAE names are based on the spectral distance between the DPOAEs and the zwuis complex. **A:** The single tone is below the zwuis complex (f_s<f_z). **B:** The single tone is above the zwuis complex (f_s>f_z).

was placed inside the ear-canal. For stimulus delivery, two speakers (Visaton FRS 7) were connected to the probe via flexible tubing.

The stimulus is illustrated in Fig. 1. One of the primary tones was replaced by a so-called zwuis-complex. This is defined as a set of components for which all the difference and sum frequencies are unique. The *M*-tone zwuis-complex ($f_{z(1)}, f_{z(2)} \ldots f_{z(M)}$) may either replace the higher primary (f_s<f_z; Fig. 1A) or the lower one (f_s>f_z; Fig. 1B). Interaction of the single primary with a unique pair of stimulus frequencies from the zwuis-complex results in the depicted 3rd-order DPOAEs.

3 Results

As an example. Fig. 2 shows an amplitude spectrum of the ear-canal pressure when the higher primary was replaced by a 9-tone zwuis-complex. The aforementioned 3rd-order DPOAEs are readily identified, Each of these DPOAE frequencies results from a unique set of three stimulus frequencies.

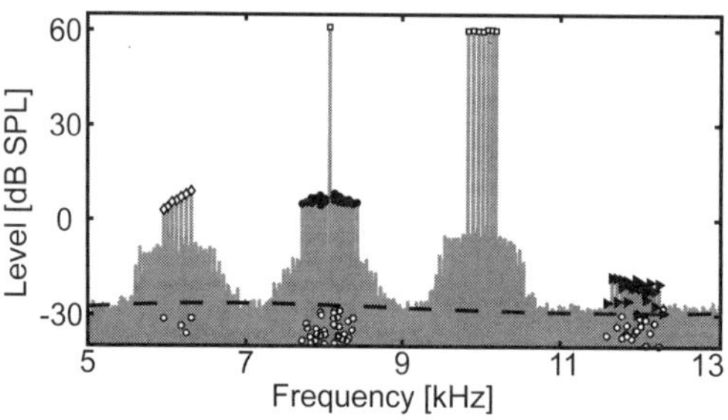

Figure 2. Amplitude spectrum of the recorded microphone signal. Besides the stimulus tones, three groups of 3rd-order OAEs can be identified. The dashed line indicates the estimated noise floor, while the white circles give the recorded system distortion at the OAE frequencies. See Fig. 1 for description of remaining symbols.

The noise floor, estimated by considering spectral components incommensurate with the stimulus, is indicated by the dashed line. System distortion at the DPOAE-frequencies, determined from control experiments without a gerbil, is also shown (white circles). The *M*-tone zwuis-complex $f_{z(1)}, f_{z(2)} \ldots f_{z(M)}$ may either replace the higher primary (f_z>f_s) or the lower one (f_z<f_s). The resulting three groups of 3rd-order DPOAEs are:

1. M different DPOAEs replacing the $2f_s$-f_z: the "far group"
2. $\frac{1}{2}M(M+1)$ different DPOAEs replacing the $2f_z$-f_s: the "near group"
3. $M(M-1)$ different DPOAEs surrounding f_s: the "suppression group"

where the naming is based on the spectral distance between DPOAEs and the zwuis complex.

3.1 *The far group*

These DPOAEs occur spectrally at the same side as, but beyond the single component, *i.e.* below f_s when $f_s<f_z$ (Fig. 1A) and above f_s when $f_s>f_z$ (Fig. 1B). Their frequencies are given by $f_{\mathrm{far}}(a)=2f_s$ -$f_{z(a)}$, $a=1...M$. Thus, each of these emissions results from the interaction of only two primaries, f_s and $f_{z(a)}$, completely analogous to the conventional stimulus paradigm where only the two primaries f_s and $f_{z(a)}$ are presented.

We tested this analogy by first evoking emissions with a 6-tone zwuis-complex replacing the higher primary (Fig. 3), followed by 6 consecutive recordings in which each $f_{z(a)}$ was presented separately in combination with f_s. The multi-tone stimulus resulted in somewhat attenuated emission. However, for the DPOAE phase, no difference was observed between the two paradigms; the calculated group delays (slope of phase *vs.* frequency curves) were in both cases 7.0 ms and were within 30 μs of each other.

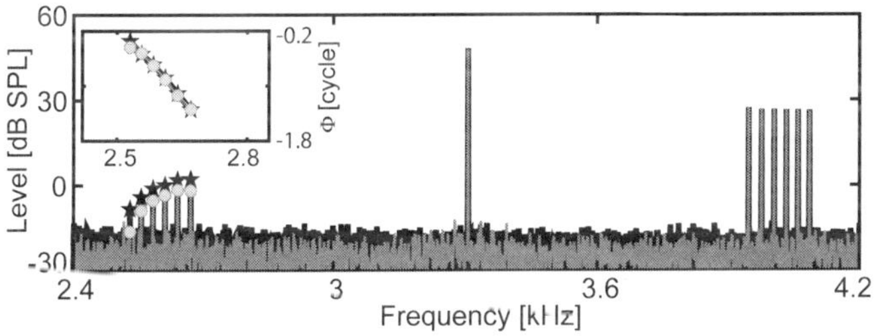

Figure 3. Comparison between amplitude spectra of far group emissions (gray line; circles) and DPOAEs evoked with six two-tone stimuli (black line; stars). The inset shows corres-ponding DPOAE phases.

3.2 *The near group*

The frequencies of the near group are given by $f_{\mathrm{near}}(a,b)=f_{z(a)}+f_{z(b)}-f_s$, $a\leq b$. They occur below f_z when $f_s>f_z$ (Fig. 4A) and above f_z when $f_s<f_z$ (not shown). The M-component subset for which $a=b$ results from the nonlinear interaction of only two primaries, but in general *three* primaries interact to generate each f_{near} component. In Fig. 4A, the $a=b$ subset is marked with white circles. Their amplitudes are ~6 dB lower than the remaining $a\neq b$ components. This is a combinatorial effect analogous to the coefficient 2 of the ab term in the expansion of $(a+b)^2$. Apart from this combinatorial effect, the $a=b$ subset has no special status within the group of f_{near} components.

Fig. 4B shows DP phase as a function of DP frequency for the f_{near} group. The bandwidth of the f_{near} group is twice the bandwidth of the zwuis complex.

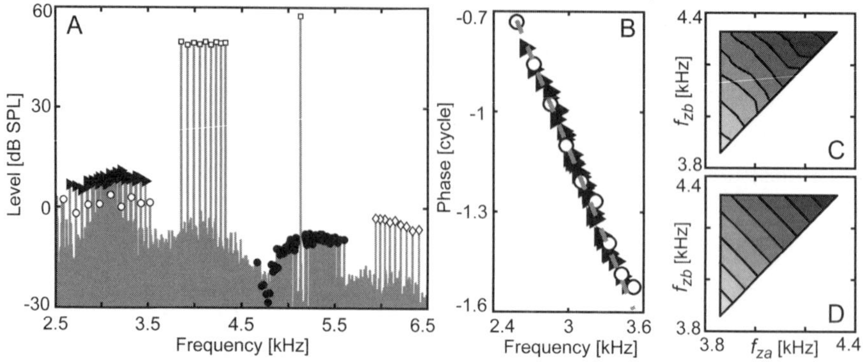

Figure 4. **A:** Amplitude spectrum showing the three 3rd-order DPOAE groups. In the near group (leftmost group), white circles indicate DPOAEs involving $f_a=f_b$. Other symbols are as in Fig. 1. **B:** Corresponding phase for the near group. A straight line (Eq. 1; gray dashed line) was fitted to these data: $\Phi_{near} = 1.66 - 0.92*f_{near}$. **C:** Same phase data, but presented in two-dimensional format. Contour lines are drawn with 0.1 cycle intervals. **D:** Same as **C**, but for the model ($\Phi=\Phi_0+\alpha f_{z(a)}+\beta f_{z(b)}$; see text) fitted to the phase data.

Over this limited domain, DP phase is well described by a straight line with intercept Φ_0 and slope $-\tau$:

$$\begin{aligned}\Phi_{\text{near}}(a,b) &= \Phi_0 - \tau f_{near}(a,b) \\ &= \Phi_0 - \tau(f_{z(a)} + f_{z(b)} - f_s)\end{aligned} \tag{1}$$

DP phase is here expressed as a function of DP frequency, so the two interacting primaries $f_{z(a)}$ and $f_{z(b)}$ occur only in the fixed combination $f_{z(a)}+f_{z(b)}$. But the data allow a more comprehensive evaluation of the dependence of Φ on $f_{z(a)}$ and $f_{z(b)}$, because each possible pair $f_{z(a,b)}$ ($a\leq b$) gives rise to a unique and identifiable DP. The complete, two-dimensional dependence of Φ on the primary pair ($f_{z(a)}, f_{z(b)}$) is shown as a contour plot in Fig 4C. The contours are perpendicular to the main diagonal, showing that Φ is determined by $f_{z(a)}+f_{z(b)}$. We fitted Φ to the model $\Phi=\Phi_0+\alpha f_{z(a)}+\beta f_{z(b)}$, yielding values of α and β that were not significantly different (Fig. 4D). Thus at hindsight, the phase data of Fig. 4C are exhaustively described by the intercept Φ_0 and group delay τ of Eq. 1. We found this to be the case for both spectral arrangements $f_s<f_z$ and $f_s>f_z$.

3.3 The suppression group

This group is novel; it does not occur in a two-primary stimulus paradigm. The DPs of this group occur at $f_{\text{supp}}(a,b)=f_s+f_{z(a)}-f_{z(b)}$. Each pair ($a\neq b$) produces a unique DP component. We refer to this new group as "suppression" DPOAEs for the following reason. The single primary f_s is modulated ("partially suppressed") by the envelope of the zwuis-complex. This envelope is dominated by difference frequencies $f_{z(a)}-f_{z(b)}$, resulting in the occurrence of sidebands at $f_s \pm(f_{z(a)}-f_{z(b)})$. These sidebands constitute the suppression group. Fig. 5 shows the suppression group of DPOAEs, both for $f_s>f_z$ (upper panels) and $f_s<f_z$ (lower panels). The phase curves are shown in Figs. 5B,E. As before, DP phase is plotted against DP frequency and we fitted these curves by straight lines.

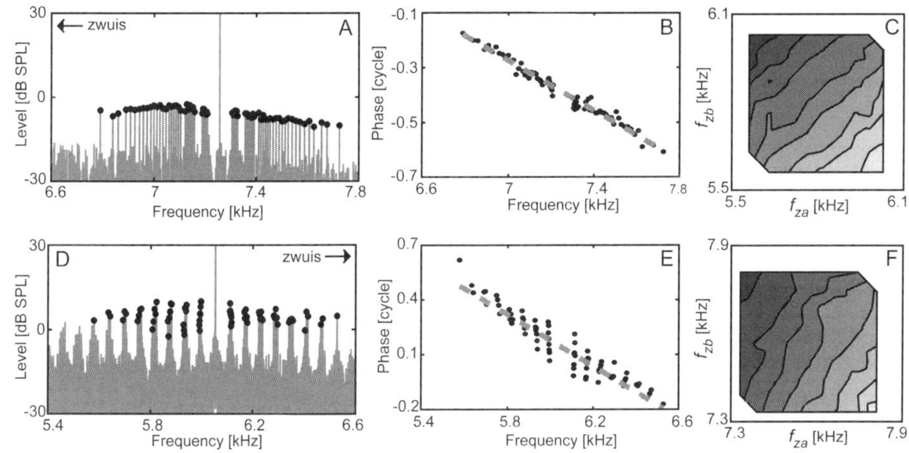

Figure 5: A: Amplitude spectrum for suppression group ($f_s>f_z$). **B:** Corresponding DPOAE phase. A straight line (Eq. 2; gray dashed line) was fitted to these data: $\Phi_{\text{supp}} = 2.89 - 0.45*f_{supp}$. **C:** Same phase data as in **B**, but presented in two-dimensional format. Contour lines are drawn with 0.1 cycle intervals. **D-F:** same as **A-C**, but for $f_s<f_z$. The straight line fitted to the phase data is given by: $\Phi_{supp} = 4.41 - 0.71*f_{supp}$.

In analogy with Eq. (1),

$$\begin{aligned}\Phi_{\text{supp}}(a,b) &= \Phi_0 - \tau f_{\text{supp}}(a,b) \\ &= \Phi_0 - \tau(f_s + f_{z(a)} - f_{z(b)})\end{aligned} \tag{2}$$

The phase patterns in the $f_s<f_z$ case (Fig. 5E) suggest that Eq. 2 is not an exhaustive description of DP phase as a function of primary frequencies. This is further analyzed by the full, two-dimensional phase dependence on ($f_{z(a)}, f_{z(b)}$) shown in Figs. 5C,F for the two cases $f_s>f_z$ and $f_s<f_z$. When $f_s>f_z$, (Fig. 5C) the contour lines are parallel to the main diagonal, in agreement with Eq. (2). But when $f_s<f_z$ (Fig. 5F), the contour lines are not parallel to the diagonal, and Eq. (2) fails to describe the frequency-dependence of the DP phase. This was confirmed by fitting $\Phi_{\text{supp}}=\Phi_0+\alpha f_{z(a)}+\beta f_{z(b)}$, yielding nonzero $\alpha+\beta$.

3.4 Model attempt

In a first attempt to understand the phase data, we evaluated a simple model based on the following assumptions: 1) scaling invariance [3]: within the frequency range of a single recording, cochlear phase $\Phi(x,f)$ as a function of cochlear position x and frequency f is given by $\Phi(x,f) = \Phi_0 - \tau(F_x - f)$, with F_x the characteristic frequency at position x; 2) DP generation can be approximated by distortion at a single location; 3) DPOAEs travel from their generation site to the ear canal via a backward traveling wave. Denoting the CF of the generation site by F_G, a straightforward calculation yields:

$$\Phi_{\text{supp}} = 2(\Phi_0 - \tau(f_{\text{DP}} - F_{\text{G}})) \tag{3}$$

where f_{DP} is the frequency of the measured DP component. Reasonable choices of F_G, such as f_s when $f_s>f_z$ and $\frac{1}{2}(f_{z(a)}+f_{z(b)})$ when $f_z>f_s$, led to poor agreement between model and data. The failure of Eq. (3) to describe our data was most apparent when pooling multiple datasets having a fixed set of f_z and differing in f_s. Dropping assumption (3) in favor of instantaneous backward propagation [4] (which merely removes the factor 2 in Eq. 3) did not improve the fit.

4 Discussion

Replacing one of the primaries of a two-tone stimulus by a zwuis group of tones results in a rich spectrum of 3rd-order DPOAEs. Two groups of DPOAEs are analogous to the two cubic difference tones evoked by the usual two-tone stimulus, and their phase data indicate that the analogy is a close one. The fact that so many DPOAEs can be measured in a single recording may provide a significant improvement over the traditional method of varying, across recordings, one or both primary frequencies. Apart from the obvious gain in measurement efficiency, simultaneous measurement of multiple components eliminates temporal effects that may plague sequential measurements.

On a more fundamental level, the three-tone interaction underlying most of the novel DPs allows a more comprehensive analysis of DP amplitude and phase than does the two-tone paradigm. Moreover, a whole new family of DPOAEs ("suppression group") arises, which possesses an interesting asymmetry in phase patterns depending on the relative frequencies of the single component and the tone complex.

Based on assumptions commonly used when analyzing DPOAEs evoked by two tones, we formulated a simple model for the description of DPOAE phase. It did not describe our phase data well. We consider this model as preliminary, but we do expect that the new type of DPOAE data will also provide more stringent tests of existing, more elaborate models of DPOAE generation and propagation. Last but not least, we are happy that the complex stimuli that we need for our neural recordings also generate DPOAEs that are amenable to systematic analysis. This means we will be able to combine the different types of measurements.

Acknowledgments

Part of this work was supported by NWO-VENI grant 863.08.003 to SWFM.

References

1. Probst, R., Lonsbury-Martin, B.L., Martin, G. K., 1991. A review of otoacoustic emissions. J. Acoust. Soc. Am. 89, 2027-2067.
2. van der Heijden, M., Joris, P.X., 2003. Cochlear phase and amplitude retrieved from the auditory nerve at arbitrary frequencies. J. Neurosci. 23, 9194-9198.
3. Shera, C.A., Talmadge, C.A., Tubis, A., 2000. Interrelations among distortion-product phase-gradient delays: Their connection to scaling symmetry and its breaking. J. Acoust. Soc. Am. 108, 2933-2948.
4. Ren, T. 2004. Reverse propagation of sound in the gerbil cochlea. Nat. Neurosci. 7, 333-334.

REMOVAL OF THE DPOAE SECOND GENERATION SOURCE WITH A PULSED PARADIGM METHOD IMPROVES HEARING THRESHOLD ESTIMATION IN HUMANS

DIANA TURCANU, ALES VETESNIK, ERNST DALHOFF, ANTHONY W. GUMMER

Section of Physiological Acoustics and Communication, Eberhard-Karls-University, Elfriede-Aulhorn-Str. 5, Tübingen, 72076, Germany

Objective hearing threshold estimation in humans by means of extrapolation of the input-output function of the cubic distortion product of otoacoustic emissions is negatively influenced by the secondary source of the DPOAE generation mechanism. To separate the primary source from the secondary source, a pulsed f2 paradigm was used in normal hearing subjects. The separated primary-source component of the DPOAE was used to estimate hearing threshold. The results show that the removal of the DPOAE second generation source significantly improves the estimate of hearing threshold.

1 Summary

A novel method of objective hearing threshold estimation in humans uses the extrapolation of the input-output (I/O) function of the cubic distortion product of otoacoustic emissions (DPOAE) measured with continuous tones [1]. Recently, it has been shown that the method is strongly influenced by the secondary source of the DPOAE generation mechanism [2]. The latter is partially responsible for deficiencies encountered in the threshold estimation method [2].

Using normal hearing subjects, we investigate the possibility of separating the primary source from the secondary source contribution in the time domain during the DPOAE onset period. For this purpose, a pulsed f2 primary paradigm is used [3]. Employment of this paradigm in a nonlinear hydrodynamic model of the human cochlea for describing the primary DPOAE source generation mechanism in the time domain showed that, while the DP component at the primary generation site is near its maximum, the DP component of the basilar membrane response at the secondary generation site is only starting to build up [4].

DPOAE I/O functions at the cubic difference frequency, 2f1-f2, were measured with 20-Hz frequency resolution in four normal hearing subjects. The primaries, f1 and f2, were presented continuously for f1 and pulsed for f2 (100 ms in a time-window of 200 ms). The responses of the primary tones were removed by varying the phases of the primaries by 90° for f1 and 180° for f2 and ensemble averaging of 4x50 time-windows of 200 ms each [3]. Stimulus parameters were: f2 = 1.8-2.5 kHz, f2/ f1 = 1.2, L2 = 25-65 dB SPL and L1 = 0.4L2 + 39 dB [5].

To demonstrate experimentally the outcomes of the cochlear model and to establish objectively the time necessary for the DPOAE primary source to reach its maximum before being influenced by the secondary generator, a time-windowing method [6] was applied to the DPOAEs measured by the pulsed f2 paradigm. First, the time signal of the

DPOAE was sampled in 1-ms steps from the onset of f2 and DP-grams were constructed for these time instants. Each constructed DP-gram was time-windowed to extract the primary and secondary source components. We show that the DPOAE primary source reaches its steady-state within 8-9 ms after the f2 onset, before the secondary source starts to contribute significantly. This suggests that the primary source component can be obtained by just taking the time-signal measured with the pulsed paradigm 8 ms after the f2 onset. Additionally, comparison is made between the decomposition method using the pulsed f2 primary paradigm and the method using time-windowing [6] of the steady-state DPOAEs.

Using the DPOAE primary-source component, separated by the pulsed paradigm method, I/O functions were constructed and hearing thresholds were estimated by extrapolating these functions to the L2 axis. For comparison, thresholds were also estimated from the steady-state DPOAEs. We found that the hearing threshold estimated using the pulsed paradigm mirrors the Békésy threshold and shows less variability between adjacent frequencies than thresholds estimated by the steady-state DPOAEs. Moreover, the standard deviation of the estimates obtained with the pulsed primary paradigm method (6 dB) was smaller than that obtained for the steady-state DPOAEs (14 dB).

Acknowledgments

This work is supported by the German Research Council DFG Gu 194/8-1.

References

1. Boege, P., Janssen, T., 2002. Pure-tone threshold estimation from extrapolated distortion product otoacoustic emission I/O functions in normal and cochlear hearing loss ears. J. Acoust. Soc. Am. 111, 1810–1818.
2. Mauermann, M., Kollmeier, B., 2004. Distortion product otoacoustic emission (DPOAE) input/output functions and the influence of the second DPOAE source. J. Acoust. Soc. Am. 116, 2199–2212.
3. Whitehead, M.L., Stagner, B.B., Martin, G.K., Lonsbury-Martin, B.L., 1996. Visualization of the onset of distortion-product otoacoustic emissions, and measurement of their latency. J. Acoust. Soc. Am. 110, 1663–1679.
4. Vetesnik, A., Nobili, R., Gummer, A., 2006. How does the inner ear generate distortion product otoacoustic emissions? ORL 68, 347–352.
5. Kummer, P., Janssen, T., Hulin, P., Arnold, W., 2000. Optimal L1-L2 primary tone level separation remains independent of test frequency in humans. Hear. Res. 146, 47–56.
6. Kalluri, R., Shera, C.A., 2001. Distortion-product source unmixing: A test of the two-mechanism model for DPOAE generation. J. Acoust. Soc. Am. 109, 622–637.

HARD X-RAYS CAN BE USED TO VISUALIZE COCHLEAR SOFT TISSUE DISPLACEMENTS IN A CLOSED COCHLEA

C.-P. RICHTER, A. FISHMAN, L. FAN., S. SHINTANI, C. RAU

Department of Otolaryngology, Northwestern University
Chicago, IL, USA

Experiments were made at the Advanced Photon Source (APS), Argonne National Laboratory. The APS is a synchrotron radiation source of the third generation, for which the particular characteristic is the highly coherent X-ray radiation. X-rays are generated with an undulator, inserted in a straight section of the storage ring. Images taken with hard X-rays at full field. A video flow algorithm by Lucas and Kanade was used to determine and quantify cochlear soft tissue displacements. The results show that displacements as low as 100 nm could be visualized.

1 Introduction

The dearth of experimental data on the inner workings of the organ of Corti is a simple consequence of technical problems related to making micromechanical measurements. Starting with Békésy's pioneering observations, much that has become available is based on some version of *in vitro* measurements. Exceptions are the somewhat limited *in vivo* measurements on tectorial membrane or reticular lamina vibrations obtained in the opened cochlear apex [1-4]. Again, in an *in vitro* cochlea preparation, Mammano and Ashmore [5] compared reticular lamina and basilar membrane motion using electrical stimulation. All other relevant work was performed either on isolated cochlear segments [6-10] or on the isolated cochlea preparation [11, 12]. Essentially all of the more recent work, except the experiments of Reuter et al. [6] Morioka et al. [7] Karavitaki et al. [13] and Richter et al. [14, 15] is based on laser interferometry, meaning that the measurements provide a motion component that is in line with the laser beam. By doing multiple measurements with controlled angle changes between the preparation and the beam, it is possible, at least in theory, to describe the three-dimensional motion of the target; this multi-dimensional approach, however, has not been used extensively and the reliability of the method is uncertain. Confocal laser interferometry permits measurements to be taken without placement of reflective material on the vibrating tissue. This latter technique would theoretically allow measurements of the motion of cellular elements within the organ of Corti [12, 16].

The limitation of all methods is their invasive character. Recently, efforts have been made towards developing novel methods to measure cochlear mechanics in an intact cochlea and at mid cochlear section. One promising approach is optical coherence tomography (OCT). At present, however, the method still requires opening of the cochlea [17-19].

We have introduced for imaging of soft tissue using hard X-rays and inline phase contrast, which does not require opening of the intact cochlea [20]. With the present experiments we show that soft tissue displacements can be quantified.

2 Methods

2.1 The radiation source

Experiments were conducted at 34-ID-C beamline of the Advanced Photon Source (APS), Argonne National Laboratory. The APS is a synchrotron radiation source of the third generation, for which the particular characteristic is the highly coherent X-ray radiation. X-rays are generated with an undulator, inserted in a straight section of the storage ring. The full width half maximum (FWHM) source size is 600 μm by 40 μm (horizontal x vertical) and the beam divergence is 40 μrad by 12 μrad (horizontal x vertical). The liquid nitrogen-cooled mirror rejects higher undulator harmonics above 15 keV at an incidence angle of 5 mrad. The fixed-exit double crystal monochromator is water-cooled and yields an energy bandwidth of $\Delta E/E=10^{-4}$ using Si (111) crystals over an energy range of 6-30 keV. The beam size is ~1 mm^2 at a distance of 55 m from the source and the monochromatic flux is about 10^{13} Photons/second.

The imaging setup was at 55 m from the source, consisting of a sample stage and a CCD-based detector system both placed on a vibration isolation table. The distance between sample stage and detector can be easily selected for absorption and/or in-line phase contrast imaging. For this study, all images were captured at a distance d = 40 mm. All alignment stages are fully motorized and are also chosen to meet the requirements for high-resolution tomography. The camera system consists of a scintillation screen coupled via an optical microscope to a CCD detector. The scintillation screen transforms the X-rays into visible light, which was then projected with an optical microscope onto the chip of a CCD camera (Kodak KX2e). The camera has 14-bit resolution with a 1560x1024 pixel array and 9x9 μm^2 size of each pixel. Using a 20x objective lens (Mitutoyo) the effective pixel size on the screen is 0.45 μm. The scintillation screen was a Yttrium Aluminum Garnet (YAG) single crystals with an 6 μm-thick europium-doped active layer. This imaging system provided a resolution of 2.4 μm. The resolution was determined by taking the FWHM of the line-spread function, which was measured using a sharp edge.

2.2 Data acquisition

A series of images was captured at 2 seconds exposure time as a function of the pressure gradient across the cochlear partition. To generate the pressure gradient, a small opening of approximately 100 μm in diameter was drilled into scala tympani of the basal turn. A similar hole was then made in the cochlear apex. A small tubing was inserted through the hole in the cochlear base and was fixed with acrylic. The tubing was attached to a syringe and fluids were infused continuously through the cochlea at flow rates between

0.6 and 1.5 μl/minute. Additional images were taken for all projections: the "flat" beam profile without the sample (flat field) and the dark counting of the CCD without X-rays on the detector system (dark field). Inhomogeneities of the beam profile were corrected with these additional images captured at each pressure gradient. The corrected images were used as frame sequence for motion analysis of soft tissue using optical flow technique.

2.3 Optical flow analysis

The displacement of soft tissue was quantified by optical flow analysis. Optical flow estimates were generated using the Lucas and Kanade algorithm [21] implemented in the FlowJ plugin for ImageJ (http://rsb.info.nih.gov/ij/). The Lucas and Kanade algorithm is a gradient-based technique that estimates two-dimensional velocity fields by using spatio-temporal derivatives of the image intensity field. The parameters used to determine the spatial and temporal properties of the analysis were as follows: σ_s, the dimensionless space constant for velocity determination; σ_t, the dimensionless time constant for velocity determination; τ, the threshold for velocity detection (pixel/s); σ_W, the dimensionless space constant over which velocities were constrained to be regular; and ρ, the saturation point of the pseudocolor scale (pixel/s). The resulting flow fields were visualized using DCM 2D format also implemented by the FlowJ plugin.

3 Results

Optical-flow analysis revealed that increasing of pressure displaced cochlear soft tissue. At flow rate of 40%, velocity range was between 0.11 to 0.37 pixel per frame, which corresponds to 50 nm and 167 nm per frame. For the comparison, the data was obtained at the same location for each of the three flow rates shown in Fig. 1. The error bars show the standard deviation. The lowest average velocity is 110 nm in this measurement.

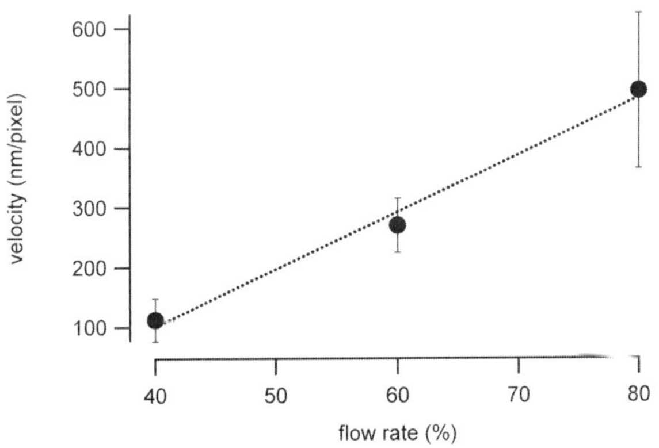

Figure 1. Displacements for different flow rates were determined (100% = 1.5 μl/min). Increasing the flow rate increased the displacement of the structures between the frames as could be determined by the intensity of color plot (blue lines) on the right. The smallest displacement detected in the pilot experiment was ~100nm.

4 Discussion

To our knowledge, these results are the first demonstration of how X-ray phase contrast can complement existing imaging methods to investigate movement of soft tissue of the inner ear. The present approach allowed us to measure movement of soft cochlear tissue at a high spatial resolution without opening the cochlea. Phase contrast X-ray imaging techniques can provide the information imparted by conventional radiography, with additional information on soft tissue. The photon flux, photon energy, sample-detector distance, image acquisition time, and orientation of the cochlea relative to the beam are important factors for visualizing soft tissues structures using in line phase-contrast imaging technique.

When hard X-rays are used, it is important to ensure that the tissue is not damaged by the radiation dosage applied. When determining the radiation dose, several factors have to be considered: the particle type, the beam energy spectrum, and the properties of the absorbing medium. The problem is generally complex for photons because the dose is delivered by recoiling charged particles in the medium, which have a spread of energies and directions. For the Advanced Photon Source at Argonne National Laboratories, the thickness of the stop-shutter has been estimated assuming that the remaining radiation dose for humans is less than 1.25 μSv/h (0.125 mrem/h). The estimate for the size of the stop-shutter also provides information that can be used to estimate the maximum dosage that would be applied to tissue when exposed to the beam [22]. The dose is about 28 mrem/s. For comparison, the dose of a conventional X-ray image is ~14 mrem. Keep in mind, however, that the dose applied during the experiments, is less than the maximum dose because we used a monochromatic beam (whereas the dose calculation was done for the full spectral beam). Furthermore, a fast shutter minimized the exposure time, and consequently the radiation dose and risk for tissue damage.

Acknowledgments

The authors thank all members of UNICAT for their support, in particular K. M. Peterson and P. R. Jemian. CR would like to thank T.C. Chiang for his support. The UNICAT facility at the Advanced Photon Source (APS) is supported by the U.S. DOE under Award No. DEFG02-91ER45439, through the Frederick Seitz Materials Research Laboratory at the University of Illinois at Urbana-Champaign, the ORNL (U.S. DOE contract DE-AC05-00OR22725 with UT-Battelle LLC), NIST (U.S. Department of Commerce) and UOP LLC. The APS is supported by the U.S. DOE, Basic Energy Sciences, Office of Science under contract No. W-31-109-ENG-38. CPR is supported by a grant from the NSF (IBN-0415901).

References

1. Rhode, W. S., and Cooper, N. P. 1996. Nonlinear mechanisms in the apical turn of the chinchilla cochlea. In: Lewis, E. R., Long, G. R., Lyon, R. F., Narins, P. M.,

Steele, C. R., and Hecht-Poinar, E., (Eds.), Diversity in auditory mechanics, 1 ed. World Scientific, Singapore//New Jersy//London//Hong Kong. pp. 318-324.
2. Hao, L. F., and Khanna, S. M., 1996. Reissner's membrane vibrations in the apical turn of a living guinea pig cochlea. Hear. Res. 99, 176-189.
3. Zinn, C., Maier, H., Zenner, H., and Gummer, A. W., 2000 Apr. Evidence for active, nonlinear, negative feedback in the vibration response of the apical region of the in-vivo guinea-pig cochlea. Hear. Res. 142, 159-183.
4. Dong, W., and Cooper, N. P., 2002. Three dimensional, in vivo measurements of the tectorial membrane's vibratory responses to sound. Assoc. Res. Otolaryngol., Vol. 25. 905.
5. Mammano, F., and Ashmore, J. F., 1993. Reverse transduction measured in the isolated cochlea by laser Michelson interferometry. Nature 365, 838-841.
6. Reuter, G., Gitter, A. H., Thurm, U., and Zenner, H. P., 1992. High frequency radial movements of the reticular lamina induced by outer hair cell motility. Hear. Res. 60, 236-246.
7. Morioka, I., Reuter, G., Reiss, P., Gummer, A. W., Hemmert, W., and Zenner, H. P., 1995. Sound-induced displacement responses in the plane of the organ of Corti in the isolated guinea-pig cochlea. Hear. Res. 83, 142-150.
8. Nowotny, M., and Gummer, A. W., 2006. Nanomechanics of the subtectorial space caused by electromechanics of cochlear outer hair cells. Proc. Natl. Acad. Sci. USA. 103, 2120-2125.
9. Scherer, M. P., and Gummer, A. W., 2004. Vibration pattern of the organ of Corti up to 50 kHz: evidence for resonant electromechanical force. Proc. Natl. Acad. Sci. USA. 101, 17652-17657.
10. Chan, D. K., and Hudspeth, A. J., 2005. Ca2+ current-driven nonlinear amplification by the mammalian cochlea in vitro. Nat. Neurosci. 8, 149-155.
11. Decraemer, W. F., Khanna, S. M., and Funnell, W. R., 1989 Mar. Interferometric measurement of the amplitude and phase of tympanic membrane vibrations in cat. Hear. Res. 38, 1-17.
12. Fridberger, A., Boutet de Monvel, J., and Ulfendahl, M., 2002. Internal shearing within the hearing organ evoked by basilar membrane motion. J. Neurosci. 22, 9850-9857.
13. Karavitaki, K. D., Mountain, D. C., and Cody, A. R. 1996. Electrically evoked micromechanical movements from the apical turn of the gerbil cochlea. In: Lewis, E. R., Long, G. R., Lyon, R. F., Narins, P. M., Steele, C. R., and Hecht-Poinar, E., (Eds.), Diversity in auditory mechanics, 1 ed. World Scientific, Singapore//New Jersy//London//Hong Kong. pp. 392-397.
14. Richter, C.-P., and Dallos, P. 2003. Micromechanics in the gerbil hemicochlea. In: Gummer, T., (Ed.), Meeting on cochlear mechanics, Titisee (Germany), 1 ed. pp. 97-109.
15. Richter, C.-P., and Dallos, P. 2001. Multiple modes of vibration detected in the gerbil hemicochlea. In: Breebaart, D. J., Houstma, A. J. M., Kohlrausch, A., Prijs, V. F., and Schoonhoven, R., (Eds.), Physiological and Psychophysical Bases of Audiory Function, Vol. 1, 1 ed. Shaker Publishing, St. Maateslaan. pp. 44-50.
16. Fridberger, A., and de Monvel, J. B., 2003 May. Sound-induced differential motion within the hearing organ. Nat. Neurosci. 6, 446-448.

17. Choudhury, N., Song, G., Chen, F., Matthews, S., Tschinkel, T., Zheng, J., Jacques, S. L., and Nuttall, A. L., 2006. Low coherence interferometry of the cochlear partition. Hear. Res. 220, 1-9.
18. Chen, F., Choudhury, N., Zheng, J., Matthews, S., Nutall, A. L., and Jacques, S. L., 2007. In vivo imaging and low-coherence interferometry of organ of Corti vibration. J. Biomed. Opt. 12, 021006.
19. Hong, S. S., and Freeman, D. M., 2006. Doppler optical coherence microscopy for studies of cochlear mechanics. J. Biomed. Opt. 11, 054014.
20. Rau, C., Robinson, I. K., and Richter, C. P., 2006. Visualizing soft tissue in the mammalian cochlea with coherent hard X-rays. Microsc. Res. Tech. 69, 660-665.
21. Lucas, B., and Kanade, T., 1981. An iterative image resgistration technique with an application to stereo vision. Proc. DARPA Image Understanding Workshop. 121-130.
22. Job, P. K., and Micklich, B. J., 2005. Design calculations for the Advanced Photon Source safety shutters. ANL/APS/LS. -309.

SECTION IV

MICROMECHANICS: BM, TM AND SUB-TECTORIAL SPACE

A.W. Gummer and W.E. Brownell

FREQUENCY-SELECTIVE RESPONSE OF THE TECTORIAL MEMBRANE IN THE FROG BASILAR PAPILLA

R.L.M. SCHOFFELEN, J.M. SEGENHOUT, P. VAN DIJK

Dpt of Otothinolaryngology/Head and Neck Surgery, University Medical Center Groningen, University of Groningen, Groningen, The Netherlands

The frog's basilar papilla is a useful study object for cochlear mechanics, because of it's relatively simple anatomy and functionality. We investigated the displacement amplitudes of the basilar papilla's tectorial membrane in response to stimulation of the oval window at various frequencies within the auditory range of the Northern leopard frog. From our measurement data we find that the tectorial membrane exhibits a frequency selective response. The peak response was found to occur at 1500Hz in correspondence with known data for the response of auditory nerve fibers from the organ.
From these data we conclude that mechanical tuning contributes significantly to the frequency selectivity of the frog's basilar papilla

1. Introduction

The frog inner ear is unique among vertebrates in that it contains two dedicated auditory end organs, the amphibian papilla (AP) and the basilar papilla (BP). The animal's auditory range is divided over these two organs. The AP detects the lower part of the frequency range, from 100Hz up to approximately 1000Hz [1], in most species. The higher frequencies are detected in the BP. The range of the latter organ is species dependent [2].

Both the AP and the BP lack a basilar membrane (BM), but do have a tectorial membrane (TM) connecting directly to the stereovilli of the hair cells. Presumably, the AP is the functionally more complex organ of the two. It is partially electrically tuned [3], it exhibits tonotopy, and it emits spontaneous otoacoustic emissions. The hair cell orientation and TM structures are far more complex and neural responses are more sharply tuned in comparison with the BP's. The BP seems to function in a simpler manner. It lacks efferent innervation, electrical tuning, spontaneous otoactoustic emissions, and has a rather simple hair-cell orientation map [4]. Responses of the auditory nerve fibers connecting to the BP are broadly tuned with a single characteristic frequency [5].

Because of the BP's anatomy and functionality, we hypothesize that the frequency-selectivity of this organ must be the result of mechanical tuning. This hypothesis has been put forward and investigated before. Earlier investigations focused on models of the basilar papilla [6], or the entire inner ear [7], or on measurements of associated structures [8]. We investigated the TM by means of optical recordings of its response to stimulation of the oval window.

2. Methods

Animals and preparation: In our experiments we used Northern leopard frogs, *Rana pipiens pipiens*. The animals were obtained through a commercial supplier

(Charles D. Sullivan Co. Inc, Nashville (TN), USA, via Exoterra Schaudi GmbH, Holzheim, Germany), and housed in the university's laboratory-animal facility until the day of the experiment. All procedures were approved in advanced by the Institutional Animal Care and Use Committee of the University of Groningen.

The animal was terminated prior to the experiment through a double pith procedure. The ear was excised from the skull. The outside of the otic capsule was cleaned of any soft tissue, and the tympanic membrane and the lateral part of the columella were removed. The perilymphatic space was opened at the round window to allow a direct view onto the BP. The endolymphatic labyrinth remained closed. A cylindrical perspex prothesis was glued on the operculum to facilitate positioning of the stimulator.

The specimen was submerged in amphibian ringer solution (Carolina Biological Supply Company, Burlington (NC), USA) throughout the preparation and further experimental procedures.

Measurement set-up: The heart of the measurement set-up is a trifocal microscope (Olympus Corporation, Japan) which sits on a vibration-isolation table inside a darkened sound isolating booth. Any equipment that is not mounted directly on the microscope is kept outside of the booth, except for the stimulator's amplifier.

The microscope has 5x (NA=0.10) and 40x (NA=0.80) water-immersion objectives. The 40x objective is connected to the microscope's revolver with a piezo-electric positioner (Physik Instrumente GmbH & Co., Karlsruhe, Germany), which is used to remotely control the focal plane. The microscope's lighting system has been adapted to house a green Luxeon power LED (Lumileds Lighting, San Jose (CA), USA). A scientific grade digital camera (DVC Company, Austin (TX), USA) is mounted on the microscope with a 0.5x video adapter.

A closed-loop piezo actuator (Physik Instrumente) is used as a stimulation device. The actuator is set in a micro-positioner (Kanetec Co., Ltd., Kyoto, Japan) which sits on the microscope table. A custom needle is connected to the piezo stack.

A computer (National Instruments, Austin (TX), USA) is used to automate the measurement control and data-acquisition. The system runs on Windows XP (Microsoft Corp., Redmond (WA), USA); signal control and image acquisition are programmed in LabView v8.2 (National Instruments). An external trigger signal is used to synchronize the onset of the various signals.

Finally, an analogue oscilloscope (Hameg Instruments GmbH, Mainhausen, Germany) is connected to the loop-back output of the stimulator's amplifier.

Procedure: The reduced specimen was positioned under the microscope and clamped firmly into place; its position was adapted to optimize a direct view of the BP's TM from the round-window side. The stimulator needle was applied to the center of the operculum prothesis.

Stimulation signals were sinusoids of varying frequency, applied to the oval window. After the stimulus was switched on the LED was used to stroboscopically illuminate the specimen. The illumination frequency was equal to the stimulation frequency. The illumination duty cycle was 10%, with the 'on' phase of the LED centered around a specified phase of the stimulation signal. Images were taken with the camera; the acquisition time and voltage over the LED were adjusted for each specimen to optimize the use of the camera's output range. For each tested frequency, images were taken at 8 equidistant phases of the stimulus signal at a 20x magnification.

For each frequency this procedure was repeated for at least 30 focal planes within the TM, spaced $1\mu m$ apart. These were combined to generate a 3D image of the motion of the TM under stimulation of the oval window.

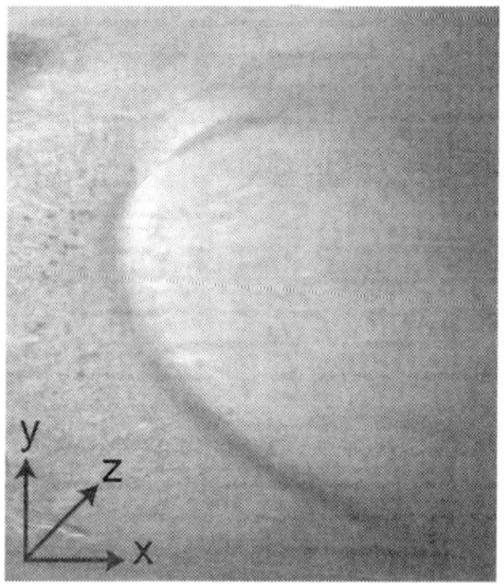

Fig. 1. Individual photo taken from data set, with axis-convention definition.

Data analysis: Data-analysis was performed on a Linux workstation, using the nD image analysis package developed at the Massachusetts Institute of Technology's department of Electrical Engineering and Computer Science. This software uses optical-flow analysis for the calculation of the displacement of structures within a sequence of images [9].

Two separate analysis procedures were used. The first method was used to achieve an initial overview of the motion in the membrane. For each pixel location in the y-direction of the data set, an xz-cross section was taken (axis convention defined in Fig. 1). We defined 800 voxels in each 2D image: 80 in the x-direction by 10 in the z direction. The nD software was used to trace the motion of each voxel. The displacements were averaged in the z-direction. An averaging procedure in the y direction was used for initial smoothing.
Next, we fitted a sinusoidal function to the displacement data. This procedure resulted in a response-amplitude field that can be superposed on the data image and evaluated. The amplitude field was then smoothed with a gaussian kernel (FWHM = 0.83 voxels) in order to increase insight and presentability.

The second method uses regions of interest (ROIs) to gain a more detailed view of the motion of the membrane. Six rectangular ROIs were defined within the TM in the 3D image. Each ROI was approximately between one fifth and one quarter of the TM size in same direction along the x and y axes, and approximately one third of the TM size in the z direction. The 3D motion of the membrane was then traced by the nD software and output as a x, y, and z amplitudes. We evaluated only the amplitude data for the fundamental frequency. The amplitude of the 3D displacement vector was calculated for each ROI at each stimulation frequency. The ROIs were such that there were two ROIs in each of the following three zones: the TM's free edge, the center, and the area close to the connection with the hair-bundles. After computation of the displacement amplitude the ROIs were averaged per zone.

3. Results

In Fig. 2 an overview of the response-amplitude fields for an individual specimen is displayed. The images were constructed using the first analysis method discussed above. The amplitude data were corrected for variations in the stimulation signal due to the response of the piezo stimulator.

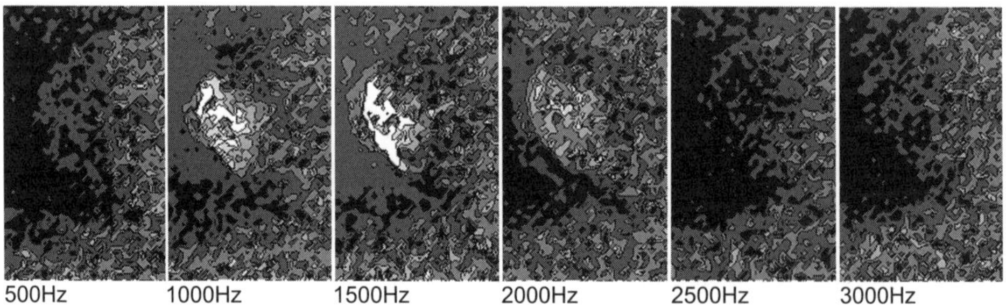

Fig. 2. Response-amplitude fields for the x direction at various frequencies, data from an individual specimen. (Displacements in de z direction show a similar pattern). The grayscale-code runs from black for no displacement, to white for maximum displacement; gray values are linearly scaled. Measurement data were linearly corrected for differences in stimulation amplitude; correction to the maximum applied amplitude of 0.2μm.

These images clearly show differences in the response amplitude of the TM for different frequencies. The TM displacement is low at stimulation frequencies of 500Hz, 2500Hz, and 3000Hz; at 1000Hz, 1500Hz, and 2000Hz there is considerably more displacement of the TM. The areas surrounding the TM show hardly any increase in response. The peak response appears to occur at 1500Hz.

Closer comparison with original images yields that, within the TM, the areas closest to the lumen's boundary show the largest responses, while the TM's free edge is hardly displaced at all.

The second analysis method described, gives us a quantification of the observations from the response-amplitude fields. The average displacements of the ROIs in the different zones of the TM are displayed in Fig. 3. In this measurement the stimulus amplitude was approximately 48nm; we applied a linear correction for variations in the stimulus amplitude to match this amplitude.

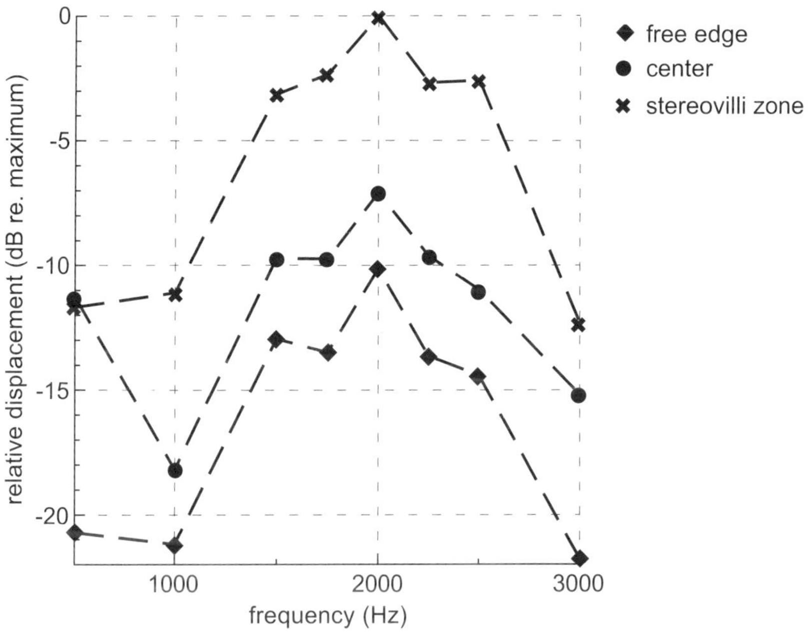

Fig. 3. Relative frequency dependence of responses in three zones of the TM: squares mark the zone near the TM's edge in the lumen, circles the central zone between edge and hair bundle connections, and crosses the near the hair bundle attachments. All three zones contain 2 ROIs of which the displacement measurements are averaged in this graph. All displacements are given in dB relative to the maximum displacement measurement in the graph.

In this specimen, the peak response occurs at 2000Hz for all zones. The notion that maximum response is found near the hair bundle connections is confirmed. The TM's amplitude range is approximately 12dB across the measured frequencies for a constant stimulation signal. On the decibel scale the response curves are remarkably congruent: the difference between the TM edge and the hair-bundle zone is approximately 10 to 12dB, while the central zone's response is 6 to 8dB below the hair-bundle zone's for most measurement points.

4. Discussion

The data presented above proof that the frequency-selective response of the frog's BP can be observed in its tectorial membrane. In our measurements, the peak responses were observed at 1500Hz and 2000Hz for two different specimens. This corresponds with the organ's tuning, as observed when measuring distortion product oto-acoustic emissions [10] and auditory nerve-fiber responses [5].

The amplitude distribution within the TM shows that the displacement of the membrane is largest near its coupling to the stereovilli. The response of this area surpasses the response of the TM's free edge by approximately 10dB. Comparing our results with Van Bergeijk's [6] observations on mechanical models of the BP's TM, we must conclude that we found no experimental evidence for tonotopy or traveling waves to occur in the membrane.

Based on our current data we can not claim that the TM itself tunes the response. The tuning may be the result of the properties of the contact membrane and the lumen, as suggested by Purgue and Narins [7], or of the hair-bundle stiffness. However, the presence of a frequency-selective response in the tectorial membrane, without apparent central control of the organ, provides evidence for a mechanical basis of the tuning of the BP.

Acknowledgments

We would like to thank D.M. Freeman and A.J. Aranyosi for making the nD image analysis software available to us, and for their support during the start up of the experiments.

This study was supported by the Heinsius Houbolt Foundation and the Netherlands Organisation for Scientific Research, and is part of the research program of our department: Communication through Hearing and Speech.

References

1. Wever, E. G., *The amphibian ear* (Princeton University Press, USA, 1985).
2. Schoffelen, R. L. M., Segenhout, J. M. and van Dijk, P., 2008. Mechanics of the exceptional anuran ear. J Comp Physiol A. 194, 417–428.
3. Pitchford, S. and Ashmore, J. F., 1987. An electrical resonance in hair cells of the amphibian papilla of the frog *Rana temporaria*. Hear Res. 27, 75–83.
4. Lewis, E. R. and Narins, P. M., The acoustic periphery of amphibians; anatomy and physiology, in *Comparative hearing: Fish and Amphibians*, eds. Fay, R. R. and Popper, A. N., Springer Handbook of Auditory Research, Vol. 11 (Springer, 1999).
5. Ronken, D. A., 1990. Basic properties of auditory-nerve responses from a 'simple' ear: the basilar papilla of the frog. Hear Res. 47, 63–82.
6. Van Bergeijk, W. A., 1957. Observations on models of the basilar papilla of the frog's ear. J Acoust Soc Am. 29, 1159–1162.
7. Purgue, A. P. and Narins, P. M., 2000. A model for energy flow in the inner ear of the bullfrog (*Rana catesbeiana*). J Comp Physiol A. 186, 489–495.
8. Purgue, A. P. and Narins, P. M., 2000. Mechanics of the inner ear of the bullfrog (*Rana catesbeiana*): the contact membranes and the periotic canal. J Comp Physiol A. 186, 481–488.
9. Davis, C. and Freeman, D., 1998. Using a light microscope to measure motions with nanometer accuracy. Opt Eng. 37, 1299–1304.
10. Meenderink, S. W. F., Van Dijk, P. and Narins, P. M., 2005. Comparison between distortion product otoacoustic emissions and nerve fiber responses from the basilar papilla of the frog. J Acoust Soc Am. 117, 3165–73.

Comments and Discussion

Aranyosi: In the anatomy SEM image you showed, the TM looks somewhat like a pendulum. Are your results consistent with it moving like one as well?

Schoffelen: From the motion data I showed, the impression may be that the TM moves in a pendulum-like manner. However, at this point that seems to be too simple as a working model.

Several questions and limitations that underlie this this statement:

(1) I only showed the amplitude distributions. For pendulum-like motion to exist, no phase differences between the different regions of the TM may exist. I am looking at those data as we speak, but the results aren't conclusive yet.

(2) If we ignore the first point for now. The peak displacement of a point on a pendulum should be linearly related to the distance to the rotational axis. Let's position the rotational axis of a hypothetical pendulum at the 'free edge' of the TM, and assume the amplitude of to be measured (on average) at the center of the zones. The difference in max. displacement between zones 1 and 3 (edge and hair bundles, respectively) would in that case be 14dB; a value we only approximately achieved in one specimen (see conference paper) – other specimens show lower response differences. Furthermore, we would expect the amplitude of zone 2 (center), to be about 10dB higher than the amplitude of zone 1, and about 4dB lower than that of zone 3. We cannot consistently find that ratio in our data.

We could achieve the measured dB–value ratios between any two zones by moving the rotational axis, of course. However, that would not solve the problem of third zone's response not being in line with the expectations.

(3) The biggest problem with a simple pendulum model is anatomical: the connection of the TM to the hair bundles is away from the rotational axis of the pendulum, therefore the motion would be damped at the "bottom" of the pendulum. The stiffness of the TM would, most likely not be high enough to behave like a (rigid) pendulum in that situation. (This may account for some of the problems mentioned under point 2.)

So at the moment we cannot simplify this TM any further than "a membrane", and our concept of it is somewhere between a pendulum or door and Van Bergeijk's (1957) model of a drum-like membrane [6].

MECHANICAL RESPONSE OF THE BASILAR MEMBRANE TO LATERAL MICROMANIPULATION

S. O. NEWBURG[1*], A. ZOSULS[1], P. E. BARBONE[2], D. C. MOUNTAIN[1]

[1] *Department of Biomedical Engineering and Hearing Research Center, Boston University*
[2] *Department of Aerospace and Mechanical Engineering, Boston University*
[*] *44 Cummington Street, Boston, MA 02215, USA*

The role of wave propagation in the cochlea can be founded on detailed knowledge of the mechanical properties of the basilar membrane. In this study, measurements of lateral point stiffness and longitudinal coupling were made on gerbil basilar membranes *in vitro*. We designed an integrated optical imaging system and basilar membrane manipulator, in which a calibrated glass micropipette is used to displace the basilar membrane laterally. Because the tissue deformations occur in the optical plane of the microscope objective, detailed images of the movement of the basilar membrane ultrastructure are obtained. The motion is measured by matching features in successive images. It was found that the lateral point stiffness in the arcuate zone is 0.2–0.5 N/m, in the pectinate zone is 2–3 N/m, and increases to over 10 N/m near the spiral ligament. To compute the longitudinal coupling, the displacement field was measured quantitatively using image registration. It was found that the tissue in compression has a space constant of 7.6 μm and the tissue in tension has a longer space constant of 10.5 μm. The correlation of mechanical properties to anatomically distinct regions of the basilar membrane supports the conclusion that these regions serve different functional roles.

1. Introduction

A striking feature of the ultrastructure of the basilar membrane (BM) is the presence of radially oriented bundles of collagen fibers. In the arcuate zone, fibers are uniformly distributed, and in the pectinate zone fibers are organized into two strata of radially oriented bundles interspersed in a homogeneous ground substance [1, 2]. These bundles are readily observed using conventional light microscopy.

The collagen bundles were proposed to reinforce the stiffness of the BM and act independently of each other [3], like a set of piano strings, supporting the idea that traveling wave propagation is coupled predominantly by the cochlear fluids. However, the BM exhibits intrinsic longitudinal coupling that acts to broaden the tuning of the cochlear amplifier, especially at the high frequencies in the basal turn, where the shortest wavelengths are observed [4].

The objective of the present work is to measure the mechanical properties of the component materials of the BM and its composite structure in order to deduce its functional role in micromechanics. By utilizing high-resolution imaging and applying deformations to the tissue laterally in the optical plane, the displacements of individual collagen bundles are measured. Image analysis yields measurements of the lateral point stiffness and longitudinal coupling in the tympanic surface of the BM.

2. Methods

Measurements were made *in vitro* on cochleae excised from 25–45 g female Mongolian gerbils (*Meriones unguicilatus*). Intra-peritoneal injection of ketamine and xylazine anesthetized the gerbils prior to decapitation in accordance with guidelines approved by the Boston University IACUC. Micro-dissection isolated the basal turn and provide an unobstructed path to both sides of the cochlear partition. The preparation was mounted on the stage of an inverted microscope and bathed in oxygenated L-15 Leibovitz culture medium at room temperature. The cochlear partition was imaged from below through the scala vestibuli so the BM faced upward providing the force probe access from above.

The microscope was fitted with a 20× water-immersion objective and was configured for transmitted-light bright field illumination. Still images were recorded using a digital SLR camera with a 6.0 megapixel CCD sensor mounted on the photographic port of the microscope, giving a system resolution of 156 nm per pixel.

The lateral force probe consists of a piezoelectric bimorph to which a pulled borosilicate glass micropipette was affixed such that the micropipette tip undergoes lateral displacement when voltage is applied. The bending stiffness of the micropipette was calibrated using the string instrument method of Zwislocki and Cefaratti [5]. The measured stiffnesses of the micropipettes range from 1.6 to 3.6 N/m.

The probe was positioned over the BM and inserted to penetration depth of 5–8 μm, which was determined by visual observation of the micropipette tip and the BM fiber bundles. The probe was then used to apply a series of lateral static displacements, 630 nm to 3.0 μm, to the base of the micropipette, and images of the micropipette tip and surrounding tissue were recorded at each displacement. When the micropipette deflected the BM, its measured bending and calibrated stiffness were used to determine the point stiffness of the tissue.

When displacements were applied, deflection of the fiber bundles surrounding the probe tip was observed. The displacement field was measured quantitatively using image registration [6]. In this finite element method, a mesh is generated to divide the raw images into 10,000 square elements, each 10 pixels wide. An iterative feature matching algorithm, with regularization, estimated the displacement of each element between a pair of images.

3. Results

The illustration of the organ of Corti in Fig. 1A depicts the cross-section of the basal turn of gerbil and is based on images from the hemicochlea preparation [8], the isolated turn preparations, and histological sections. It is provided for comparison with the radial positions of the lateral point stiffness measurements shown in Fig. 1B. These measurements were made near the round window in the basal turn, where, in the gerbil, the density of collagen bundles is greatest. The micropipette tip was inserted into the BM just enough to pierce the surface and firmly engage the first bundle layer. The increasing stiffness from the spiral lamina to the spi-

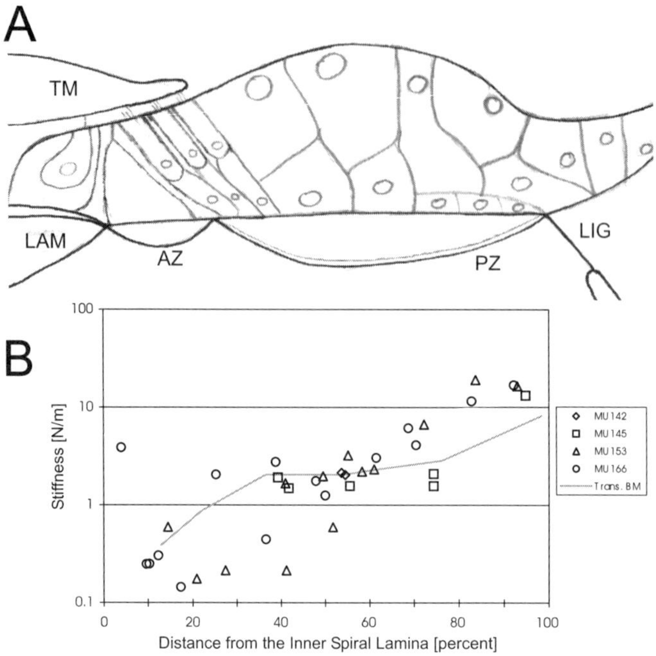

Fig. 1. (A) Cross sectional view of the organ of Corti in the basal turn. TM, tectorial membrane; LAM, osseous spiral lamina; AZ, arcuate zone; PZ, pectinate zone; LIG, spiral ligament. (B) Lateral point stiffness of the BM. The radial locations are normalized by the BM width for each experimental animal, represented by open symbols. The transverse point stiffness of the BM with cellular components stripped away [7] is overlaid for comparison.

ral ligament corresponds to anatomical features of the BM. The arcuate zone has the lowest stiffness of 0.2–0.5 N/m, the pectinate zone has a stiffness of 2–3 N/m, and the point stiffness increases to over 10 N/m near the spiral ligament. Previous measurements of the transverse point stiffness of the BM [7] with the cellular components of the organ of Corti stripped away show a similar trend. This, along with visual observations that the lateral strain does not couple through the thickness of the BM to the interior fiber bundle layer, nor to the adjacent cell bodies, leads to the conclusion that the lateral point stiffness is dominated by collagen bundles at the surface of the BM.

The photomicrograph in Fig. 2A shows the BM with the sharp probe tip located in the pectinate zone. The radially-oriented collagen bundles of the pectinate zone are evident. They run vertically through the image and provide stable, high contrast features for pattern matching. The collagen bundles disperse into smaller individual fibers in the arcuate zone, seen as the dark region at the top of the frame.

The propagation of strain in the BM is highlighted in Fig. 2B by subtracting the mean of the image sequence to remove image features that remain stationary. The resulting image shows the BM fibers with the probe tip at the center, and what appears to be a red blood cell drifting at the surface of the BM on the lower right.

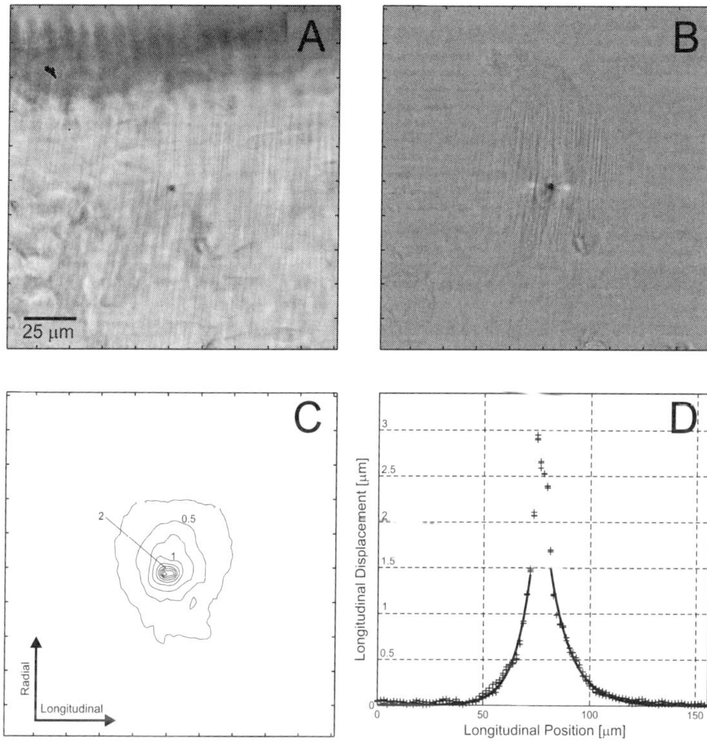

Fig. 2. Lateral displacement of the probe tip induces a displacement field in the surrounding tissue. (A) Photomicrograph of the BM with the probe tip embedded in the collagen fiber bundle layer at the tympanic surface. The 25 μm scale bar applies to all four panels. (B) Subtracting a motion blurred mask highlights features that undergo displacement, i.e., the probe tip and surrounding radial bundles. (C) Contour plot showing the result of elastic imaging analysis. The isolines represent the measured longitudinal displacement, and are labeled in micrometers. (D) The displacement falls off with increased distance from the probe tip. This relationship is characterized as an exponential decay.

The contour map in Fig. 2C shows the lateral displacement field measured using biomechanical imaging analysis. The probe was initially applying strain to the tissue, and then stepped to the right, returning the tissue to the initial unstrained condition displayed in panel A. This measurement was repeated three times and averaged. Contours of equal displacement magnitude are roughly elliptical in the immediate vicinity of the probe, and form more circular patterns as the strain propagates out into the tissue. Figure 2D is a plot of the lateral displacements from each of the three measurements with respect to longitudinal position, passing through the probe tip. In order to avoid deformation effects at the probe-tissue interface, the tails of the displacement profile are least squares fit with exponential curves to find the space constant for longitudinal coupling. It is found that tissue to the left of the probe experiencing compression has a space constant of 7.6 μm, and the tissue to the right, in tension, has a longer space constant of 10.5 μm.

4. Discussion

The variation in lateral point stiffness across the width of the BM corresponds to its anatomical features and indicates their functional differences. The collagen fibers in the arcuate zone are evenly distributed and show a relatively low stiffness when displaced laterally. As the BM transitions into the pectinate zone, these fibers become organized into bundles. These bundles, concentrated at the upper and lower surfaces of the BM, have a substantially higher stiffness than that seen in the arcuate zone. Fibroblasts in the spiral ligament anchor and are believed to apply tension to the BM [9]. The lateral point stiffness of the BM increases as the probe is brought closer to the spiral ligament. This increased stiffness is most likely caused by geometric changes in loading conditions rather than material properties. The collagen bundles are secured at the boundary of the spiral ligament. The point stiffness measured near the support of the spiral ligament is higher than that measured far from the spiral ligament. Thus the strain in the bundles increases because the applied deflection causes more elongation per unit length in the bundles. It is conceivable that there is a gradient in material properties across the width of the pectinate zone. However, the BM morphology provides no indication to support this assertion, and we feel that it is unlikely.

The radial orientation of the collagen bundles suggests the composite material would be transversely isotropic. This symmetry would lead to similar material properties in the longitudinal and transverse directions. The transverse and lateral point stiffness measurements in Fig. 1, while not directly comparable, show a trend of increasing stiffness from lamina to ligament that generally supports this hypothesis.

The layered structure of the pectinate zone, with stiff surfaces separated by a soft ground substance, presumably increases its flexural rigidity. This is analogous to an engineered material, like a foam-core–fiberglass laminate, that efficiently maximizes the stiffness of the structure by concentrating material away from the neutral bending axis. The arcuate zone lacks this structural organization, and accordingly shows a much lower stiffness.

The pectinate zone could be serving as an anchor for the contractile motion of the outer hair cells. It has been shown that OHC contractions pump fluid from the spaces of Nuel into the tunnel of Corti [10]. The lower stiffness of the arcuate zone indicates that it serves a different function, possibly to provide an elastic wall to alter the characteristic impedance for fluid flow in the tunnel.

These experiments yield estimates for longitudinal coupling that show a slightly longer space constant for the tissue in tension than in compression. The fact that strain is apparent in the tissue in tension is a clear indication that there is coupling among the collagen bundles. If the BM behaved like a set of uncoupled strings, strain would be observed only when the material is in compression, and there would be no displacement on the side opposite the probe motion. The combined effect of the homogeneous ground substance surrounding the fiber bundles and anastomoses between them could support greater tension than compression resulting in greater strain and a longer space constant in tension.

These estimates of longitudinal coupling can be compared with published estimates derived using different techniques. Naidu and Mountain [4] used a force probe to apply transverse displacements to the BM in the isolated turn preparation, and using imaging, measured the lateral deflection of features in the surrounding tissue, yielding space constants $\lambda = 40$ μm in the middle turn and $\lambda = 15$ μm in the basal turn. Emadi *et al.* [11] used a force probe to measure the point stiffness gradient near the cut edge of the hemicochlea preparation, yielding $\lambda = 21$ μm in the middle turn. With the lateral force probe, the space constant for the tissue in tension in the basal turn is $\lambda = 10$ μm. It is more appropriate to compare the space constant in tension than compression, because the transverse deflection used in the other two methods apply tension to the fiber bundles. The smaller space constant measured using the lateral force probe is likely the result of the deformation being localized to a single layer of collagen bundles rather than the entire thickness of the BM.

The lateral force probe technique contributes measurements of the localized interactions in the BM ultrastructure and helps unify its mechanical properties with observations from histology. The BM has mechanical properties that arise from distinctly organized structural elements, suggesting there are likely functional differences between the arcuate and pectinate zones.

Acknowledgments

This work was funded by NIDCD R01 DC000029.

References

1. Iurato, S., 1962. Functional implications of the nature and submicroscopic structure of the tectorial and basilar membranes. J Acoust Soc Am. 34, 1386–1395.
2. Dreiling, F. J., Henson, M. M. and Henson, O. W., Jr., 2002. The presence and arrangement of type II collagen in the basilar membrane. Hear Res. 166, 166–80.
3. Voldřich, L., 1978. Mechanical properties of basilar membrane. Acta Otolaryngol. 86, 331–5.
4. Naidu, R. C. and Mountain, D. C., 2001. Longitudinal coupling in the basilar membrane. J Assoc Res Otolaryngol. 2, 257–67.
5. Zwislocki, J. J. and Cefaratti, L. K., 1989. Tectorial membrane II: Stiffness measurements in vivo. Hear Res. 42, 211–27.
6. Richards, M. S., 2007. Quantitative three dimensional elasticity imaging, PhD thesis, Boston University, USA.
7. Naidu, R. C. and Mountain, D. C., 1998. Measurements of the stiffness map challenge a basic tenet of cochlear theories. Hear Res. 124, 124–31.
8. Edge, R. M., Evans, B. N., Pearce, M., Richter, C. P., Hu, X. and Dallos, P., 1998. Morphology of the unfixed cochlea. Hear Res. 124, 1–16.
9. Henson, M. M. and Henson, O. W., Jr., 1988. Tension fibroblasts and the connective tissue matrix of the spiral ligament. Hear Res. 35, 237–58.
10. Karavitaki, K. D. and Mountain, D. C., 2007. Evidence for outer hair cell driven oscillatory fluid flow in the tunnel of corti. Biophys J. 92, 3284–93.
11. Emadi, G., Richter, C. P. and Dallos, P., 2004. Stiffness of the gerbil basilar membrane: radial and longitudinal variations. J Neurophysiol. 91, 474–88.

Comments and Discussion

Puria: Your measurements are excellent. However, I wonder if your measurements are from the non-linear region of the stress-strain curve, especially given the ~1 μm displacements you used. I am not convinced that the ~0.5% harmonic distortion you showed during discussion are sufficient.

Aranyosi: Following up on Sunil's question — your maximum displacements are quite large (several micrometers), but you also have measurements for much smaller, sub-micrometer displacements. Do you see any nonlinearity in the force–displacement relation over this range of displacements?

Newburg: For displacements below 2 μm the force and displacement show a linear relationship. The force–displacement relation for larger applied displacements is nonlinear. The measured force is 10 to 20 percent greater for displacements of 4 to 8 μm than what would be predicted from a linear fit to the data measured at displacements below 2 μm. For this reason, we will use only the data in the linear regime to estimate the point stiffness of the BM.

de Boer: Are fibres stiff by themselves or by being stretched?

Newburg: When displacements were applied, fiber bending was observed. The displacement of an individual fiber does not decrease linearly to zero at its ends. Therefore, it is not behaving like an ideal string that can only support tension. The fibers do get stretched, but the observed bending indicates there are factors contributing reaction forces in addition to the tension in the fibers. These factors may include intrinsic bending stiffness of the collagen bundles and the stiffness of the surrounding ground substance.

TRAVELING WAVES... ON THE TECTORIAL MEMBRANE

ROOZBEH GHAFFARI[1,2], A. J. ARANYOSI[2], DENNIS M. FREEMAN[1,2]

[1]*Speech and Hearing Bioscience and Technology, Harvard-MIT Division of Health Sciences and Technology, Cambridge, MA. 02139 USA*

[2]*MIT Research Laboratory of Electronics, Cambridge, MA. 02139 USA*

We show that isolated sections of the TM suspended in artificial endolymph support traveling waves of radial motion that propagate longitudinally. The velocities of the TM traveling waves are remarkably similar to those of basilar membrane (BM) near the best place. A simple model shows that these waves are intrinsic to the viscoelastic structure of the TM. Analysis of models of cochlear loads suggests that TM traveling waves are little affected by the stiffness of hair bundles and the limbal attachment or by viscous damping in the subtectorial space. TM waves are inconsistent with a fundamental assumption in "classical" cochlear macromechanical models: that adjacent longitudinal sections of the cochlea are coupled through the fluid alone. Also, cochlear micromechanics, which are conventionally envisioned as local processes that depend on just neighboring structures, must be rethought as a global process extending over significant longitudinal distances.

1. Introduction

The tectorial membrane (TM) is a gelatinous matrix that overlies the mechanosensory bundles of hair cells in the cochlea. Based on this anatomical configuration, the TM is believed to play a key role in stimulating hair cells. However, the mechanical processes by which the TM stimulates hair cells remain unclear, largely because the important dynamic mechanical properties of the TM have proved difficult to measure. Consequently, there are a variety of cochlear models that describe the mechanical function of the TM [1, 2, 3, 4]. These "classical" models assume that adjacent longitudinal sections of the cochlea are uncoupled except for energy propagation through the fluid via the basilar membrane (BM) traveling wave [5]. Within this theoretical framework, the important mechanical feature of the TM is its point impedance.

However, recent measurements of TM mechanical properties do not agree with the models. Instead, these measurements indicate that the TM is viscoelastic and can spatially couple motion over significant longitudinal distances [6, 7]. Here we show that radial shear forces applied to isolated TMs launch waves of radial motion that propagate longitudinally along the TM. These waves introduce a new mechanism for longitudinal coupling at both macromechanical and micromechanical scales.

2. Methods

TM segments were excised from the cochleae of adult male mice (strains B6129F1 and CD-1, 4–10 weeks old, Taconic) and were placed in a fresh bath of artificial endolymph using a previously published surgical technique [8]. TM segments were classified as basal or apical based on the cochlear turn from which they were excised. A wave chamber was developed to study the dynamic properties of the TM at audio frequencies. This chamber consisted of two parallel supports separated by 300–480 μm. One support was attached with epoxy to a piezo-electric actuator (resonance frequency: 138 kHz, Thorlabs, Inc.) and loosely coupled to the underlying glass slide. Mechanical shear forces were delivered in the radial direction through this vibrating support (Fig. 1A). The second support was firmly fixed to the underlying glass slide. The surfaces of the supports were coated with tissue adhesive and perfused with artificial endolymph. The TM was injected into this chamber with a glass-tip pipette. A sterilized eyelash was used to help position the TM between the supports (Fig. 1B). As an alternative to the vibrating support TM waves were also generated with microfabricated shear probes (Fig. 1A). Each probe consisted of a shear plate and a base, which was coupled to a piezo-electric actuator. The shear plate was carefully placed in contact with the surface of the suspended TM near Hensen's stripe (Fig. 1B) and used to apply point displacements in the radial direction at audio frequencies. An optical system was used to track displacement magnitude and phase of TM wave motion. Motions were captured at high frequencies using stroboscopic illumination and analyzed with computer vision algorithms as described previously [8, 9].

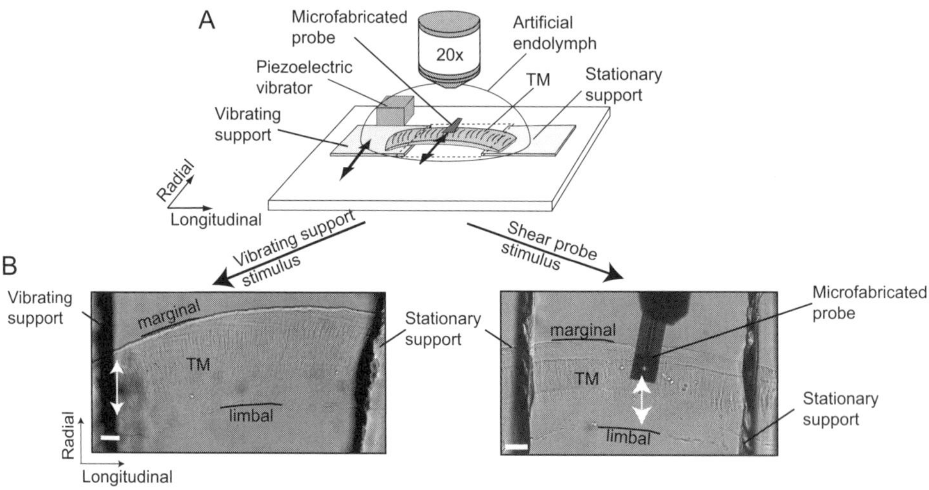

Fig. 1. (A) Schematic drawing of TM segment suspended between two supports in a wave chamber. Double-sided arrows indicate sinusoidal displacement applied to the TM along the radial direction through a vibrating support (left) or a microfabricated shear probe (center). (B) Images of suspended TM taken with a light microscope. (Left) Forces were delivered by vibrating one of the supports (scale bar, 50 μm). (Right) Forces were delivered through a probe engaged on the surface of suspended TM (scale bar, 30 μm).

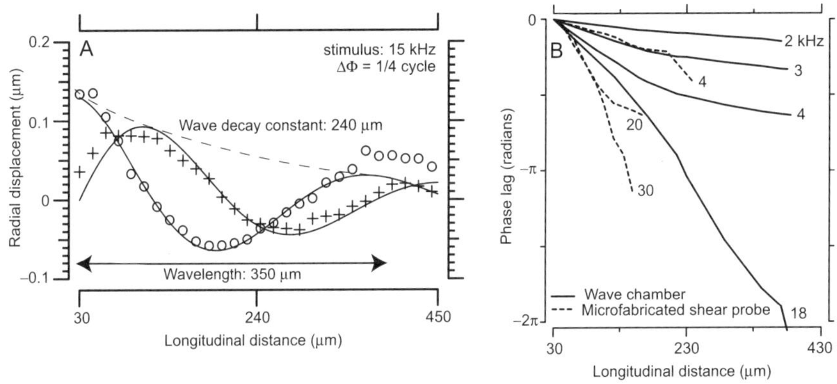

Fig. 2. (A) TM radial displacement vs. longitudinal distance in response to 15 kHz stimulation of vibrating support. Solid lines represent exponentially decaying sinusoid fit to the measurements at sixteen equally spaced stimulus phases (two shown separated by 1/4 cycle). (B) Phase vs. longitudinal distance for stimulus frequencies 2–30 kHz for measurements with vibrating support (solid) and microfabricated probe (dashed). Phase lag decreased monotonically with distance and became steeper with increasing frequency. Phase was measured relative to a point on TM surface ~ 30 μm from the stimulus source.

3. Measurements and Model

Fig. 2A shows the spatial pattern of radial displacement of a typical basal TM segment in response to 15 kHz motion of the vibrating support. The measurements were well fit by an exponentially decaying sinusoid, corresponding to a wave with wavelength of 350 μm and wave decay constant of 240 μm. TM wave motion scaled linearly with applied displacement at the vibrating support and wavelengths were constant over the range of displacements applied to the TM ($\sim 90 - 340$ nm).

The phase of radial displacement varied with longitudinal distance in a frequency-dependent manner. In the vibrating support configuration (Fig. 2B), phase accumulation vs. longitudinal distance was frequency dependent, with little or no phase accumulation at low frequencies (below ~ 2 kHz) but more than a full cycle ($> 2\pi$ radians) at 18 kHz.

The velocity of wave propagation v_s was computed for basal and apical TM segments as the product of frequency and wavelength for each stimulus frequency. TM wave velocity values ranged from 2–10 m/s and increased with stimulus frequency (Fig. 3A). The propagation velocities of the TM wave are strikingly similar to those of the BM wave near the best place (Fig. 3B).

Waves propagated towards the stationary support when either the basal or apical end of the TM segment was attached to the vibrating support. Furthermore, TM waves propagated in both directions when excited by a probe attached at the midpoint of the span between the supports. These results show that both forward and reverse traveling waves propagate through the TM.

Previously the material properties of the TM were shown to be viscoelastic over a broad range of frequencies [6, 7]. For pure shear waves in an infinite, isotropic,

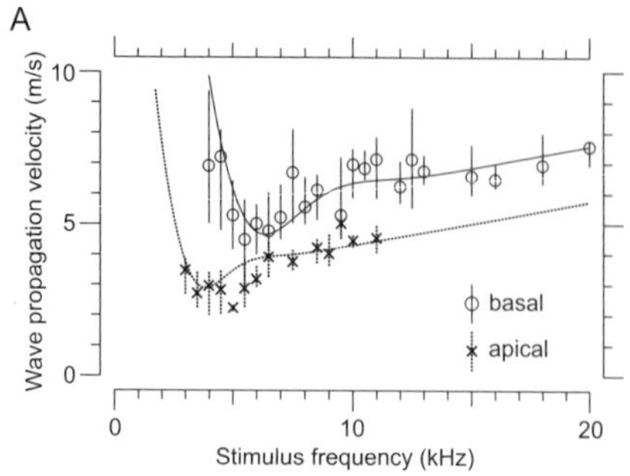

B

Species	Distance from base, mm (%Length)	CF, kHz	Velocity, m/s
Gerbil	1.6 (14)	25	22.5
Guinea pig	1.7 (9)	31	28
Guinea pig	4 (22)	15	10
Guinea pig	5-6.5 (27-35)	5-9	5-6
Guinea pig	12.8 (69)	1	1.55
Cat	2.6 (10)	31	19
Chinchilla	2.6 (14)	12	6
Chinchilla	2.7 (15)	10	6
Chinchilla	14 (76)	0.62	0.76
Squirrel monkey	9 (39)	6.5	12

Compiled by Robles and Ruggero, 2001

Fig. 3. Propagation velocity of TM traveling waves compared to BM waves. (A) Symbols represent the median values of wave propagation velocity (v_s) measured across multiple frequencies for basal (○) and apical (×) segments. Interquartile ranges are represented with vertical lines. Fit lines represent model predictions of v_s vs frequency for typical basal (solid) and apical (dotted) TM samples. (B) BM wave propagation velocities near CF locations across different species (Table adapted from [12]). TM waves propagate with velocities (2–10 m/s) that are comparable to those of the BM traveling wave.

viscoelastic material, v_s is related to the shear storage modulus G' and the shear viscosity η by $v_s = \sqrt{2(G'^2 + \omega^2\eta^2)/\rho(G' + \sqrt{G'^2 + \omega^2\eta^2})}$, where ω is the angular frequency of vibration and ρ is the density of the material, assumed to equal that of water [10]. This equation suggests that the velocity should increase with frequency, as was seen in Fig. 3A. Furthermore, best fits of G' and η to the data in Fig. 3 are comparable to those determined using an independent method [7], implying that the interaction of TM inertial and shear properties give rise to traveling waves.

To account for the finite cross-section of the TM, the finite length of the wave chamber, and the effect of surrounding fluid, we analyzed a finite difference model of the TM [8]. The model consisted of a longitudinally distributed series of masses coupled by viscous and elastic elements. The radial displacements of the TM at the vibrating and stationary supports were constrained in the model as they were in the wave chamber. The measurements of v_s were fit by this model for both basal and apical TM segments. The best fit of this model to TM wave displacements predicted material properties for basal ($G' = 47 \pm 12$ kPa; $\eta = 0.19 \pm 0.07$ Pa·s, $n = 5$) and apical ($G' = 17 \pm 5$ kPa; $\eta = 0.15 \pm 0.05$ Pa·s, $n = 3$) TM segments that are remarkably similar to those determined using an independent method [7].

The finite difference model also showed that the TM can support waves in the presence of cochlear loads, such as hair bundle stiffness, the limbal attachment, and damping in the subtectorial space [8]. We assumed that a nominal hair bundle stiffness of 3.5 mN/m was evenly distributed across an 8 μm extent of the TM for each of the three rows of OHCs. Adding hair bundle stiffness increased wave decay constants by 1%. We modeled the limbal attachment as a local stiffness load. The effect of the limbal attachment was apparent only for sufficiently large stiffnesses, which caused the wavelength and wave decay constant to increase by 10–15%. Damping in the subtectorial space was estimated assuming that fluid flow was Couette. Viscous

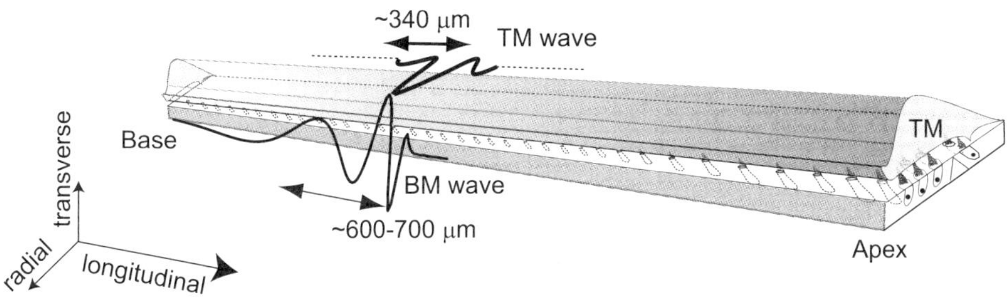

Fig. 4. Schematic drawing illustrating the concept that the cochlea supports two traveling waves. TM wave (top) and BM wave (bottom) both propagate longitudinally with comparable wavelengths and velocities near CF. The observed TM waves are longitudinally propagating waves of radial motion; the BM waves are longitudinally propagating waves of transverse motion. These two waves can be coupled through the OHCs and cochlear fluids.

damping in the subtectorial space did not cause significant dissipation even for gaps as small as 1 μm. This result is striking, since such damping has been modeled as the primary loss mechanism in the cochlea [11].

4. Discussion

The waves reported in this study provide evidence for significant spatial coupling through cochlear structures. This finding counters a fundamental assumption in "classical" cochlear macromechanical models: that adjacent longitudinal sections of the cochlea are coupled through the fluid alone [13]. This finding also counters a fundamental assumption in classical micromechanical models: that micromechanics are local, depending on motions of only adjacent structures. Instead the presence of both BM and TM waves in the cochlea (Fig. 4) offers entirely new ways to think about cochlear mechanics. We briefly describe three possibilities:

(1) **Otoacoustic emissions** are thought to arise from retrograde BM traveling waves or compressive fluid waves. Retrograde TM waves [14] could also contribute to emissions.

(2) **Traveling wave amplification** of two coupled anterograde traveling waves has been described theoretically as a mechanism for cochlear amplification [15]. Anterograde TM waves are a logical substrate for the second wave.

(3) **Recirculating amplification** is made possible by anterograde propagation of BM waves driving retrograde propagation of TM waves. This idea provides a natural mechanism by which amplification basal to the best place can be coupled into a feedback system. In this regard, it is important to note that decay constants of the TM wave span regions of the cochlea with best frequencies that vary by 0.5 to 1 octave.

These possibilities illustrate some ways by which TM waves can fundamentally change the way that we think about cochlear mechanics.

Acknowledgments

This research was supported by grant R01-DC00238 from the National Institutes of Health. R. Ghaffari was supported by a training grant from the National Institutes of Health to the Speech and Hearing Bioscience and Technology Program in the Harvard-MIT Division of Health Sciences and Technology. The main results of this work have been published before in [8].

References

1. Davis, H., 1958. A mechano-electrical theory of cochlear action. Ann Otol Rhino Laryngol. 67, 789–801.
2. Allen, J. B., 1980. Cochlear micromechanics — a physical model of transduction. J Acoust Soc Am. 68, 1660–1670.
3. Zwislocki, J. J., 1979. Tectorial membrane: a possible sharpening effect on the frequency analysis in the cochlea. Acta Otolaryngol. 87, 267–269.
4. Mammano, F. and Nobili, R., 1993. Biophysics of the cochlea: linear approximation. J Acoust Soc Am. 93, 3320–3332.
5. von Békésy, G., *Experiments in Hearing* (McGraw-Hill, New York, 1960).
6. Abnet, C. C. and Freeman, D. M., 2000. Deformations of the isolated mouse tectorial membrane produced by oscillatory forces. Hear Res. 144, 29–46.
7. Gu, J. W., Hemmert, W., Freeman, D. M. and Aranyosi, A. J., 2008. Frequency-dependent shear impedance of the tectorial membrane. Biophys J. in press.
8. Ghaffari, R., Aranyosi, A. J. and Freeman, D. M., 2007. Longitudinally propagating traveling waves of the mammalian tectorial membrane. Proc Nat Acad Sci USA. 104, 16510–16515.
9. Aranyosi, A. J. and Freeman, D. M., 2004. Sound-induced motions of individual cochlear hair bundles. Biophys J. 87, 3536–3546.
10. Chen, S., Fatemi, M. and Greenleaf, J. F., 2004. Quantifying elasticity and viscosity from measurement of shear wave speed dispersion. J Acoust Soc Am. 115, 2781–2785.
11. Allen, J. B., 1977. Cochlear micromechanics — a mechanism for transforming mechanical to neural tuning within the cochlea. J Acoust Soc Am. 62, 930–939.
12. Robles, L. and Ruggero, M. A., 2001. Mechanics of the mammalian cochlea. Physiol Rev. 81, 1305–1352.
13. de Boer, E., 1996. Mechanics of the cochlea: modeling efforts, in *The Cochlea*, eds. Dallos, P., Popper, A. N. and Fay, R. R., Springer Handbook of Auditory Research, Vol. 8 (Springer-Verlag, New York)
14. Aranyosi, A. J., Ghaffari, R. and Freeman, D. M., Tectorial membrane traveling waves: a new mechanism for longitudinal coupling. This volume.
15. Hubbard, A. E., 1993. A traveling-wave amplifier model of the cochlea. Science. 259, 68–71.

Comments and Discussion

Fulton: Would you agree that the propagating waves you describe are surface acoustic waves of the "Lamb" type, i.e., energy is propagated at a slow velocity (3-10 m/sec) with the amplitude of the waves in the plane of the surface of the substrate (gelatinous matrix)?

Your figure 4 can be compared to figure 4 in Fulton (this volume). Fulton suggests the waves traveling along Hensens stripe (the hump in the TM surface labeled only TM in your figure) consists of a Lamb wave as you suggest and a Rayleigh wave (traveling at the same substrate-controlled velocity but with amplitude orthogonal to the surface of the substrate) instead of the largely conceptual traveling wave described in the review by Robles and Ruggero.

The waveform you attribute to Robles and Ruggero, 2001, could not be located in that review but appears to be an earlier conceptual drawing from von Bekesy, 1960, or in a more artistic form in Tonndorf, 1960. It has appeared frequently in the pedagogical literature. In either case, the illustrated waveform represents a pulse traveling through a dispersive liquid (Lighthill, 1978) under highly controlled conditions to avoid the closely associated discontinuity described by Lighthill. The method of achieving the associated slow velocity, and avoiding the discontinuity, was not described by these early authors as highlighted in the review by Robles and Ruggero.

Can you comment on this?

Ghaffari: The waves we have measured appear to be bulk shear waves rather than surface acoustic waves. Our technique does not currently allow us to test whether the TM supports transverse surface acoustic waves.

Braun: The idea that the TM may greatly enhance cochlear sensitivity and tuning appears to be in conflict with important recent data. Russell et al. (2007) and Lukashkin et al. (2008) reported that *Tectb*$^{-/-}$ mice, which have a reduced longitudinal elastic coupling along the TM, have unchanged neural sensitivity above ca 20 kHz compared to wild-type mice. Do these findings not indicate that TM elasticity has no role at hearing threshold above 20 kHz in mice? These two papers further reported the absence of a low-frequency tail in neural tuning curves in *Tectb*$^{-/-}$ mice, even with stimuli reaching 100 dB SPL. Do both findings, taken together, not suggest that the normal TM in wild-type *Tectb*$^{+/+}$ animals has a mechanical function at high sound levels, rather than at low ones?

References

Russell IJ, Legan PK, Lukashkina VA, Lukashkin AN, Goodyear RJ, Richardson GP, 2007. Sharpened cochlear tuning in a mouse with a genetically modified tectorial membrane. Nat Neurosci 10, 215-223.

Lukashkin AN, Lukashkina VA, Richardson GP, Russell IJ, 2008. Does the cochlea compromise on sensitivity and frequency selectivity? This volume.

Ghaffari: The Russell and Lukashkin studies support our conclusions. They see a small ($\sim$ 10 dB) loss of sensitivity associated with sharper tuning in *Tectb*$^{-/-}$ mice, indicating that the TM plays a significant role in determining both sensitivity and tuning at low sound levels. In light of TM waves, their findings suggest that the extent of wave coupling determines the tradeoff between these two fundamental properties of the cochlea. Moreover, $Tecta^{\Delta ENT/\Delta ENT}$ mice, in which the TM is completely detached from the organ of Corti, have significantly elevated thresholds and broad tuning compared to wild types, further highlighting the importance of the TM for normal cochlear function at low sound levels.

Braun: Lukashkin et al. (2008) reported in Table 1 and Fig. 2A that $Tectb^{-/-}$ mice had lower (!) hearing thresholds at 36 and 54 kHz than wild type mice. Basilar membrane (BM) threshold responses at the CF place of 53 kHz, however, were "about 10 dB less sensitive" in $Tectb^{-/-}$ mice. The most parsimonious interpretation of these findings would be that the (side) effect of OHC motility upon BM motion is smaller in $Tectb^{-/-}$ mice than in wild type mice. Evidently, the longitudinal elastic coupling of the normal TM does not support hearing sensitivity at these frequencies.

In animals where the TM is widely separated from the reticular lamina, as in the $Tecta^{\Delta ENT/\Delta ENT}$ mice, signal transfer between OHCs and IHCs is severely impeded or even totally interrupted, according to all extant cochlear hypotheses. Data from these animals are therefore not informative concerning the present question.

de Boer: In order to include the influence of the sub-tectorial space you could consider letting the specimen float on a glass plate. Would that be possible?

Ghaffari: A technically simpler approach is to position a plate above the free-floating TM sample at various distances comparable to the thickness of the sub-tectorial space. We are currently investigating this approach.

THE ANISOTROPY OF THE TECTORIAL MEMBRANE GUIDES STEREOCILIA DEFLECTION

R. GUETA[1], D. BARLAM[2], R. Z. SHNECK[3], I. ROUSSO[1]

[1] *Dept. of Structural Biology, Weizmann Institute of Science, Israel*

[2] *Mechanical Engineering, Ben-Gurion University of the Negev, Israel*

[3] *Materials Engineering, Ben-Gurion University of the Negev, Israel*

The extreme performance of mammalian hearing is due to the cochlea's remarkable amplification of sound-induced mechanical stimuli. The mechanical coupling between the tectorial membrane (TM) and outer-hair cells' stereocilia leads to a crucial step in hearing mechanics: the conversion of outer-hair cells (OHC's) vertical displacement motion to the lateral motion of the stereocilia. Nonetheless, the mechanism which governs stereocilia motion is not known. Here we present a mechanism for this motion conversion by demonstrating that the relations between material properties of the TM along its three axes act as a "mechanical guide". By using lateral force spectroscopy measurements combined with finite-element simulations, we show that along the entire length of the TM, the normal stiffness modulus (Ez) is significantly larger than the lateral modulus values (Ex, Ey). Our "mechanical guiding" model is supported by finite-element simulations of the interactions between the TM and the stereocilia. When the TM is simulated as an isotropic material (Ex=Ey=Ez), no lateral deflection of the stereocilia is generated. In contrast, when our experimentally obtained mechanical properties (Ez>Ex,Ey) were incorporated into the simulated TM, a clear lateral motion is observed. Our results provide a mechanism for cochlear amplification induced by TM-stereocilia coupling.

1 Introduction

During acoustic stimulation, fluid flow inside the cochlea induces transverse displacement of the basilar membrane (BM). This displacement results in the outer hair cells moving against the tectorial membrane (TM). During this motion, the stereocilium bundles at the end of the hair cells must be laterally deflected as they are pushed against the TM, and only then the shearing motion between adjacent stereocilia will cause the mechanosensitive ion-channels at their tips to open. The resulting influx of cations changes the polarization of the outer-hair cells (OHCs) [1], which then exert mechanical force through an electromechanical transduction process [2, 3]. This force is fed back into the BM's motion and amplifies the initial mechanical input. A fundamental question associated with this process is why stereocilia, which are more than 4 orders of magnitude stiffer than the TM (~3 GPa [4] 12 compared to 30-300 kPa [5, 6]), are deflected by it, rather than piercing it.

A possible reason for deflection is that the mechanical properties of the TM are not spatially uniform but differ along different axes. We propose a "mechanical guiding" model by which lateral deflection of the stereocilia is governed by the anisotropic properties of the TM. Specifically, we submit that the stiffness normal to the membrane surface of the TM (represented as the normal Young's modulus, Ez) is higher than its lateral stiffness (Ex, Ey). Upon stimulation, the properties of the TM coupled with the orientation of the stereocilia restrict their vertical more than their lateral motion, and thus the vertical movement of the OHCs is lateralized at the stereocilia.

In this study, the normal and lateral stiffness values of TM samples isolated from mice were analyzed through indentation type and lateral force experiments conducted with an atomic force microscopy (AFM). The results of this study show the role of the TM material properties in the micromechanical process that couples basilar membrane motion to hair-bundles deflection motion.

2 Results and Discussion

To test our "mechanical guiding" model, the material properties of the TM were measured using normal-indentation and lateral-force spectroscopy conducted by an AFM. For each position on the TM, we derive the normal Young's modulus by analyzing the force-distance (FD) curve using a modified Hertz model. From the FD curve we can also determine the depth to which the TM is indented by a given normal loading force (Fig. 1 A). Subsequently, we perform a lateral force measurement at the same TM position (Fig. 1 B) for a specific TM indentation depth (achieved by applying the appropriate normal loading force, as determined from the FD curve). For this study, all lateral-force measurements were carried out with loading forces which give rise to 150-200 nm indentation depths. This level of indentation is similar to the initial depth (~100 nm) at which stereocilia are naturally embedded within the TM, as determined by cryo-Electron Microscopy [7].

Similarly to our normal force results [6], we find a gradual increase in the lateral stiffness of the TM going from the apical to the basilar region of the cochlea, independent of radial orientation. The lateral stiffness of the TM at its apical end (parallel: 1.9±0.2

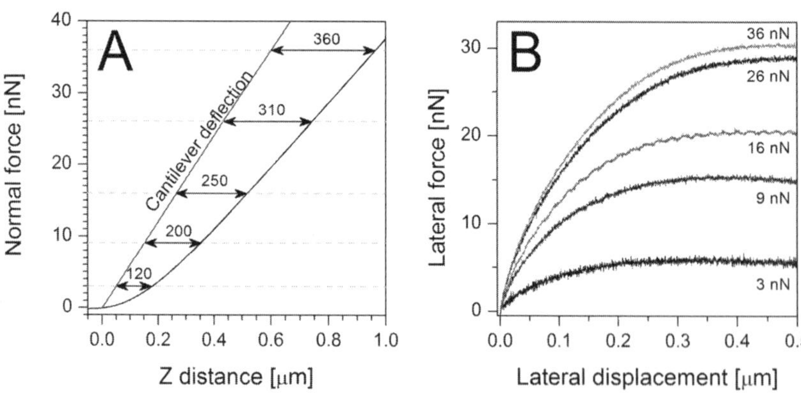

Figure 1. Normal and lateral force spectroscopy. (A) An averaged force-distance curve for a TM sample and the corresponding cantilever deflection. Double-headed arrows indicate the depth (nm) to which the TM is indented by each of five different normal loading forces (3, 9, 16, 26, and 36 nN). (B) Averaged lateral-force versus lateral-displacement curves acquired at the same TM position as was used for the force-distance curve (A) under five different normal loading forces denoted adjacent to the curves. The lateral force grows as the normal loading force and, consequently, the indentation depth increases.

N/m, n=18; perpendicular: 2.4±0.2 N/m, n=27) is only about one-fifth of its stiffness at its basal end (parallel: 10.5±1.1 N/m, n=19; perpendicular: 12.4±1.0 N/m, n=36). In the mid-turn region of the cochlea, the lateral stiffness of the TM is 4.4±0.3 N/m (n=36) for an orientation parallel to the radial direction, and 4.8±0.3 N/m (n=34) for an orientation perpendicular to it. Earlier material property measurements made using magnetic beads attached to the surface of the TM found it to be 3-fold stiffer in the radial (x) direction than in the longitudinal (y) direction [8]. However, we find the TM to be slightly (10-20%) softer along its radial direction compared to the longitudinal direction, with the observed differences probably arising from the arrangement of the TM's collagen fibres along its radial direction [9]. This discrepancy between the findings of the two research groups may point to differences between the properties of the TM surface (as measured in the earlier study) and its bulk properties (as assessed using our constant indentation approach). Alternatively, it may result from the different measurement methods applied.

To evaluate the inter-relations between the mechanical properties of the TM along its different axes, the corresponding Young's moduli, Ex, Ey (parallel and perpendicular to the radial direction, respectively) and Ez, should be compared. The normal Young's modulus (Ez) can be reliably obtained by analysing the indentation experiments using the well-established Hertz model. However, no model is available to describe lateral force measurements, which prevents their being used to directly estimate lateral moduli.

To obtain these moduli, we simulate our lateral-force measurements in the framework of continuum mechanics using the finite-element (FE) method. The TM is modelled as an anisotropic material characterized by three independent Young's moduli (Ex, Ey, and Ez). While Ex, and Ey are free parameters, Ez is a constant which is determined experimentally from the normal FD curve. At the end of the FE calculation, the simulated lateral-force curve is superimposed over the experimental curve. The differences between the two curves are minimized by optimising the Ex, and Ey values of the modelled TM. The model properties that best fit our force measurements are considered the estimated TM lateral moduli.

The FE analysis results (Table 1) indicate that, along the entire length of the TM, both lateral moduli (Ex, Ey) are significantly lower than the normal modulus (Ez). The lateral modulus Ex (parallel to the TM's radial direction) is about 6.7, 4, and 2.5 times lower than the normal modulus in the basilar, mid-turn, and apical regions, respectively.

It is possible, however, that the observed differences in stiffness between the lateral and normal directions are affected by the fact that different experimental methods were

Table 1. Summary of the mechanical properties of the TM.

Longitudinal position	Young's modulus (kPa)		
	E_x	E_y	E_z
Apical	11	16	28
Midturn	18	31	73
Basal	45	75	300

used to obtain them (indentation vs. lateral force). To exclude this possibility, we characterized the isotropic mechanical properties of polydimethyl siloxane (PDMS). Analysis of the lateral moduli showed that the FE-model-derived Ex was equal to Ey, and they were both slightly larger than the Hertz-model-derived Ez.

Finally, we simulated the mechanical interactions between the TM and embedded stereocilia using the FE method. The TM was modeled as an elastic three-dimensional plate. Given that a stereocilium is markedly stiffer than the TM, it was modeled as a non-deformable rod (Fig. 2 A-C). The diameter of the stereocilium (300 nm), its initial position inside the TM (100 nm), and angle relative to the TM's surface (20°) were obtained from electron microscopy and histological data [7]. To simulate in vivo motion, the stereocilium was pushed against the TM along its longitudinal axis (Fig. 2 A, white double-headed arrow), while rotational movements were unconstrained. In the control simulation, the TM is modeled as an isotropic material. As the stereocilium is pushed against the isotropic TM, it pierces it while maintaining its initial orientation (Fig. 2 B)[1]. By contrast, when the TM is modeled as an anisotropic material which has mechanical properties as per the findings of this study (Table 1), the vertical motion of the base of the stereocilium is converted into a lateral deflection of its tip (Fig. 2 C, D)[a]. One of the common models suggests that shear motion between OHCs and the TM induce stereocilium deflection. If we include such a relative shear motion and model the TM as an isotropic material, the stereocilium must be pushed by a 42° angle relative to its longitudinal axis in order to yield deflection angle that is similar to those obtained in Table 2.

The mechanosensitive ion-channels at the stereocilia tips open in response to the shearing motion of adjacent stereocilia. This motion depends on the lateral movement of the stereocilium tip rather than on its deflection angle. The length of the stereocilia is smallest at the basal end of the cochlea, and it gradually increases along the length of the cochlea until it is maximized at the apex [10]. If the stereocilia deflection angle were constant throughout the entire length of the cochlea, the resulting lateral displacements of their tips would be smallest at the basal end and largest at the apex. We find that when maximally deflected, the stereocilium is tilted by an angle that perfectly matches the corresponding stereocilium length [10], such that the resulting lateral movement of the stereocilium is kept constant (Table 2). The lateral movement and the deflection angle of the stereocilium are predicted by our model to be slightly reduced when the higher estimated spring constant value (1.25 pN/nm [11]) for the stereocilia bending stiffness is modeled. Strikingly, these predicted deflection angles and lateral movements agree extremely well with the experimental observations [12]. Our results indicate that the motion of the stereocilium can be controlled by the anisotropic properties of the TM.

---

[1] Full animations of the finite-element simulation for the mechanical interactions between the TM and stereocilia can be obtained from our laboratory webpage at: http://www.weizmann.ac.il/sb/faculty_pages/Rousso/Gallery_anim.html

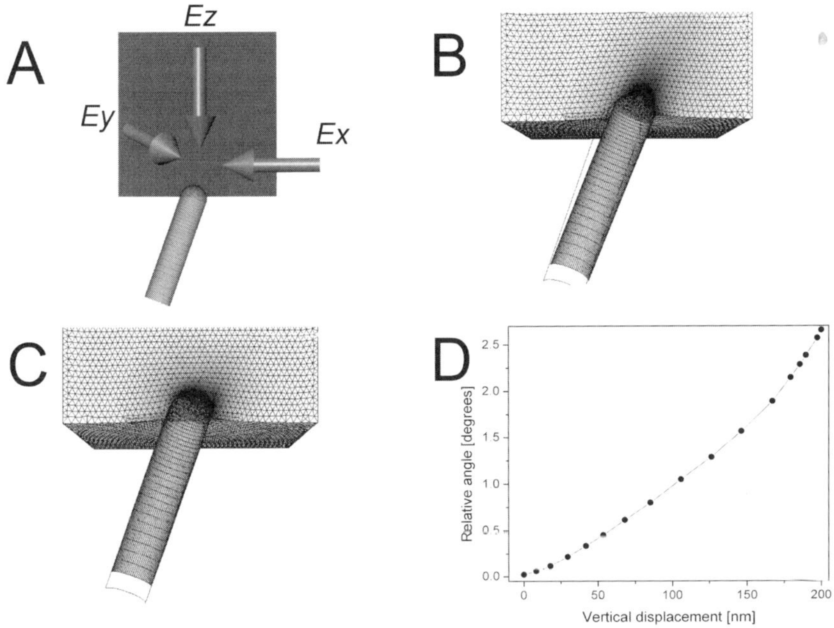

Figure 2. Finite element simulation for the mechanical interactions between the TM and stereocilia. (A) The TM is modeled as an elastic 3D plate made of anisotropic homogenous material. The stereocilium is modeled as a non-deformable rod capped by a sphere at its tip, implanted 100 nm into the TM. The base of the stereocilium is moved vertically along its longitudinal axes (double-headed white arrow). (B) (B) The penetration of the stereocilium as it is pushed against an isotropic TM model with Ex=Ey=Ez=300 KPa. (C) The deflection of the stereocilium as it is pushed against an anisotropic TM model with Ex=45, Ey=75, and Ez=300 KPa. The white outline shows the contour of the initial stereocilium orientation. (D) The relative angle of the stereocilium as a function of the vertical displacement of its base.

Table 2. Summary of the finite-element simulations results for the mechanical interactions between the TM and embedded stereocilia.

Longitudinal position	Stereocilia length[a] (μm)	Stereocilia bending stiffness			
		No spring		1.25[b] [pN/nm]	
		Deflection (nm)	Displacement (degrees)	Deflection (nm)	Displacement (degrees)
Apical	5	1.3	113	1.1	96
Midturn	3.75	1.8	115	1.6	105
Basal	2.5	2.6	113	2.5	109

[a] Taken from Wright [10]; [b] Taken from Crawford and Fettiplace [11].

Acknowledgments

This work was supported in part by the Israel Science Foundation, the Minerva Foundation with funding from the Federal German Ministry of Education and Research, a grant from the Jean-Jacques Brunschwig Fund for the Molecular Genetics of Cancer, the Kimmelman Center for Macromolecular Assemblies, and the Helen and Martin Kimmel Center for Nanoscale Science.

References

1. Hudspeth, A. J. 1989. How the ear's works work. Nature 341:397-404.
2. Brownell, W. E. 1990. Outer hair cell electromotility and otoacoustic emissions. Ear Hear. 11:82-92.
3. Brownell, W. E., C. R. Bader, D. Bertrand, and Y. de Ribaupierre. 1985. Evoked mechanical responses of isolated cochlear outer hair cells. Science 227:194-196.
4. Duncan, R. K., and J. W. Grant. 1997. A finite-element model of inner ear hair bundle micromechanics. Hear. Res. 104:15-26.
5. Ghaffari, R., A. J. Aranyosi, and D. M. Freeman. 2007. Longitudinally propagating traveling waves of the mammalian tectorial membrane. Proc Natl Acad Sci U S A 104:16510-16515.
6. Gueta, R., D. Barlam, R. Z. Shneck, and I. Rousso. 2006. Measurement of the mechanical properties of isolated tectorial membrane using atomic force microscopy. Proc. Natl. Acad. Sci. U. S. A. 103:14790-14795.
7. Tsuprun, V., and P. Santi. 1998. Structure of outer hair cell stereocilia links in the chinchilla. J. Neurocytol. 27:517-528.
8. Abnet, C. C., and D. M. Freeman. 2000. Deformations of the isolated mouse tectorial membrane produced by oscillatory forces. Hear. Res. 144:29-46.
9. Gueta, R., E. Tal, Y. Silberberg, and I. Rousso. 2007. The 3D structure of the tectorial membrane determined by second-harmonic imaging microscopy. J. Struct. Biol. 159:103-110.
10. Wright, A. 1984. Dimensions of the Cochlear Stereocilia in Man and the Guinea-Pig. Hear. Res. 13:89-98.
11. Crawford, A. C., and R. Fettiplace. 1985. The mechanical properties of ciliary bundles of turtle cochlear hair cells. The Journal of physiology 364:359-379.
12. Fridberger, A., I. Tomo, and J. B. de Monvel. 2006. Imaging hair cell transduction at the speed of sound: Dynamic behavior of mammalian stereocilia. Proc. Natl. Acad. Sci. U. S. A. 103:1918-1923.

Comments and Discussion

Fulton: The reported experiments assume the TM is an elastic three-dimensional anisotropic plate (a physical solid) with three independent Young's moduli that are independent of the stimulus displacement frequency. An alternate proposal is that the surface of the TM in contact with the stereocilia, e.g. Kimura's membrane, is a liquid crystalline material (a physical gel) with surface properties highly dependent on the frequency of any excitation. Ghaffari, Aranyosi & Freeman (this volume) use the term gelatinous matrix for this structure. Under such a proposal, the surface properties of the liquid crystalline material are highly frequency dependent. Can you provide any information on the spatial displacement frequencies used in your experiments?

Rousso: Our lateral displacements are carried out at 1-2 Hz frequencies. We therefore consider these measurements as pseudo-static ones.

Aranyosi: In our measurements the TM is about twice as stiff in the radial direction as in the longitudinal or transverse directions. Gavara and Chadwick (this volume, and at the ARO meeting) report similar findings, as do Richter et al. (2007). You report the opposite, that the TM is stiffest transversely and least stiff radially. How do you explain these differences? Is it possible that you're measuring the local properties of Kimura's membrane rather than the bulk properties of the TM?
Also, can you comment on how properties such as surface roughness and sliding friction affect the lateral AFM measurements?

Roussso: Your question is a very important one, but also quite a challenging one to answer. The mechanical properties of the TM have been measured by several groups often using different methods. We do not have yet a decisive answer but we can propose a possible explanation:

The most plausible explanation is that in our studies (described in this proceeding and in Gueta et al., Biophys. J., 2008) we have focused solely on the Kimura's membrane anisotropic properties. This ~300 nm thick structure is important as it is where the OHC stereocilia are attached and interact. Our results are unlikely to represent the bulk properties of the TM but rather those of the Kimura's membrane. In a previous study (Gueta et al., PNAS, 2006) we have shown that this region is significantly different than all other regions of the TM. Specifically, it has the largest structural and mechanical differences along the longitudinal direction. To the best of our knowledge, all other studies were not limited to this region.

In addition, the variance could be attributed to differences in experimental methods used to analyze the TM properties. Mainly, the different probe size used in the various measurements. We have used a relatively small probe (2 μm), while Richter et al. used a needle tip which was 20 μm in diameter, Gavara and Chadwick as well as Shoelson et al. used 10 μm beads, and previous measurements in your lab, have utilized 12 μm magnetic beads to mechanically perturb the TM. We have chosen a smaller bead since according to our FE simulation (Gueta et al., 2006), such a bead will deform an area which is similar to that of a single OHC bundle.

When using a nanometer scaled probe (standard AFM tip) there is a large effect of surface roughness to the extent that it is impossible to carry out these measurements. Using a 2 μm spherical bead and applying relatively small displacements, the local roughness of the TM is being averaged by the blunt probe. Using standard nanoscale probes, it was impossible to obtain relatively smooth and spike-less lateral force curves.

TECTORIAL MEMBRANE TRAVELING WAVES: A NEW MECHANISM FOR LONGITUDINAL COUPLING

A. J. ARANYOSI[1], ROOZBEH GHAFFARI[1,2], DENNIS M. FREEMAN[1,2]

[1]*MIT Research Laboratory of Electronics, Cambridge, MA. 02139 USA*
[2]*Speech and Hearing Bioscience and Technology, Harvard-MIT Division of Health Sciences and Technology, Cambridge, MA. 02139 USA*

The ability of the TM to support traveling waves suggests that its motion in response to sound may be more complex than previously believed. To investigate this possibility, we have studied responses of a wave model of the TM to distributed pressures predicted from a passive long-wave cochlear model. The resulting TM motions have a peak near the best frequency of the local BM drive, a notch at a lower frequency, and negative group delays at frequencies from well below to slightly above the best frequency. The negative group delays suggest that TM motion below the best frequency is dominated by the contribution of backward-traveling TM waves. Above the best frequency, forward-traveling TM waves dominate the response. These features are robust against variations in the wavelength and decay constant of the TM waves. Taken together, the notches in the TM magnitude response, negative group delays in the phase response, and increased sharpness of tuning are inconsistent with classical TM models. These phenomena can be explained in terms of TM waves propagating in both the apical and basal directions, highlighting the importance of modeling TM motion as a global wave phenomenon.

1. Introduction – The TM supports traveling waves

We have recently shown that the tectorial membrane (TM) in isolation supports traveling waves of radial motion that propagate longitudinally [1, 2]. These waves can propagate over hundreds of micrometers with velocities comparable to those of the basilar membrane (BM) traveling wave near its peak. Because the TM is in contact with the hair bundles of outer hair cells (OHCs), TM waves can have a significant effect on cochlear function. Since the TM supports waves that propagate in both the apical and basal directions in response to point forces, its response to distributed forces may be quite complex. This paper explores the response of a wave model of the TM to distributed pressures proportional to BM motion determined from a long-wave model. The resulting motions differ in significant ways from those predicted by resonant [3, 4] and inertial [5] point-impedance TM models.

1.1. *A closed-form approximation for TM waves*

TM waves can be approximately characterized by two parameters: the wavelength λ and the decay constant α. These parameters depend on the density ρ, the shear storage modulus $G'(x)$, and the loss modulus $\eta(x)$ of the TM, as well as on the effective storage modulus $\kappa_{\text{hb}}(x)$ of the hair bundles and limbal attachment and the effective loss modulus $\eta_{\text{sts}}(x)$ of the subtectorial space. The spatial gradient of pressure $\partial/\partial x(p_{\text{TM}}(x,\omega)) = Z(x,\omega)v_{\text{TM}}(x,\omega) + p_{\text{BM}}(x,\omega)$, where $v_{\text{TM}}(x,\omega)$ is TM velocity, ω is frequency, $Z(x,\omega) = j\omega\rho(x) + \kappa_{\text{hb}}(x)/j\omega + \eta_{\text{sts}}(x)$ is the local TM impedance

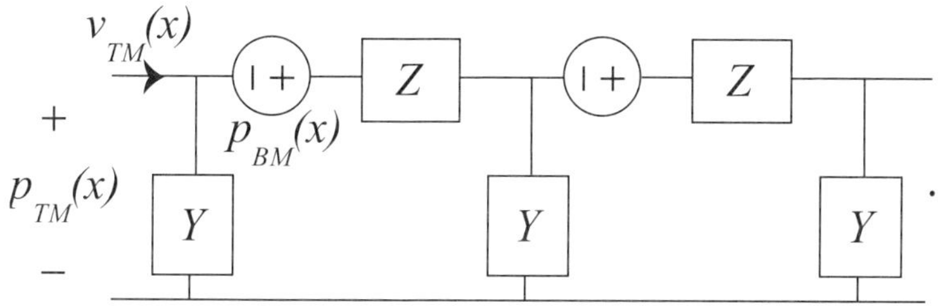

Fig. 1. Schematic diagram of the TM wave model. The velocity $v_{\rm TM}(x,\omega)$ flows through the local impedance $Z(x,\omega)$. The coupling admittance $Y(x,\omega)$ determines the pressure $p_{\rm TM}(x,\omega)$ in proportion to the spatial gradient of TM velocity. Pressure applied by the BM at each location is represented by a pressure source $p_{\rm BM}(x,\omega)$.

(i.e., the part that senses local velocity) [1], and $p_{\rm BM}(x,\omega)$ is the pressure applied by the BM. The spatial velocity gradient $\partial/\partial x(v_{\rm TM}(x,\omega)) = Y(x,\omega)p_{\rm TM}(x,\omega)$, where the coupling impedance $Y^{-1}(x,\omega) = G'(x)/j\omega + \eta(x)$ senses velocity gradients. In other words, spatial pressure gradients move the mass of the TM while spatial velocity gradients shear the tissue. These two phenomena combined give rise to a traveling wave on the TM. Figure 1 illustrates the interconnection of the model components. If we assume that the model parameters do not vary significantly over the distances for which TM waves propagate, we can represent TM waves at any cochlear location using a linear transmission line model. Under these conditions $v_{\rm TM}(x,\omega) = v_0 e^{j(\omega t - x/\zeta)}$, where the complex wavelength $\zeta(x,\omega)$ is related to the wavelength λ and decay constant α by $1/\zeta(x,\omega) = 2\pi/\lambda(x,\omega) - j/\alpha(x,\omega)$. The complex wavelength $\zeta(x,\omega) = 1/\sqrt{Z(x,\omega)Y(x,\omega)}$.

2. Driving TM waves with a cochlear model

We assume that the TM is driven by a spatially distributed pressure $p_{\rm BM}(x,\omega)$ applied by the BM. In point-impedance models of the TM, the relation between $p_{\rm BM}(x,\omega)$ and $v_{\rm TM}(x,\omega)$ is determined by the local impedance $Z(x,\omega)$. In the wave model this relation is more complex, and depends on the local impedance $Z(x,\omega)$, the coupling admittance $Y(x,\omega)$, and the spatially distributed pressure $p_{\rm BM}(x,\omega)$ applied by the BM. The extent of coupling is determined by $Y(x,\omega)$; if $Y(x,\omega) \to \infty$ the model degenerates to a resonant or inertial point-impedance model, depending on the formulation of $Z(x,\omega)$. We drive the TM model with a pressure $p_{\rm BM}(x,\omega)$ predicted from a long-wave cochlear model as described by de Boer (1996). Since the interaction of waves on the TM depends critically on their relative phase, the long-wave model was chosen because of its ability to accurately match the phase of motion of a passive cochlea. The pressure applied to the TM at each location was proportional to the acceleration of the BM as predicted by the long-wave model. This pressure launched waves in both the apical and basal directions, so that the net pressure applied to the TM at any location was a sum of pressures from waves launched throughout the cochlea. These waves attenuated as they propagated, so the response at any location was dominated by waves propagating from nearby cochlear locations. No attempt was made to couple motions of the TM back into the long-wave model of BM motion. There was also no attempt to model OHC somatic [6] or bundle motility [7, 8]. Rather, this study focuses on differences between classical models of TM motion (with local interactions only) and wave models of TM motion when the TM is driven by distributed pressures.

3. Results

3.1. *TM responses are inconsistent with classical models*

Figure 2 shows the magnitude and phase of BM velocity vs. frequency as predicted by the long-wave model and TM velocity predicted by the wave model in response to distributed pressures proportional to BM motion.

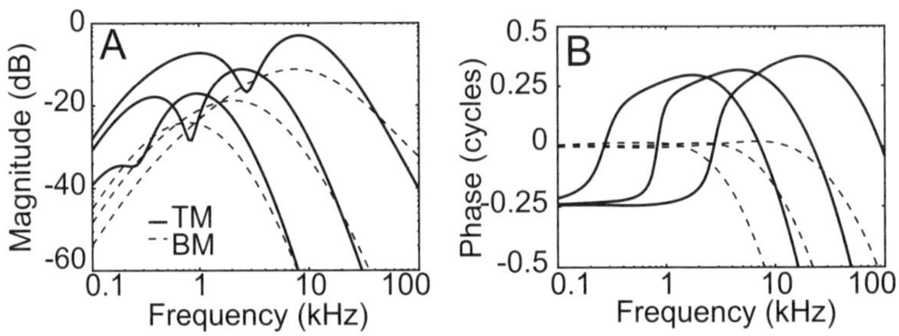

Fig. 2. Magnitude (left) and phase (right) of BM motion (dashed) predicted from a long-wave model and TM motion (solid) predicted from the TM model in response to distributed pressures proportional to BM motion. The multiple curves represent multiple cochlear locations. In comparison to the BM, the TM is more sharply tuned, has a notch below CF, and a phase that increases with frequency over much of the frequency range.

Several significant differences between TM and BM velocity are evident:

- The TM is more sharply tuned than the BM;
- The best frequency of the TM is slightly higher than that of the BM;
- Below the best frequency, the TM magnitude exhibits a prominent notch;
- Above the best frequency, the TM magnitude falls more quickly than that of the BM;
- The TM phase increases with frequency for frequencies up to and slightly greater than the TM best frequency.

These features are not consistent with classical models of the TM. For example, the resonant TM model predicts sharper tuning of the TM, but this sharp tuning corresponds to a phase lag rather than a lead. Neither the resonant nor inertial models predict the presence of a notch. This notch is similar to the notch in BM motion attributed to the TM in other studies [10, 11]. Over most of the low frequency range, the group delay of the TM response is negative, suggesting that TM waves are propagating back toward the stapes.

3.2. *Effect of wave parameters on TM tuning*

Figure 3 shows the magnitude and phase of the ratio of TM to BM velocity at one cochlear location as wave parameters were increased and decreased by factors of two and three.

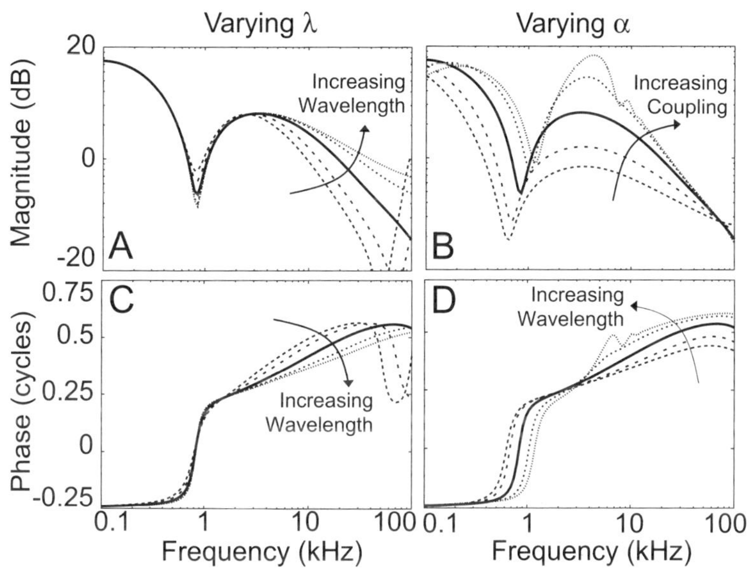

Fig. 3. Magnitude (top) and phase (bottom) of the ratio of TM to BM velocity vs. frequency. The left column shows the effect of varying λ, the right column shows the effect of varying α.

Increasing wavelength λ:

- Decreases the sharpness of tuning of the TM peak by increasing high-frequency TM motion;
- Increases the depth of the notch;
- Extends the phase increase to higher frequencies;
- Does not change the low-frequency gain.

Increasing decay constant α:

- Increases the overall sensitivity of the TM to applied pressures;
- Slightly increases the sharpness of tuning of the TM;
- Increases the frequency of the notch;
- Increases the low-frequency gain, except at the lowest frequencies;
- Increases the slope of phase vs. frequency.

4. Discussion

4.1. *Bidirectional TM traveling waves*

The wave model response to BM-like distributed forces is unlike that of passive point-impedance models of the TM. The magnitude of the TM response exhibits a notch in addition to a magnitude peak. The phase exhibits a negative group delay, suggesting that waves are propagating backwards along the TM. These responses have a simple interpretation in terms of TM waves. At any observed location, the motion of the TM is due to the sum of waves coming from nearby apical and basal segments of the TM. At low frequencies, TM motions are in phase throughout the cochlea and simply sum. As the frequency increases, apical phase accumulation in the BM wave introduces an increasing phase lag in TM waves propagating backward from apical regions. These waves interfere destructively with waves from more basal regions (propagating both apically and basally) that have little phase lag, creating a notch in TM velocity. At higher frequencies, the response is dominated by TM waves propagating basally from near the peak of the BM response. As the frequency approaches the local best frequency, these basal-going waves have a shorter distance to travel and therefore undergo less attenuation, giving rise to the magnitude peak. Above the best frequency, the peak of the BM response moves basal to the response location. In this frequency range the apically propagating TM waves contribute, but their net contribution decreases with frequency as the peak moves further away. In this region the magnitude of TM motion falls with frequency and the slope of the phase vs. frequency curve reverses sign.

4.2. *The effect of model parameters*

To distinguish the relative contributions of spatial phase variation and longitudinal coupling to the overall response of the TM, we need to look at how changes in λ and α affect TM motion. With α constant (i.e., the extent of the TM contributing to the response at a particular location remains constant), the bandwidth of the peak in TM velocity increases with increasing λ. Since an increased λ means that the phase changes more slowly with longitudinal position, this result means that the contribution of waves from multiple locations can sum constructively up to higher frequencies, so the width of the response peak grows. Similarly, the magnitude of the peak depends strongly on α, since more energy will propagate from each cochlear location at which waves are generated. However, the presence of the notch and magnitude peak and the dominance of backward-traveling TM waves are not fundamentally altered by changes in λ and α.

4.3. *Implications for cochlear tuning*

In conventional cochlear models, the TM is represented as a point impedance whose motion depends only on motions of nearby structures. The notches in the TM magnitude response and negative group delays in the phase response seen in this simple model are inconsistent with point-impedance models of the TM. Moreover, the increased sharpness of tuning compared to the BM is inconsistent with models that include longitudinal coupling without phase accumulation (i.e., space constants alone), which inherently broaden tuning. These phenomena can be explained in terms of TM waves propagating in both the apical and basal directions. These results highlight the importance of modeling TM motion as a global wave phenomenon rather than a local point-impedance phenomenon.

Acknowledgments

Supported by NIH grant R01-DC00238. R.G. was supported in part by an NIH grant to the Speech and Hearing Bioscience and Technology program in the Harvard-MIT Division of Health Sciences and Technology.

References

1. Ghaffari, R., Aranyosi, A. J. and Freeman, D. M., 2007. Longitudinally propagating traveling waves of the mammalian tectorial membrane. Proc Nat Acad Sci USA. 104, 16510–16515.
2. Ghaffari, R., Aranyosi, A. J. and Freeman, D. M., 2009. Traveling waves ... on the tectorial membrane. This volume.
3. Allen, J. B., 1980. Cochlear micromechanics — a physical model of transduction. J Acoust Soc Am. 68, 1660–1670.
4. Zwislocki, J. J., 1979. Tectorial membrane: a possible sharpening effect on the frequency analysis in the cochlea. Acta Otolaryngol. 87, 267–269.
5. Mammano, F. and Nobili, R., 1993. Biophysics of the cochlea: linear approximation. J Acoust Soc Am. 93, 3320–3332.
6. Brownell, W. E., Bader, C. R., Bertrand, D. and de Ribaupierre, Y., 1985. Evoked mechanical responses of isolated cochlear hair cells. Science. 227, 194–196.
7. Chan, D. and Hudspeth, A., 2005. Ca^{2+} current-driven nonlinear amplification by the mammalian cochlea in vitro. Nature Neurosci. 8, 149–155.
8. Kennedy, H., Crawford, A. and Fettiplace, R., 2005. Force generation by mammalian hair bundles supports a role in cochlear amplification. Nature. 433, 880–883.
9. de Boer, E., Mechanics of the cochlea: modeling efforts, in *The Cochlea*, eds. Dallos, P., Popper, A. N. and Fay, R. R., Springer Handbook of Auditory Research, Vol. 8 (Springer-Verlag, New York, 1996)
10. Russell, I. J., Legan, P. K., Lukashkina, V. A., Lukashkin, A. N., Goodyear, R. J. and Richardson, G. P., 2007. Sharpened cochlear tuning in a mouse with a genetically modified tectorial membrane. Nature Neurosci. 10, 215–223.
11. Lukashkin, A. N., Lukashkina, V. A., Richardson, G. P. and Russell, I. J., 2009. Does the cochlea compromise on sensitivity and frequency selectivity? This volume.

Comments and Discussion

Fulton: The first reference describes a "Lamb" type surface acoustic wave moving longitudinally along the TM. Is this the type of slow wave assumed in this paper?

Aranyosi: We do not believe that the TM waves we have measured are Lamb waves, but transverse (shear) waves (n.b. "transverse" here means orthogonal to the direction of wave propagation, which includes the radial as well as the transverse cochlear axis for longitudinally propagating waves). Lamb waves propagate through compression and rarefaction of the tissue near the surface. We see no evidence of such compression in our measurements, nor do we expect the tissue to be significantly compressible at audio frequencies. Shear waves, on the other hand, can easily propagate on an incompressible tissue. The model represents shear wave motion of the TM.

MODELING HENSEN'S STRIPE AS A TOPOGRAPHIC WAVEGUIDE THAT DEFINES THE ROLES OF THE OHC AND IHC

JAMES T. FULTON

Hearing Concepts, 1106 Sandpiper Drive, Corona Del Mar, California 92625, USA

The surgical literature suggests a direct acoustic connection between the oval window and Hensen's stripe on the gelatinous surface of the tectorial membrane (TM). I propose that Hensen's stripe can act as a topographic waveguide. The curvature of such a topographic waveguide is known to control the dispersion of the energy as a function of frequency. Describing this curvature by a Hankel function, the propagation velocity by the data of Ghaffari et al., and employing the dispersion equation of Marcatili, provides a complete description of the place-frequency-delay function of hearing for all species of mammals. It may also explain the very high rate of attenuation versus frequency (exceeding 500 dB/octave) on the high frequency side of the best frequency in both OHC and IHC signaling channels. This attenuation is described by "an exponential of an exponential" function. The defined mechanism may explain the unique sensory roles of both the OHC and IHC.

1 Introduction

It is difficult to find one comprehensive description of the end-to end-processes between the oval window of the labyrinth and the mechanical excitation applied to the inner and outer hair cells (IHC/OHC). The goal of this paper is to present a contiguous model of the hearing process that is compatible with the surgical literature, describes the dispersion of energy with frequency, explains the explicitly different sensory roles played by the inner and outer hair cells, and provides a novel explanation for the unique attenuation characteristic associated with the high frequency roll-off of the neural signals derived from the cochlear partition. An expanded version of this paper is available on-line at www.hearingresearch.net/pdf/MoHpaper.pdf. A series of animations is available in the structural model section of www.hearingresearch.net/files/animation.htm.

2 The operation of the vestibule of the labyrinth and the cochlear partition

The surgical literature describes the relationship between the elements of the labyrinth, including the crucial geometric, hydraulic and acoustic arrangements between the oval window and the cochlear partition. This literature includes many photographs and physical models of the relevant structures obtained using lost wax techniques [1]. Figure 1 stresses the close coupling between the oval window and the initial segment of the scala media that extends into the vestibule of the labyrinth. The scala media has a flattened surface of nearly 1 mm^2 aligned perpendicularly to the energy waves generated in the perilymph by the oval window. This energy is traveling as a compression wave at a nominal 1500 meters/sec. As shown in Inset I of Fig. 1, I propose that the energy passes through the outer surface of the scala media and impinges on the initial segment of the TM membrane. I hypothesize that the energy passes through the acoustically transparent

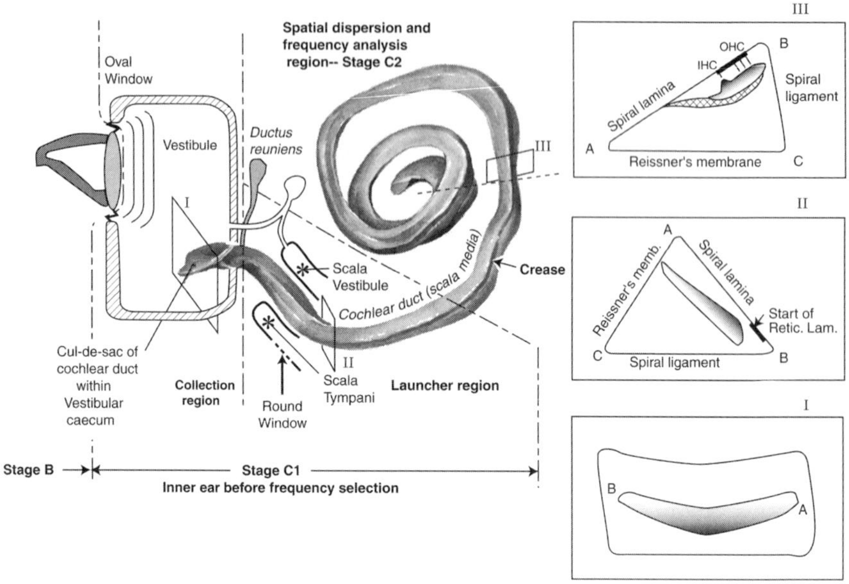

Figure 1. Detailed configuration of the vestibule and cochlear duct. Inset I; the gel surface of TM is the only distinct feature within the duct. Inset II; the initial section of the cochlear duct twists while condensing the energy into an area near the future Organ of Corti. Inset III; the gel surface achieves its final shape and location with the bulk of the TM acting as an inertial mass.

first surface and interacts with a putative density difference encountered at the second surface, as illustrated in Fig. 2. This encounter generates a new Rayleigh type surface acoustic wave traveling along the TM membrane at a nominal 6 meters/sec. The energy then travels along the membrane and is concentrated in the region of Hensen's stripe (Insets II & III of Fig. 1) as it travels toward the Organ of Corti.

To understand the operation of hearing as it relates to the above figure, it is necessary to be familiar with the unusual characteristics of the liquid-crystalline layer (LCL) (colloquially a gel) of the TM facing the reticular lamina. Freeman et al. have provided a recent summary in the hearing literature [2]. A broader discussion including other technological environments can be found in Fulton [3]. Ghaffari et al. have corroborated the low propagation velocity discussed below [4] However, they focused on the propagation of Lamb waves on the above defined surface. They did not address either Rayleigh waves or the more relevant modified Rayleigh waves. A primary feature of the liquid-crystalline surface is its radically different surface tension characteristics under dynamic conditions (excitation above 200 Hz) compared to static conditions.

Figure 2 illustrates how the initial portion of the LCL might operate as an acoustic Rayleigh wave launcher. Such launchers are common elements in engineering applications such as the acousto-mechanical filters in televisions, cell phones and other electronic devices. The launcher shown is a second surface launcher. The incident wave

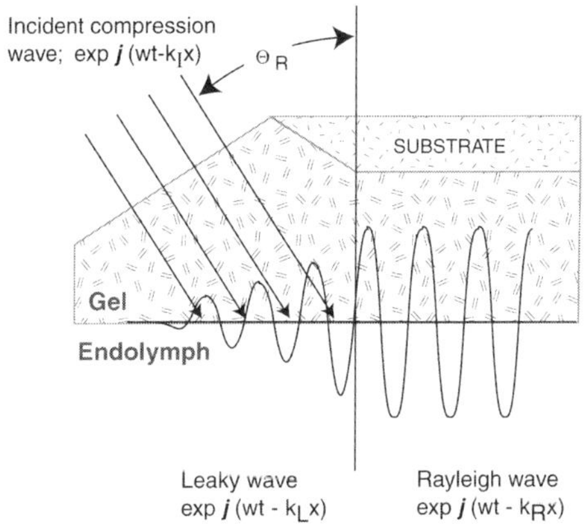

Figure 2. A second-surface SAW launcher as used in hearing. The density of the endolymph opposite the gel surface must be of lower density than the gel.

is shown perpendicular to the first surface and at an acute angle to the second surface. The second angle is quite small with an angle described by the ratio of the velocity of the Rayleigh wave to the velocity of the compression wave involved, 6/1500 or 0.004 radians in the case of human hearing.

In the portion of the scala media towards the right of Figure 2, the energy propagates as a Rayleigh wave but begins to propagate as a modified Rayleigh wave (a combination of a Rayleigh wave and a Lamb wave) at it encounters the topographic feature identified as Hensen's stripe. This feature can be described as a topographic waveguide. As discovered by Marcatili, a topographic waveguide will shed its energy as a function of frequency as it begins to curve [5]. This is the condition encountered by the energy propagating along Hensen's stripe as it enters the right hand side of Figure 2.

3 The mathematical description and performance of the cochlear partition

Earlier investigators have attempted to describe the curvature of the cochlea, and Hensen's stripe, as either an equiangular or logarithmic helix. A closer examination suggests Hensen's stripe (based primarily on measurements related to the overall tectorial membrane) is more appropriately described by a Hankel function, a member of the Bessel function family. Figure 3 shows the Hankel function as an overlay to the data of Bredberg from a human cochlea [6]. For purposes of this work, a "Standard Human Cochlear Partition" will be defined that is 2 7/8 turns long with a frequency of 66 Hz at a distance of 38 mm from the origin of the cochlear partition as a dispersive device. This origin is defined functionally by the beginning of the in-plane curvature of Hensen's stripe (although it is commonly defined morphologically as near the wall between the labyrinth and the cochlea).

To optimally track the curvature of Hensen's stripe in the area of the apical hook, an additional term has been added to the basic Hankel function in Figure 3 The long-dashed

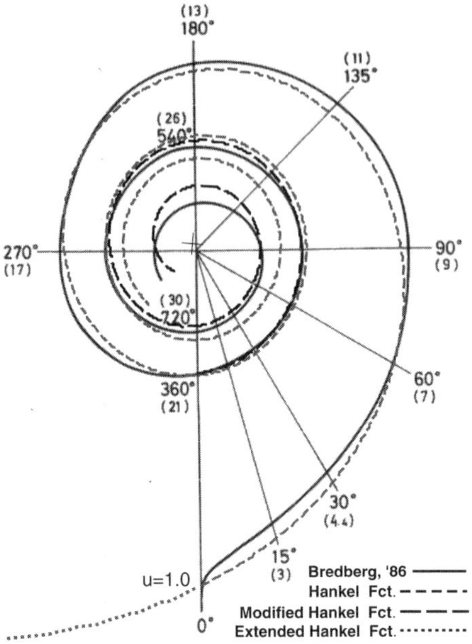

Figure 3. Overlaying a Hankel function to Hensen's stripe. The function is of the first kind and zero order. Its center is indicated by the small cross. The Hankel function is plotted for 0.97< u < 19. The modified Hankel function overlays a figure from Bredberg, 1968. The dotted line shows the extended Hankel function for u<0.97. The numbers in brackets represent the distance from u=0.97.

line in the figure represents a Hankel function modified by an arbitrary factor, $(1-(u/22)^4)$. This modified Hankel function fits the original Bredberg curve to well within the variation in the measured data. The resulting equation for the active portion of the cochlear partition (Hensen's stripe) as a function of distance, u, is given by;

$$HS = 7.5 \cdot H_0^1(u) \cdot \left(1 - \left(\frac{u}{22}\right)^4\right) \qquad \text{for } 1.0 < u < 19. \tag{1}$$

where *HS* is the local radius in millimeters from the center of the cochlea (corresponding to the axis of the modiolus), and the figure is rotated until the value at u = 1.0 lies along Bredberg's negative vertical axis.

Marcatili found the energy shed by a curved topographic waveguide was a function of the curvature and the frequency of the energy involved in 1969. The key equation is (13) in Marcatili & Miller [7]. They express the argument of an exponential as:

$$\alpha_r = c_1 \exp(-c_2 R) \tag{2}$$

where R is the radius of curvature (not the local radius, HS) of the guide.

Their equation can be rewritten to include a critical radius, R_c ,where this critical radius is frequency sensitive. With different values for the constants, the argument can be rewritten as:

$$\alpha_r = c_1 \exp(-c_2 R/R_c(f)) \tag{3}$$

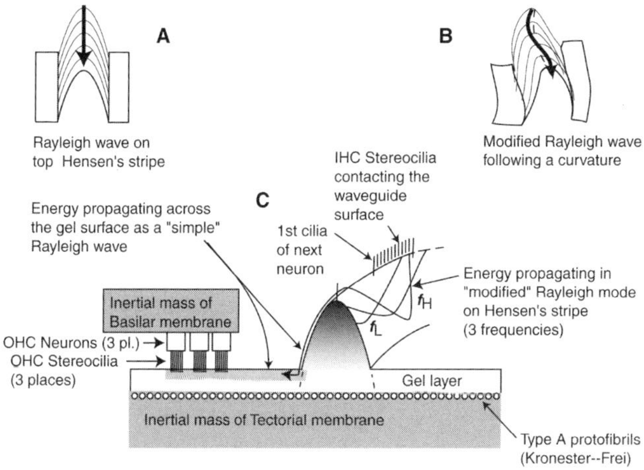

Figure 4. Profiles of the acoustic waveguide of hearing. A; the Rayleigh wave oscillates in the vertical plane and propagates along the top of a straight wave-guide. B; with curvature of the waveguide, a modified Rayleigh wave results (with both Rayleigh & Lamb components). C; the energy of a given frequency travels farther down the sides of the waveguide as curvature increases, until it sheds onto the gel layer surface, where it propagates in a straight line.

This formulation of an argument of an exponential that is itself an exponential is extremely rare in physics. Many authors have expounded on this mechanism in recent years (see the supporting paper).

The proposed operational situation is related to the findings of Marcatili in Figure 4. With increasing waveguide curvature, energy travels down the side of the topographic waveguide and eventually is shed onto the plane of the gel surface, where it propagates in a straight line at an angle to the centerline of Hensen's stripe at that location.

Figure 4 also shows the proposed sensing of the propagated energy by the IHC and the OHC. The IHC sense a broadband signal truncated at the critical frequency, R_c. The OHC sense a signal concatenating the broadband signal with a narrower spatial filter resulting from the dispersion of the energy and a narrow collection window formed by the echeloned OHC.

Figure 5 shows the predicted spectral dispersion compared to the measured data of Schuknect [8]. The predicted performance is extended to include the performance of other animals, including dolphins at up to 150 kHz (using a different abscissa)

The three broad band-pass frequency responses shown in Figure 6 illustrate the predicted response for the human cochlea at three distances along Hensen's stripe. This predicted response is obtained by performing a line integration using Marcatili's argument along Hensen's stripe. The result predicts the 10 dB/decade attenuation at low frequencies and the astounding "exponential of an exponential" decay at high frequencies. Also shown is the frequency response of the OHC channel associated with the broadband response at 15,000 Hz. This predicted responses may provide a foundation for several features of the OHC responses associated with Kiang. For signals recorded within the neural system prior to the cochlear nucleus, an inflection point is observed on each side of the narrowband response due to the concatenation (a multiplication) of filter responses described above. These inflection points have been labeled seams by Kiang. For peripheral measurements of the residual response of the

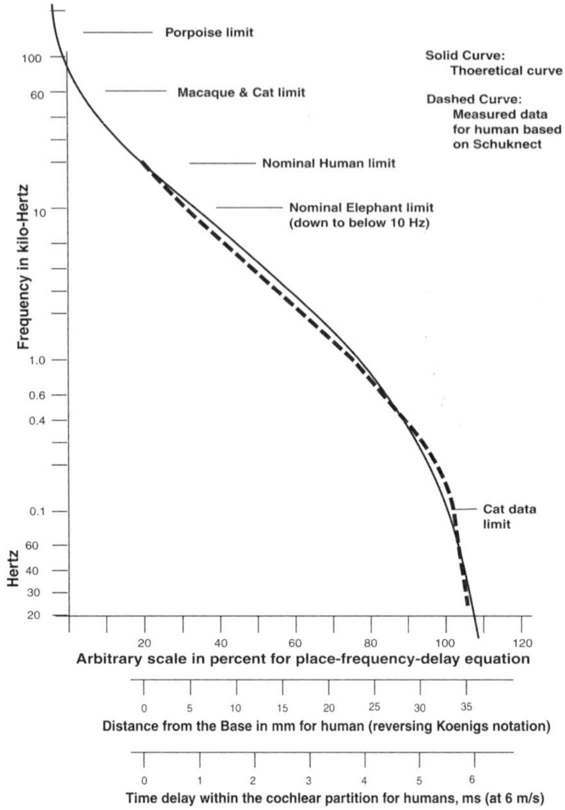

Figure 5. Proposed place-frequency-delay characteristic of the cochlear partition based on a propagation velocity of 6 meters/sec (solid line). Dashed line, measured human data from Schuknect. Also shown, the limits of hearing in various animals as they relate to the theoretical curve.

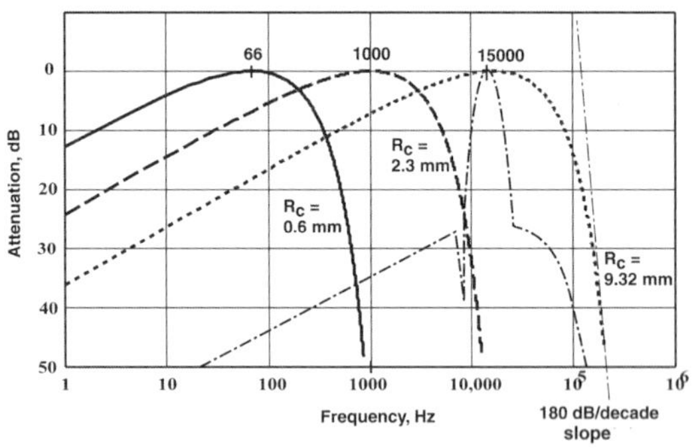

Figure 6. Predicted attenuation for the human cochlea at three radii of curvature along Hensen's stripe. Solid line: 0.6 mm radius with optimum performance at 66 Hz (ignoring the effect introduced by the apical hook). The construction line at the right falls at 180 dB/decade. The rate of attenuation is still increasing at the 50 dB down point for all responses. The complex dash-dot line represents the attenuation of the OHC channel formed at $R_c = 9.32$ mm.

basilar membrane, acting as an inertial mass, a local minimum is frequently observed on the low frequency side of the narrowband response. In the model, this feature represents a null in the summation of the basilar membranes motion reacting to the forces applied to it by both the OHC and IHC.

4 Conclusion

This paper has introduced a paradigm that is beyond the purview of de Boer's discussion in [9]. The surgical literature describes an acoustic path involving the tectorial membrane, and particularly Hensen's stripe, as the principle element of acoustic energy propagation within the cochlea, with the bulk of the tectorial membrane and the basilar membrane playing ancillary roles as inertial masses. Analyses based on Hensen's stripe as the principle element of acoustic energy propagation lead to:-· unique and specific descriptions of separate sensory roles for the OHC and the IHC; a theoretical description of the complete place-frequency-delay characteristic of human hearing; and a novel theoretical explanation for the extremely high attenuation at acoustic frequencies higher than best frequency for the OHC and IHC signaling channels.

Propagation from the oval window to the Organ of Corti via Hensen's stripe negates the need for a mechanical amplifier to raise the amplitude of basilar membrane vibrations. The spatial dispersion from a curved waveguide achieves increasingly high attenuations per octave with increasing frequency above the peak response frequency. The fact that this mechanism predicts both the low pass characteristic and the extremely high attenuation per octave at high frequencies measured in the cochlea is strong support for the validity of this theory of cochlear operation.

References

1. Donaldson, J. Lambert, P. Duckert, L., & Rubel, E., 1992. Surgical Anatomy of the Temporal Bone, 4th Ed. NY: Raven Press, pages 71, and 267-277.
2. Freeman, D. Masaki, K. McAllister, A. et al., 2003. Static material properties of the tectorial membrane: a summary. Hear Res 180, pp 11-27
3. Fulton, J. T., 2008. Hearing: A 21st Century Paradigm. Victoria, BC, CA: Trafford
4. Ghaffari, R. Aranyosi, A. & Freeman, D., 2007. Longitudinally propagating traveling waves on the mammalian tectorial membrane. PNAS 41, 16510-16515
5. Marcatili, E., 1969. Bends in optical dielectric waveguides. BSTJ 48, pp 2103-2132
6. Bredberg, G., 1968. Cellular pattern and nerve supply of the human organ of Corti. Acta Oto-laryng Suppl 236, 37-39
7. Marcatili, E. & Miller, S., 1969. Improved relations describing directional control in electromagnetic wave guidance. BSTJ 48, 2161-2181
8. Schuknect, H., 1974. Pathology of the Ear. Cambridge, MA: Harvard University Press, Table 3.2
9. de Boer, E., 2006. Cochlear activity in perspective. In: Nuttall, A.L. et al. eds. Auditory Mechanisms: Processes and Models. World Scientific, Singapore, pp. 393-409.

MEASUREMENT OF ANISOTROPIC MECHANICAL PROPERTIES OF THE TECTORIAL MEMBRANE*

N. GAVARA, R. S. CHADWICK

Auditory Mechanics Section, National Institute on Deafness and Other Communication Disorders, National Institutes of Health, 10 Center Drive MSC 1417
Bethesda, MD 20892, USA

The tectorial membrane (TM) in the cochlea is an anisotropic tissue with a key role in hearing. The TM's structural and mechanical anisotropy is provided by oriented collagen bundles about 1 micron thick. Here we report the three elastic moduli that characterize the TM, as well as the novel technique used to measure the mechanical properties of an anisotropic material. We have measured mechanical anisotropy by combining Atomic Force Microscopy (AFM) and optical tracking of microspheres. The surface Green's tensor for an incompressible anisotropic material was then used to compute the elastic moduli from imposed forces and the resulting surface deformations. Our results suggest a critical role of TM's strong anisotropy by enhancing the cochlear amplifier.

1 Introduction

The mammalian TM has a highly organized anisotropic structure at the microscopic scale. This acellular matrix contains two main groups of components, collagen fibrils and glycoproteins. Glycoproteins compose a striated-sheet matrix surrounding the collagen fibrils [1]. Collagen fibrils are organized in heavy bundles that run radially across the TM. A number of studies have measured the mechanical properties of TM [2-5]. Nevertheless, these studies modeled the TM as an isotropic homogeneous material, reporting only one elastic modulus. A recent study has used AFM to estimate the effective Young's modulus of the TM along different orthogonal directions [6]. However, the lack of a model to understand the data does not provide knowledge of the anisotropic mechanical properties of the TM. The TM can be described using the simplest model of anisotropy, the transversely isotropic model, in which there is a single family of parallel fibres embedded in a matrix whose elastic properties are the same in any direction perpendicular to the fibres. The fibres correspond to the collagen bundles and the matrix corresponds to the glycoproteins. When the material is incompressible, its mechanical properties can be described with three elastic moduli, a fibre modulus (E_f) and two shear moduli, one parallel to the fibres (μ_L) and another perpendicular or transverse to them (μ_T).

We have measured the anisotropic mechanical properties of the TM by combining AFM to exert point-like forces using a spherical tip, and optical tracking of microspheres to measure surface displacements. We previously described the relationship between point-like forces applied to the surface of a material with transverse isotropy and the resulting surface displacements [7]. Such relations are described by a surface Green's

---

* The source material for this paper is from a manuscript submitted to *Nature Materials*.

tensor, whose elements represent surface displacements in the *i*th direction that result from a unit point force at the origin acting in the *j*th direction. The Green's tensor contains the three elastic moduli of the material. Although most elements of the tensor are very complex, along special directions some elements are given by simple functions that involve only one elastic modulus. Such elements are then useful to estimate the elastic moduli of the material. This novel approach has enabled us to measure for the first time the three anisotropic elastic moduli of an anisotropic material.

2 Methods

2.1 Sample isolation

Tissue samples were acquired from 12 juvenile female pigmented guinea pigs (weight 150-200 g). All animals included in the study tested positive for pinna (startle) response. TMs were isolated as previously outlined [3]. Samples were laid onto coverslips coated with Cell-Tak for subsequent measurement. We discarded samples that appeared to be folded, damaged in any way or attached with the cover net facing up. TM segments were classified as basal or apical based on the measured distance from the edge of the marginal band to the ridge associated with the attachment of the TM to the spiral limbus [8]. All animal procedures were conducted according to approved National Institutes of Health (NIH) animal protocol number 1186-07.

2.2 Setup

Measurements were performed on the stage of a Bioscope II AFM (Veeco Metrology Inc., Santa Barbara, CA), coupled to an inverted fluorescence microscope (Fig 1). The AFM was used to exert controlled forces applied to the surface of the TM. A latex bead (10 μm diam.) glued to a tipless cantilever (nominal spring constant 0.58 N/m) was used to indent the TM. The vertical displacement of the tip was controlled by the Z-piezo of the AFM head. Forces applied to the TM were measured by the system's photodiode. Lateral displacements of the sample with respect to the tip were generated by the X-Y piezos located on the stage. At the beginning of the measurements, the angle between the fibres of the TM and the X-Y piezos of the stage was established. All sample displacements and force impositions were subsequently performed on the orthogonal coordinate system defined by the vector parallel to the fibres, the vector perpendicular to the fibres in the plane of the surface, and the vector normal to the surface. To enable the AFM to exert a force component on the plane defined by the TM surface, a custom-made wedge was positioned between the stage and the AFM head (10 degrees tilting). The head ensemble could also be rotated, enabling the force on the surface plane to be directed parallel or perpendicular to the TM fibres. Fluorescent polystyrene beads (1 μm diam.) deposited onto the surface of the TM were used as markers to track TM deformations.

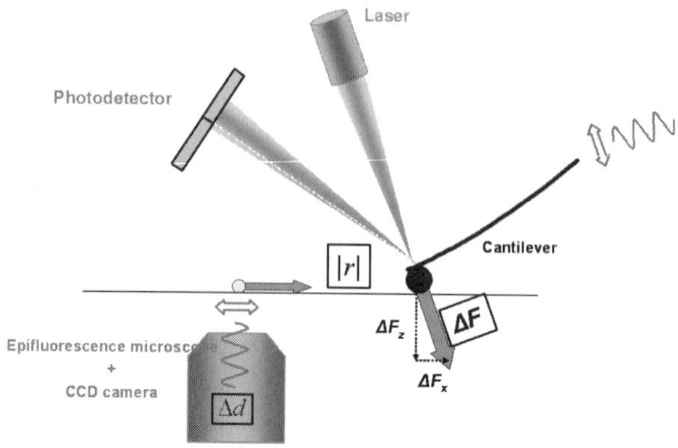

Figure 1. Sketch of the setup to measure mechanical anisotropy.

2.3 Protocol

Before measurements, the relationship between photodiode signal and cantilever deflection was calibrated. Then a fluorescent bead located near the Hensen's stripe was chosen and a phase contrast image was acquired to compute the angle of alignment of the fibres below the bead. During the experimental protocol five subsets of measurements were performed. On the first 2 measurement subsets the head was not tilted, and the sample was moved away from the tip in directions parallel and perpendicular to the fibres. Subsequently, the head was tilted so that a component of force was exerted in parallel to the fibres and the sample was laterally moved away in parallel and perpendicular directions. Finally, the head was tilted to exert a component of force perpendicular to the fibres in the plane defined by the TM surface, and the sample was moved away only parallel to the fibre. A measurement subset consisted of 5 indentation measurements performed at increasing tip-bead distances, starting at 8 μm using 1 μm increases. Each time the sample was moved, the contact point was visually estimated using a force-displacement curves. Indentation measurements were performed at 1 μm initial indentation, with a 1 μm amplitude triangular ramping wave at 0.3 Hz frequency. A force-displacement curve was recorded for one cycle at 512 points per cycle, whereas images of the fluorescent bead were recorded at 25 frames/sec using a 40× objective.

2.4 Data processing

Applied force was calculated from AFM recordings, using the cantilever bending and its spring constant. Amplitude of force changes (ΔF) were computed by measuring the slope of the force-displacement curve and multiplying it by the length of half cycle. Sample deformation was measured by computing changes in the position of the centre of mass of the fluorescent using a Gaussian mask algorithm. After correcting for slow drifts of the system, the position of the bead over time resembled the triangular pattern of the applied force. The slope of the resulting curve times the length of one half cycle was used as the amplitude of bead displacement (*Δd*). In order to assert sample fibre deformation due solely to forces applied on the surface plane, displacements measured on the non-tilting configuration were subtracted from displacements measured on the tilting configuration. The ratio of applied forces in the surface plane (ΔF_S) to resulting displacements (Δd_S) was computed as follows:

$$\frac{\Delta d_S}{\Delta F_S} = \left(\Delta d_T - \Delta d_{NT} \cdot \frac{\Delta F_T \cos\alpha}{\Delta F_{NT}} \right) \cdot \frac{1}{\Delta F_T \sin\alpha} \tag{1}$$

where α is the angle by which the head of the AFM system was tilted, and the *T* and *NT* subscripts indicate tilting and non-tilting configuration, accordingly.

2.5 Modelling

The Green's tensor $G_{ij}(x,y,z)$ resulting from the Boussinesq-like solution for an anisotropic material provides 9 relationships between point forces, elastic moduli and resulting displacements along each axis [7]. In our experiments, the direction of applied forces and the selected component of the resulting displacements were paired to reproduce the elements of the Green's tensor that provided direct estimates of the elastic moduli. The following relations were used to compute μ_L, μ_T and E_f:

$$G_{11}(0,y,0) = \frac{\Delta d_1}{\Delta F_1} = \frac{1}{4\pi\mu_T}\frac{1}{y} \tag{2a}$$

$$G_{12}(x,y,0) = \frac{\Delta d_1}{\Delta F_2} = \frac{\mu_T}{2\pi\sqrt{E_f \mu_L}}\frac{y}{x^2} \tag{2b}$$

$$G_{22}(0,y,0) = \frac{\Delta d_2}{\Delta F_2} = \frac{1}{2\pi\mu_L}\frac{1}{y} \tag{2c}$$

3 Results

Ramping indentations resulted in oscillatory forces with amplitudes ranging 10-120 nN. Consistently, the position of the fluorescent beads also displayed oscillatory movements, with amplitudes ranging 20-200 nm. As the tip was moved away from the bead during a set of measurements, forces did not change markedly, but resulting bead displacements were found to decrease. Therefore, $\Delta d_i/\Delta F_j$ was observed to decrease with increasing tip-bead distances. In addition, decreases in $\Delta d_i/\Delta F_j$ were well-fitted by a linear function of $1/|r^n|$. On additional measurements, we verified that the slope of the fit did not depend markedly on the initial indentation (data not shown). The slopes of the fits were then used to compute the three elastic moduli. The fibre modulus was found to be on the order of a MPa, whereas shear moduli were four orders of magnitude smaller (Table 1). All elastic moduli displayed larger values at the base of the TM than at the apex.

Knowledge of the elastic moduli and the expression of the surface Green's tensor can be used to estimate other mechanical properties of the tissue. We have derived

Table 1. Elastic moduli at the base and apex.

	Apex	Base
μ_L (Pa)	181 ± 66	334 ± 450
μ_T (Pa)	72 ± 8	205 ± 170
E_f (MPa)	0.46 ± 0.08	2.6 ± 0.1

Table 2. Other mechanical properties of the TM.

	Apex		Base	
	this work	others [a]	this work	others [a]
Young's Modulus (KPa)	0.78	0.53 [5] ~ 4.5 [3] ~ 25 [4]	1.93	1.9 [5] ~ 4.5 [3] ~ 215 [4]
Wave speed (m/s)	9.4	~ 4 [8]	18.0	~ 7 [8]

[a] Numbers in superscript identify referenced paper.

expressions that relate the propagation velocity of travelling waves or effective Young's modulus to the three elastic moduli that characterize a transversely isotropic material. Table 2 shows the estimates of these two quantities, as well as values previously reported in the literature. With one exception, our estimates lie in the range of the prior findings and reproduce qualitatively the gradient reported in the mechanical properties from base to apex in the TM.

4 Discussion

Unlike common AFM or magnetic bead techniques to estimate stiffness, which measure sample deformation through cantilever bending or bead movement, our setup uses microscopic beads resting over a fibre to track surface deformation. By uncoupling the probe imposing force onto the surface from the probe tracking sample deformation we can measure the third elastic modulus, that is, the fibre modulus. Estimation of this modulus would not be possible with a standard AFM alone or a rheometer. During force imposition, the temporal displacement of the bead followed a triangular pattern, parallel to the vertical ramping of the cantilever. This clearly indicates that the cantilever tip was not slipping on the surface, as is the case on AFM's lateral force measurements [6]. Therefore, the ratio of bead displacement to applied forces permitted real estimates of the elastic modulus. The main role of the TM in hearing comprises its mechanical interaction with the tips of the stereocilia of the OHC. The diameter of a single stereocilium or its initial indentation into the TM is in the hundreds of nanometres range [9]. By contrast, the amount of indentation and the effective contact area of the probe used in our experiments was on the micron range. To test the possible implications of this scale difference, we applied force ramps using different initial indentations and contact areas. Interestingly, we did not observe marked changes on the mechanical properties of the TM (data not shown). Therefore, our results reflect the anisotropic mechanical properties of the TM in the spatial scale relevant to hearing.

We have previously shown that radial motion of the TM is responsible for sound-induced OHC stereocilia shearing and subsequent deflection [10]. It should be noted that the upper end of the tip-link, the structure that mediates mechanotransduction, does not extend to the top of the stereocilium. Therefore, tip-link stretching and resulting opening of ion-channels depends not on the lateral movement of the stereocilium tip but rather on its deflection angle. Since OHC stereocilia display different translational stiffness and lengths from base to apex [11-12], different shear forces are required to induce the same amount of stereocilia angular deflection at different cochlear locations. Using the values reported in the previous references, we estimate that the shear force needed to deflect a single stereocilium by a fixed angle is ~3 fold larger in the base than in the apex. Interestingly, shear forces required to deform the TM in the radial direction scale as $(\mu_L E_f)^{1/2}$, yielding a striking 3.2 fold difference from base to apex. As a result, the same radial motion of the TM induces the same angular deflection of stereocilia anywhere in the cochlea. Therefore, the mechanical properties of the TM seem to be finely tuned to guarantee that despite the varying morphology of the stereocilia bundle, sound-induced tip-link stretching remains constant along the length of the cochlea. Because of TM's strong anisotropy, deformations of the TM's surface decay very fast as we move perpendicular to the fibres, but very slowly in the direction parallel to them. Importantly, the direction of the fibres is consistent with the direction of stereocilia bundle deflection that leads to maximal tip-link stretching. Therefore, this strong radial coupling between neighbouring cells of different rows ensures that the stereocilia of the three OHCs deflect in concert. Thus, the strongly anisotropic TM may help coordinating somatic and bundle contributions to the cochlear amplifier.

Acknowledgments

We thank T. B. Friedman, K. H. Iwasa, G. P. Richardson and I. A. Belyantseva for critical input. This work was supported by National Institutes of Health Intramural Research Program Project Z01-DC000033-10.

References

1. Goodyear, R.J., Richardson, G.P., 2002. Extracellular matrices associated with the apical surfaces of sensory epithelia in the inner ear: molecular and structural diversity. J Neurobiol. 53, 212-227.
2. Abnet, C.C., Freeman, D.M., 2000. Deformations of the isolated mouse tectorial membrane produced by oscillatory forces. Hear Res. 144, 29-46.
3. Shoelson, B., Dimitriadis, E.K., Cai, H., Kachar, B., Chadwick R.S., 20004. Evidence and implications of inhomogeneity in tectorial membrane elasticity. Biophys. J. 87, 2768-77.
4. Gueta, R., Barlam, D., Shneck. R.Z., Rousso, I., 2006. Measurement of the mechanical properties of isolated tectorial membrane using atomic force microscopy. Proc Natl Acad Sci U.S.A. 103, 14790-5.
5. Richter, C.P., Emadi, G., Getnick, G., Quesnel, A., Dallos, P., 2007. Tectorial membrane stiffness gradients. Biophys J. 93, 2265-76.

6. Gueta, R., Barlam, D., Shneck, R., Rousso, I., 2008. Sound-evoked deflections of outer hair cells stereocilia arise from tectorial membrane anisotropy. Biophys. J., biophysj.107.125203.
7. Chadwick, R.S., Shoelson, B., Cai, H.X., 2004. Surface Green's functions for an incompressible, transversely isotropic elastic half-space. SIAM J Appl Math. 64, 1186-97.
8. Ghaffari, R., Aranyosi, A.J., Freeman, D.M., 2007. Longitudinally propagating traveling waves of the mammalian tectorial membrane. Proc Natl Acad Sci U.S.A. 104, 16510-5.
9. Tsuprun, V., Santi, P., 1998. Structure of outer hair cell stereocilia links in the chinchilla. J Neurocytol. 27, 517-28.
10. Cai, H., Shoelson, B., Chadwick, R.S., Evidence of tectorial membrane radial motion in a propagating mode of a complex cochlear model. Proc Natl Acad Sci U.S.A. 101, 6243-8.
11. Kros, C.J., 1996. Physiology of Mammalian Cochlear Hair Cells. In: Dallos, P., Popper, N., Fay, R.R. (Eds.), The cochlea. Springer-Verlag, New York, pp. 319-85.
12. Wright, A., 1984. Dimensions of the cochlear stereocilia in man and the guinea pig Hear Res. 13, 89-98.

Comments and Discussion

Rousso: In table 1, the Young's modulus of the collagen fibers has a different value in the apex and in the base. Young's modulus is an inherent material property that does not depend on the size and geometry of the material, and therefore should be the same. Could you comment what is the implication that your analysis yielded different Young's moduli for collagen at the two regions? Could it imply that these moduli, derived from your model, do not represent the "pure" modulus of collagen but rather they represent some sort of bulk moduli which include the collagen fibers ultrastructure?

Chadwick: Yes that's correct, the fiber modulus that we have measured should be interpreted as the modulus of a composite fiber formed from many fibrils. The fact that the fiber modulus is larger at the base than the apex implies that the fiber structure is not uniform. This is also borne out by our morphological analysis of DIC images of fibers, which shows that the fiber width increases and the gaps between them decrease from apex to base.

Fulton: In an earlier paper, Cai, Shoelson & Chadwick treated the TMs as linear homogeneous isotropic solid domains. In this paper, the TMs are modeled "using the simplest anisotropic model," the tranversely isotropic model where the parallel fibres are embedded in a non fibre matrix. You speak of the radial motion of the TM as if it is a bulk effect. Is it possible that only the surface of the TM facing the endolymph is mobile? Is it possible that this layer is supporting the propagation of a surface acoustic wave?...[editors note: the full text of this comment is available in an on-line forum at www.mechanicsofhearing.com.]

DEFLECTION OF IHC STEREOCILIA IN RESPONSE TO SOMATIC OHC ELECTROMOTILITY

CAIO CHIARADIA, MANUELA NOWOTNY, ANTHONY W. GUMMER

Section of Physiological Acoustics and Communication, University of Tübingen Elfriede-Aulhorn-Str. 5, Tübingen, 72076, Germany

Nowotny and Gummer [1] have recently shown that for stimulus frequencies less than about 3 kHz, the somatic electromotility of the outer hair cells (OHCs) leads to anti-phasic motion of the reticular lamina (RL) and tectorial membrane (TM) in the region of the inner hair cells (IHCs). This motion is predicted to cause displacement of the fluid inside the subtectorial space, the radial component of which is believed to deflect the IHC stereocilia. We performed experiments in an *in-vitro* preparation of the guinea-pig cochlea to measure IHC stereocilia motion in response to intracochlear electrical stimulation. The present results show that 1) stereocilia deflection amplitude is high-pass filtered relative to OHC transversal displacement, attaining on average 6 dB and 2) OHC contraction causes IHC stereociliary deflection in the excitatory direction. In summary, in addition to high-frequency tuned amplification, there exists a second "amplifying" process in the cochlea, operating below about 3 kHz, which positively couples somatic electromotility directly to the IHC stereocilia.

1 Introduction

The high sensitivity and precise frequency discrimination of the mammalian cochlea is based on electromotility of the outer hair cells (OHCs), which amplifies, cycle by cycle, the sound-induced mechanical vibration of the cochlear partition. Coupling of electromotility to the inner hair cells (IHCs), the true sensory cells of the cochlea, still remains poorly understood. Nowotny and Gummer [1] have recently shown that for stimulus frequencies less than about 3 kHz, somatic electromotility of the OHC leads to anti-phasic motion of the reticular lamina (RL) and the tectorial membrane (TM) in the region of the IHC. That is, the subtectorial space in this region is synchronously contracted and enlarged due to the anti-phasic motion of the RL and TM. This motion causes displacement of the fluid inside the subtectorial space, the radial component of which is expected to deflect the IHC stereocilia.

The objective of this work was to determine whether the anti-phasic motion, elicited by electromechanical force from the OHC, really does deflect IHC stereocilia and if so, to determine the radial displacement of the stereocilia relative to the transversal displacements of the RL in the regions of the IHC and OHC.

2 Methods

It is *rotation* of the IHC stereocilia about their rootlets that modulates the open probability of their mechanoelectrical transduction channels. Therefore, in order to correctly evaluate stereocilia deflection, one must measure not only the displacement of the tip of the hair bundle but also motion of the whole cell, that is, its translation and rotation. Subtraction of these motion components from that of the tip of the stereocilia bundle yields the true deflection of the stereocilia.

The influence of cell rotation on the apparent motion of the tip of the stereocilia bundle is perhaps not so evident, but can be significant. The RL rotates around a pivot point located close to the joint of the pillar cells [1, 2]. In this case, (the apex of) the IHC would experience the same rotation and this rotation would be expressed as displacement of the tip of its stereocilia bundle. When the distance between the pivot point and the root of the hair bundle of the IHC is similar to the height of the hair bundle itself, any displacement of the cuticular plate of the IHC in the transverse direction, driven by the rotation of the RL, leads to a similar displacement of the tip of the stereocilia. Therefore, a dedicated system was built to measure both the radial displacement of the stereocilia tip and also the two dimensional motion of its IHC.

Radial motion was measured with a fast line camera adapted onto an upright microscope (Zeiss Axioskop FS2). The camera was able to resolve different structures in the organ of Corti, such as the stereocilia bundle or the border of the hair cells. Each pixel corresponded to 32 nm. The calculation of the displacement of these structures was performed with an autocorrelation-based procedure, with the frames generated by the camera as input. With this procedure, we could detect displacements with a background noise level smaller than 2 nm up to at least 4 kHz.

Rotation of the RL, as well as transversal motion of the TM, was evaluated by measuring transversal displacements at different radial positions with a laser Doppler vibrometer (LDV; Polytec). The laser beam was coupled into the microscope in such a way that the focus plane of the beam was in the same plane as that of the eyepieces. Hence, one could readily discern the position of the focused laser spot on the measurement surface. The diameter of the focussed laser beam was 0.5 μm and the depth of focus was ± 1.5 μm. Consequently, vibration measurements could be made at either the RL or opposing TM surface with little optical interference between them.

Adult pigmented guinea pigs were decapitated after anaesthesia with CO_2. The temporal bone was removed and stored in cooled Hanks' balanced salt solution (HBSS), adjusted to pH 7.2 - 7.4 and osmolality 300 - 310 mOsm. Parts of the bony cochlear wall of the first, second and fourth turns were removed. The remaining bony septum covering the third turn was opened to obtain a panoramic view of the organ of Corti through the still intact Reissner's membrane. The bone of the "floor" of this turn was partially removed to allow fluid exchange between scala tympani and the bath solution. The preparation was accommodated into a two-compartment chamber, totally filled with artificial perilymph of low chloride concentration (NaCl 30 mM, sodium gluconate 110 mM, KCl 3 mM, $MgCl_2$ 0.9 mM, $CaCl_2$ 1.30 mM, pyruvate 2 mM, glucose 5 mM, Hepes 10 mM). The preparation was inserted into an opening between the two compartments. Fluid leak across the interface between the chamber and the irregular cochlear wall was prevented with modelling clay. A drop of cyanoacrylate glue was used to block the natural communication provided by the cochlear turn itself. In that way, the upper and the lower compartments of the preparation chamber were completely isolated from each other. Finally, Reissner's membrane was removed and the solution of the upper compartment was substituted with artificial endolymph (KCl 150 mM, $CaCl_2$ 30 μM, pyruvate 1 mM, glucose 5 mM, Hepes 10 mM). To electrically stimulate the organ of Corti, two pairs of platinum electrodes were placed in the upper and lower compartments. The (*in-vivo*) characteristic frequency of the recording location was estimated to be 800 Hz.

Measurements of the transversal displacement of the RL were performed at the outer and inner border of the IHC, at the joint of the pillar cells and at the first and second rows of the OHCs. With respect to the TM, transversal vibration measurements were performed at its lower surface in the regions over the IHC, the apical joint of the pillar cells and the second row of OHCs. Displacements of the tip of the stereocilia and of the outer border of the IHC were measured in the radial direction with the line camera. A complete set of measurements with a multi-tone stimulus of 12 frequencies required typically 20 min. For positive current into artificial endolymph, transversal phase was positive in the direction of scala vestibuli and radial phase was positive for stereocilia deflection in the direction of the longest stereocilia (the excitatory direction).

3 Results

Radial motion of the tips of the stereocilia, relative to the cuticular plate of their IHC, was indeed detected in response to the electrical stimulation. This motion was evaluated after subtracting the translational and rotational components of the RL displacement. The amplitudes of the stereocilia deflection were in the range 5 to 30 nm, depending on stimulus frequency and also on the preparation (N = 8). The transversal motions of the TM and RL at the IHC were 180° out of phase (data not illustrated), in agreement with Nowotny and Gummer [1].

We define a stereocilia deflection gain as the ratio between radial displacement of the tips of the IHC stereocilia and transversal displacement of the RL at the OHC. This gain is plotted in Fig. 1 for a single preparation. Notice that this amplitude response is high-pass with corner frequency near the *in-vivo* CF of 800 Hz.

On average (N = 8 preparations), the amplitude of the gain had slope 4 dB/oct and attained 6 ± 3 dB at 3 kHz. The average phase of the gain was $-113^{\circ} \pm 33^{\circ}$ and exhibited little frequency dependence over the measurement range of 30 Hz to 4 kHz. This phase difference means that radial displacement of the IHC stereocilia leads OHC somatic displacement by approximately 90° (180° – 113° = 77°).

Application to the artificial perilymph of 9-AC – a chloride-channel blocker known to block OHC electromotility [1] –, caused a 20-30 dB attenuation of the vibration responses (N = 3). This observation suggests that the electrically induced vibration responses are mainly induced by electromechanical force from the OHC soma. It also demonstrates that the responses are not due to some artefact such as direct electrical stimulation of surface charge on the stereocilia.

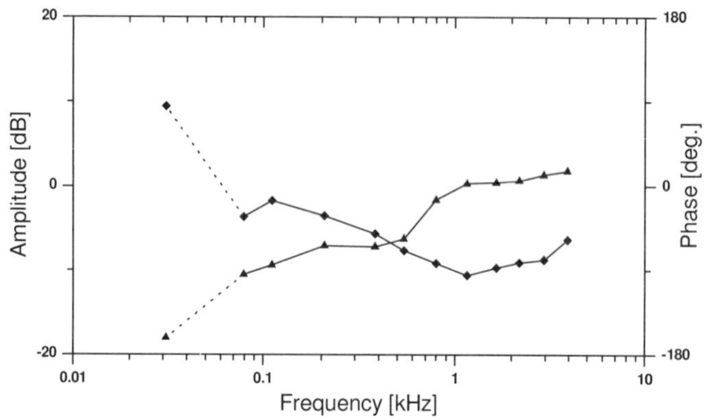

Figure 1. Stereocilia deflection gain, defined as the radial displacement of the tip of the IHC stereocilia relative to the transversal displacement of the RL at the first row of OHCs. Triangles: Amplitude; Diamonds: Phase. The dashed lines depict the region of the frequency spectrum where the measurements were dominated by noise.

4 Discussion

The results support the hypothesis, postulated in [1], that OHC electromotility can be coupled directly to the IHC stereocilia via "pulsatile" fluid forces caused by the anti-phasic motion of the TM and the RL at the IHC. Moreover, the coupling is positive (excitatory) because the observed phases indicate that OHC contraction, known to be in-phase with OHC depolarization, causes deflection of the IHC stereocilia in the excitatory direction. The 90° phase lead of radial deflection of the IHC stereocilia relative to OHC somatic displacement indicates that the force which drives the bundle is in-phase with the OHC transversal velocity. In other words, this phase relation is in agreement with the notion that a fluid force proportional to radial velocity of endolymph in the subtectorial space, generated by counter-phasic transversal motion of the TM and the RL at the IHC, deflects IHC stereocilia.

Clearly, there are still a number of open questions. For example, does this motion occur for acoustic stimulation in the *in-vivo* cochlea? Is the observed deflection of the stereocilia actually driven by forces associated with pulsatile fluid motion?

With respect to the first issue, *in-vivo* measurements of the acoustically driven vibration pattern of the basilar membrane along its radial dimension [2] present a phase difference of 180° between the regions corresponding to the IHC and the foot of the outer pillar cells. Indeed, these authors suggest electromechanically driven rotation of the RL in a very similar fashion to that described here.

Regarding the second question, there are two components of the combined motion of the TM and RL that can drive fluid flow in the subtectorial space: 1) pulsatile motion driven by counter-phasic transversal motion of the TM and the RL in the region of the IHC and 2) "classical" shearing motion, driven by relative radial motion of the TM and the RL. It is unlikely that shearing motion is the main driving force for stereocilia deflection in our preparation because for all frequencies tested, this deflection occurs in the opposite phase to that expected. This is not the case for the pulsatile motion, which was predicted to drive the hair bundle with the observed phase [1].

In summary, up to at least 4 kHz, the somatic electromechanical force can be coupled directly to the IHC stereocilia: OHC contraction causes excitatory deflection of IHC stereocilia. The unequivocal relationship between the anti-phasic transversal motion of the RL and the TM at the IHC and the radial deflection of the IHC stereocilia bundle can only be shown when measurements of fluid flow in the subtectorial space are performed.

Acknowledgments

Supported by DFG Gu 194/7-1

References

1. Nowotny, M., Gummer, A. W., 2006. Nanomechanics of the subtectorial space caused by electromechanics of cochlear outer hair cells. Proc. Natl. Acad. Sci. U.S.A. 103: 2120 - 2125.
2. Nilsen, K. E., Russell, I. J., 2000. Spatial and temporal representation of a tone on the guinea pig basilar membrane. Proc. Natl. Acad. Sci. U.S.A. 97: 11751 - 11758.

Comments and Discussion

Braun: The detailed experimental demonstration of a direct coupling between OHC motility and IHC stereocilia motion, without any mechanical involvement of the basilar membrane (BM) is a major breakthrough. It demands a more specific definition of the cochlear amplifier than the one that has been in wide use during the past 25 years. The genuine cochlear amplifier need not be a BM amplifier. It only needs to be an amplifier of forces that excite the IHCs.

FLUID MECHANICS IN THE SUBTECTORIAL SPACE

J. BAUMGART[◇], C. CHIARADIA[♮], M. FLEISCHER[‡], Y. YARIN[§], R. GRUNDMANN[◇], A.W. GUMMER[♮]

[◇]*Institute for Aerospace Engineering, Faculty of Mechanical Engineering, Technische Universität Dresden, 01062 Dresden, Germany*

[♮]*Section of Physiological Acoustics and Communication, University of Tübingen, Elfriede-Aulhorn-Str. 5, 72076 Tübingen, Germany*

[‡]*Institute for Solid Mechanics, Faculty of Mechanical Engineering, Technische Universität Dresden, 01062 Dresden, Germany*

[§]*Department of Otorhinolaryngology, Faculty of Medicine, Technische Universität Dresden, 01062 Dresden, Germany*

In the subtectorial space, momentum is transported by the fluid to the stereocilia of the inner hair cell (IHC), resulting in bending of the hair bundle. The fluid must pass through the v-shaped arrangement of the outer hair cell (OHC) stereocilia and is "squeezed" between the tectorial membrane (TM) and the reticular lamina, especially due to the Hensen's stripe. Here, we analyze the flow field by means of numerical discretization. Since the geometry is complex, the finite-element-method is employed. An implementation is developed which allows computation in the frequency domain, based on a pressure stabilised velocity formulation. Matrix coupling with the structure is accomplished and a single solution step of the whole system matrix is sufficient to retrieve the displacement field of the fluid and structural parts. The proposed method allows a solution of the fluid-structure interaction in the subtectorial space within a reasonable time. For a two-dimensional mesh with 120,000 degrees of freedom of a complete cross-section of the organ of Corti, including stereocilia, the solution time is less than 20 seconds on a convential PC.

1. Introduction

Here, a novel technique is proposed to model fluid-structure-interaction problems which are common in the field of inner-ear mechanics. It is assumed that the non-linear term in the Navier-Stokes equation can be neglected. In other words, the Reynolds number is smaller than unity, so that a computation of acoustic streaming is not necessary [1]. Although the dimensions are small, we assume that a continuum approach is still valid for the geometrical dimensions of the order of one micrometer. The no-slip condition is used at the boundaries, since the Debye-length is about one nanometer [2] and, therefore, much smaller than the geometry.

Furthermore, we perform a discretisation of the linear viscous fluid with near-incompressibility. In this respect, the only assumptions are that the amplitudes are small and the wavelength long. Estimations about the flow field are not necessary, as they are in lumped models for the fluid in a gap [3, 4]. The linear approach allows a much faster solution than with staggered techniques, such as those used by Givelberg and Bunn [5].

After presenting the method, it is applied to a two-dimensional cross-section model of the guinea-pig at a characteristic place-frequency of 0.8 kHz. Coiling effects of the cochlea are neglected [6, 7], because we are examining vibration responses from the boundaries of the subtectorial space, directly in response to somatic electromechanical forces from the OHCs.

2. Methods

2.1. *Fluid-Structure-Interaction*

For small amplitudes the non-linear term of the Navier-Stokes equations can be neglected. If the problem is isothermal and variations of pressure, p, and density, ϱ, are small compared to their mean value, a nearly incompressible fluid can be assumed [8]. The mass-conservation equation

$$\frac{1}{K}\frac{\partial p}{\partial t} + \nabla \cdot \vec{v} = 0 \tag{1}$$

relates the time derivative of the pressure divided by the compressibility K to the divergence of the velocity vector $\vec{v}$. The linear part of the momentum equation without volume forces

$$\varrho \frac{\partial \vec{v}}{\partial t} = -\nabla p + \nabla \cdot \tau \tag{2}$$

relates the acceleration to the gradient of pressure and stresses. A Stokes-fluid is described by

$$\tau = \mu \left(\nabla \vec{v} + (\nabla \vec{v})^{\mathrm{T}} - \frac{2}{3} \nabla \cdot \vec{v} \right) \tag{3}$$

where μ is the dynamic viscosity. The unknown variables are the pressure, p, and the velocities, $\vec{v}$.

For an efficient coupling with the solid structure, the velocity is substituted by the time derivative of the displacement $\vec{u}$

$$\vec{v} = \frac{\partial \vec{u}}{\partial t}\ . \tag{4}$$

Applying a finite element discretisation to Eqns. (1,2,4) yields a set of linear equations for the unknown degrees of freedom:

$$\begin{bmatrix} \mathbf{M} & \mathbf{0} \\ \mathbf{0} & \mathbf{0} \end{bmatrix} \begin{Bmatrix} \ddot{\mathbf{u}} \\ \ddot{\mathbf{p}} \end{Bmatrix} + \begin{bmatrix} \mathbf{A} & \mathbf{0} \\ \mathbf{0} & \mathbf{0} \end{bmatrix} \begin{Bmatrix} \dot{\mathbf{u}} \\ \dot{\mathbf{p}} \end{Bmatrix} + \begin{bmatrix} \mathbf{0} & \mathbf{C} \\ \mathbf{C}^{\mathrm{T}} & -\mathbf{V} \end{bmatrix} \begin{Bmatrix} \mathbf{u} \\ \mathbf{p} \end{Bmatrix} = \begin{Bmatrix} \mathbf{f} \\ \mathbf{0} \end{Bmatrix} \tag{5}$$

for given forces $\mathbf{f}$. Here $\dot{()}$ denotes the first time derivative $\partial()/\partial t$ and $\ddot{()}$ the second. The matrix $\mathbf{A}$ describes the shearing and is proportional to the dynamic viscosity μ. $\mathbf{V}$ ensures the nearly incompressibile material and depends on the compressibility K. $\mathbf{C}$ is the coupling between pressure and velocity and $\mathbf{M}$ is the inertia and depends on the fluid density ϱ. Details are given by Zienkiewicz *et al.* [9]. After elimination of the pressure variable, the system

$$\mathbf{M}\,\ddot{\mathbf{u}} + \mathbf{A}\,\dot{\mathbf{u}} + \mathbf{C}\,\mathbf{V}^{-1}\,\mathbf{C}^{\mathrm{T}}\,\mathbf{u} = \mathbf{f} \tag{6}$$

describes a fluid simply in terms of displacement. This representation allows a direct implementation in a finite element solver for structure problems and leads to a monolithic fluid-structure approach. A time-harmonic Ansatz is possible for the linear system. Thus, for excitation with a single frequency, the system given by Eqn. (6) must be solved only once. Here, this fluid element is based on quadratic shape functions for displacement and

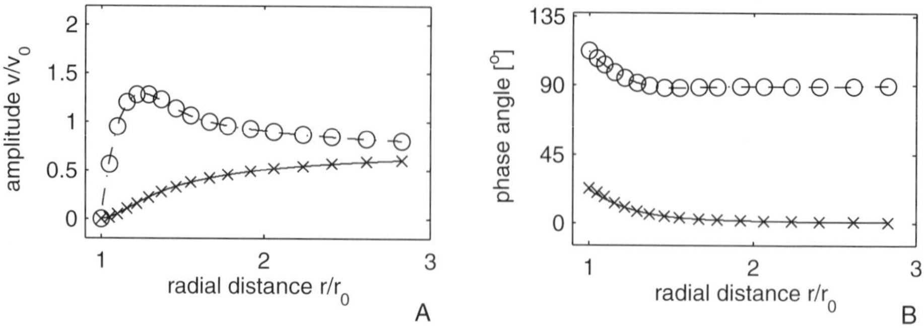

Fig. 1. Comparison of the velocities around an oscillating cylinder normal to its axis with a Womersley number of 10. Velocity profile in radial direction, 45° off the main axis. Analytical solution for the peripheral (— · —) and radial (——) direction and numerical results at the nodes ∘ and ×, respectively. (A) Amplitude. (B) Phase angle.

geometry. Linear shape functions are used for the pressure to ensure stability. The element is implemented as an user-defined element in Ansys [10].

The correct implementation is validated by the analytical solution for the oscillating motion of an infinite cylinder normal to its axis [11]. This problem involves inertia and viscous forces on a curved geometry. There is good agreement in the entire flow field, provided the unsteady boundary layer is resolved. The comparison is plotted in Fig. 1 for the amplitude and phase of the circumferential and radial velocity component.

2.2. *Geometry*

The numerical model is built on geometrical data from the guinea-pig cochlea at a characteristic place-frequency of 0.8 kHz. Data derive from live-stained (FM 1-43, Invitrogen) and TM-stained (Lectin SBA Conjugates, Invitrogen) preparations and stacks of images obtained from laser scanning microscope Leica SP5 (Fig. 2). The geometry is manually constructed and represented by splines. Each cell type and the fluid spaces are grouped into closed areas. A set of mechanical properties is assigned and then meshed with finite elements (quadratic shape functions).

2.3. *Properties*

The elasticity of the material is based on data used by Steele and Puria [4] and adjusted to match experimental data from different sources. The measurements of Cooper [12] give an indication for the stiffness distribution of the basilar membrane. The OHCs are matched to data from Frank *et al.* [13] and recent impedance measurements [14]. The stiffness of the hair bundle is based on values from Strelioff and Flock [15]. The impedance along the reticular lamina is similar to values measured by Scherer and Gummer [16]. The fluid damping was not sufficient to match these data and additional material damping in the Hensen and sulcus cells was necessary. According to the experiments of Ghaffari *et al.* [17], the TM is highly viscous. The vibrational measurements at the boundary of the subtectorial

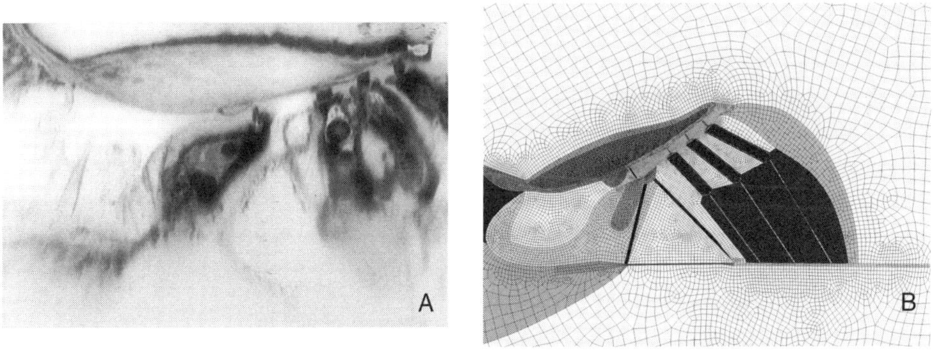

Fig. 2. Cross section of the guinea-pig organ of Corti at a characteristic place-frequency of 0.8 kHz. (A) Live staining with FM 1-43 (Invitrogen). Image is extracted from a stack. (B) Geometry and mesh for the computation. Here, the different grey levels just indicate different areas and do not correspond to physical properties.

space by Nowotny and Gummer [18] serve here as reference for the motion around the IHC stereocilia. The motion of the cross-section is compared with measured displacements [19, 20].

The dynamic viscosity of the fluid is 1 Pa ms[a], as measured at room temperature for Hanks solution [1]. The density for all materials is set to 0.001 ng/μm^3. The OHCs are modelled with a piezoelectric wall and are filled with incompressible fluid. The elasticity of the basilar membrane is set to yield a resonance frequency of 0.8 kHz for a constant pressure load radially across the membrane. Comparisons are made with experimental vibration data from the TM and reticular lamina in an in-vitro preparation in which the TM was intact, with the longest OHC stereocilia embedded in the TM [18].

3. Results

The numerical method was applied to solve a fluid-structure interaction in the organ of Corti. The two-dimensional mesh has about 60,000 nodes. Ansys [10] was used as finite element software. The approach is based on non-moving meshes and has no limitations due to small amplitudes.

To model the motion of the organ of Corti in response to somatic electromechanical force from the OHCs, a sinusoidal voltage is applied across the OHC wall. Due to the assigned piezoelectric property of the cell wall, the applied voltage sinusoidally modulates the wall thickness. In turn, the axial length of the cell is modulated because the intracellular fluid is incompressible. This excitation causes a pattern of motion, as shown in Fig. 3.

The inner and outer pillar cells are not connected to each other in the longitudinal direction. Thus, there can be fluid flow through the rows of OHCs and pillar cells. The fluid spaces are connected to model this here in two dimensions.

[a]The following basis units are used: mass in nanogram (ng), length in micrometer (μm), time in millisecond (ms). Thus, the pressure is in Pa.

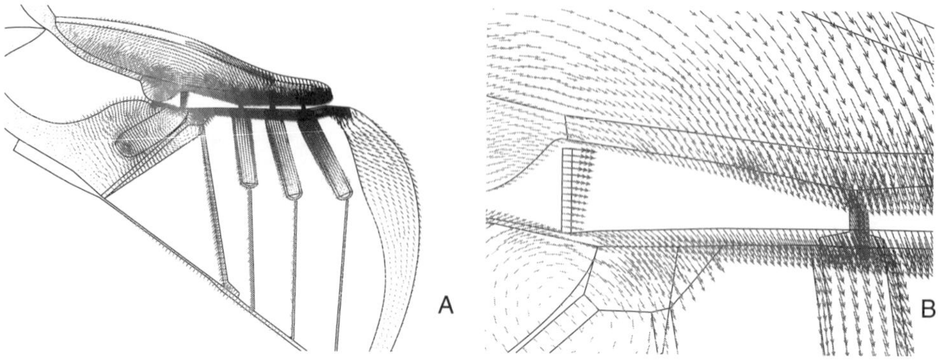

Fig. 3. Displacement of the organ of Corti at a characteristic place-frequency of 0.8 kHz (A) and a close up around the IHC stereocilia (B), at the time instant of maximum OHC contraction.

The IHC stereocilia bending amplitude and direction depend on the size of the gap between Hensen's stripe and the reticular lamina. The motion of the subtectorial space at the IHC is characterized by an anti-phasic transversal motion of the reticular lamina and TM, as also found experimentally [18]. For a sufficiently small gap (here $\approx 2\mu m$), motion of the reticular lamina towards the TM at the IHC causes deflection of the IHC stereocilia in the excitatory (depolarizing) direction.

4. Discussion

Fluid flow in the subtectorial space cannot be investigated without considering the motion of the neighbouring structures. The proposed method allows fast computation of strongly coupled fluid-structure interaction problems. For a two-dimensional model of the organ of Corti with around 120,000 degrees of freedom – that is, with two degrees of freedom for each of the 60,000 nodes –, the solution of the problem requires less than 20 seconds for one frequency on a conventional PC (AMD Athlon 64 3500+, 2GB RAM, SuSE Linux 10.3).

Application of the proposed method to a two-dimensional model of the organ of Corti reveals deflection of the IHC stereocilia in the excitatory direction for contraction of the OHC. This motion is caused by a squeezing flow between Hensen's stripe and reticular lamina. This phase relation is found experimentally by Chiaradia *et al.* [21] and the influence of Hensen's stripe has been addressed by Steele and Puria [4].

Acknowledgments

This work was supported by DFG Gr 1388/14 (J.B. and R.G.), DFG Gu 194/7 (C.C. and A.W.G.), DFG Ha 2075/9 (M.F.), DFG Za 249/4 (Y.Y.) and by DAAD D/07/46200 (J.B.).

We thank Manuela Nowotny for experimental data, together with Charles Steele and Sunil Puria for helpful discussion. We thank Roland Gärtner for a three dimensional model of the organ of Corti.

References

1. J. Baumgart and R. Grundmann, 2007. Fluid motion in the organ of corti. In 19th International Congress on Acoustics.
2. J. Baumgart, M. Fleischer, Y. Yarin, and R. Grundmann, 2007. Fluid flow around the stereocilia. In 8th International Symposium on Experimental and Computational Aerothermodynamics of Internal Flows.
3. P.J. Kolston, 1999. Comparing in vitro, in situ, and in vivo experimental data in a three-dimensional model of mammalian cochlear mechanics. Proc Natl Acad Sci U S A, 96(7), 3676–3681.
4. C.R. Steele and S. Puria, 2005. Force on inner hair cell cilia. International J. Solids and Structures, 42(21-22), 5887–5904.
5. E. Givelberg and J. Bunn, 2003. A comprehensive three-dimensional model of the cochlea. Journal of Computational Physics, 191(2), 377–391.
6. H. Cai, D. Manoussaki, and R. Chadwick, 2005. Effects of coiling on the micromechanics of the mammalian cochlea. Journal of The Royal Society Interface, 2(4), 341–348.
7. D. Manoussaki, R.S. Chadwick, D.R. Ketten, J. Arruda, E.K. Dimitriadis, and J.T. O'Malley, 2008. The influence of cochlear shape on low-frequency hearing. Proc Natl Acad Sci U S A, 105(16), 6162–6166.
8. O.C. Zienkiewicz, R.L. Taylor, and J.Z. Zhu, 2006. *The Finite Element Method: For Fluid Dynamics*. Elsevier.
9. O.C. Zienkiewicz, R.L. Taylor, and J.Z. Zhu, 2006. *The Finite Element Method: Its Basis & Fundamentals*. Elsevier.
10. Ansys Inc, 2007. *Ansys ver.11*. www.ansys.com.
11. C.-Y. Wang, 1968. On high-frequency oscillatory viscous flows. Journal of Fluid Mechanics, 32(1), 55–68.
12. N.P. Cooper, 1999. Radial variation in the vibrations of the cochlear partition. In Recent Developments in Auditory Mechanics. H. Wada *et al.* (eds.) World Scientific, Singapore, pp. 109–115.
13. G. Frank, W. Hemmert, and A.W. Gummer, 1999. Limiting dynamics of high-frequency electromechanical transduction of outer hair cells. Proc Natl Acad Sci U S A, 96(8), 4420–4425.
14. T. Eckrich, M. Nowotny, C. Harasztosi, M. Scherer, and A.W. Gummer, 2008. Impedance measurements of isolated outer hair cells. 31st MidWinter Meeting, Assoc. Res. Otolaryngology.
15. D. Strelioff and A. Flock, 1984. Stiffness os sensory-cell hair bundles in the isolated guinea pig cochlea. Hearing Research, 15(1), 19–28.
16. M.P. Scherer and A.W. Gummer, 2004. Vibration pattern of the organ of corti up to 50 kHz: evidence for resonant electromechanical force. Proc Natl Acad Sci U S A, 101(51), 17652–17657.
17. R. Ghaffari, A.J. Aranyosi, and D.M. Freeman, 2007. Longitudinally propagating traveling waves of the mammalian tectorial membrane. Proc Natl Acad Sci U S A, 104(42), 16510–16515.
18. M. Nowotny and A.W. Gummer, 2006. Nanomechanics of the subtectorial space caused by electromechanics of cochlear outer hair cells. Proc Natl Acad Sci U S A, 103(7), 2120–2125.
19. K.D. Karavitaki and D.C. Mountain, 2007. Imaging electrically evoked micromechanical motion within the organ of corti of the excised gerbil cochlea. Biophys J, 92(9), 3294–3316.
20. D.K. Chan and A.J. Hudspeth, 2005. Mechanical Responses of the Organ of Corti to Acoustic and Electrical Stimulation In Vitro. Biophys. J., 89(6), 4382–4395.
21. C. Chiaradia, M. Nowotny, and A.W. Gummer, 2009. Deflection of IHC stereocilia in response to somatic OHC electromotility. This volume.

SECTION V

MODELLING THE COCHLEAR AMPLIFIER AND THE COCHLEA'S DYNAMICS

J.P. Wilson and I.J. Russell

IS STEREOCILIA VELOCITY OR DISPLACEMENT FEEDBACK USED IN THE COCHLEAR AMPLIFIER?

SHAN LU

Boston University Center for Hearing Research, Dept. of Electrical and Computer Engineering, Boston University, 8 Saint Mary's Street, Boston, MA 02215, USA

DAVID MOUNTAIN

Boston University Center for Hearing Research, Dept. of Biomedical Engineering, Boston University, 44 Cummington Street, Boston, MA 02215, USA

ALLYN HUBBARD

Boston University Center for Hearing Research, Dept. of Electrical and Computer Engineering, Dept. of Biomedical Engineering, Boston University, 8 Saint Mary's Street, Boston, MA 02215, USA

Outer hair cells (OHC) play an important role in cochlear amplification. The OHC senses stereocilia motion and creates a force feedback to the organ of Corti. It is largely accepted that the stereocilia displacement drives the OHC apical conductance change, which, in turn, drives somatic motility. Recent research shows that the tension gated OHC current exhibits fast adaptation in response to stereocilia displacement. Such an adaptation process resembles a high-pass filter or differentiator, at least for the inward current. Since velocity is the derivative of displacement, fast adaptation may indicate that it is the stereocilia velocity, rather than displacement is the more important driver of the OHC apical conductance. We changed our multi-compartment, piezo-electro-mechanical model to sense stereocilia velocity rather than displacement. This new model can well match measured basilar membrane velocity and our own cochlear microphonic data.

1 Introduction

Since outer hair cell (OHC) somatic electromotility was discovered [1], it is largely accepted that OHC voltage-dependent length changes observed *in vitro* form the basis for the cochlear amplifier (CA) *in vivo*, although there is evidence suggesting that hair bundle motility could also contribute to the CA [2]. Our previous research suggests that a nonlinear, multicompartment model with a piezoelectric OHC feedback system can partially cancel the voltage drop across the cell wall in the high frequency range [3]. The model enables us to explore the cochlear mechanical responses and the electrical potential changes simultaneously. In this paper we present a model based on our previous one, but with its OHC tension-gated channel sensing reticular laminar (RL) velocity, rather than displacement. Simulation results show that this new model can overcome the OHC low-pass filtering problem, and can replicate cochlear responses at both low and high frequencies.

2 Method

Figure 1 shows the structure of one section of the model. The left portion is the hydromechanical model as described in previous publications [4-6]. The formulation of this model includes scala vestibule (SV), scala tympani (ST) and organ of Corti (OC) fluid compartments. The basilar membrane (BM) and RL impedances separate these compartments vertically.

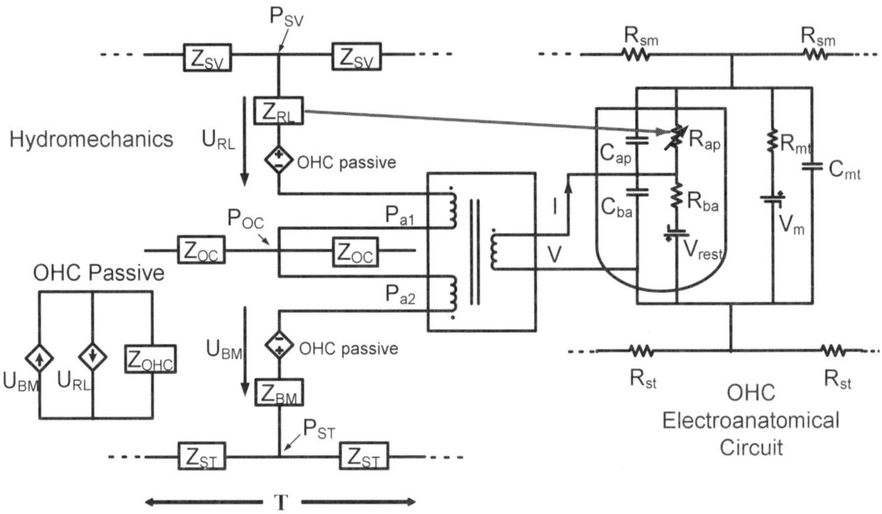

Figure 1. The schematic drawing of one section of the model. The left part is the hydromechanical system. The right part is the OHC anatomical circuit. A bi-directional piezoelectric feedback system connects the mechanical part and the electrical part. The adjacent sections of the OHC circuit model are coupled by resistors in both SM and ST.

The right part of Figure 1 depicts one section of the OHC electro-anatomical model. In the apical part of the OHC circuit, the tension-gated conductance changes nonlinearly in response to the stereocilia motion, which is assumed to be proportional to RL motion. The asymmetrical nonlinearity of the tension-gated conductance is usually described by a Boltzmann function [7]. In our model a similar asymmetrical nonlinear Boltzmann curve was used:

$$g_{ap} = \frac{g_{\max}}{1+e^{-\delta v/w}(1+e^{-\delta v/w})} + g_0 \tag{1}$$

where δv is the motion of stereocilia, w is the saturation constant, g_0 is the conductance constant representing the minimum conductance of the channel, and g_{max} is the maximum conductance of the stereocilia channel ($g_0 << g_{max}$). The gain of the piezoelectric feedback system is controlled by the transformer ratio N. In the model, N is set to be about 30,000 in the basal part of the cochlea when the cochlea is active. N is about 10 times smaller in the apical part of the cochlea than it is in the basal part. The force generated is about 2nN per OHC per mV change in the OHC basal lateral transmembrane potential. In our previous models that implemented the piezo-electrical feedback system with OHC tension gated channel sensing the RL displacement, we were unable to get the correct phase relationship between the sensed RL motion and the active force that was produced. Therefore we were unable to get enough gain to boost BM vibration to match the physiological data.

Existing experimental results suggest that the mechanoelectrical channel is fast-adaptive [8-10]. This fast adaptive channel may be approximately described as being related to the derivative of displacement, which is velocity. So in this work, we used RL velocity to drive the conductance change of the OHC tension gated channel, instead of using RL displacement.

Table 1. Parameters used in the OHC electroanatomical model.

Name	Meaning	Value	Unit
V_{rest}	OHC Resting potential	-70	mV
Vm	Stria vascularis battery	110	mV
g_{max}	Max conductance of OHC tension gated channel	110	nS
g_0	Apical leakage conductance	11	nS
w	Saturation constant in OHC tension gated channel	70	μm/s
R_{ap}	OHC Apical resistance	$1/g_{ap}$	Ω
C_{ap}	OHC Apical capacitance	21	pF
R_{ba}	OHC Basal resistance	1.2e6	Ω
C_{ba}	OHC Basal capacitance	209	pF
R_{sm}	Fluid resistance in SM	$R_{sm} = R_f \frac{T}{A_{sm}}$	Ω
C_{mt}	Capacitance between SM and ST	383.3	pF
R_{mt}	Source resistance of stria vascularis battery	2e5	Ω

3 Results

Cochlear microphonic (CM) is the AC potential generated by the hair cells in response to acoustic stimulation. Figure 2 compares CM input/output responses of the model with experimental data at low frequencies. The experimental CM data were recorded from a second-turn location in the gerbil cochlea where the CF is around 2.5 KHz [11]. In all four panels, the input/output responses from both the model and the experimental data are nearly linear when the input level is below 60 dB SPL. Both the model and experimental data enter saturation at about 60dB SPL. The comparison of experimental and model results shows a good fit, which suggests the model has realistic parameter settings in OHC electroanatomical model.

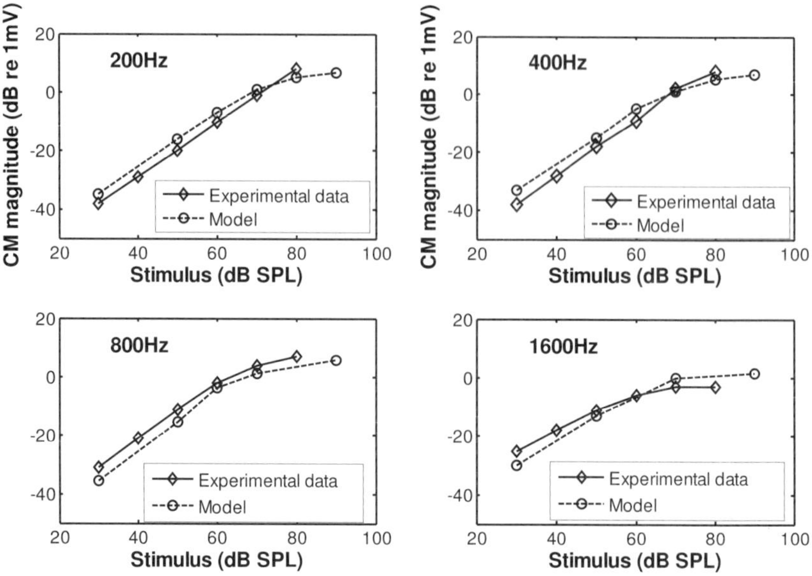

Figure 2. Comparison of Cochlear Microphonic results. The solid lines are experimental data from gerbil from our laboratory. They are recorded from a location where the CF is about 2.5 KHz. The input frequencies are: 200 Hz, 400 Hz, 800 Hz and 1600 Hz. The dashed lines are the simulation results of our model at the same cochlear location.

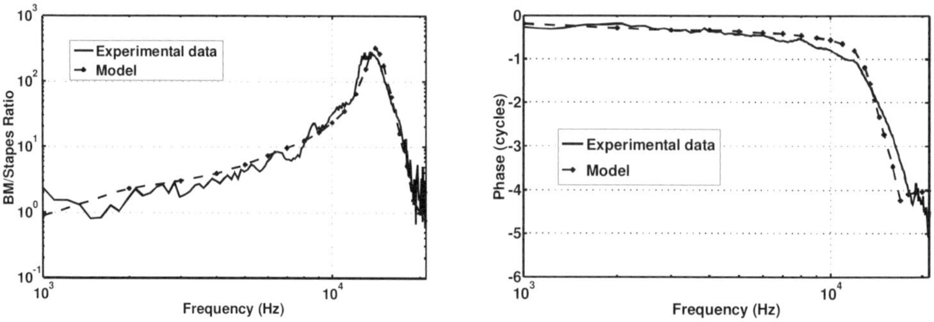

Figure 3. BM velocity ratio (BM/stapes) comparison. The left panel is magnitude. The right panel is phase. The solid line is experimental data for gerbil. It is recorded from a location where CF is about 14 KHz. The dashed line is the simulation result of our model.

Figure 3 shows the BM velocity ratio comparison with experimental data [12]. The data were recorded at the location where CF is about 14 KHz. The figure shows the BM velocity ratio when the input was 40 dB SPL in the ear canal. The magnitude of BM velocity ratio (left panel) from the model agrees well with the experimental results over the entire range of comparison. The model shows smaller phase changes before the CF and larger phase changes after the CF (right panel) as compared with experimental data.

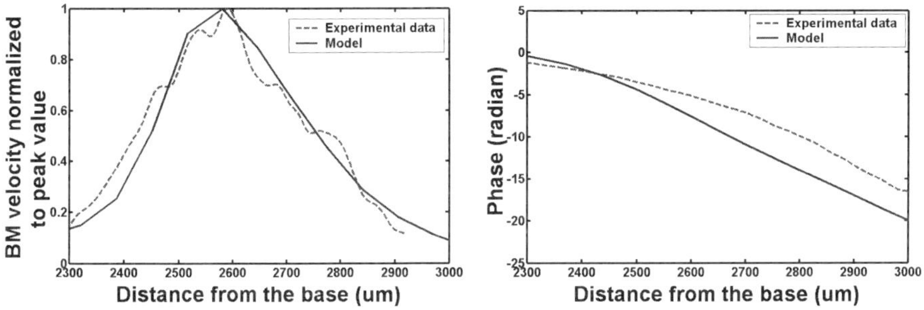

Figure 4. Longitudinal BM velocity profiles. The solid line is experimental data from the gerbil. It is recorded from a location where CF is about 16 KHz. The dashed line is the simulation result of our model.

Figure 4 compares the longitudinal velocity pattern of the model with experimental data obtained from the gerbil in a region where the CF is about 16 KHz [13]. The input stimulus is 50 dB SPL in the ear canal. The recording range is from 2.2 mm to 2.9 mm. The left panel shows the normalized BM velocity magnitude. The data is normalized by its peak value.

4 Discussion

The phase angle between the active force generation and the RL motion is critical to the CA. If the phase is not correct, the active force will not inject sufficient power into the hydromechanical system before the system becomes unstable due to increasing gain (N) [14]. The model with RL displacement driving the OHC tension-gated channel cannot produce sufficient cochlear amplification to match the experimental data, due to incorrect phase delay between motion sensing and force generation. When we used RL velocity to drive the apical conductance of OHC, the system was stable and the gain used is large enough to match the experimental data and the system is far from becoming unstable.

The results from our model suggest that the stereocilia fast adaptation may play an important role in OHC-mediated cochlear active amplification. Current research shows that the tension gated OHC current exhibits a fast adaptation in response to stereocilia motion and this fast adaptation causes a 90-degree phase lead [15]. Such an adaptation process, at least for inward current, resembles a high pass filter or a derivative of the stereocilia displacement. Since velocity is the derivative of the displacement, fast adaptation may underlie the reason that, velocity driven apical conductance, rather than displacement, works well in our model. Our study suggests that both hair bundle fast adaptation and somatic motility are essential to the active response of the cochlea. The hair bundle fast adaptation moderates the phase between RL-stereocilia motion and active force generated by the OHC (mechano-electro transduction), while somatic motility decides the magnitude of the active force (reverse transduction).

Acknowledgments

This work was supported by NIH. We thank Tianyin Ren for providing us with the experimental data and giving us useful feedback.

References

1. Brownell, W.E., C.R. Bader, D. Bertrand, and Y. de Ribaupierre, 1985. Evoked mechanical responses of isolated cochlear outer hair cells. Science, 227, 194-6.
2. Hudspeth, A.J., 1997. Mechanical amplification of stimuli by hair cells. Curr Opin Neurobiol, 7, 480-6.
3. Lu, S., J. Spisak, D.C. Mountain, and A.E. Hubbard, 2006. A New Multicompartment Model of the Cochlea, in Auditory Mechanics: Processes and Models, A.L. Nuttall et al., eds., World Scientific: Singapore, pp. 508-509.
4. Lu, S., A.E. Hubbard, F. Chen, and D.C. Mountain, 2008. Is Longitudinal Fluid Flow in The Organ of Corti Fundamental to Cochlear Function? Submitted.
5. Hubbard, A.E., D.C. Mountain, and F. Chen, 2003. Time-domain responses from a nonlinear sandwich model of the cochlea, in Biophysics of the Cochlea: From Molecules to Models, A.W. Gummer, ed., World Scientific: Singapore, Hong Kong. pp. 351-358.
6. Hubbard, A.E., Z. Yang, L. Shatz, and D.C. Mountain, 2000. Multi-mode Cochlear models, in Recent Developments in Auditory Mechanics, H. Wada, ed., World Scientific: Singapore, pp. 167-173.
7. Holton, T. and A.J. Hudspeth, 1986. The transduction channel of hair cells from the bull-frog characterized by noise analysis. J Physiol, 375, 195-227.
8. Kennedy, H.J., M.G. Evans, A.C. Crawford, and R. Fettiplace, 2003. Fast adaptation of mechanoelectrical transducer channels in mammalian cochlear hair cells. Nat Neurosci, 6, 832-6.
9. Strassmaier, M. and P.G. Gillespie, 2003. Fast adaptation in the mammalian cochlea: a conserved mechanism for cochlear amplification. Nat Neurosci, 6, 790-1.
10. Ricci, A.J., A.C. Crawford, and R. Fettiplace, 2000. Active hair bundle motion linked to fast transducer adaptation in auditory hair cells. J Neurosci, 20, 7131-42.
11. Nakajima, H.H., E.S. Olson, D.C. Mountain, and A.E. Hubbard, 1994. Electrically evoked otoacoustic emissions from the apical turns of the gerbil cochlea. J Acoust Soc Am, 96, 786-94.
12. Ren, T. and A.L. Nuttall, 2001. Basilar membrane vibration in the basal turn of the sensitive gerbil cochlea. Hear Res, 151, 48-60.
13. Ren, T., 2002. Longitudinal pattern of basilar membrane vibration in the sensitive cochlea. Proc Natl Acad Sci U S A, 99, 17101-6.
14. Markin, V.S. and A.J. Hudspeth, 1995. Modeling the active process of the cochlea: phase relations, amplification, and spontaneous oscillation. Biophys J, 69, 138-47.
15. Rabbitt, R.D., H.E. Ayliffe, D. Christensen, K. Pamarthy, C. Durney, S. Clifford, and W.E. Brownell, 2005. Evidence of piezoelectric resonance in isolated outer hair cells. Biophys J, 88, 2257-65.

CELLULAR BASIS OF THE COCHLEAR AMPLIFIER

K. H. IWASA, B. SUL, J. FANG, G. P. SINHA
Biophysics Section, NIDCD, NIH,
Rockville, Maryland 20850, USA

Three issues regarding the cellular basis for cochlear amplifier are discussed. First, regarding the mechanism of electromotility, we present our recent experimental data that demonstrate that conformational transitions of prestin-based membrane motor involve both membrane area and thickness. Namely the conformation that has larger membrane area is thinner. Second, we show that the effectiveness of hair bundle motility is less than that of electromotility as cochlear amplifier for the mammalian ear. Third, we examine factors that could enhance electromotility. Although we found that the cochlear microphonic can enhance the effectiveness of electromotility, it alone is insufficient to counteract viscous drag beyond 10 kHz. This result suggests additional mechanisms for enhancing the efficiency of electromotility.

1. Introduction

It has been demonstrated that outer hair cells are essential for the mammalian ear for its exquisite sensitivity, frequency selectivity, and nonlinearity. Since this effect is manifested in the movement of the basilar membrane, it must be based on motile mechanisms of outer hair cells. One motile mechanism is in the hair bundles and another in the cell body. The former depends on Ca^{2+} gradient across the plasma membrane for energy source and is triggered directly by mechanical stimulation. The latter, electromotility, uses electrical energy associated with the receptor potential, which is generated by the transducer current in response to mechanical stimulation at the hair bundle. In the mammalian ear the relative significance of these two mechanisms has been debated [1, 2].

In the first part, we describe our recent finding on electromotility and then discuss the effectiveness of these motile mechanisms. The effectiveness of these mechanisms depends on the details of how they contribute to enhance the vibration of the organ of Corti (cochlear amplifier). Instead of examining detailed interactions involved in a traveling wave, we examine a simple necessary condition for the cochlear amplifier. Here we assume that these motile mechanisms must counteract viscous drag [3] and try to determine upper bounds of their effectiveness for infinitesimal mechanical stimulation. We show that this approach can be applied not only for electromotility but also to hair bundle motility, enabling to compare the mammalian ear with the avian ear, in which electromotility is absent.

2. OHC Electromotility

Electromotility is based on prestin, a membrane protein in the SLC26 superfamily of anion transporters. It involves transfer of charge across the membrane, in a manner analogous to voltage-gated ion channels or ion transporters. For this reason, we can assume that motor molecules undergo conformational changes which involve charge transfer and changes in the membrane area. The electromechanical coupling can be confirmed by inhibiting charge transfer by constraining the membrane area. This motile mechanism based on conformational changes can explain experimental observations that changes in the membrane potential induces length changes in these cylinder-shaped cells and that length changes reciprocally induce charge transfer across the plasma membrane.

We recently found evidence that conformational changes in the membrane motor involves thickness changes. When we incorporated phosphatidylcholine (PCs) with short acyl chains, nonlinear capacitance shows positive shifts up to 130 mV together with small increases in the linear capacitance, indicating reduced membrane thickness. Incorporation of PCs with long chains has the opposite effects. These observations indicate that a reduction in membrane thickness favors the extended state, which is the dominant conformation in hyperpolarized potentials, and an increase in membrane thickness favors the compact state, which is dominant in depolarized potentials. Such an effect is likely the result of hydrophobic mismatch with the lipid bilayer in the plasma membrane.

Since the volume of proteins would be conserved during conformational changes, a larger membrane area must be associated with thinner profile in the membrane, which is therefore consistent with our observations.

3. Energy Balance

We assume that the major viscous drag is associated with shear motion between the tectorial membrane and the reticular lamina [4]. Because the gap d is smaller than the boundary layer thickness, for sinusoidal shear with small amplitude x and angular frequency ω, viscous drag F_d per OHC can be expressed by, $F_d = \gamma\omega x\cos(\omega t)$ where γ is the friction coefficient per hair cell. This quantity is expressed by $\eta A/d$ with the surface area A of the gap per hair cell, the viscosity η of the medium, and the gap d. This drag results in energy loss E_d per OHC per cycle expressed as $E_d = \pi\gamma\omega x^2$.

Let F_{ohc} be the force generated by an OHC and E_{ohc} the corresponding energy generated per cycle. The condition for counteracting viscous drag is expressed by either

$$F_{\mathrm{ohc}} \geq F_d \tag{1}$$

with respect to force or

$$E_{\mathrm{ohc}} \geq E_d \tag{2}$$

with respect to energy. These inequalities lead to a frequency limit ω_{lim}, up to which the motile mechanism can be effective.

4. Electromotility Frequency Limit

4.1. *Without extracellular resistance*

We have previously examined the efficiency of electromotility ignoring the resistance of the external media to the transducer current [5]. The force F_{em} produced by electromotility can be expressed by $F_{\rm em} = \phi g x V_a/(\omega C_m)$, where C_m is the membrane capacitance of the cell, g the sensitivity of the transducer current to displacement, ϕ force produced per membrane potential change, typically 0.1 nN/mV, and V_a the voltage drop at the apical membrane, typically ~150 mV.

If the phase of this force is optimal to counteract viscous drag, the inequality (1) for force leads to a frequency limit $f_{\rm em}^{\rm limit}$ for electromotility alone,

$$f_{\rm em}^{\rm limit} = \frac{1}{2\pi}\sqrt{\frac{\phi g V_a}{\gamma C_m}}. \tag{3}$$

With realistic values for the parameters, the frequency limit $f_{\rm em}^{\rm limit}$ is about 10 kHz [5], which is appreciable but does not quite cover the auditory range of mammals. This equation shows that this frequency limit is not significantly affected by a single factor because each of these factors affect the frequency by its square root.

4.2. *With extracellular resistance*

Extracellular resistance is expected to reduce the receptor potential by reducing the receptor current. However, extracellular AC potentials (cochlear microphonics) could be utilized by another cell to generate mechanical force [6]. In order to address this issue, we consider energy balance of a cell which has the highest characteristic frequency in the cochlea. We assume that the cochlear is stimulated at this frequency.

Because the extracellular potential produced by the receptor current responding to small amplitude stimulation has the opposite phase as the receptor potential, constructive effect of the extracellular field can be expected only in the location where the phase is advanced by π. From this location, energy can me transmitted back to the characteristic location by a travoling wave. It can be shown [7] that the limiting frequency under this condition cannot exceed the value given by Eq. 3. This result strongly suggests that a mechanism that enhances electromotility is required to explain the mammalian hearing range.

5. Effectiveness of Hair Bundle Motility

To evaluate the efficiency of hair bundle motility in counteracting viscous drag, we assume that an actuator displaces a hair bundle sinusoidally with small amplitude x and angular frequency ω. The behavior of the hair bundle can be determined by the work it does to the actuator. If the bundle does positive work, it functions as an amplifier and if the work is negative it functions as a damper.

Of the two proposed mechanisms of fast adaptation, "release" and "channel re-closure," we found that only the channel re-closing performs positive work, and thus qualifies as an amplifier.

The simplest model for re-closure has a single Ca^{2+} binding site and thus four states, i.e. open and closed states, each with and without Ca^{2+} binding. This is a simplified version of six state model with two Ca^{2+} binding sites [8]. We assume that the Ca^{2+} concentration in the vicinity of a binding site of the channel is close to the level of the external medium when the channel is open and that it is close to the cytosolic level when the channel is closed.

Force produced by a hair bundle with N tip-links in response to mechanical stimulation is given by

$$F_{\text{hb}}/N = -k_g(x - x_g \Delta P_o) - k_s x, \tag{4}$$

where k_g is the stiffness of the gating spring, k_s the stiffness of other springs for tip displacement x, which we assume is the same as the amplitude of the shear in the gap. P_o is the open probability of the channel.

The energy that the hair bundle produces is calculated by integrating this force with respect to displacement over a cycle. The terms in force (Eq. 4) other than the one that contains ΔP_o do not contribute to this energy. Mechanical stimulation changes transition rates in which gating is involved, and gating is determined by free energy $k_g x_g x$ for a small displacement x. Because that small change induces deviation from the steady state, the change ΔP_o in open probability is proportional to $\beta k_g x_g x$ with $\beta = 1/(k_B T)$. This response has a phase delay due to transitions involved in Ca^{2+} binding and unbinding. Thus we may have,

$$E_{\text{hb}} = \beta N (k_g x_g x)^2 \Phi \tag{5}$$

where N is the number of tip-links per cell, Φ the phase factor of the response which depends on all kinetic rates involving the MET channel. It thus depends also on Ca^{2+} concentration gradient across the plasma membrane because some of these transition rates involve binding and unbinding of Ca^{2+}. This energy can be positive, satisfying the condition of being an amplifier. The efficiency of this mechanism can be expressed by comparing with the energy loss due to viscous damping.

The inequality (2) for energy leads to a limiting frequency $f_{\text{hb}}^{\text{limit}}$ for hair bundle motility alone, which can be expressed by,

$$f_{\text{hb}}^{\text{limit}} = \beta N (k_g x_g)^2 \Phi / (2\pi^2 \gamma). \tag{6}$$

Gating force $F_g (= k_g x_g)$ measured at the top of the bundle is related to molecular gating force f_g by $F_g = \gamma_g f_g$ with a geometrical factor γ_g, which can be approximated by s/h [9], s and h being the rootlet separation and the bundle height, respectively. If we can assume that the gap d of the subtectorial space can be approximated by the bundle height h in the definition of the viscous drag γ, we can break down the expression for limiting frequency into three factors, i.e.

$$f_{\text{hb}}^{\text{limit}} = \frac{1}{2\pi^2 k_B T} \Phi F_{\text{morph}} f_g^2. \tag{7}$$

They are the phase factor Φ, and molecular gating force f_g, and a morphological factor F_{morph}, which is given by,

$$F_{\text{morph}} = \frac{N s^2}{h A}. \tag{8}$$

The phase factor Φ depends on the kinetics of the channel. In a model with a single Ca^{2+} binding site, the optimal value for this factor for open probability of 0.1 is about 0.05 for the 100-fold Ca^{2+} gradient across the apical membrane of the mammalian hair cells. It is about 0.07 for the 1000-fold Ca^{2+} gradient of avian hair cells. A striking feature of this expression is that each factor is important for the limiting frequency, quite different from the expression for the electromotility-based limiting frequency. Gating force has a particularly significant effect because it affects the limiting frequency on the second power.

5.1. *Mammalian ear*

The number of tip-links for a apical OHC is about 60 [10], and the drag coefficient per OHC is 0.3×10^{-7} Ns/m if the gap $d(= h)$ is 5 μm, the same as the bundle height of apical cells [5]. If we use a value 500 fN per channel for gating force of OHCs, the frequency limit $f_{\rm hb}^{\rm limit}$ is ~300 Hz. At the basal turn, the limit is about 2 kHz because $f_{\rm hb}^{\rm limit}$ is inversely proportional to the bundle height (see Eq. 8) and N, the number of tip-links, is about 70. The range obtained would be compatible with 500 Hz resonance in an in vitro prepatation [2]. The limit of 2 kHz is, however, much lower than the limit of 10 kHz, our estimate based on electromotility.

5.2. *Avian ear*

The avian basilar papilla has strikingly different appearances being packed with hair cells with large hair bundles. Despite these cells are classified into tall and short hair cells, the tallest cilia in each of these hair cells are firmly anchored to the tectorial membrane. For this reason, hair bundles of all hair cells are expected to contribute to the limiting frequency.

Because the basal papilla of the chicken has hair cells at a higher density, the area A of the gap per hair cell is smaller. This difference gives a factor 3 or 4. The gap and hair bundle height in the basal part of the papilla is about 2 μm, about twice larger. The number N of tip-links in a single hair cell of basal cells is about 200 for chicken [11], about 3 times as many as in basal OHCs. The morphological factor $F_{\rm morph}$ is therefore 5 or 6 times higher in the chicken. In addition, the optimal value for the phase factor Φ is higher than the mammalian value due to the larger Ca^{2+} concentration gradient across the apical membrane of avian hair cells. If a single MET channel produces the same force as mammalian one, the frequency limit $f_{\rm hb}^{\rm limit}$ predicted would be between 12 and 15 kHz, exceeding the chicken's auditory limit of about 5 kHz. This frequency limit is, of course, subject to the uncertainties of each factor that contributes to $F_{\rm morph}$, the morphological factor. Such uncertainties include, for example, the shrinkage of samples used in scanning electromicroscopic studies. In addition, it is particularly sensitive to the value of gating force, which appears to be difficult to determine in avian hair cells [12].

6. Conclusions

The observed dependence of electromotility on membrane thickness indicates that conformational changes of prestin-based motor undergoes thickness changes. This observation supports that electromotility is based on area changes of the membrane motor.

In order to assess the effectiveness of electromotility and hair bundle motility, comparing the mechanical energy that these motile mechanisms produce with the viscous loss appears to be useful. We showed that electromotility can be more effective than hair bundle motility as the basis for the cochlear amplifier in the mammalian ear. However, this conclusion may depend on a phase factor of the hair bundle response that depends on the kinetics of the MET channel, which has not been experimentally determined. This conclusion can be justified by an analysis of the avian ear. Namely, assuming the same kinetics can explain the auditory frequency range for the chicken, which does not have electromotility in hair cells. Our analysis also suggests the existence of an additional factor that improves the efficiency of electromotility to explain the auditory range of the mammalian ear.

Acknowledgment

This work was supported by the Intramural Research Program of NIDCD, NIH.

References

1. R. Fettiplace and C. M. Hackney, 2006. The sensory and motor roles of auditory hair cells. Nat. Rev. Neurosci. 7, 19–29.
2. D. K. Chan and A. J. Hudspeth, 2005. Ca^{2+} current-driven nonlinear amplification by the mammalian cochlea in vitro. Nature Neurosci. 8, 149–155.
3. T. Gold, 1948. Hearing II. The physical basis of the action of the cochlea. Proc. Roy. Soc., B 135, 492–8.
4. J. Allen, 1980. Cochlear micromechanics–a physical model of transduction. J. Acoust. Soc. Am. 68, 1660–70.
5. M. Ospeck, X.-. X. Dong and K. H. Iwasa, 2003. Limiting frequency of the cochlear amplifier based on electromotility of outer hair cells. Biophys. J. 84, 739–49.
6. P. Dallos and B. N. Evans, 1995. High-Frequency Outer Hair Cell Motility: Corrections and Addendum Science 268, 1420.
7. K. H. Iwasa and B. Sul, 2008. J. Acoust. Soc. Am. in press.
8. Y. Choe, M. O. Magnasco and A. J. Hudspeth, 1998. A model for amplification of hair-bundle motion by cyclical binding of Ca^{2+} to mechanoelectrical-transduction channels. Proc. Natl. Acad. Sci. USA 95, 15321–26.
9. R. A. Jacobs and A. J. Hudspeth, 1990. Ultrastructural correlates of mechanoelectrical transduction in hair cells of the bullfrog's internal ear. Cold Spring Harb Symp Quant Biol 55, 547–61.
10. D. J. Lim, 1980. Cochlear anatomy related to cochlear micromechanics. A review. J. Acoust. Soc. Am. 67, 1686–95.
11. L. G. Tilney and J. C. Saunders, 1983. Actin filaments, stereocilia, and hair cells of the bird cochlea. I. Length, number, width, and distribution of stereocilia of each hair cell are related to the position of the hair cell on the cochlea. J. Cell Biol. 96, 807–21.
12. D. Bozovic, personal communication.

Comments and Discussion

Bell: You talk about area changes, but it is unclear what effect that will have on the volume of the OHC. This would be useful to spell out, as volume changes are relevant to the detection and generation of fast pressure waves.

Iwasa: Thank you for your comments. The volume I am referring to is the volume of membrane proteins and that has nothing to do with the cell volume.

TILT OF THE OUTER HAIR CELL LATTICE: ORIGIN OF DUAL TUNING TIPS AND COCHLEAR BANDWIDTH

ANDREW BELL, TED MADDESS

Centre for Visual Sciences and ARC Centre of Excellence in Vision Science, Research School of Biological Sciences, The Australian National University, Canberra, ACT 0200, Australia

Using spatial autocorrelation analysis of published outer hair cell stereovilli positions in a primate, this work measured average spacing between cells. We find that the OHCs occupy a nearly rectangular unit cell that, from base to apex, exhibits a consistent tilt of about 4°. The implications of this "tilted box" geometry in terms of the squirting wave model of the cochlear amplifier are examined, and two major conclusions emerge. (1) The minimum and maximum wavelengths accommodated by the short and long diagonals of the tilted box (directions reinforced by the 'V'-shaped arms of the stereovilli) allow two distinct frequencies to map to every place. (2) The amount of tilt dictates the difference between these two frequencies – i.e., the relative bandwidth or Q. Thus, a 4° tilt leads by simple geometry to a Q of about 12 at 1 kHz and 25 at 8 kHz, figures in line with some otoacoustic and psychophysical results.

1 Introduction

The geometrically precise arrangement of outer hair cells in the cochlea – typically in three rows – is remarkable. This distinctive feature, which escapes standard explanation, was the initial stimulus for formulating a micromechanical model of the cochlea in terms of a surface acoustic wave (SAW) resonator [1, 2]. Subsequent modelling demonstrated how the three rows could, via wave interaction, form a positive feedback loop [3]. These publications strengthened the supposition by Bell and Fletcher [4] that the cochlear amplifier could be understood as a standing wave formed between the rows. The key conjecture of [4] was that the feedback was carried by 'squirting waves' which propagate slowly – at some mm/s – in the subtectorial space. Independent work by Elliott [5] and Gummer et al. [6] has elaborated the distinctive properties of squirting waves, particularly their extremely slow phase velocity and high dispersion [$f \propto (1/\text{wavelength})^3$], and has strengthened the idea that these waves play a key role in cochlear tuning. If a whole wavelength standing wave were to exist between rows of OHCs, this would deliver sharp tuning and high sensitivity to the IHCs, provided the gain of the system were set just below oscillation threshold. The net result might be that the distance between the rows could, through control of the propagation delay and hence the feedback delay, effectively govern the radial tuning of the cochlear partition from 20–20 kHz (see Fig. 1 of [4]).

Given the theoretical feasibility of the squirting wave model of cochlear tuning, we were encouraged to look at the spacing of OHCs in more detail. To our knowledge, the most comprehensive set of published data on OHC positions is the surface view tracings of Lonsbury-Martin et al. [7] for the primate cochlea. To examine whether the observed spacing is fully consistent with the squirting wave model, we applied spatial autocorrelation analysis to this data, a method by which the average spacing between hundreds of cells can be measured. The results show that the row spacing is consistent with the squirting wave model. They also reveal another distinctive property: that the OHC unit cell exhibits a tilt of about 4°. Below we show how this small tilt is important in governing the Q achievable at any point on the partition. The "tilted box" geometry of the OHC lattice allows two frequencies to be tuned at every cochlear location and this dual tuning in turn dictates the maximum Q.

2 Methods

Lonsbury-Martin et al. [7] contains data on the relative positions of the OHCs from 4 rhesus monkeys. The most complete data – thousands of cells in all – is for monkey 84-098, which is presented as a map of all the 'V's of the stereovilli over 24–60% of the length of the right cochlea and 24–59% for the left. Here, for brevity, we confine ourselves to presenting data from this monkey's data. These authors used surface view microscopy in the plane of the plateau of Corti after preparing the cochleas. The specimens for 84-098 were broken into 15 segments (right ear) or 11 segments (left ear) and the outline of the OHC stereovilli traced. A sample segment is shown in Fig. 1, which depicts nearly 200 'V's from 44–46% of the right cochlea length. Note that the specimens were fixed, alcohol dried, embedded in plastic, and baked, so that the possibility of distortion should be kept in mind. Notwithstanding, data from all four monkeys (eight cochleas) fell into a similar pattern, giving some confidence that the measurements, which are reported at face value, may not deviate too far from in vivo ones.

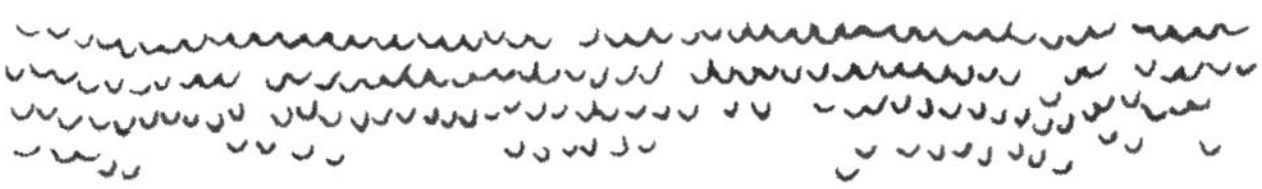

Figure 1. A segment of the stereovilli pattern of outer hair cells from the right ear of monkey 84-098 (from [8]). It represents a region between 44% (1.4 kHz, right) and 46% (1.6 kHz, left) of the distance from the apex.

The stereovilli layout in Fig. 1 shows irregularity – more so than seen in other laboratory animals – and the underlying pattern is not obvious. To reveal regularity in cell spacing, a spatial autocorrelation analysis [8] was brought to bear on the traced images. This method allows the average distances between cells to be measured and presented as a two-dimensional map in which recurring patterns appear as correlation peaks.

Using Matlab (The Mathworks, Natick, MA), a program was written which used as input the published black-and-white images, which were scanned at 600 or 1200 dpi (depending on original size). The program performed autocorrelation analysis by first using a fast two-dimensional Fourier transform to create a power spectrum; it then created an autocorrelation image from an inverse Fourier transform of the real part of the power spectrum. This method is hundreds of times faster than directly computing the autocorrelation function (but in a cross-check gave the same results). The program divided each of the published segments into equal steps of about 10 cell-widths across and then averaged the results from multiple steps to give an average autocorrelation across the segment. This subdivision appeared about optimum in terms of balancing the requirements of high spatial resolving power, good signal-to-noise ratio, and assimilating deviations in segment straightness.

3 Results

Figure 2 shows the average autocorrelation function corresponding to the segment shown in Fig. 1. It is displayed as a 3-dimensional plot, and radial and longitudinal slices through it are also shown. For ease of interpretation, a contour map of the same data is shown in Fig. 3 (middle), which clearly shows the regularity in cell positioning that underlies the three OHC rows – OHC1 at the top, OHC2 in the middle (and offset longitudinally by half an intercell distance), and OHC3 at the bottom. This regularity can be described in terms of an average spacing of the cells along the row, a, and of a radial separation between the rows, $b/2$ (i.e., the distance from OHC1 to OHC3 is b).

Using this labelling scheme, it is then straightforward to see that the entire base-to-apex array of OHCs is made up of repeating unit cells laid side by side. The unit cell is almost rectangular, and is outlined in Fig. 3, which shows contour maps of regions ranging from near the apex (left) and near the base (right). A similar display for the right ear is given in Fig. 4. They are all parallelograms with sides a and b. There is one other parameter needed to describe the unit cell, and this expresses the extent to which the unit cell deviates from an upright form. This deviation is given in terms of an angle, θ, and measures the tilt of the unit cell from the normal (the radial direction).

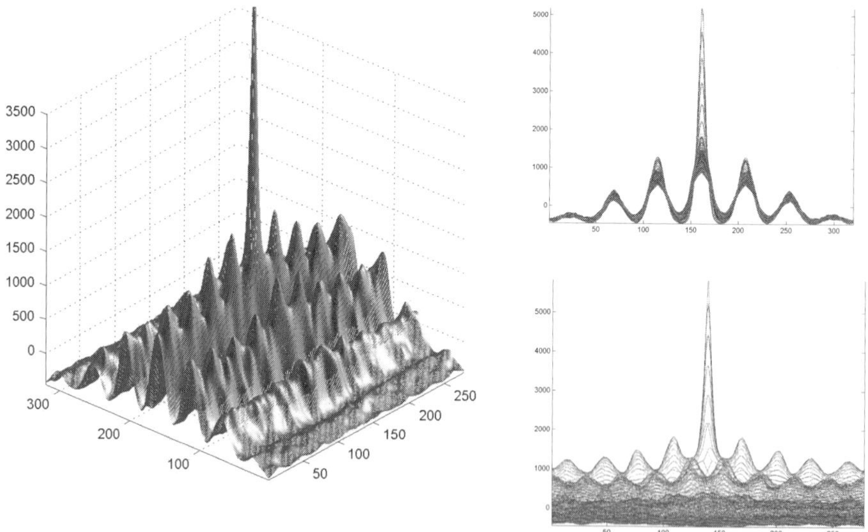

Figure 2. *Left*: Autocorrelation image of Fig. 1 after dividing the strip into 8 segments horizontally, performing autocorrelation, and averaging the result. *Right*: radial slices through the image (top) and longitudinal slices (bottom). Vertical axis is correlation magnitude, other axes are distance in pixels.

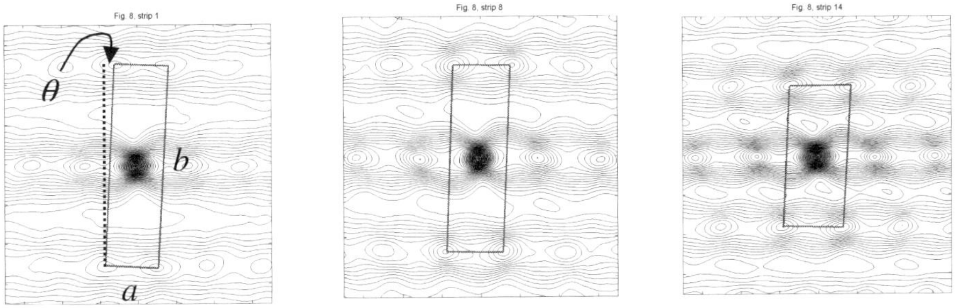

Figure 3. Contour maps derived from autocorrelation analysis of three chosen OHC strips (out of 15 available) from the right ear of monkey 84-098. At left is that from 24–27% from the apex (the most apical available); at centre is that derived from analysis of Fig. 1 (44–46% from apex); and at right is that from a strip of cells 56–58% from the apex. The corresponding CFs are about 0.5, 1.6, and 3.5 kHz. The unit cells are outlined.

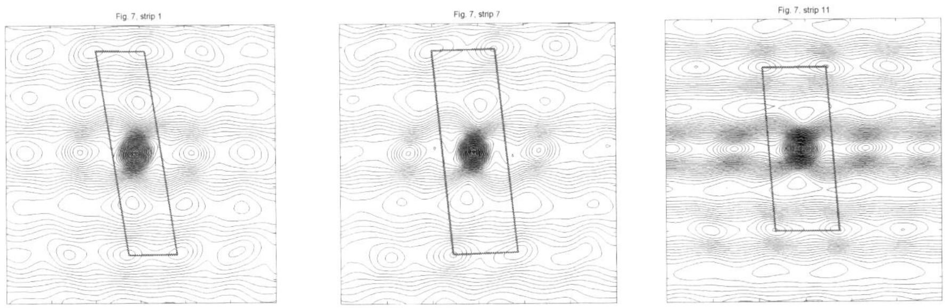

Figure 4. As for Fig. 3, but for 3 strips (selected from 11 available) for the left ear of monkey 84-098. At left is that for the strip 24–26% from the apex (the most apical available); at centre is that derived from 40–44% from apex); and at right is that from the strip of cells 54–59% from the apex. The CFs are similar to before. Again the unit cell is outlined. Note the opposite direction of tilt compared with Fig. 3.

The section below reports the autocorrelation analyses of all the strips from the left and right ear of monkey 84-098 in terms of a, b, and θ. As we subsequently show, when b/a and θ are interpreted in terms of a squirting-wave-based SAW system, they play powerful roles in controlling the frequencies and bandwidths achievable. Moreover, a bank of resonators extending along the cochlea will, if simultaneously excited, give rise to Q cycles of delay for each resonator and in this way produce an apparent traveling wave from base to apex.

3.1 Apex-to-Base Trends

As apparent from Figs 3 and 4, there are clear trends in cell spacing from apex to base. Analysis shows that a is effectively constant from base to apex. Hence, the ratio b/a can be used to describe the height of the unit cell. In this fashion, the major apex-to-base trend in row spacing can be summarised as shown in Fig. 5.

The trend in b/a ranges linearly from 2.4 at the 60% position to 3.7 at the 25% position. A 1.5-fold row spacing change translates, by cubing, to a nearly 4-fold change in squirting wave frequency, consistent with the CFs attributed by [7] to these regions.

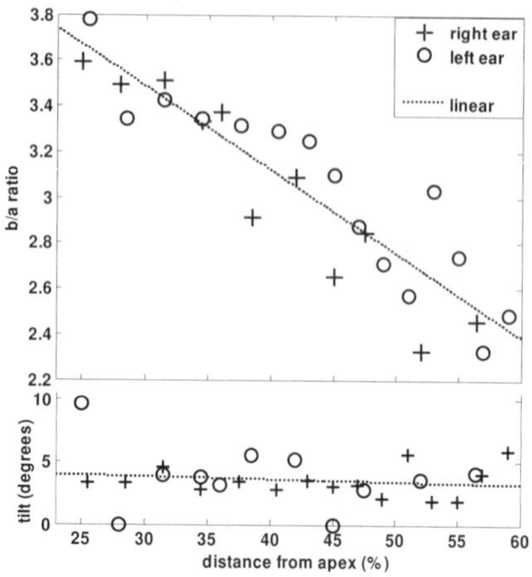

Figure 5. The *b/a* ratio of the unit cell for right and left ears, plotted as distance from the apex (top), together with the corresponding tilt (bottom). The CF ranges from about 500 Hz at left to 4 kHz at right. The trends are fitted with straight lines.

Similarly, spot data from the 80% region showed a *b/a* ratio of about 1.5, giving a 2.5-fold changein spacing compared to the 25% position and hence a 15-fold frequency shift, congruent with [7]. Notably, the tilt remains nearly constant at 3.6 ± 1.9° (mean ± SD) throughout the range. (The other three monkeys showed greater tilts, from 5–11°, although the data for them were sparse). The tilt (OHC1 relative to OHC3) was always towards the left in the left ear and towards the right in the right; that is, the tilt is in the direction of the apex.

4 Discussion and Conclusion

We have shown that the distance between the OHC rows decreases progressively from apex to base, allowing standing waves of increasing frequency to be accommodated. That is, row spacing largely controls cochlear tuning (as shown in [4]). In this situation, tilt of the OHC lattice is important, for it delivers two distinct frequencies to each unit cell, thereby establishing its Q. The derivation is as follows. First, tilt causes the two diagonals of the unit cell to differ in length. Now, in the SAW model, standing waves are formed between OHC1 and OHC3, and in a tilted case there will be a minimum possible standing wave frequency corresponding to the longest diagonal (length L) and a highest possible frequency corresponding to the shortest diagonal (S). Of course, this dual-frequency cavity will require each arm of the stereovilli (the ‘V’) on each OHC to operate independently, a hypothesis yet to be confirmed. Significantly, fine-structure studies often find distinctive features at frequency ratios of 1.04–1.10 (e.g., the dual tuning tips of [10]). Let us now consider a unit cell at 1 kHz, which for monkeys (and humans) has a *b/a* ratio of about 3.0. Simple geometry shows that for a constant 3.6° tilt the ratio of the diagonals (L/S) will be 1.026. But squirting wave dispersion means that the associated

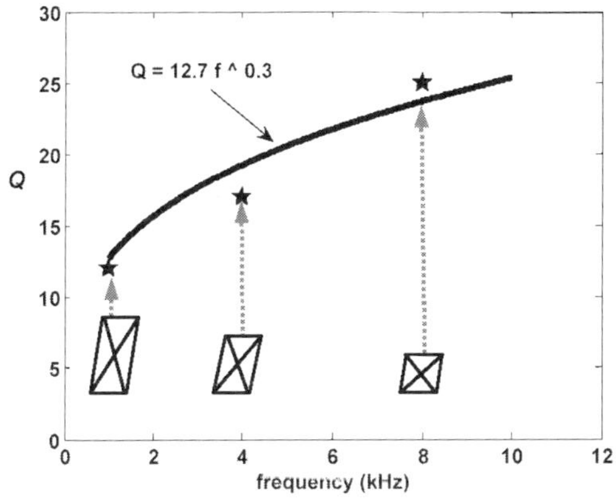

Figure 6. How tilt determines Q. Tilt of the unit cell causes the two diagonals to differ in length and therefore in the standing wave frequencies they support. The difference in frequencies can be expressed as Q. For 3 typical unit cells (a, b, 3.6°) as shown, calculated Q values (stars) match the empirical function of [9] (curve).

ratio of frequencies will be $(1.026)^3 = 1.08$ (1.3 semitone). Expressed another way, the difference in frequencies compared to the average frequency will be 1/0.08, giving a bandwidth, or Q, of 12. Similarly, at a CF where distance b is halved, b/a will now be 1.5 and CF will be about 8 times (2^3) higher or 8 kHz (assuming other squirting wave parameters stay about the same). Geometry now shows that for an invariant 3.6° tilt and b/a of 1.5, the ratio of diagonals, L/S, has now reduced to 1.013, and the corresponding frequency ratio is $(1.013)^3 = 1.04$ (0.7 semitone). Expressed as Q, we see that 1/0.04 = 25. Likewise, at about 4 kHz, geometry dictates a frequency ratio of 1.06 (1.0 semitone) and a Q of 17.

These Q values match the human data of [9], who found, psychophysically and oto-acoustically, the empirical relation $Q = 12.7 f^{0.3}$. These congruences are shown in Fig. 6. In short: $\boxed{\text{tilt} \rightarrow \text{L/S};\ (\text{L/S})^3 = (f_S/f_L);\ 1/(f_S/f_L - 1) \approx (\text{av.}\ f_S \,\&\, f_L)/(f_S - f_L) = Q.}$

Acknowledgments

We thank Dr Brenda Lonsbury-Martin for permission to use the monkey data and Elsevier Science for copyright permission (Fig. 1). This work reports research undertaken as part of the first author's PhD [2], and the receipt of an ANU PhD Scholarship and supplementary funding from the University of Tübingen is gratefully acknowledged.

References

1. Bell, A., 2006. Sensors, motors, and tuning in the cochlea: interacting cells could form a surface acoustic wave resonator. Bioinsp. Biomim. 1, 96–101.
2. Bell, A., 2005. The Underwater Piano: A Resonance Theory of Cochlear Mechanics. PhD thesis, Australian National University. <http://thesis.anu.edu.au>

3. Bell, A., 2007. Tuning the cochlea: wave-mediated positive feedback between cells. Biol. Cybern. 96, 421–438.
4. Bell, A. and Fletcher, N.H., 2004. The cochlear amplifier as a standing wave: "squirting" waves between rows of outer hair cells? J. Acoust. Soc. Am. 116, 1016–1024.
5. Elliott, S.J., Pierzycki, R., and Lineton, B., 2005. Incorporation of an active feedback loop into the "squirting wave" model of the cochlear amplifier. Proceedings, Twelfth International Congress on Sound and Vibration, Lisbon.
6. Gummer, A.W., et al., 2006. Pulsating fluid motion and deflection of the stereocilia of the inner hair cells due to the electromechanics of the outer hair cells. Auditory Mechanisms: Processes and Models. World Scientific, New Jersey, pp. 17–25.
7. Lonsbury-Martin, B.L., et al., 1988. Spontaneous otoacoustic emissions in a non-human primate. II. Cochlear anatomy. Hear. Res., 33, 69–94.
8. Bracewell, R.N., 1986. The Fourier Transform and Its Applications, McGraw-Hill.
9. Shera, C.A., Guinan, J.J., and Oxenham, A.J., 2002. Revised estimates of human cochlear tuning from otoacoustic and behavioral measurements. Proc. Nat. Acad. Sci. U.S.A., 99, 3318–3323.
10. Abdala, C., et al., 2007. Distortion product otoacoustic emission suppression tuning and acoustic admittance in human infants: birth through 6 months. J. Acoust. Soc. Am., 121, 3617–3627.

Comments and Discussion

Aranyosi: You present an intriguing idea. I've got a few questions:
(1) How would this idea work in species with more than three rows of outer hair cells? Is it necessary to have three rows, or is two enough?
(2) Can your proposed mechanism explain the level dependence of cochlear tuning?
(3) The counterphase contraction of different rows of OHCs would generate significant fluid flow in both the subtectorial space and the spaces of Nuel. How would this fluid motion affect the tuning you predict?

Bell: (1) As with any SAW device, 3 rows are the minimum required for full-wavelength mode resonance. However, 4 or more rows work in a similar way: alternating polarity between the rows is retained and there is half a wavelength between each pair. That means the frequency of tuning - the resonance frequency - is the same but additional gain is available. There is a trade-off between number of rows and the speed of response to a transient. Two rows might be enough, or even only one, although the location of nodes and antinodes might be more difficult to arrange anatomically and quarter-wavelength modes would then seem the most easily excited. [...]

(2) In its simplest form, no. The highest achievable Q is geometrically fixed, although reduced gain of the positive feedback loop between the rows might well reduce the tilted system's Q, as it does in the untilted case (e.g., Fig. 12 of Bell (2007)). However, whether this reduction is sufficient to match observations would require additional modeling. At high SPLs, passive responses would also begin to affect the total system.

(3) Since the squirting waves are conjectured to give rise to a standing wave, there is no net fluid flow. Waves are proceeding in both directions equally, so there is no net transfer of fluid in the subtectorial space. Similarly, counterphasic activity of the rows of outer hair cells is not expected to give any net fluid flow in the spaces of Nuel over a cycle. Whether some rectification of this alternating fluid flow takes place is a case for experiment to decide. Note, however, that over time frames less than a cycle, OHC motion (two rows up and one down and vice versa) should mean that some vibration will be imparted to surrounding structures, including in particular the basilar membrane.

Gummer: Your model requires wave-motion on the reticular lamina (RL) and also counter-phasic motion at the lower surface of the tectorial membrane (TM). However, vibration measurements at these surfaces in response to intracochlear electrical stimulation show absolutely no evidence of wave motion (Nowotny and Gummer, 2006). Moreover, theoretical calculations (Elliott, 2007; Nowotny and Gummer, 2006) support these experimental observations. [Although those experiments were conducted in an in-vitro preparation, the system appeared to be in "excellent" mechanical condition, as judged by attachment of the outer hair cell (OHC) stereocilia to the TM and the absence of any visible TM retraction or OHC swelling (Nowotny and Gummer, 2007).] How do you reconcile your assumptions with experimental evidence? Do you feel free to neglect the results of those electrical stimulation experiments because you believe that a change of transmembrane potential is not the stimulus for generation of mechanical force by the OHC?

Bell: It is not unexpected that a dead cochlea will behave differently to a live one. In particular, the efferent system has been severed in an in vitro system, and thus control of the resting membrane potential of the outer hair cells has been lost. My model calls for antiphasic motions of OHC2 compared to OHC1 and OHC3 (in response to a common pressure stimulus), and this can be achieved if the resting membrane potential of OHC2 is higher (or lower) than that of OHC1 and OHC3. The resting membrane potential is known to be labile; moreover this potential determines whether certain voltage-sensitive ion channels are open or closed. Thus, via efferent control, the resting membrane potentials between the rows can be made to differ, and in this way antiphasic responses to the same primary stimulus can result.

There is some experimental evidence that the resting membrane potential varies between the rows. Thus, Tanaka et al. (1980) looked at d.c. gradients and found (their Fig. 3) that 4 cells from OHC1 had a gradient of 100-120 mV compared to 20-90 mV for 10 cells from OHC2 and OHC3. Clearly, more definitive work is needed in this area, but a general argument can be made that inter-row potential differences are important in controlling cochlear gain. It is known that, in vivo, cutting the nerve supply to the cochlea results almost immediately (within minutes) in loss of active cochlear properties, such as evoked emissions. This is puzzling in that, in vitro, isolated outer hair cells continue to respond normally for tens of minutes after extraction. An explanation [see Chapter 9 of

Bell (2006)] is that each row requires a different membrane potential, set by the efferent system, in order to join OHC1, -2, and -3 into a functional differential amplifier. [...]

In summary, a common electrical stimulus applied to an in vitro preparation does not mimic the behaviour of a live cochlea because it results in a uniform response of all the rows. In the SAW model, it is the differential response of the rows that is the crucial input signal. The analogy is with a differential amplifier in which common-mode signals are rejected. Perhaps that factor might explain the limited vibration gain of your preparations when the cochlear amplifier typically displays a gain of 60-80 dB.

Braun: [in response to Bell's reply, above] Liberman and Brown (1986) reported unequivocal anatomical data from the cat showing that one efferent fiber typically makes synaptic contacts with clusters of OHCs across all three rows of OHCs. Row-specific efferent input upon OHCs was therefore anatomically impossible. (B) Cutting of efferent innervation of OHCs does not affect threshold of hearing (e.g. see Scharf et al., 1997). This observation also disproves the hypothesis of an "antiphasic setting" of adjacent rows of OHC via efferent input.

Liberman MC, Brown MC, 1986. Physiology and anatomy of single olivocochlear neurons in the cat. Hear Res 24, 17-36.

Scharf B, Magnan J, Chays A, 1997. On the role of the olivocochlear bundle in hearing: 16 case studies. Hear Res 103, 101-122.

ACOUSTIC STREAMING IN THE COCHLEA

F. BÖHNKE
*ENT Department, Technische Universität München, Ismaningerstr. 22
81675 München, Germany*

M. SCHARFF
*Inst. of Fluid Mechanics (FLM), Technische Universität München
85747 Garching, Germany*

A first description of Acoustic Streaming (AS) in air was given by Rayleigh in 1884. Nyborg (1952) describes AS due to attenuated waves and cited some examples, e.g. the presence of small eddies in a protoplasm. In connection with nonlinear cochlear mechanics the effects were not explicitly denoted as AS though the nonlinearity should induce eddies which were demonstrated experimentally at high pressure levels by Bekesy and suggested theoretically by e.g. Tonndorf (1), although Zwicker questioned the existence of AS at normal auditory pressure levels. Later Lighthill (2) gave an analytical approach of AS in the inner ear but a numerical evaluation failed to appear. Because of computer limitations and the lack of appropriate models AS could not be verified for the cochlea up to now. This work is a numerical approach to confirm or refute AS in the cochlea.

1 Introduction

One effect of AS is the deviation from harmonic behaviour caused by nonlinear viscous effects of the lymph with sinusoidal excitation. To simulate the acoustics of the inner ear we use a mathematical formulation developed for the simulation of 3d microfluidic biochips (3). Here the propagation of surface acoustic waves (SAW) is used to mix particles in the fluid for medical applications.

2 Mathematical Formulation

The compressible Navier-Stokes equations are used to describe the fluid flow. The model consists of two separate systems of partial differential equations to describe the flow variables at pressure p and velocity v on two different time scales. Acoustic streaming effects are a result of the nonlinearity of the Navier-Stokes equations in combination with viscosities. The derivation is done according to (4) which provides a rigorous treatment of boundary conditions often neglected in other works. Because of the low pressure levels of acoustical signals with hearing we expect *Rayleigh streaming* associated with boundary layers of a fluid near solid surfaces instead of the *Quartz wind* which requires high pressure levels in the main body of the fluid. For the simulation of acoustic streaming we use macroscale variables $\mathbf{v}^{(2)}$, $p^{(2)}$ which require macroscale forcing terms. To compute these, temporal averaging is applied using microscale, quickly oscillating terms $\mathbf{v}^{(1)}$, $p^{(1)}$.

2.1 Acoustic subproblem

In a first step, the acoustic subproblem to obtain $\mathbf{v}^{(1)}$, $p^{(1)}$ in the fluid must be solved. Because of a dimensionless formulation, the density ρ does not appear and viscosity $\tilde{\nu}_1 = \eta / \rho^{(0)}$ and bulk viscosity $\tilde{\nu}_2 = (\eta + \zeta/3)/\rho^{(0)}$ are kinematic viscosities. Constants are: $a = \frac{VT}{L}$, $b = \frac{VT^2}{L^3}$, and $d = \frac{L}{c_0^2 VT^3}$, where V is velocity scale, L is length scale, T is time scale, and c_0 is sound speed in the lymph. η, ζ are standard and bulk viscosity coefficients. $\rho^{(0)}$ is the time and space independent constant density. The bulk viscosity ζ or second viscosity appears in processes which are accompanied by a change in volume (i.e. in density) of the fluid.

$$
\begin{aligned}
a\frac{\partial \mathbf{v}^{(1)}}{\partial t} - b\left(\tilde{\nu}_1 \Delta \mathbf{v}^{(1)} + \tilde{\nu}_2 \nabla(\nabla \cdot \mathbf{v}^{(1)})\right) + \nabla p^{(1)} &= 0 \\
d\frac{\partial p^{(1)}}{\partial t} + \nabla \cdot \mathbf{v}^{(1)} &= 0 \\
a\mathbf{v}^{(1)} &= \frac{\partial \mathbf{u}_{FP}}{\partial t}
\end{aligned}
\tag{1}
$$

The last equation is the boundary condition at the stapes footplate. With the solution of equation 1 the time averaged (symbols $\langle\ \rangle$) right hand sides in equation 2 are used to solve the acoustic streaming subsystem (Equation 3) for $\mathbf{v}^{(2)}$, $p^{(2)}$

$$
\begin{aligned}
g &= -\left\langle [\nabla \mathbf{v}^{(1)}]\, \mathbf{u}_{FP} \right\rangle \\
\mathbf{f}_v &= a^2 \left\langle (\nabla \cdot \mathbf{v}^{(1)})\mathbf{v}^{(1)} + [\nabla \mathbf{v}^{(1)}]\, \mathbf{v}^{(1)} \right\rangle \\
f_p &= -ad\left\langle \nabla \cdot (p^{(1)} \mathbf{v}^{(1)}) \right\rangle
\end{aligned}
\tag{2}
$$

Note that the term g depends on the product of the stimulating displacement $\mathbf{u}_{FP}$ and the acoustic field $\mathbf{v}^{(1)}$ and not on the scaling constants *a*, *b* or *d*. Therefore the streaming field $\mathbf{v}^{(2)}$, $p^{(2)}$ is proportional to the time averaged value of this product, shown in Eq. 3. The backward coupling of the fluid to the elastic basilar membrane is neglected preliminary.

2.2 Acoustic streaming subproblem

$$
\begin{aligned}
\mathbf{v}^{(2)} &= g \\
-b\tilde{\nu}_1 \Delta \mathbf{v}^{(2)} - b\tilde{\nu}_2 \nabla(\nabla \cdot \mathbf{v}^{(2)}) + \nabla p^{(2)} &= \mathbf{f}_v \\
\nabla \cdot \mathbf{v}^{(2)} &= f_p
\end{aligned}
\tag{3}
$$

Equation 3 is a Stokes system with non-homogeneous divergence constraint and can be solved using a stable finite element discretization such as the Taylor-Hood element. Preliminary numerical calculations with two-dimensional models show the effect of AS with dimensions and frequencies (e.g. $f = 10$ kHz) in case of the inner ear.

Acknowledgments

We gratefully acknowledge the disposal of mathematical C-Code by D. Köster.

References

1. J. Tonndorf, 1970. Cochlear Mechanics and Hydro-dynamics, in Foundations of Modern Auditory Theory, ed. J. V. Tobias, Academic Press, pp. 205-254.
2. J. Lighthill, 1993. Acoustic Streaming in the Ear Itself, Contemp. Math., Fluid Dynamics in Biology, Am. Math. Society, ed. A.Y. Cheer, C.P. van Dam, pp. 187-236.
3. D. Köster, 2006. Numerical Simulation of Acoustic Streaming on SAW-driven Biochips, PhD Dissertation, Universität Augsburg, Germany, 1-148.
4. Bradley, 1996. Acoustic streaming field structure: The influence of the radiator. J. Acoust. Soc. Am., 100(3) 1399-1408.

Comments and Discussion

Gummer: To summarize this discussion, do you agree that at functionally relevant stimulus frequencies the cochlear fluid velocities and, therefore, the Reynold's number, are too small for there to be streaming of any significance, as illustrated in the talk by Baumgart et al. (this book)?

Böhnke: No, I don't agree. The value of the Reynolds number is not a criterion for the presence of acoustic streaming. As shown in the presentation the product of the space gradient of velocity of the primary acoustic field times the displacement of the radiator (boundary displacement) determines the streaming velocity independent of viscosity and therefore the Reynolds number. This result was newly derived by Bradley (JASA, 1996) and former by Schlichting (1932). Therefore we expect relevant streaming velocities as a consequence of high basilar membrane displacement gradients. If we assume a fluid particle velocity gradient of the primary acoustic field of 1 μm/s per μm behind the basilar membrane maximum displacement and a (footplate) displacement of 10nm, the resulting streaming velocity would be 10nm/s, which is (only) 100 times smaller than the assumed particle velocity of the primary acoustic field.

Gummer and Baumgart: Since our own paper (Baumgart et al., Fluid mechanics in the subtectorial space, this book) is not really detailed concerning acoustic streaming, we would like to add some comments to the present discussion. First, we regard acoustic streaming as a steady-flow field which is induced by sinusoidal excitation. Second, the traveling wave on the basilar membrane can produce significant streaming (Lighthill, 1992; J. Fluid Mechanics 239, 551-606). Third, the relative magnitude of the steady-flow field increases with stimulus amplitude. Thus, acoustic streaming becomes increasingly important with increasing stimulus amplitude. Forth, the non-linear convective term in the Navier-Stokes-equations (conservation of momentum in Eulerian notation) is responsible for acoustic streaming. It is the product of a convective velocity times a spatial gradient of velocity. At the boundary, where the velocity gradients are highest in the field, this should be the boundary velocity (as pointed out by Dr. Böhnke in his discussion). Finally, in our present work, we are interested in the cochlear amplifier. As presented theoretically by Lighthill (1991; ASME: J. Vibration and Acoustics 113, 1-13) and demonstrated experimentally by others, increasing amplification with decreasing excitation amplitude cannot be explained by fluid mechanics. Indeed, we like many others maintain that acoustic streaming can be neglected for the small amplitudes associated with cochlear amplification (velocities below 10 mm/s up to 20 kHz and 80 dB SPL). A fluid-based change of velocity amplitude of, say, 1% due to the nonlinear convection term, as might be expected at sound pressures of about 80 dB SPL for a footplate displacement amplitude of about 10 nm up to 1 kHz (and 13 dB/oct less at higher frequencies; Aibara et al., 2001; Hearing Research 152, 100-109), can therefore be regarded as functionally irrelevant below 80 dB SPL. Moreover, uncertainties in parameters, such as geometry and material properties, are still too large to allow a meaningful description of second-order effects such as acoustic streaming. Nevertheless, since acoustic streaming becomes increasingly important for higher amplitudes, it is certainly of interest to investigate possible protection mechanisms in the cochlea that might be based on acoustic streaming.

HOOK REGION REPRESENTED IN A COCHLEAR MODEL

CHARLES R. STEELE, NAMKEUN KIM, SUNIL PURIA

Mechanical Engineering and Department of Otolaryngology-Head and Neck Surgery, Stanford University, Durand Building, Stanford, CA 94305, USA

The present interest is in discontinuities. Particularly the geometry of the hook region, with the flexible round window nearly parallel with the basilar membrane, is not represented by a standard box model, in which both stapes and round window are placed at the end. A better model represents the round window by a soft membrane in the wall of scala tympani, with the end closed. This complicates the analysis considerably. Features are that the significant compression wave, i.e., the fast wave, is of negligible magnitude in this region, and that significant evanescent waves occur because of the discontinuities at the beginning and end of the simulated round window. The effect of this on both high frequency, with maximum basilar membrane response in the hook region, and lower frequencies are determined.

1 Introduction

Work is progressing on the full three-dimensional, organ of Corti model of the cochlea. In comparison with measurements from various laboratories, the recent results include: (1) The seeming discrepancy between point load measurements and the basilar membrane deformation observed is attributable to the longitudinal stiffness of pillar heads. (2) The model with outer cell properties from isolated hair cell measurements agrees well with results for electrical excitation of the intact cochlear section. (3) The general deformation agrees well with results from confocal optical measurements in the apex of the guinea pig. However, many measurements are made in the vicinity of the hook region at the base of the cochlea, such as by Ren and Nuttall [1] and Olson [2]. The standard "box" model of the cochlea is indicated in Figure 1(a), with the stapes and round windows located at the end. So the motivation of the present work is to generalize the models to consider a more realistic location of the round window. More generally, we wish to develop the capability for including any discontinuity in geometry and properties, such as with a pressure probe, a short electrode, or opening in the wall. For this we work with the model containing the basilar membrane but excluding the organ of Corti. However, the procedure can be generalized to include the organ of Corti as outlined by Steele [3].

2 Mathematical Analysis

The full Navier-Stokes equations for the viscous, compressible fluid in the scalae is used, with the simplification that the amplitude is small, so the convective nonlinear term is negligible. Peterson and Bogart [4] discuss the following two types of solutions for the box model of the cochlea.

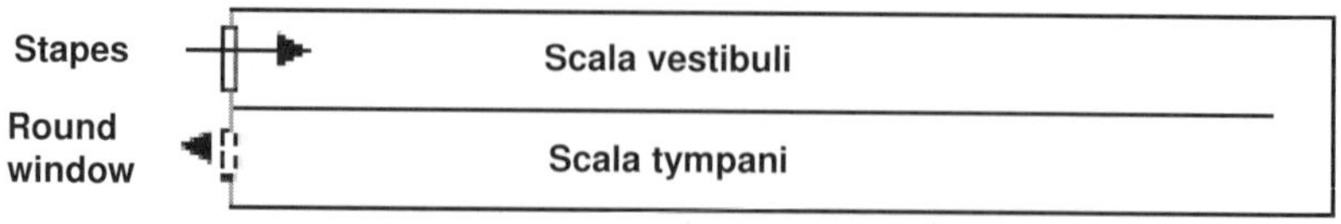

Figure 1(a). Standard box model of the cochlea. For an inward displacement of the stapes, there is an equal outward volume displacement of the round window.

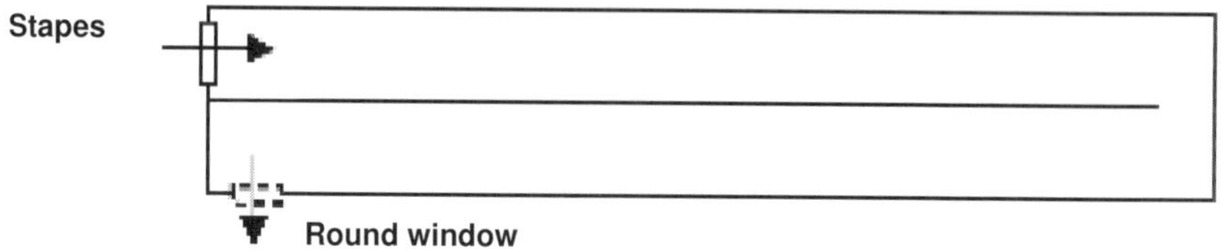

Figure 1(b). Better approximation to the hook region, in which the round window is located in the wall of scala tympani, rather than at the end.

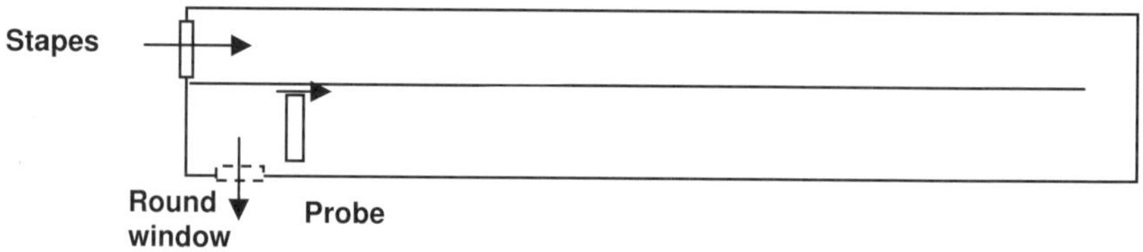

Figure 1(c). A reduction in the height of scala tympani, as a rough approximation to the effect of an insertion of a probe.

2.1 Fast wave

The pressure is equal on both sides of the basilar membrane, so there is no displacement of the basilar membrane. So this is simply an acoustic wave traveling along the cochlea. Generally the effect of this wave has been ignored, but the pressure measurements of Olsen [2] clearly show its presence. When there are changes in the dimensions of the scala then this wave is necessary to satisfy the continuity conditions. The important feature is that the pressure is the same in both scala, but the total volume displacement depends on the cross sectional area of each scala. In the symmetric model of Peterson and Bogart [4] this is the *even* solution.

2.2 Slow traveling wave

For this the pressure is of opposite sign in the two scalae, so there is a resultant pressure on the partition causing a displacement of the basilar membrane. Thus there is a fluid-elastic interaction that yields a traveling wave that has phase and group velocities much slower than the acoustic wave. In the symmetric model this is the *odd* solution. Because the stiffness of the basilar membrane changes by orders of magnitude along the length, there is no closed form solution, even for the one-dimensional model. Because the wavelength becomes very short in the interesting region, direct numerical methods require a very fine mesh. However the problem is suitable for the approximate WKB method as used by Steele and Taber [5] and recently by Yoon, et al. [6].

The key is to solve for the local wave number as a function of the input frequency. For the rectangular scalae, a Fourier series is used in the lateral direction to provide the 3-dimensional solution for the viscous fluid. Each term satisfies the condition of no slip on the partition, where the basilar membrane is located and no slip on the opposite wall. Then the compatibility of the fluid and the partition yields the dispersion relation, i.e., the relation of the local wave number and the frequency. This is a nonlinear relation that must be solved numerically. Subsequently, the amplitude function is determined. For inviscid fluid, the amplitude function is determined from the condition that the power in at one section must be equal to the power out at another section. For the viscous fluid this is extended to a consideration of the reciprocity of the forward and backward waves. The effective pressure on the cross section is determined so that the power is equal to the effective pressure multiplied by the volume velocity of the fluid.

2.3 Evanescent wave

The traveling wave is characterized by a wave number whose real part (giving the phase) is much greater than the imaginary part (giving an exponential decrease in the total wave amplitude). In addition, there is a doubly infinite spectrum of solutions of the dispersion relation given by wave numbers whose imaginary parts are dominant. These correspond to fluid modes on the cross section necessary to solve prescribed distribution of pressure or velocity on a given fluid cross section. However, for the closed walls, as in Figure 1(a), these modes have zero resultant fluid displacement, decay rapidly with the distance from the end, and consequently are generally ignored. For a wall with an opening, as in Figure 1(b), the situation is different. The fast wave with constant pressure cannot occur in such a region. Fortunately, the boundary condition of zero pressure on the wall yields nearly imaginary wave numbers that provide a net displacement of the fluid. In the present analysis the first such evanescent wave replaces the fast wave in the region with the wall opening.

2.4 Computational procedure

In the current calculation, the cochlea is divided into some fifteen sections, in each of which forty points along the length are used. In each of the sections, the matrix of solutions consists of the forward and backward slow waves and the forward and backward fast waves. The latter are replaced by the evanescent waves in the section with an opening of the wall. The dependent variables are the volume velocity and effective pressure in each scala, so the system is fourth order. Discontinues in the wall condition, geometry or stiffness at the section boundaries are permitted, as in Figure 1(c). However, the effective pressure and the volume displacement must be continuous in each scala. Significant backward waves are generated by discontinuities.

The procedure is to start at the helicotrema, where the impedance is known, and work toward the stapes with a computation of the total impedance at the boundary of each section. Then the input conditions at the stapes are used to evaluate the magnitude of the solution, and the system reverse solved for the solution in each section. The calculation requires some five seconds per frequency on a laptop computer (Mac G4).

3 Results

The geometry of the hook region is not simple and is not well represented by the standard box model in Figure 1(a) with the stapes and round window at the end. It is well known that the round window is more parallel to the partition. From micro-CT images in gerbil, it seems that the round window in the cross section is oriented at an angle of about 60° to the partition. Our initial approximation in Figure 1(b) has the round window located on the wall of the scala tympani, i.e., parallel with the partition in both longitudinal and transverse directions.

Figure 2 shows results for the stapes driven at a frequency of 5kHz with a unit magnitude of velocity. Figure 2(a) shows the components of the volume velocities in the

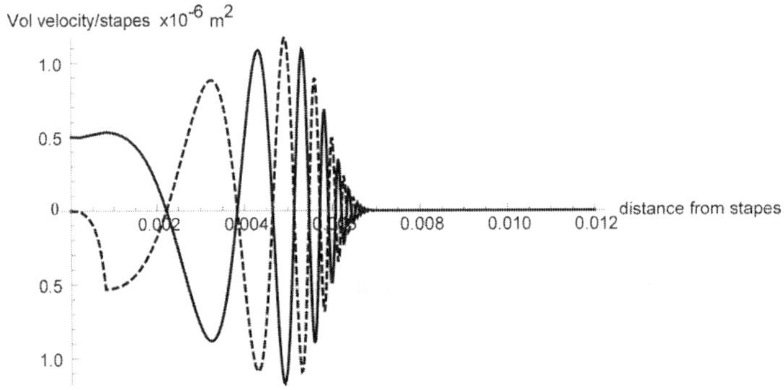

Figure 2(a). Volume velocity in scala tympani (solid) and in scala vestibuli (dashed) at 5 kHz. A unit velocity of the stapes is prescribed, so the scala tympani has a volume velocity equal to the area of the stapes. The scala vestibuli has zero velocity at the stapes, but along the round window region, becomes equal and opposite in direction to that in scala vestibuli. At distances past the maximum, the velocities become zero.

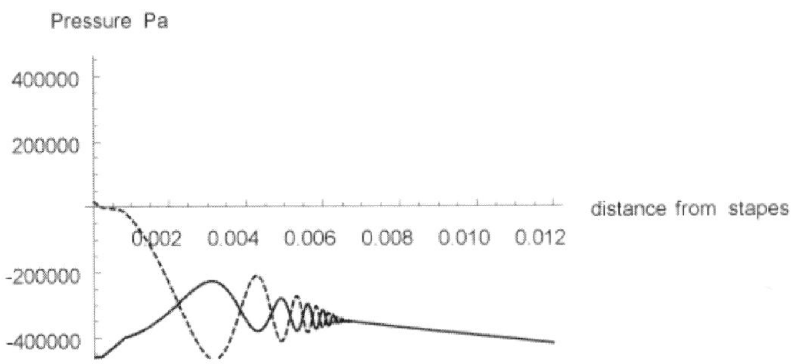

Figure 2(b). Effective pressure in scala tympani (solid) and in scala vestibuli (dashed) at 5 kHz. The scala vestibuli has a high value at the stapes, while that in scala vestibuli is low. Then come the oscillations of the traveling wave. After these diminish, only the fast wave remains, for which the pressure in both scalae is the same and no forcing of the basilar membrane occurs.

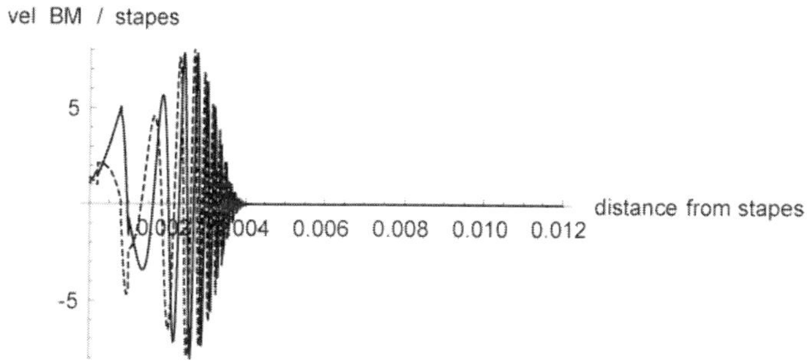

Figure 3. Response of the basilar membrane for 20 kHz with a reduced height of scala tympani in the region from 0.8 to 1 mm. This is an exaggerated representation of the presence of a pressure probe, which causes a reduction in maximum amplitude.

scalae that are in phase with the stapes. At the stapes end, the scala vestibuli has the input value, but the scala tympani has zero velocity corresponding to the closed end. In the region of the round window, from 0.2 to 0.8 mm, the scala tympani volume velocity rapidly increases in magnitude, corresponding to the fluid flow out of the opening in the wall. Subsequently, the velocities are equal in magnitude and opposite in sign, as in the usual model Figure 1(a). After the maximum, the volume velocities become very small, corresponding to the apical region where there is only the fast wave. This is clear in Figure 2(b) where the effective pressures in the scalae are shown. The pressure on the basilar membrane is much larger than the effective pressure and produces the usual response. For higher frequencies, there is more reduction in amplitude for the configuration Figure 1(b) compared to Figure 1(a).

An additional feature is in the geometry shown in Figure 1(c), in which the height of scala tympani is reduced substantially in the region from 0.8 to 1 mm. This condition may approximate to some degree the effect of inserting the pressure probe of Olson [2] . However, this model is more severe, since the flow around the probe is not permitted. The consequence is a further reduction in the basilar membrane response as shown in Figure 3.

4 Conclusion

The procedure of simultaneously using the slow and fast waves, both forward and backward, works well for discontinuities in properties along the cochlea. The extension for the open wall of scala tympani is to replace the fast wave with the first evanescent wave that has a net volume velocity. Thus the full three-dimensional behavior of the viscous fluid is taken into consideration for little cost in computational time.

Acknowledgments

This work is supported by NIDCD of NIH Grant No. DC007910.

References

1. T. Ren and A.L. Nuttall, 2001. Basilar membrane vibration in the basal turn of the sensitive gerbil cochlea, Hear. Res. 151: 48-60.
2. E.S. Olson, 2001. Intracochlear pressure measurements related to cochlear tuning, J. Acoust. Soc. Am., 110: 249–367.
3. C.R. Steele, 1999. Toward three-dimensional analysis of cochlear structure, ORL 61: 238-251 (1999).
4. L.C. Peterson and B.P. Bogart, 1950. A dynamic theory of the cochlea. J. Acoust. Soc. Am., 22: 369-381.
5. C.R. Steele and L.A. Taber, 1979. Comparison of WKB calculations and experimental results for three-dimensional modeling of cochlear mechanics, J. Acoust. Soc. Am., 65: 1007-1008.
6. Y.J. Yoon, S. Puria, and C.R. Steele, 2007. Intracochlear pressure and derived quantities from a three-dimensional model, J. Acoust. Soc. Am., 122 (2): 952–966.

Comments and Discussion

Bell: Since pressure is a scalar quantity, I don't understand how Fig. 1a differs physically from 1b.

Puria: In Fig. 1a, the end condition at the stapes ($x = 0$) is that the effective pressure in scala tympani is zero, the approximation for a very compliant round window membrane at the end of the box. For Fig. 1b, the condition at $x = 0$ for scala tympani is zero volume displacement in the x-direction. Along the wall of scala tympani the condition used is that the region simulating the round window, from $x = 0.2$ to $x = 0.8$ mm, is that all the three-dimensional stress components, i.e., the normal component (pressure) and the two shear components, are zero. Remember that we use ten to forty harmonics of the fluid displacement across the width of each scala, and the exact solution for each harmonic in the direction normal to the basilar membrane is used, to obtain the local solution of the phase-frequency equation. In the region of the zero wall stress components, there can be no fast wave, while for the configuration in Fig. 1a, the fast wave is present everywhere.

Fulton: The simple box model has been used in pedagogy for a very long time. It is appropriate that you develop a more detailed representation of the round window area of the scala tympani.

Have you considered updating the relationship of the stapes/oval window to the scala vestibuli in your model? The physiological, and particularly the surgical anatomy, literature would suggest there is no low impedance path between these two structures.

Puria: The current model simply moves the round window to the wall of scala tympani as a significant first step. As you suggest, an improved model is desirable for the relation of oval window, vestibule, semi-circular canals and scala vestibuli, not to mention the sharp bend of the hook region. Each of these geometric features must have an effect on the details of the basilar membrane response.

de Boer: Would it be possible to reformulate the problem by defining a few types of Green's functions? That would at least allow you to use more longitudinal sections.

Puria: We consider the actual variation of the height and width of each scala, for which a usual Green's function is not convenient. Particularly to handle the sharp discontinuities, the cochlea is divided into some fifteen sections, on each of which a discontinuity in properties is permitted. The procedure is such that the computational time increases linearly with the number of sections, so there is not much restriction. Between sections is a "large element", in which the property variation is "smooth" and forward and backward waves are included. Within the large element, the exact three-dimensional equations for the viscous fluid and the interaction with the basilar membrane is computed on forty points. From this phase-frequency relation at each point there is a doubly infinite set of roots. The simplification comes in using only those with a significant volume displacement of fluid in the scalae, i.e., the usual slow traveling wave and the fast acoustic wave. In a region of an opening of the wall of scala tympani, the latter is replaced with the first evanescent wave.

COCHLEAR MODELING USING "TIME-AVERAGED LAGRANGIAN" METHOD: COMPARISON WITH V_{BM}, P_{ST}, AND Z_C MEASUREMENTS

Y. YOON, N. KIM, S. PURIA, C.R. STEELE

Mechanical Engineering, Stanford University, 496 Lomita Mall, Durand Rm 224 Stanford, CA 94305, USA

In this work, basilar membrane velocity (V_{BM}), scala tympani intracochlear pressure (P_{ST}), and cochlear input impedances (Z_c) for gerbil and chinchilla are implemented using a three-dimensional hydro-dynamic cochlear model using 1) time-averaged Lagrangian, 2) push-pull mechanism in active case, and 3) the complex anatomy of cochlear scalae by micro computed tomography (μCT) scanning and 3-D reconstructions of gerbil and chinchilla temporal bones. The objective of this work is to compare the calculations and the physiological measurements of gerbil and chinchilla cochlear such as V_{BM} (Ren and Nuttall [1]), P_{ST} (Olson [2]), and Z_C (Decraemer *et al.* [3], Songer and Rosowski [4], Ruggero *et al.* [5]) with present model. A WKB asymptotic method combined with Fourier series expansions is used to provide an efficient simulation. V_{BM} and P_{ST} simulation results for the gerbil cochlea show good agreement both in the magnitude and the phase for the physiological measurements without larger phase excursion. Z_C simulation from the gerbil and chinchilla model show reasonably good agreement with measurement.

1 Introduction

In the previous work, V_{BM}, P_{ST}, and organ of Corti impedance for gerbil cochlea were calculated and compared with physiological measurements (Lim and Steele [6], Yoon *et al.* [7]). However, the model V_{BM} and P_{ST} results had phase excursion issue since the amplitude function took the averaged value of Laplace's equation. To resolve the larger phase excursion issues in the previous model, we uses the "time-averaged Lagrangian" for the conservative system (Steele and Taber [8]). In addition to this, the gerbil and chinchilla cochlear anatomy from the micro computed tomography (μCT) scanning and 3-D reconstructions of temporal bones are used to construct current model. In the active case, the push-pull mechanism from the OHC and phalangeal process is implemented in the new model. Through this new model, V_{BM} and P_{ST} area calculated and the simulation results show good agreements with animal measurements both in its magnitude and phase. Finally, the acoustic cochlear input impedance (Z_C) are calculated and compared with recent animal measurements which were measured from the gerbil (Decraemer *et al.*[3]) and chinchilla (Songer and Rosowski [4], Ruggero *et al.* [5]) cochlea.

2 Method and Results

Overall model features and dimensions of the three-dimensional hydro-dynamic model were described in the previous work (Yoon *et al.* [7]). In addition to this, the current model has new features of: 1) push-pull mechanism which consists of the downward push from an expansion of the outer hair cell (OHC) and an upward pull through the phalangeal process from an expansion of the OHC, and 2) scala areas by the μCT scanning 3-D reconstruction. The most important change comparing to previous model is that the equation for wave propagaion in current model is obtained by equating the time-averaged kinetic and potential energies of the conservative system (Steele and Taber [8]). By using the "time-averaged Lagrangian" for the conservative system the larger phase excursion issue in V_{BM} and P_{SV} of the previous research was resolved.

2.1 BM velocity with feed-forward (push) / feed-backward (pull) mechanism

The details of BM velocity calculation using "time-averaged Lagrangian" are in the Steele and Taber [8]. The boundary conditions of matching the averaged fluid volume displacement at the stapes (Eq. (1)) and zero pressure at the helicotrema are taken into account for V_{BM}.

$$Q\,|_{stapes} = A_{st}\delta_{st} = \overline{U} L_1 L_2\,|_{x=0} \tag{1}$$

where Q is fluid volume displacement, A_{st} is stapes area, δ_{st} is stapes displacement, U is the fluid displacement, and L_1, L_2 is width and height of the fluid chamber.

For the push-pull active model, the arrangement of the OC is indicated in Figure 1. If the OHCs expand, then the Deiters rod at the horizontal distance $x + \Delta x_1$ will be pushed down by the OHC with the apex at x, and pulled up by the phalangeal process connected to the OHC with the apex at $x + \Delta x_2$. The force in an OHC is proportional to the shear force on the cilia at the apex, which is approximately equal to the total force acting on the BM. This is quantified by writing the total force on the BM as:

$$F_{PZ} = 2F_{BM}^{f} + F_{BM}^{C} \tag{2}$$

in which F_{BM}^{C} is the force due to the OHCs and F_{BM}^{f} is the force due to the fluid from one side.

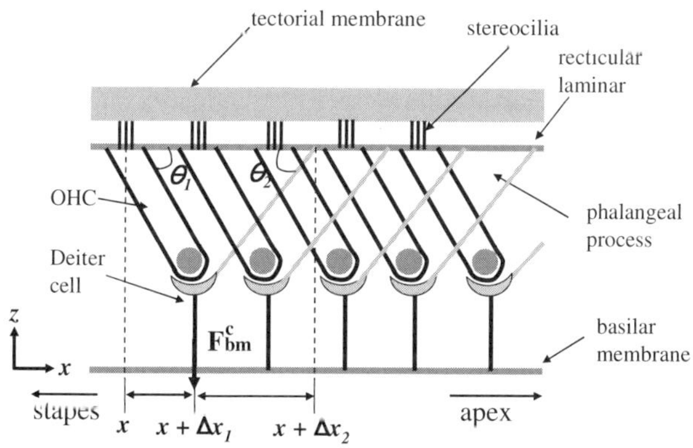

Figure 1. Schematic of the longitudinal view of organ of Corti, showing the tilt of the outer hair cells. For one hair cell with apex at the distance x, the base is at $x + \Delta x_1$, while the phalangeal process connected at the base attaches to the upper surface at $x + \Delta x_2$. θ_1: OHC angle, θ_2: phalengeal process angle with recticular laminar. The force on the BM through the Deiters' rod is F^c_{bm}, which consists of the downward push from an expansion of the hair cell at x and an upward pull through the phalangeal process from an expansion of the hair cell at $x + \Delta x_2$.

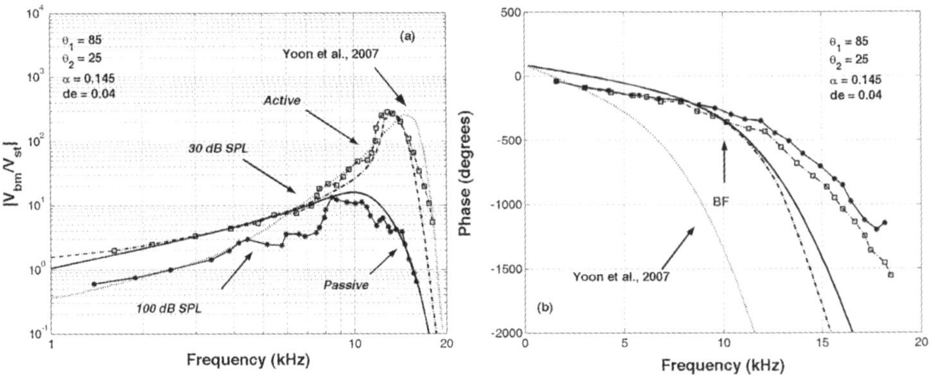

Figure 2. Basilar membrane (BM) velocity relative to the stapes (a) magnitude and (b) corresponding phase for the gerbil cochlea at 4.2 mm from the base (BF = 10 kHz). For the active model, 0.145 feed-forward and backward gain factor was used. Experimental data (expt.) are included for comparison (Ren and Nuttall [1]).

The gerbil cochlear BM velocity magnitude and phase for 4.2 mm from the base (BF = 10 kHz) relative to the stapes displacement are computed up to 20 kHz (Figure 2(a) and 2(b)). Model results are compared with the gerbil experimental data (Ren and Nuttall [1]). Active model results are calculated at a low stimulus level (30 dB SPL at the ear canal); θ_1 = 85° (OHC angle), θ_2 = 25° (phalangeal process angle), α = 0.145 (feed-forward and backward gain factor). The passive and active model shows qualitatively and quantitatively good agreement with data (Figure 2).

In the relative BM velocity magnitude plot (Figure. 2(a)), BF place shifts from 10 kHz (passive model) to 13 kHz (active model) both in the simulation and measurement. Overestimated BF place shift in the previous feed-forward active model (Yoon *et al.* [7]) is resolved by using push-pull mechanism in active model. From the parameter study, we find that the BF place shifts to the low frequency by using small phalangeal process angle in the active case.

Figure 2(b), the phase shows good agreement with measurement up to BF place. This is the most impressive improvement in the current time-averaged Lagrangian model, when we compare to the 2.5 cycle phase difference in the previous model (Yoon *et al.* [7]).

2.2 *Intracochlear pressure*

For the intracochlear pressure calculation, the boundary conditions of zero total pressure (p_{total}) at round window and the averaged total pressure at x=0 has to be equal to the prescribed total pressure at the stapes (p_{stapes}) (Eq. (3)).

$$\overline{p_{total}}\,|_{x=0} = p_{stapes}\,. \tag{3}$$

Gerbil scala tympani intracochlear pressures for the passive and active case at 2.6 *mm* from the stapes (BF = 15 kHz) and 10 μm away from the BM are shown in

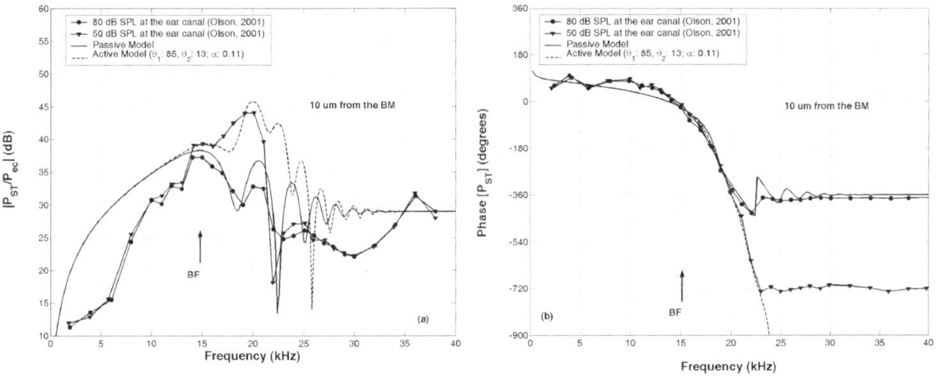

Figure 3. Combined slow and fast wave intracochlear pressure in the scala tympani (ST) of the gerbil. (a) Intracochlear pressure magnitude and (b) corresponding phase at 2.6 mm from the stapes (BF = 15 kHz) and 10 μm away from the BM (80 dB SPL: passive & 50 dB SPL: active case). Data are from Olson [2], expt. 9-8-98-I-usual.

Figure 3. Model results are compared with Olson's experimental data [2] of expt. 9-8-98-I-usual. Active model results are calculated at a low stimulus level (50 dB SPL at the ear canal); $\theta_1 = 85^\circ$ (OHC angle), $\theta_2 = 13^\circ$ (phalangeal process angle), $\alpha = 0.11$ (feed-forward and backward gain factor). In the magnitude (Figure 2(a)), the model results and measurement data shows good agreement in (i) nonlinearity with 10 dB gain from the OHCs, (ii) 2/3 octave BF shift in the active case, and (iii) less peaks and valleys than previous results in Yoon *et al.* [7]. From Figure 2(b), phase from the model shows good agreement with experimental data. The phase excursion issue in the previous model (Lim and Steele [6]; Yoon *et al.* [7]) is resolved by taking the time-averaged Lagrangian method.

2.3 *Cochlear input impedance*

With the present model, the acoustic input impedance (Z_c) of the cochlea can be obtained by taking the ratio of P_{sv} which consist both the fast and the slow wave to the averaged $\mathring{U}$ in the SV fluid behind the footplate (Eq. (4)). Z_c for the gerbil and chinchilla cochlea are calculated and compared with measurements (Decraemer *et al.* [3], Songer and Rosowski [4], Ruggero *et al.* [5]) and theoretical result from Puria and Allen [9] (Figure (4)).

$$Z_c = \left(\frac{P_{SV}}{\dot{U}} \right)_{x=0} \tag{4}$$

Z_c magnitude from the gerbil and chinchilla (Figure 4(a) and 4(b)) show good agreement with recent animal measurements. Z_c phase from the model remains near zero while the measurements shows above zero degree. However, Figure 4(b) measurements from Decraemer et al. [3] show that the angle is mostly equal to values close to 0, 1, or 2 cycle which represents zero degree in phase. For stimulus frequencies below 10 kHz, it is observed that the gerbil Z_c is 5 dB larger than the chinchilla Z_c in magnitude. For the most part, the chinchilla scalae area is greater than gerbil and this result corresponds to the inverse relationship between Z_c magnitude and the scale area (Puria and Allen [9]).

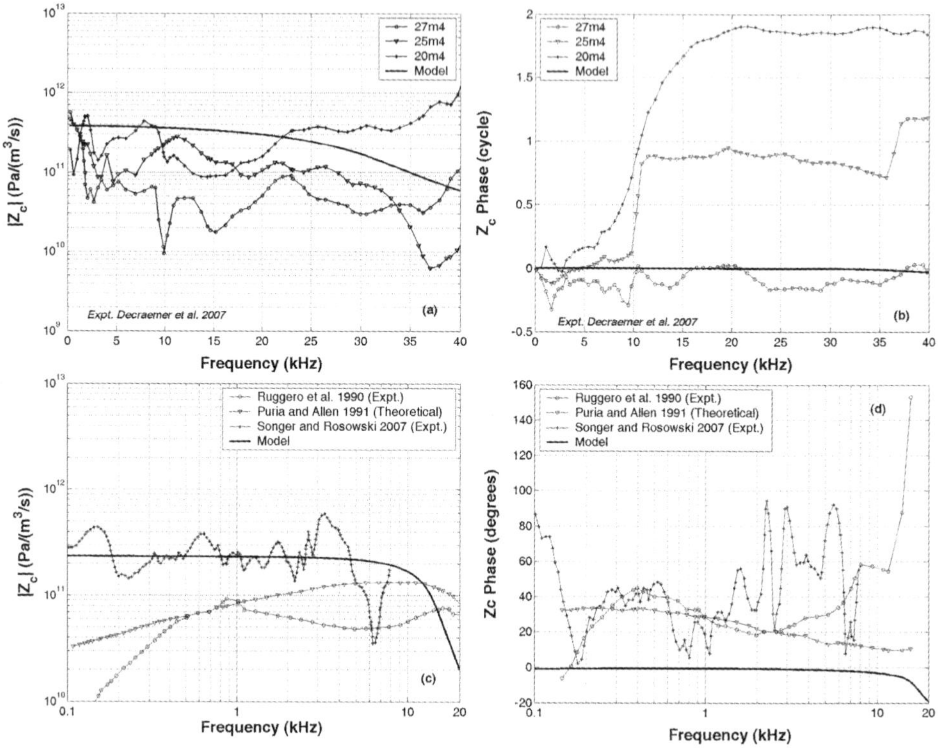

Figure 4. Cochlear input impedance Zc of gerbil [(a) amplitude; (b) phase] and chinchilla [(c) amplitude; (d) phase]. Model results (solid) were compared with experimental and theoretical Zc. Gerbil data are from Decraemer *et al.* [3] and Chinchilla experimental data are from Songer and Rosowski [4] and Ruggero *et al.* [5] and theoretical results are from Puria and Allen [9].

3 Discussion

New anaomically based three-dimensional hydro-dynamic cochlear model resolved the phase excursion issue by taking the time-averaged Lagrangian method and BF shift discrepancy in active model by implementing push-pull mechanism. The intracochlear pressure and cochlear input impedance for the gerbil and chinchilla cochlea were calculated and showed good agreements with animal experiments. This gives support to our proposition that the present model is close to the actual behavior of the gerbil and chinchilla cochlea.

Acknowledgments

This work is supported by grant no. R01 - DC007910 from the NIDCD of NIH.

References

1. Ren, T., Nuttall, A. L., 2001. Basilar membrane vibration in the basal turn of the sensitive gerbil cochlea. Hear. Res. 151, 48-60.
2. Olson, E. S., 2001. Intracochlear pressure measurements related to cochlear tuning. J. Acoust. Soc. Am. 110 (1), 349-367.
3. Decraemer, W. F., Khanna, S. M., de La Rochefoucauld, O., Dong, W., Dirckx, J. J. J., Olson, E. S., 2007. Scala vestibuli pressure and three-dimensional stapes velocity measured in direct succession in gerbil. J. Acoust. Soc. Am. 121 (5): 2774–2791.
4. Songer, J. E., Rosowski, J. J., 2007. Transmission matrix analysis of the chinchilla middle ear. J. Acoust. Soc. Am. 122 (2), 932–942.
5. Ruggero, M. A., Rich, N. C., Robles, L., Shivapuja, B. G., 1990. Middlear response in the chinchilla and its relationship to mechanics at the base of the cochlea. J. Acoust. Soc. Am. 87 (4), 1612–1629.
6. Lim, K. M., Steele, C. R., 2002. A three-dimensional nonlinear active cochlear model analyzed by the WKB-numeric method. Hear. Res. 170, 190-205.
7. Yoon, Y. J., Puria, S. Steele, C. R., 2007. Intracochlear pressure and derived quantities from a three-dimensional model. J. Acoust. Soc. Am. 122 (2), 952–966.
8. Steele, C. R., Taber, L. A., 1979. Comparison of WKB calculations and experimental results for three-dimensional cochlear models. J. Acoust. Soc. Am. 65 (2), 1007–1018.
9. Puria, S., Allen, J. B., 1991. A parametric study of cochlear input impedance. J. Acoust. Soc. Am. 89 (1), 287–309.
10. Lynch, T. J., Nedzelnitsky III, V., Peake, W. T., 1982. Input impedance of the cochlea in cat. J. Acoust. Soc. Am. 72, 108–130.
11. Aibara, R., Welsh, J. T., Puria, S., Goode, R. L., 2001. Human middle-ear sound transfer function and cochlear input impedance. Hear. Res. 152 (1-2), 100-109.

Comments and Discussion

Olson: It would be nice to have a bit of an "intuitive" explanation for why the different computational methods give substantially different phase excursions. The phase represents wavelength (or wavenumber) which is mainly determined by BM + OC stiffness and fluid mass. Comparing the previous and present work, is the stiffness the same? And if yes, is it appropriate to say the previous computational method found a fluid mass that was too big (leading to large wavenumber, small wavelegth, and too much phase excursion)?

Puria: The huge part of the phase problem is that the amplitude function developed by previous work (Lim and Steele, 2002, Yoon et al. 2007) was not the best. They took the averaged value of Laplace's equation, while it is better to use a consideration of work for a conservative system. This leads to the time-averaged Lagrangean. The previous method had a simple derivation which provides the 2nd order wave equation without an assumption of sinusoidal waveforms. What they used is similar to a Gelerkin procedure, but with the only weight function of 1. The amplitude function has little dependence on the viscosity, and should provide the energy balance, i.e., work in at one section should be equal to the work out at another section.

Shera: In addition to the intuitive understanding, it would be nice to know, at a more technical level, where and why the previous computational method went wrong. The previous method makes certain assumptions and approximations - which one(s) turn out not to be good enough for this problem, and why?

Puria: As you know, the WKB solution for the wave equation consists of the rapidly-varying exponential term with a "phase function" and a slowly-varying "amplitude function". Without damping, the total amplitude is given by the amplitude function, but with damping, the phase acquires a small imaginary part which has a dominant effect on the total amplitude, i.e., the apical roll-off of cochlear response for a given frequency. However, the slowly-varying amplitude function still has an effect on the location of the peak response. These phase and amplitude functions can be obtained by a formal asymptotic expansion procedure for the usual wave equation. However, Whitham (Jimenez and Whitham 1976) developed the "time-averaged Lagrangean" for the conservative system (without damping) which yields a variational formulation from which the amplitude function can be obtained more readily in terms of the derivative with respect to wavenumber of the phase-frequency relation (the eikonal equation). This result for the amplitude function satisfies the basic condition of conservation of energy, i.e., over an arbitrary distance, the power input at one end is equal to the power taken out at the other end. When damping is included, energy is not conserved, but the amplitude function from the derivative of the eikonal equation remains a good approximation. Recently, however, we observed that a direction integration of the three-dimensional Navier-Stokes equation yields the attractive result of a one-dimensional wave equation without the direct assumption of the exponential form (Lim and Steele 2002). Consequently, a direct numerical integration was used in the very long wavelength region and the WKB approximation used for the rest of the cochlea.

More recently we realized that a significant difference in the amplitude function exists with the problem getting worse as one moves apically, and the time-averaged Lagrangean gives a better result. A substantial theoretical improvement is obtained by using the Maxwell-Betti reciprocity relation which can be extended to the system with feedback (feed-forward and feed-backward) of energy by using the adjoint wave. So far it seems that the amplitude function from the reciprocity and from the time-averaged Lagrangean are nearly the same. Another view is that we wish for one condition to obtain one amplitude function. In a Gelerkin procedure the exact equation is multiplied by weighting functions to obtain approximations. In the one case the one weighting function is just unity, while the reciprocity provides a more complex weighting and seems to be better. For the same BF, there is a small shift in stiffness, and we find that additional mass corresponding to the thickness of the gerbil basilar membrane is needed for best agreement with the measurements.

A LUMPED-ELEMENT MODEL OF THE APICAL COCHLEA AT LOW FREQUENCIES

TORSTEN MARQUARDT
UCL Ear Institute, University College London, 332 Gray's Inn Road, London, WC1X8EE, UK

JOHANNES HENSEL
Physikalisch-Technische Bundesanstalt, WG 1.61, Bundesallee 100, D-38116 Braunschweig

1 Summary

Below a certain stimulus frequency, the traveling wave reaches the end of the basilar membrane (BM) and differential pressure across the BM is shunted through the helicotrema. Dallos [1] showed that in cat and chinchilla this oscillatory perilymph flow through ducts and helicotrema is inertia-dominated. In contrast, in cochleae of guinea pig and kangaroo rat, which exhibit more turns, more tapering, and a smaller helicotrema, this perilymph movement is dominated by viscous friction. Recently, we developed a non-invasive DPOAE-based technique to measure the forward middle-ear transfer function (fMETF) for low frequencies, and were able to compare guinea pig and human cochleae [2]. The results support Dallos' anatomy-based conjecture that the human cochlea should be inertia dominated at lowest frequencies. In the frequency range of transition between resistive traveling wave and shunting helicotrema, we observed pronounced resonances in both species, indicating oscillatory interactions between BM stiffness and perilymph inertia. Although Dallos' fMETF of all four species also showed such resonances, his discussion was focused on their slopes. His 1st-order model did not include compliance, and therefore did not account for the observed oscillation. Similar irregularities in fMETF of guinea pigs were also observed by Décory *et al.* [3]. Because the cochlear impedance of guinea pigs is apparently not inertia-dominated Décory's first-order model had no mass element, and yields only a step in the fMETF, instead of a resonance. Puria and Allen's transmission line model [4] shows resonances, which they interpreted as standing waves along the BM. Here, we show that also a simple lumped-element model can account for the resonance present in both, viscosity-dominated and inertia-dominated fMETFs, given it is of second order, i.e. incorporating compliance and mass elements.

Figure 1 shows the electrical equivalent circuit of the apical cochlear acoustics. The middle ear is modeled by a compliance element (C_{ME}) because the frequencies considered here are well below its resonance frequency. R_{BMI} represents the resistive traveling-wave impedance. C_{BMI} sets its lower cutoff frequency. M_{peri} represents perilymph inertia, and prevents acoustic energy from traveling beyond the characteristic place. During low-frequency stimulation, however, M_{peri} impedes less than C_{BMI}, and acoustic energy

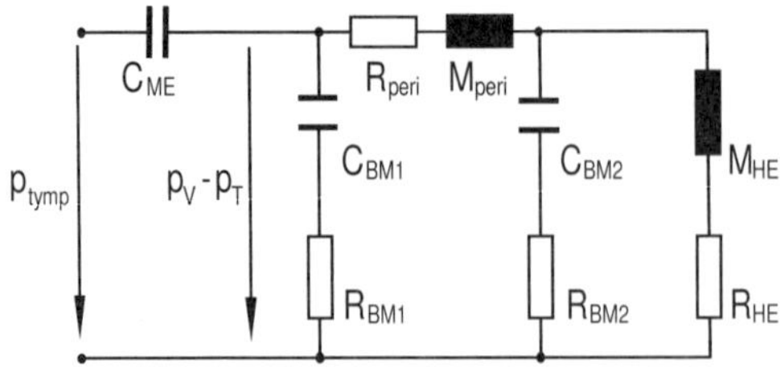

Figure 1. Model of the apical cochlea.

	Guinea pig	Human
C_{ME}	$8.3 \cdot 10^{-16}\ m^5N^{-1}$	$5 \cdot 10^{-16}\ m^5N^{-1}$
C_{BM1}	$1.3 \cdot 10^{-14}\ m^5N^{-1}$	$3.3 \cdot 10^{-14}\ m^5N^{-1}$
R_{BM1}	$7 \cdot 10^{10}\ Nsm^{-5}$	$13 \cdot 10^{10}\ Nsm^{-5}$
R_{peri}	$50 \cdot 10^{9}\ Nsm^{-5}$	$1 \cdot 10^{9}\ Nsm^{-5}$
M_{peri}	$7 \cdot 10^{7}\ Ns^2m^{-5}$	$30 \cdot 10^{7}\ Ns^2m^{-5}$
C_{BM2}	$4 \cdot 10^{-14}\ m^5N^{-1}$	$10 \cdot 10^{-14}\ m^5N^{-1}$
R_{BM2}	$3 \cdot 10^{9}\ Nsm^{-5}$	$15 \cdot 10^{9}\ Nsm^{-5}$
M_{HE}	$5 \cdot 10^{7}\ Ns^2m^{-5}$	$12 \cdot 10^{7}\ Ns^2m^{-5}$
R_{HE}	$9 \cdot 10^{9}\ Nsm^{-5}$	$1 \cdot 10^{9}\ Nsm^{-5}$

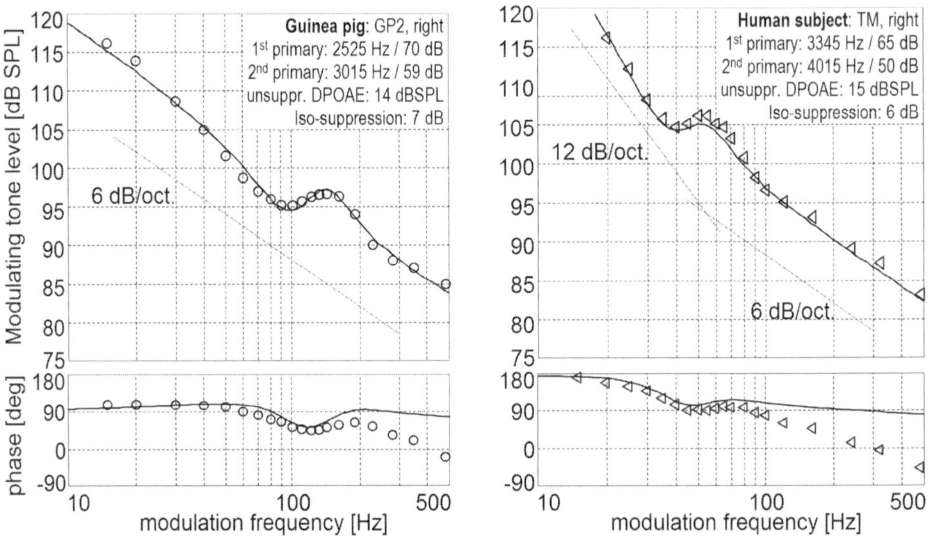

Figure 2. Inverted fMETF obtained from guinea pig (o) and human cochlea (◁) by iso-suppression of DPOAE [2]. Magnitudes of the model results (solid) are vertically offset to provide alignment with the physiological data.

reaches the apical BM (C_{BM2} and R_{BM2}). For lowest frequencies, even the most compliant apical end of the BM is too stiff (C_{BM2}), and perilymph is forced through the helicotrema. The ratio between inertia (M_{peri} , M_{HE}) and viscous friction (R_{peri} , R_{HE}), related to this periymph movement, is species-dependent and determines the fMETF slope at those frequencies [1].

Inverted fMETFs were obtained experimentally [1,2] by adjusting the stimulus level so as to evoke constant BM displacement amplitude at each stimulation frequency. BM displacement was measured far basal from the helicotrema, where it is stiffness-dominated to low-frequency stimulation, and therefore proportional to the pressure difference across the BM (p_{SV} - p_{ST}). The fMETF is here defined as the ratio between this pressure difference and the pressure at the tympanum (p_{tymp}). Examples of *1/fMETF* from human and guinea pig cochleae are shown together with our model results in figure 2.

The fact that the damping of the resonance can be independently adjusted (R_{BM2}) from the slope of the fMETF below the resonance (R_{peri}) indicates that only the apical BM is part of the resonating system.

Acknowledgment

The presentation was kindly supported by a Deafness Research UK Travel Grant.

References

1. Dallos, P., 1970. Low-frequency auditory characteristics: Species dependence. J. Acoust. Soc. Am. 48, 489–499.
2. Marquardt, T., Hensel, J., Mrowinski, D., Scholz, G., 2007. Low-frequency characteristics of human and guinea pig cochleae. J. Acoust. Soc. Am. 121,
3. Décory, L., Franke, R.B., Dancer, A.L., 1990. Measurements of the middle ear transfer function in cat, chinchilla and guinea pig. In: Dallos et al. (Eds.), The mechanics and biophysics of hearing, Springer, New York, pp.
4. Puria, S., and Allen, J. B., 1991. A parametric study of cochlear input impedance. J. Acoust. Soc. Am. 89, 287–309.

COCHLEAR MECHANICS: A SIDEWAYS LOOK

A. R. GARDNER-MEDWIN
Dept. Neuroscience, Physiol. & Pharmacol., University College London, London WC1E 6BT, UK

This short paper sets out some simple issues and questions that arise in viewing the cochlear mechanics literature. Given the complexity of the literature and of many current models, the approaches here may risk seeming simplistic. But the challenge of trying to dismiss them or produce more clear answers may be worthwhile.

1 The extent of intrinsic damping

Gold's model [1] of viscous damping of a miniature piano string in aqueous fluid is not very appropriate for the basilar membrane (BM). Passive damping of the resonance of a small elastic region separating two aqueous chambers may actually be much more consistent with observed critical bandwidths [2]. This raises the possibility that active mechanisms, combined with what looks rather like an enhancement of viscous damping due to the structural features of the organ of Corti, may serve to modulate damping and provide a trade-off between frequency selectivity and transient resolution, rather than simply to counter viscosity. It is interesting in this context that a raised kinematic viscosity was found necessary in a physical model of the cochlea to eliminate unphysiological standing waves [3].

2 The nature of the inertial mass

The density of most BM structures cannot greatly exceed that of the fluids, so the effective inertial mass in BM resonance is substantially a function of the pattern of surrounding fluid flow. Contributions to effective mass are proportional to the square of local fluid velocity, and for divergent flow are therefore dominated by the varying size and shape of the local zones of maximum displacement around the BM and within a travelling wave [2]. This introduces non-linearities and means that local effective mass at a point on the BM is a variable depending on local wavelength. It falls (and the local maximally sensitive frequency rises) if there is a travelling wave slowing down as it approaches resonance at a slightly more apical point. This approach suggests a possible way to handle effective mass in a 1-D model, complementary for example to the more thorough analysis in 2-D models by de la Rochefoucauld and Olson [4].

3 The consequence of Outer Hair Cell (OHC) depolarisation

Though BM displacement towards scala media is generally agreed to cause OHC depolarisation, it seems unclear whether the result of this depolarisation and consequent shortening of OHCs will shift fluid in the same or opposite direction to the stimulus.

Even an attempt to answer this directly [5] seems equivocal, since imposed current across the BM into scala media will (as observed) have depolarised the apical OHC membranes, but will simultaneously have hyper-polarised the basal portions to an uncertain extent.

4 The role of active negative stiffness

It is tempting to regard negative stiffness of BM structures (in which an active internal mechanism produces force in the same direction as displacement [6,7]) as providing amplification. The internal force adds to the effect of external force at frequencies below resonance. However, without a phase shift this internal force (in phase with the displacement) does not do work and add energy to the system: the increased energy of displacement comes from greater work done by the external force. This is the opposite of the active energy feed into a travelling wave that is required to produce cochlear amplification. A phase delay is in fact inevitable as soon as such a mechanism operates at increased frequency, but with a negative active stiffness this leads to extraction of energy from the system, similar to viscous energy dissipation. To achieve negative damping and energy feed into a BM disturbance (true amplification) an active force must act with positive stiffness and a phase delay $\leq \pi$ (optimally $\pi/2$). Active negative stiffness, where it occurs, may have a different physiological role. The benefits from positive active stiffness (which - subject to the uncertainties of the previous paragraph - may be generated by OHC prestin mechanisms) can arise in two ways [8]: firstly by reducing the energy loss from a travelling wave as this passes over basal regions of the BM that are actively stiffened for frequencies below local resonance, and secondly by feeding energy into the wave as the resonant point is approached, due to an increasing phase lag.

5 The significance of OHC membrane capacitance

In considering the role of OHC membrane potential (V_m) and prestin in cochlear amplification, attenuation of the changes of V_m at audio frequencies (due to membrane capacitance) is often considered a problem. However, if the effect of OHC motility is to oppose BM displacement at low frequencies (positive stiffening), then it is this capacitance that gives the phase lag needed to add energy to the travelling wave and counter viscous damping at frequencies approaching resonance. If a local passive resonance has frequency f and intrinsic damping with rate constant k for energy dissipation (with $Q_{passive}=2\pi f/k$), then for OHCs to give rise to high gain and high Q, their ratio of active stiffness to local passive stiffness (β) at low frequencies must be at least $2/Q_{passive}$. In addition they must have a membrane capacitance sufficient that the time constant (τ_m =RC) is approximately β/k for large β. Given these conditions, less capacitance would risk spontaneous oscillation while lower membrane resistance (as may result from large stimuli or efferent activation) lowers gain and Q. The roll-off of OHC response at audio frequencies above the 3dB cut-off frequency ($1/(2\pi\tau_m)$) is not critical, since the optimal operating conditions may be well above this, by a factor $\beta Q_{passive}$.

References

1. Gold, T., 1948. Hearing II. The physical basis of the action of the cochlea. Proc. Roy. Soc. B, 135, 492-8.
2. Gardner-Medwin, A.R., 2006. Viscous damping of acoustic resonance with a restricted zone of wall compliance. Proc. Physiol. Soc. 3, DC2 [also 9].
3. White, R.D., & Grosh, K., 2005. Microengineered hydromechanical cochlear model. PNAS 102, 1296-1301.
4. de la Rochefoucauld, O., & Olson, E.S., 2007. The role of organ of Corti mass in passive cochlear tuning. Biophys. J. 93,3434-50.
5. Mammano, F., & Ashmore, J.F., 1993. Reverse transduction measured in the isolated cochlea by laser Michelson interferometry. Nature 365, 838-41.
6. Martin, P., Mehta, A.D., & Hudspeth, A.J., 2000. Negative hair-bundle stiffness betrays a mechanism for mechanical amplification by the hair cell. PNAS 97, 12026-31.
7. Kennedy. H.J., Crawford, A.C., & Fettiplace, R., 2006. Force generation by mammalian hair bundles supports a role in cochlear amplification. Nature 433, 880-3
8. Gardner-Medwin, A.R., 2007. OHC motility as a means to concentrate delivery of cochlear stimulus energy. Int. J. Audiology 46, 652 [also 9].
9. see further details at: http://www.ucl.ac.uk/~ucgbarg/pubres.htm#cochlea.

NONLINEAR RESPONSES OF A NONLINEAR COCHLEAR MODEL WITH THE FUNCTION OF AN OUTER HAIR CELL MODEL

Y. MURAKAMI, M. UNOKI

School of Information Science, Japan Advanced Institute of Science and Technology
1-1 Asahidai, Nomi, Ishikawa 923–1292 Japan

To investigate how outer hair cells (OHCs) produce compressive nonlinearity in both the cochlear I/O function and tuning curve for a single tone, we present a nonlinear cochlear model with a nonlinear OHC model. In modeling cochlea filtering, we modeled somatic motility as a function of OHCs and the interaction between the basilar membrane and somatic motility through the tectorial membrane as mechanoelectrical transducer networks. The parameter values of the model were set to the estimates for human data. Signal frequencies of 0.125, 0.25, 0.5, 1, 2, 3, 4, and 6 kHz were used in the simulations of cochlear filtering. The results revealed that this model can account for the compressive nonlinearity in both the I/O functions and tuning curves of a cochlea obtained from the experiments. They also suggest that the somatic motility depending on the transducer currents produces nonlinearities in the I/O functions and tuning curves of cochlea.

1. Introduction

Cochlear nonlinearities such as compression, two-tone suppression, and distortion products have been found in physiological and psychoacoustical studies. Compression in the cochlear I/O function has especially been discussed in both physiological [1, 2], and psychological studies [1, 3, 4, 5]. These studies have shown that outer hair cells (OHCs) play a significant role in producing compressive nonlinearity. It has been considered that the somatic length changes in the OHCs depending on the membrane potential, i.e., the somatic motility of the OHCs [6], are involved in producing compression. However, it has not yet been completely revealed how the somatic motility of the OHCs produces compressive nonlinearity. Some typical cochlear models have assumed that the interaction between the basilar membrane (BM) and the somatic motility of the OHCs through the tectorial membrane (TM) produces compression [7, 8]. Although compressive nonlinearities obtained from these models are consistent with experimental results at lower input levels, broken-stick nonlinearity in the cochlear I/O functions [1, 4] and the nonlinear characteristics of the tuning curves of the BM have not yet been clearly discussed. We present a nonlinear cochlear transmission-line model in this paper, into which a nonlinear OHC model based on the findings of somatic motility has been incorporated, that enables us to account for the experimental results with regards to nonlinearities in the I/O functions and tuning curves. We then discuss how the somatic motility of the OHCs produces compression and tuning as cochlear nonlinearities by conducting simulations of the cochlea model for single-tone responses.

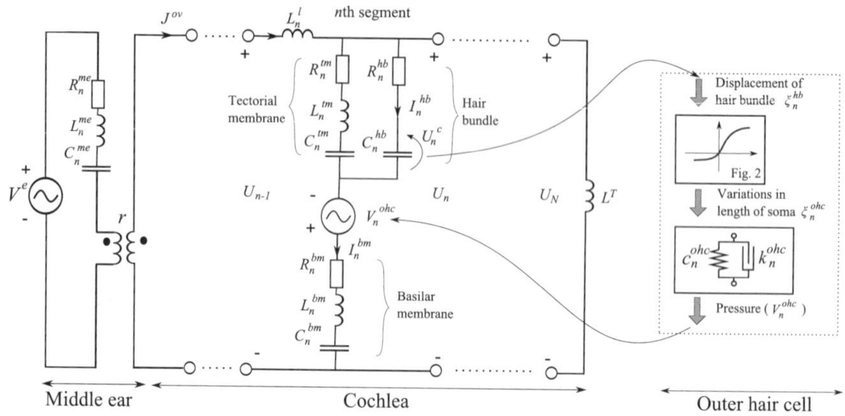

Fig. 1. Electroacoustic transmission-line networks of auditory peripheral system: middle ear, Basilar membrane, tectorial membrane, and outer hair cell models.

2. Methods

The computational auditory peripheral model we used was constructed as an electroacoustic transmission-line model, based on the implementations of Giguére and Woodland [9], our previous work [10], and Neely and Kim [11]. The overall structure of the model consists of a middle-ear model and a cochlear model, as shown in Fig. 1. Since this model is used to account for both physiological and psychoacoustical data, an outer ear model was not incorporated into this.

The sound pressure at the eardrum and the stapes volume velocity in the middle-ear model were represented as voltage source, $V^{\mathrm{e}}(t)$, as input, and current, $J^{\mathrm{ov}}(t)$, as output. The impedance of the middle ear was represented as the impedance of an RLC circuit, as shown at the left of Fig. 1. An ideal transformer represents the effective acoustic transform ratio between the eardrum and the oval window.

The cochlear model was spatially discretized into N small segments according to cochlear length. This model was transformed from Neely and Kim's [11] mechanical model into an electroacoustic transmission-line model, as shown in the middle of Fig. 1. A small segment, Δx, is provided by dividing the cochlear length l^{bm} by $N-1$. The input of the cochlear model is the current, $J^{\mathrm{ov}}(t)$, transmitted from the middle-ear model. The voltage difference at the nth segment $U_n(t)$ is analogous to the pressure difference between the scala vestibule (SV) and the scala tympani (ST). Each segment represents a micromechanical model and this is coupled by inductance L^l that represents the acoustic mass of the lymph fluid (LF). The lower parts in the segments model the BM. Shunt current $I_n^{\mathrm{bm}}(t)$ represents the volume velocity of the BM. The left and right paths in the upper part of the segments model the TM and the hair bundle (HB). Inductance L, capacitance C, and resistance R represent the acoustic mass, compliance, and resistance in each part:

$$L = \frac{m}{W\Delta x}, \qquad C = \frac{W\Delta x}{k}, \qquad R = \frac{c}{W\Delta x}, \tag{1}$$

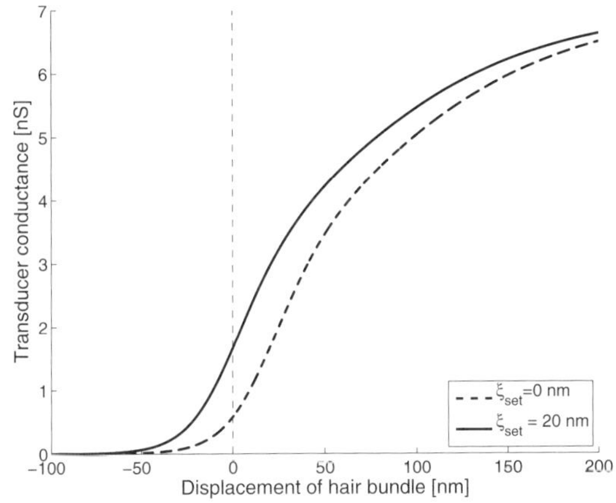

Fig. 2. Transducer conductance of OHCs as function of displacement of HB as given in [13]. Function $G_{\rm tr}$ was fitted by 2nd-order Boltzmann function. ξ_{set} is HB displacement in resting state.

where m, k, and c are the mass, stiffness and damping per unit area, n. W is the width of the scala. The apical side of the model is terminated by inductance L^T that represents the acoustic mass of the helicotrema. The suffices, "bm", "tm", and "hb", were used to represent the parameters in the respective parts.

The input of the OHC model is the displacement of the HB $\xi_n^{hb}(t)$ that is proportional to the voltage drop at the nth segment, $U_n^c(t)$, across the capacitance, $C_n^{\rm hb}$, from electroacoustic analogies in a one-dimension model. However, the two-dimensional motion of the TM models a large shearing displacement in the HB on the basal side of the cochlea rather than one-dimensional motion. Therefore, the shearing displacement in the HB at the cochlea length of x is defined as

$$\xi_n^{\rm ha} = \frac{C_n^{\rm ha} U_n^c}{W \Delta x} 40 e^{-2.5x}. \tag{2}$$

It has been assumed that the process of interaction between the BM and OHC motility is divided into five stages to enable the physiological behaviors of the OHCs [12] to be schematically illustrated. Based on this assumption, we hypothesized that the dominant nonlinearity of the cochlear would cause mechano-electro transformation. This transformation, $\rm G_{tr}$, can be approximated by the 2nd-order Boltzmann function, as shown in Fig. 2 [13]. The length changes in the OHC's soma $\xi_n^{\rm ohc}(t)$ can be represented as the proportion to the approximated transformation:

$$\xi_n^{\rm ohc}(t) = \alpha_{\rm tr} \left({\rm G_{tr}}(\xi_n^{\rm hb}(t) + \xi^{\rm set}) - {\rm G_{tr}}(\xi^{\rm set}) \right), \tag{3}$$

where $\xi^{\rm set}$ is the resting displacement of the HB and $\alpha_{\rm tr}$ is chosen as $\xi_n^{\rm ohc}(t) = \xi_n^{\rm hb}(t)$ at low vibration amplitude. Voltage source $V_n^{\rm ohc}$ that represents the pressure produced by the somatic motility of the OHC model is

$$V_n^{\rm ohc}(t) = \gamma \left(c^{\rm ohc} \dot{\xi}_n^{\rm ohc}(t) + k^{\rm ohc} \xi_n^{\rm ohc}(t) \right), \tag{4}$$

Table 1. Parameter values of electroacoustic transmission-line model. Other parameter values used for mechanical and mechano–electro transducer functions are same as in [11, 13]. x is distance from the stapes.

$L^{\rm bm} = 3 \times 10^{-3}/W\Delta x$ (H)	$L^{\rm tm} = 0.5 \times 10^{-3}/W\Delta x$ (H)
$R^{\rm tm} = 4.4e^{-1.65x}/W\Delta x$ (Ω)	$R^{\rm bm} = 6 + 670e^{-1.5x}/W\Delta x$ (Ω)
$C^{\rm bm} = W\Delta x/2.2 \times 10^{8}e^{-3x}$ (F)	$C^{\rm tm} = W\Delta x/1.4 \times 10^{6}e^{-3.3x}$ (F)
$R^{\rm hb} = 0.8e^{-0.6x}/W\Delta x$ (Ω)	$c^{\rm ohc} = 440e^{-1.5x}$ (dyn $\cdot$ s $\cdot$ cm^{-3})
$C^{\rm hb} = W\Delta x/2 \times 10^{6}e^{-3}$ (F)	$k^{\rm ohc} = 1.1 \times 10^{8}e^{-3x}$ (dyn $\cdot$ cm^{-3})
$l^{\rm bm} = 3.5$ cm	$\xi^{\rm set} = 2 \times 10^{-6}$ (cm)
$N = 512$	$\gamma = 0.72$

Table 2. Comparison of slopes of I/O functions (dB/dB) of proposed model with physiological data [1] and psychological data [3, 4, 5]. Range of input level is with minimum value for slope at each 10 dB for physiological results and this is from 30 to 80 dB for psychological results.

Signal frequency (Hz)	125	250	500	1000	2000	3000	4000	6000
Rhode (2007) [1]	-	-	-	-	-	-	-	~0.25
Proposed model (*at* CF)	0.46	0.32	0.27	0.25	0.24	0.24	0.25	0.25
Glasberg and Moore (2000) [3]	-	0.73	0.70	0.39	0.56	-	0.57	-
Unoki *et al.* (2006a) [4]	-	0.62	0.50	0.42	0.37	0.36	0.35	0.34
Unoki *et al.* (2006b) [5]	0.91	0.52	0.43	0.32	0.32	-	0.33	0.32
Proposed model (*at* BF)	0.73	0.50	0.40	0.37	0.36	0.36	0.37	0.43

where γ is a gain factor, and $c^{\rm ohc}$ and $k^{\rm ohc}$ are the damping and stiffness of the soma and the connection of the supporting cell per unit area. At low vibration amplitude ($\xi_n^{\rm ohc}(t) = \xi_n^{\rm hb}(t)$), this cochlear model is equivalent to that of Neely and Kim.

The parameter values of the transmission-line model we used were set as estimates for human data obtained from experimental animal data using Greenwood's function [14]. These are listed in Table 1. We investigated the cochlear I/O functions and the equal magnitude contours obtained from the model we presented.

3. Results

Table 2 lists the slopes of the I/O functions of the model we presented and compares the degree of compression with physiological [2] and psychological studies [3, 4, 5]. Signal frequencies of 0.125, 0.25, 0.5, 1.0, 2.0, 3.0, 4.0, and 6.0 kHz were used to calculate the common slopes of the I/O functions corresponding to the results. The slopes of the I/O functions obtained from the model were in a range from 0.24 to 0.46 dB/dB at the characteristic frequencies (CFs), while these values at the best frequencies (BFs), which are defined by the stimuli evoking the largest vibration along the cochlear length, were approximately 0.4 dB/dB at higher frequencies (over 1 kHz). These values increased as the signal frequency decreased. These results are comparable with both the physiological and psychoacoustical results.

Figure 3(A) plots the characteristics of equal-magnitude contours at CF of 6.0 kHz as a function of the sound pressure level (from 0 to 90 dB SPL). The solid lines indicate the tuning curves obtained from the model while the dashed lines

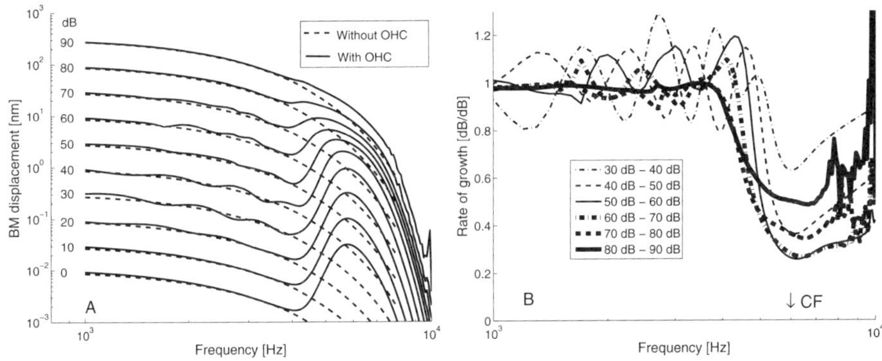

Fig. 3. Characteristics of (A) equal-magnitude contours (tuning curves) at CF= 6 kHz and (B) rate of growth function obtained from tuning curves (degree of compression) in same location.

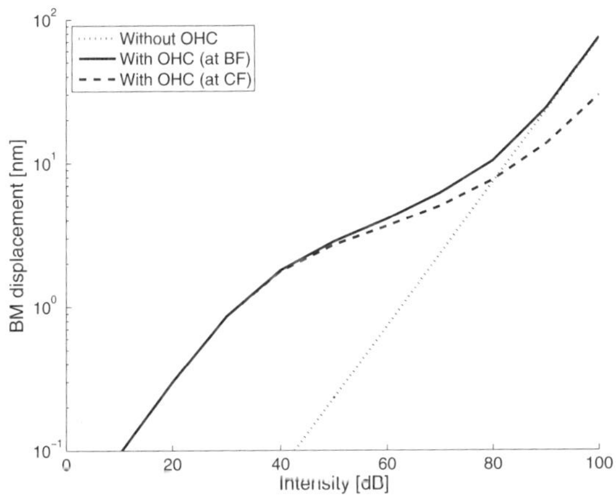

Fig. 4. Cochlear I/O functions at signal frequency of 6 kHz.

indicate those of the model without OHCs. We found that the shapes of the tuning curves obtained from the model were sharper and they were gained at lower-input levels. We also found that the tunings broadened and their gains reduced as the input level increased. Figure 3(B) plots the characteristics of the rate-of-growth (ROG) function of BM displacement. These characteristics were calculated from Fig. 3(A). At frequencies lower than the CFs, the ROG is almost equal to 1 dB/dB. The ROGs around the CFs are less than 1 dB/dB (around 0.25 dB/dB), i.e., compressive nonlinearity around the CFs. The ROGs at frequencies above the CFs are relative compressive while they are linear at frequencies lower than the CFs. This is consistent with the results observed in the physiological experiment [2].

Figure 4 plots the cochlear I/O functions obtained from the responses for a pure tone of 6 kHz. The solid and dashed lines indicate respective I/O functions at the

BFs and CFs obtained from the model. The dotted line plots the linear I/O function obtained from the model without the OHC model. The slopes of the I/O function are 1 dB/dB at lower and higher levels (under 30 dB and over 80 dB), and are lower than 1 dB/dB at moderate levels (from 30 to 80 dB). A break point in the I/O function appears at around 20–30 dB. Furthermore, the I/O functions for other frequencies have similar tendencies such as the I/O functions for 6 kHz.

4. Discussion

Figures 3(A) and 4 show how much the OHCs produced a cochlear amplifier around CFs at lower-input levels.a At these levels, the somatic motility of the OHC model behaved linearly in which the maximal displacement of the HB was less than 20 nm, as shown in Fig. 2. Furthermore, this linear behavior of the somatic motility of the OHC model resulted in producing active BM motion that passed through the TM and the supporting cells at these low levels.

Compressive nonlinearity in the I/O function appeared when the input level was just over 30 dB. At this level, the slopes of the I/O function were not completely compressive (approximately 1 dB/dB). This is because the amount of somatic motility in the OHC model had slightly reduced (maximal displacement of the HB was just over 20 nm; after this, the rise in transducer conductance increased gradually). At moderate levels (30 to 80 dB), compressive nonlinearity was seen and this was restricted around the CFs. This is because the somatic motility of the OHC model behaved nonlinearly around the location of the BM peak along the cochlear length since maximal displacement of the HB was in the range of transition, as shown in Fig. 1. At frequencies much lower than the CF, compressive nonlinearity was no longer seen since the maximal displacement of the HB was still less than 20 nm around the outside location of the BM peak so that no significant cochlear amplification (atenuation) was produced by the somatic motility of the OHC model. At higher levels (over about 80 dB), the maximal displacement of the HB was in the saturation range shown in Fig. 2 so that the somatic motility of the OHC model did not produce compressive nonlinearity and the cochlear model also then responded linearly again.

Hence, the broken-stick nonlinearity in the cochlear I/O function can be interpreted as having been produced by the transitions in linear/nonlinear behavior and the saturation of the somatic motility of the OHC model. In addition, it has been pointed out that the activity of OHCs at the base from the location of the BM peak is important for the cochlear amplifier [11].

Acknowledgments

This work was supported by a Grant-in-Aid for Scientific Research from the Ministry of Education, Culture, Sports, Science, and Technology of Japan (No. 20300064).

References

1. Bacon, S. P., Fay, R. R., and Popper, A. N., 2004. Compression, From Cochlea to Cochlear Implants, Springer, New York.
2. Rhode, S. W., 2007. Basilar membrane mechanics in the 6-9 kHz region of sensitive chinchilla cochleae, J. Acoust. Soc. Am. 121, 2792-2804.
3. Glasberg, B. and Moore, B. C. J., 2000. Frequency selectivity as a function of level and frequency measured with uniformly exciting notched noise. J. Acoust. Soc. Am., 108, 2318-2328.
4. Unoki, M., Irino, T., Glasberg, B., Moore, B. C., and Patterson, R. D., 2006. Comparison of the roex and gammachirp filters as representations of the auditory filter. J. Acoust. Soc. Am., 120(3), 1474-1492.
5. Unoki, M., Ito, K., Ishimoto, Y., and Tan, C. T., 2006. Estimate of auditory filter shape using notched noise masking for various signal frequencies. Acoust. Sci. & Tech., 27(1), 1-11.
6. Ashmore, J. F., 1987. A fast motile response in guinea-pig outer hair cells: the cellular basis of the cochlear amplifier. J. Physiol., 388, 323-347.
7. Chadwick, R. S., 1998. Compression, gain, and nonlinear distortion in an active cochlear model with subpartitions. PNAS, 95, 14594-14599.
8. Hubbard, A. E., Mountain, D. C., and Chen, F., 2003. Time-domain responses from a nonlinear sandwich model of the cochlea. In: Biophysics of the Cochlea: from Molecule to Model. A. W. Gummer, E. Dalhoff, M. Nowotny, and M. Scherer (Eds.). World Scientific, Singapore, 351-357.
9. Giguére C. and Woodland P. C. 1994. A computational model of the auditory periphery for speech and hearing research. I. Ascending path. J. Acoust. Soc. Am., 95(1), 331-342.
10. Murakami, Y. and Unoki, M. 2007. A study on the imput-output function of a nonlinear cochlear transmission-line model with the function of an outer hair cells model, J. Acoust. Soc. Am. 122(5), Pt. 2.2, 3061.
11. Neely, S. T. and Kim, D. O., 1986. A model for active elements in cochlear biomechanics. J. Acoust. Soc. Am., 79, 1472-1480.
12. Patuzzi, R. 1996. Cochlear Micromechanics and Macromechanics. In: in The Cochlea, P. Dallos, A. N. Popper, and R. R. Fay (Eds.), Springer, New York, 186-257.
13. Lukashkin, A. N. and Russell, I. J. 1998. A descriptive model of the receptor potential nonlinearities generated by the hair cell mechanoelectrical transducer. J. Acoust. Soc. Am., 103, 973-980.
14. Greenwood, D. D., 1990. A cochlear frequency-position function for several species-29 years later. J. Acoust. Soc. Am., 87, 2592-2605.

THE INFLUENCE ON PREDICTED HARMONIC GENERATION OF THE POSITION OF THE NONLINEARITY WITHIN MICROMECHANICAL MODELS

J. HOW, S. J. ELLIOTT, B. LINETON

Institute of Sound and Vibration Research, University of Southampton, Southampton, UK

1 Summary

We have compared the harmonic responses of the micromechanical lumped parameter cochlear models of Kanis and de Boer (1993) [1] and Cooper (1998) [2]. Kanis & de Boer use a Hammerstein system arrangement, incorporating a hyperbolic tangent function to represent the nonlinear action of the outer hair cells acting *after* a dynamic model of its frequency response. In contrast, Cooper's uncoupled model uses a Wiener system arrangement with the nonlinearity represented by a first order Boltzmann function, acting *before* the dynamic model of the outer hair cell frequency response. Block diagrams representing the two different micromechanical models are shown in figure 2.

We have found that the different system arrangements produce a significant difference in the spatial distributions and spectra of the harmonics in the predicted basilar membrane response, although there is little effect on the fundamental component. The results are summarized in figure 2, and have been obtained using a fully coupled cochlear model having 512 elements (figure 1), with the distribution of parameters suggested in [1]. It appears that the Wiener system arrangement produces distributions of 2nd and 3rd order harmonics that are most consistent with physiological measurements [2]. In addition we have also observed that the use of a first order Boltzmann function nonlinearity, instead of a hyperbolic tangent function, generates a 2nd order harmonic in the response, but has little effect on the fundamental or 3rd order components.

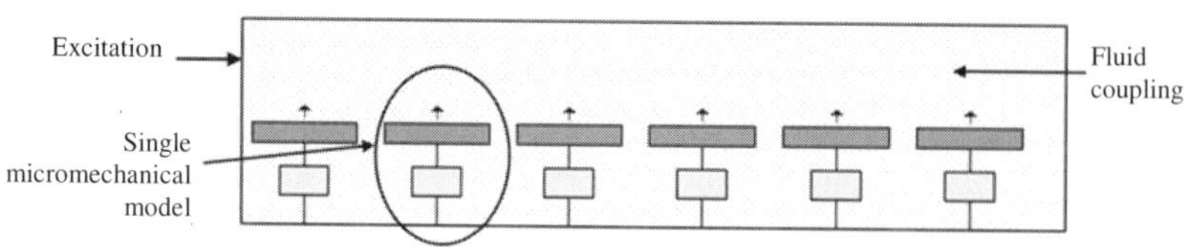

Figure 1. A discrete model of the coupled cochlea.

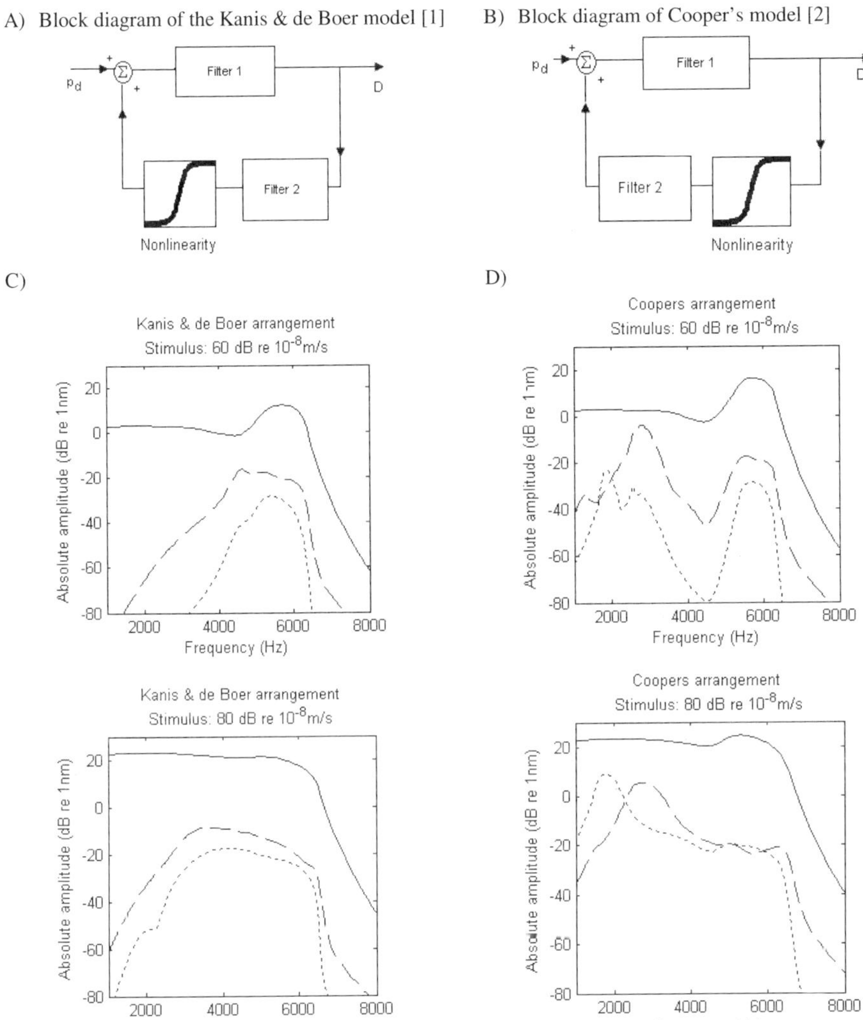

Figure 2. In A and B Filter 1 represents the passive mechanics of the basilar membrane and filter 2 is a dynamic representation of the frequency response of the active process. The nonlinearity represents the action of the outer hair cells. P_d and D correspond to the semi-difference pressure across the basilar membrane (BM) and the displacement of the BM respectively. C and D: Frequency responses of the predicted BM displacement at the 6kHz characteristic place for the arrangements of Kanis & de Boer and Cooper respectively. For each model the responses corresponding to two different stimulus levels (60 and 80 dB re 10^{-8}m/s) are shown. The solid, dashed and dotted lines correspond to the fundamental, second harmonic and third harmonic components of the basilar membrane displacement respectively.

References

1. Kanis, L. and E. de Boer. 1993. Self-suppression in a locally active nonlinear model of the cochlea: A quasilinear approach. J Acoust. Soc. Am. 94(6), 3199-3206.
2. Cooper, N.P. 1998. Harmonic distortion on the basilar membrane in the basal turn of the guinea-pig cochlea. J. Physiol. 509(1), 277-288.

BROWNIAN ENERGY DEPOT MODEL OF THE BM-OHC SYSTEM

YONG ZHANG, CHUL KOO KIM

Institute of Physics and Applied Physics, Yonsei University, Seoul 120-749, Korea

KONG-JU-BOCK LEE

Department of Physics, Ewha Womans University, Seoul 120-750, Korea
School of Physics, Korea Institute for Advanced Study, Seoul 130-722, Korea

YOUNGAH PARK

Department of Physics, Myongji University, Yongin 449-728, Korea
School of Physics, Korea Institute for Advanced Study, Seoul 130-722, Korea

In this study, we propose an energy depot model for the active oscillation of the basilar membrane which successfully describes both active and protective mechanisms. One of the main results is that thermal noise in the absence of external stimulation can be amplified leading to the spontaneous basilar membrane oscillation. Our model produces better agreement with nonlinear compressive data [1].

1. Introduction

The active amplification of the living cochlea was conjectured by Gold in 1948 [2] and is now studied widely through the Hopf equation [3, 4, 5, 6, 7]. However, the physical origin of the active oscillation remains still unclear. In mammals, the outer hair cells (OHCs) are known to be force generators for auditory sensitivity and frequency selectivity. Proposed mechanisms of the force generation are contraction of the OHC itself [8, 9, 10] and an active motion of the hair bundle [11, 12].

Energy depot models can describe the active phenomena [7, 13], because the energy supplied by the depot can induce a negative dissipation. A similar mechanism for a negative stiffness in the bullfrog's hair bundle has been reported [14]. Considering the OHC as an energy depot, we adopt the energy depot model to describe the active force generation and thus the active mechanism. By introducing a critical velocity of the basilar membrane (BM) oscillation, we also propose a braking mechanism, which protects the BM from excessive oscillations and produces better approximation in comparison with the experimental observation.

2. Energy Depot Model with Braking Mechanism

The energy depot model is described as [13]

$$\frac{dE(t)}{dt} = q - cE(t) - d(v)E(t), \tag{1}$$

where $E(t)$ is the energy density of the energy depot, q the rate of the energy pumping into the energy depot per unit area, c the rate of the internal energy dissipation, $d(v)$ the rate of the energy converting into the kinetic energy of the BM, and v the velocity of the BM. Considering that constant contribution of d(v) can be incorporated into c and that conversion rate of the energy into the kinetic energy of the BM will not be sensitive to the sign of v, we assume that $d(v)$ is an even function of v without a constant term. The energy depot model in which $d(v) = d_2 v^2$ has been extensively discussed [13]. In this paper, we include the next higher contribution to $d(v)$ and introduce a critical velocity of the BM to describe a braking mechanism. Hence the conversion rate $d(v)$ can be written as

$$d(v) = d_2 v^2 \left(1 - \frac{v^2}{v_c^2}\right), \tag{2}$$

where v_c is the critical velocity. For a finite v_c the oscillation of the BM can be braked by the negative conversion rate into the kinetic motion of the BM. Note that $v_c \to \infty$ corresponds to a system without the braking mechanism. The critical velocity will be roughly estimated from the contraction of the OHC later for our numerical calculation.

3. Results and Discussion

An adiabatic approximation in which the adaptation of the energy depot is very fast [13] yields that

$$E(t) = \frac{q}{c + d_2\left(1 - v^2/v_c^2\right)v^2}. \tag{3}$$

It should bear in mind that $E(t)$ is not constant. The oscillation of the BM is then governed by

$$\rho\dot{v} + \mu v + \kappa R = F_{passive} + F_{active}, \tag{4}$$

where ρ is the density of the BM, μ the physical damping coefficient of the BM, κ the stiffness of the BM, and R the displacement of the BM. $F_{passive}$ includes the contributions of a sound wave (external stimulation) and a noise. The noise is assumed to be a Gaussian white noise. F_{active} is the active force density acting on the BM by the OHC,

$$F_{active} = \frac{d(v)E(t)}{v} = \frac{q\frac{d_c}{c}\left(1 - \frac{v^2}{v_c^2}\right)v}{q + \frac{d_2}{c}\left(1 - \frac{v^2}{v_c^2}\right)v^2}. \tag{5}$$

Note that the Langevin equation, Eq. (4), describes the passive and the active response in a unified and natural way. It should be noted that the active force, Eq. (5), can be either positive or negative depending on the value of v. When v^2 is smaller than v_c^2, a positive active force is provided on the BM (Active mechanism). On the other hand, when v^2 is larger than v_c^2, a negative active force, hence an extra drag force, is exerted on the BM (Braking mechanism).

The density of the BM is known to be about 0.77kg/m^2 [15] and the reported damping coefficient, μ, of the BM has a large variation from 600 to 63,000kg

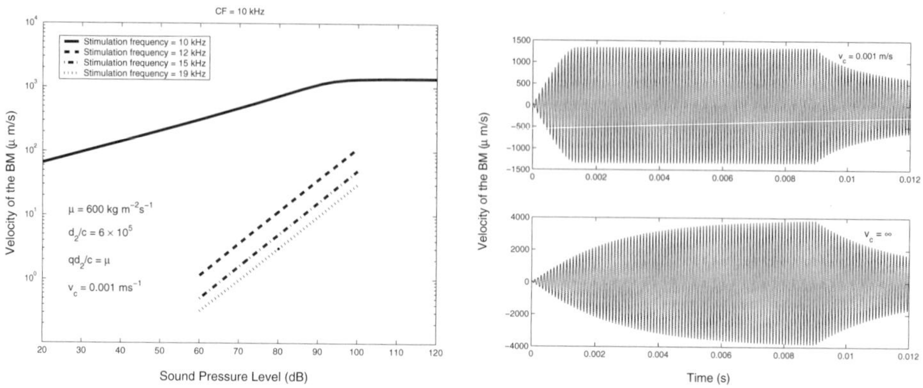

(a) (b)

Fig. 1. (a) The compressive response of the BM. The characteristic frequency is 10kHz. It shows that the response of the BM is non-linear and compressive at the characteristic frequency. (Note that the slope of the solid line is less than one.) Otherwise, the response of the BM is a linear function of the stimulation yielding that the slopes are 1. These results agree with the experimental observations [1]. (b) The dynamic response of the BM when the characteristic frequency is 10kHz and the stimulation is 100dB. The onset time is much shorter than the offset time which is not shown in full scale. This result is qualitatively consistent with the experimental observations [1]. $\mu = 600\text{kgm}^{-2}\text{s}^{-1}$, $d_2/c = 6 \times 10^5\text{m}^{-2}\text{s}^2$, and $qd_2/c = \mu$. The stimulation is unloaded at $t = 9$ms. The lower panel shows the response represents the case without the braking mechanism, showing a continuous increase as long as the stimulation is on.

$\text{m}^{-2}\text{s}^{-1}$ [15, 16] depending on the models. For a numerical calculation, we set the damping as $600 \sim 6{,}000\text{kg m}^{-2}\text{s}^{-1}$. It has been reported that the Guinea pig OHCs are able to change its length up to 5% when the transmembrane potential is varied. This corresponds to a displacement of 1μm for the OHCs of 20μm at the high-frequency end [17]. Hence, in the regime of high frequency of 10kHz, the adjustable parameter v_c is roughly estimated about 0.01m/s. By considering the OHC as a capacitor and using the observed membrane potential, membrane capacitance [18], and the scale of the OHC [19], we estimate the value of q as about 0.01kg s^{-3}. The response of the BM to the sound wave can be then obtained straightforwardly using Eqs. (4) and (5). In the following calculations, we set $qd_2/c = \mu$ and the characteristic frequency, $\omega_c = \sqrt{\kappa/\rho}$. It will be shown later that these relations arise from the conditions of the Hopf bifurcation. In the present calculations, the noise width of the distribution is chosen to be zero dB.

The compressive and dynamical responses of the BM are obtained as shown in Fig. 1(a) and Fig. 1(b), respectively. Fig. 1(a) demonstrates that the response of the BM is non-linear and compressive at the characteristic frequency, while the response is linear at other frequencies. This compressive response of the BM agrees well with the observations [20, 1] as expected. Fig. 1(b) shows the time course of the BM response, qualitatively reproducing the experimental observation that the onset time of the response is shorter than the offset time, although the onset time is somewhat larger than the experimental values [1]. It is important to notice that the velocity of the BM increases continuously as long as the stimulation is being turned on if the braking mechanism is not included ($v_c = \infty$). Hence, the braking mechanism introduced in our model is essential to explain the experimental observation at a sufficiently strong stimulation. A dependence of the BM velocity on v_c is plotted in Fig. 2(a). The results clearly show that lower critical velocity produces smaller response at the same stimulation, thus producing a super–compressive behavior qualitatively as observed in experiment [1]. Note that the saturation of the BM response disappears if the braking mechanism is not considered.

The gain of the oscillator is defined as the ratio of the displacement of the BM to the stimulation,

$$G = \frac{R}{F_{passive}}. \tag{6}$$

Using Eq. (4), we obtain the gain, which is a scaling function of the stimulation when the stimulation is not too large, $G \sim F_{passvie}^{-0.65}$. This result agrees well with the experiment [20] and the result of the Hopf bifurcation model [3], $G \sim F_{passvie}^{-2/3}$. Indeed our model includes the Hopf model naturally. To show this, we expand the active force, Eq. (5), up to the lowest nonlinear term when $d_2(1 - v^2/v_c^2)v^2/c$ is small. Eq. (4) is then approximately written as

$$F_{passive} = \rho\dot{v} + \left(\mu - q\frac{d_2}{c}\right)v + \kappa R + q\frac{d_2}{c}\left(\frac{1}{v_c^2} + \frac{d_2}{c}\right)v^3. \tag{7}$$

In the Fourier spectra space, this directly corresponds to the Hopf equation [3] with the following bifurcation conditions,

$$\omega_c = \sqrt{\frac{\kappa}{\rho}}, \tag{8}$$

$$\mu - q\frac{d_2}{c} \tag{9}$$

Hence our energy depot model with a braking mechanism reduces to the Hopf bifurcation model in a special limit.

One of the most important results of our model is an amplification of the thermal noise itself in the absence of an external stimulation. At the bifurcation point or when $qd_2/c < \mu$, the thermal noise is well suppressed. However, when $qd_2/c > \mu$, so that too much of the internal energy is converted into the kinetic energy, the thermal noise can be largely amplified as shown in Fig. 2(b). Here, we set the mean average of the noise is 0Pa and the width is 0dB. Since it can be generally assumed that the oscillators are distributed in the vicinity of the bifurcation point, the noise may be amplified incoherently in the absence of the input signal. However, when a weak input signal with the same characteristic frequency is introduced, it may phase-lock to the already existing amplified noise, thus enhancing the selectivity as discussed in Ref. [21]. Because it is believed that the spontaneous BM oscillation is crucial for understanding the spontaneous otoacoustic emission [22], our model may provide a clue to explore this interesting and important phenomenon.

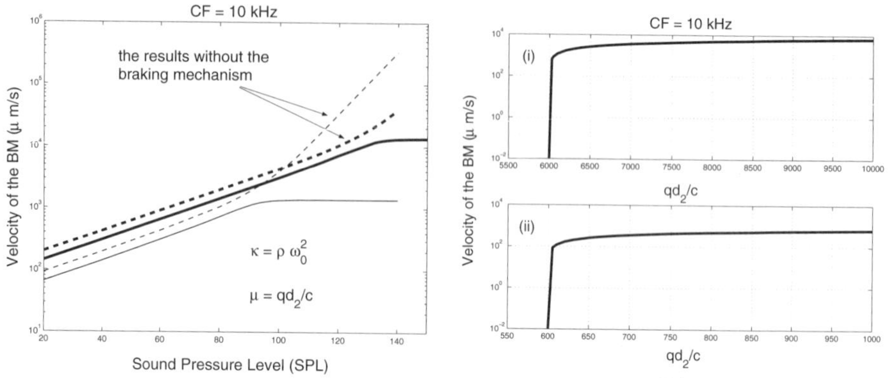

(a) (b)

Fig. 2. (a) The response of BM when $v_c = 0.001$m/s (solid thin line), $v_c = 0.01$m/s (solid thick line), and $v_c = \infty$ (dashed lines). For thin lines, $\mu = 600$ kgm^{-2}s^{-1}, $d_2/c = 6 \times 10^5$m^{-2}s^2 and for thick lines, $\mu = 6,000$kgm^{-2}s^{-1}, $d_2/c = 6,000$m^{-2}s^2. The velocity of the BM is (not) saturated when the stimulation is sufficiently large with (without) the braking mechanism. (b) The amplification of the thermal noise in the absence of external stimulation. When $qd_2/c \leq \mu$, the Brownian fluctuation is suppressed, while in the region of $qd_2/c > \mu$, the Brownian fluctuation is amplified. We use that the intensity of the thermal noise is 0Pa, the strength of the fluctuation is 0dB, in (i) $\mu = 6,000$kgm^{-2}s^{-1}, $v_c = 0.01$m/s, and in (ii) $\mu = 600$kgm^{-2}s^{-1}, $v_c = 0.001$m/s.

4. Summary

In this paper, we propose a theoretical scheme for the active and passive response of the BM based on a concept of the energy depot model with a braking mechanism. The OHC is assumed to play a role of the energy depot by supplying an active force to the BM and also by reabsorbing an excessive kinetic energy like the engine brakes in automobiles. It is shown that a single equation of motion, Eq (4), can produce all the essential passive and active behaviors. Moreover, our model shows that the Brownian noise can be sufficiently amplified under a certain condition, thus leading to the SBMO. Although we applied our model only to the mammalian cochlea, it is equally possible to be applied to the nonmammalian vertebrates because the only required physical characteristics are a supply of the active force and a self-adaptation. In this sense, we believe that the concept of the energy depot model with a braking mechanism is quite universal and can be applied to any underdamped biological systems.

Acknowledgment

This work is supported by the Korea Science and Engineering Foundation (KOSEF) (R01-2006-000-10083-0).

References

1. Ruggero, M.A., Rich, N.C., Recio, A., Narayan, S.S. and Robles, L., 1997. Basilar-membrane responses to tones at the base of the chinchilla cochlea. J. Acoust. Soc. Am. 101, 2151-2163.

2. Gold, T., 1948. Hearing. II. The Physical Basis of the Action of the Cochlea. Proc. R. Soc. London B 135, 492-498.
3. Camalet, S., Duke, T., Jülicher, F. and Prost, J., 2000. Auditory sensitivity provided by self-tuned critical oscillations of hair cells. Proc. Natl. Acad. Sci. USA 97, 3183-3188.
4. Duke, T. and Jülicher, F., 2003. Active Traveling Wave in the Cochlea. Phys. Rev. Lett. 90, 158101-158104.
5. Magnasco, M.O., 2003. A Wave Traveling over a Hopf Instability Shapes the Cochlear Tuning Curve. Phys. Rev. Lett. 90, 058101-058104.
6. Nadrowski, B., Martin, P. and Jülicher, F., 2004. Active hair-bundle motility harnesses noise to operate near an optimum of mechanosensitivity. Proc. Natl. Acad. Sci. 101, 12195-12200.
7. Stoop, R. and Kern, A., 2004. Two-Tone Suppression and Combination Tone Generation as Computations Performed by the Hopf Cochlea. Phys. Rev. Lett. 93, 268103-268107.
8. Brownell, W.E., Spector, A.A., Raphael, R.M. and Popel, A.S., 2001. MICRO- AND NANOMECHANICS OF THE COCHLEAR OUTER HAIR CELL. Annu. Rev. Biomed. Eng. 3, 169-194.
9. Liberman, M.C., Gao, J., He, D.Z.Z., Wu, X., Jia, S. and Zuo, J., 2002. Prestin is required for electromotility of the outer hair cell and for the cochlear amplifier. Nature 419, 300-303.
10. Santos-Sacchi, J., 2003. New tunes from Cortis organ: the outer hair cell boogie rules. Curr. Opin. Neurobiol. 13, 459-468.
11. Martin, P. and Hudspeth, A.J., 2001. Compressive nonlinearity in the hair bundle's active response to mechanical stimulation. Proc. Natl. Acad. Sci. USA 98, 14386-14391.
12. Lemasurier, M. and Gillespie, P.G., 2005. Hair-Cell Mechanotransduction and Cochlear Amplification. Neuron 48, 403-415.
13. Schweitzer, F., 2003. Brownian Agents and Active Particles. Springer, Berlin.
14. Goff, L.L., Bozovic, D. and Hudspeth, A.J., 2005. Adaptive shift in the domain of negative stiffness during spontaneous oscillation by hair bundles from the internal ear. Proc. Natl. Acad. Sci. USA 102, 16996-17001.
15. Jaffer, T.S.A., Kunov, H. and Wong, W., 2002. A model cochlear partition involving longitudinal elasticity. J. Acoust. Soc. Am. 112, 576-589.
16. Kim, Y. and Xin, J., 2005. A two-dimensional nonlinear nonlocal feed-forward cochlear model and time domain computation of multitone interactions. Multiscale Model Simul. 4, 664-690.
17. Holley, m.C., 1996. Outer Hair Cell Motility. In: Dallos, P., Popper A.N. and Fay, R.R.(Eds.), The Cochlea. Springer, New York, pp. 386-434.
18. Santos-Sacchi, J., 1991. Reversible inhibition of voltage-dependent outer hair cell motility and capacitance. J. Neurosci. 11, 3096-3110.
19. Furness, D.N. and Hackney, C.M., 2006. The Structure and Composition of the Stereociliary Bundle of Vertebrate hair Cells. In: Eatock, R.A., Fay, R.R. and Popper, A.N.(Eds.), Vertebrate Hair Cells. Springer, New York, pp. 95-153.
20. Robles, L. and Ruggero, M.A., 2001. Mechanics of the Mammalian Cochlea. Physiol. Rev. 81, 1305-1352.
21. Martin, P., Hudspeth, A.J. and Jülicher, F., 2001. Comparison of a hair bundle's spontaneous oscillations with its response to mechanical stimulation reveals the underlying active process. Proc. Natl. Acad. Sci. USA 98, 14380-14385.
22. Nuttall, A.L., Grosh, K., Zheng, J., de Boer, E., Zou, Y. and Ren, K., 2004. Spontaneous Basilar Membrane Oscillation and Otoacoustic Emission at 15 kHz in a Guinea Pig. J. Assoc. Res. Oto. 5, 337-348.

CONJOINED COCHLEAR MODELS: THE TWAMP AND THE SANDWICH

ALLYN HUBBARD

Electrical and Computer Engineering Department, Boston University, Boston University Center for Hearing Research, Dept. of Biomedical Engineering, Boston, MA 02215, USA

A new model of the cochlea is created by joining parts of the traveling-wave amplifier (TWAMP) and the Sandwich models. The lossy, untuned traveling-wave line of the TWAMP is retained, but the TWAMP's tuned traveling-wave line is replaced by the Sandwich's traveling-wave line that represents the reticular lamina (RL) and scala tympani. The model combines stereocilliary forces, which act between the tectorial membrane (TM) and RL, with somatic outer hair cell forces that power the Sandwich.

1 Introduction

A new cochlear model was inspired by Ghaffari et al.'s [1] reference to the TWAMP model of the cochlea [2] in a paper that describes measurements made on the TM. The data support the notion of the TM being a lossy transmission line, which is a fundamental tie to the TWAMP. Previously, the TWAMP model had been abandoned, for lack of a physical counterpart in the cochlea to the TWAMP's lossy line, despite that its pressures and volume velocities in the "classical" portion of the model were cochlea-like. The possibility that the TWAmp model could be resurrected has provided an alternative way to boost the "gain"[3] of the Sandwich model of the cochlea [4].

2 The TWAMP

A fundamental property of the TWAMP (Fig. 1) is that it amplifies when the phase velocity on the "upper" and "lower" lines approximately match. The upper line was considered to be a hydrodynamic analog, i.e. a classical one-dimensional model of the cochlea. Thus, the traveling wave slows down approaching the tuned location to match the velocity on the "lower" line, which supports slow waves. Over this region, which is on the order of 200-400 microns basal from the best place, amplification occurs. Many possibilities have been suggested for the "lower line", using either mobility or impedance

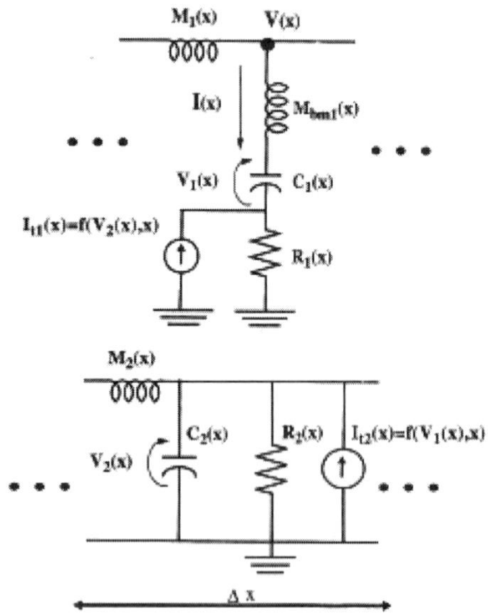

Figure 1. A circuit representation of the TWAMP cochlear model.

analogies. Whatever the ascribed analog, its mechanical or hydrodynamic counterpart behaves as computed by the electric circuit. Force or pressure in the "lower" line kicks back into the BM via a dependent source, and vice versa.

3 The Sandwich model of the cochlea

The Sandwich, too, was once abandoned: de Boer [3] declared it implausible from an efficiency standpoint. Moreover, it appeared that outer hair cells (OHCs) had nothing to "push against" and might produce no net motion of the cochlear partition (CP). But, we simulated the Sandwich model, and found that it worked well. Moreover, the model predicted longitudinal flow within the organ of Corti. RL and BM motion is amplified by both local OHC forces and a pressure wave within the OC, a mechanism that is substantially different from the mechanism of the TWAMP.

4 The conjoined models

Figure 2 shows the conjoined TWAMP and Sandwich models. The TWAMP is flipped upside down, and the previously "lower line" becomes the TM analog. The TWAMP's "upper line" is the RL hydromechanical analog. The coupling feedback forces are stereocilliary forces. The remainder of the circuitry is simply the remaining portions of the Sandwich, with active forces produced by somatic motility, represented as a

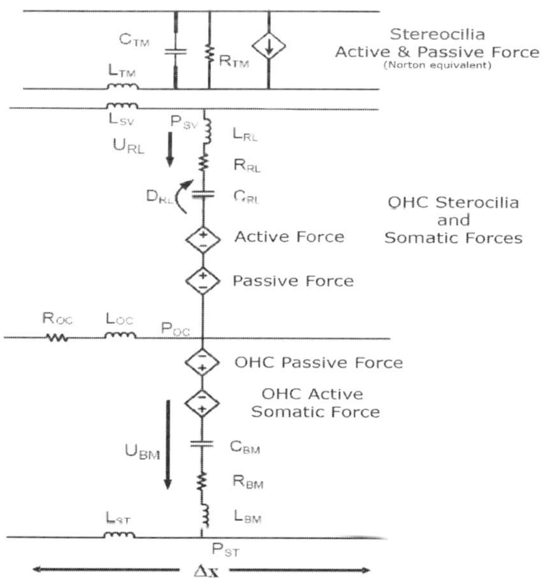

Figure 2. The TWAMP circuit representation is drawn upside-down and overlapped (conjoined) with the Sandwich model's circuitry. Numerous helper circuits that are required to switch from acoustic to mechanical analogs have been omitted. The actual simulation is computed in an analogous electrical domain using electric circuit simulation programs.

dependent source that senses RL motion. The Norton "kick-back" source in Figure 1 is combined into the Thevenin equivalent source called "OHC Active Somatic + Stereocilia Force" [see 5 for a review], with force scaled by the section's area to be pressure.

5 Discussion

This circuit or one that is topologically similar can likely work in a way that its currents and voltages are similar to analogous motions and pressures in the cochlea. The TWAMP *always* amplifies over a narrow spatial region basal to the best place, where the tuned line's velocity slows down. In the Sandwich, RL traveling wave velocity does, indeed, slow down to meet the requirement. Assuredly, refinements must be made to the circuit shown. The phase of the force contributions from the stereocilia and from the OHC soma, plus the OC pressure wave must be right. From a circuit standpoint, we can make it be right. Whether this will be physically realistic remains to be seen. The circuitry shown cannot be regarded as iron-clad at this point, and must be amended. For example, the sensed variables that produce force generation, both in the stereocilia and soma, must be related to the motion of the RL relative to the TM, instead of the way the Sandwich model currently assumes that the TM does not move. Moreover, the final model must satisfy force-balance equations; i.e. the gains and points of attachments of what are now arbitrary dependent sources must be constrained. In general, this requires additional "helper circuits" with other dependent sources to convert from mechanical to hydromechanical domains. That will have to be done for stereocilliary passive, mechanical forces also. At that point, the representation of the subtectorial space will look much like the circuitry representing flow in the organ of Corti in the Sandwich model, i.e. a fluid channel with movable floor and ceiling having interposed mechanical actuators. Fluid flows, relative motions, and pressures in the subtectorial space are quantities predicted by the model that should move us a big step closer to understanding the stimulus to the inner hair cells.

References

1. R. Ghaffari, A. Arayosi, and D. Freeman, 2007. Longitudinally propagating traveling waves of the mammalian tectorial membrane. PNAS, 104(42), 16510-16515.
2. A. Hubbard, 1993. A traveling-wave amplifier model of the cochlea. Science, 259, 68-71.
3. S. Lu, D. Mountain, and A. Hubbard, 2008. Is stereocilia velocity or displacement feedback used in the cochlear amplifier?. This volume.
4. E. deBoer, 1990. Wave-propagation modes and boundary conditions for the Ulfendahl-Flock-Khanna preparation. In: The Mechanics and Biophysics of Hearing. P. Dallos, et al. eds. Springer-Verlag, Berlin, pp. 333-339.
5. R. Fettiplace and C. Hackney, 2006. The sensory and motor roles of auditory hair cells. Nature Reviews - Neuroscience, 7, 19-29.

SECTION VI

HAIR CELLS AND ELECTRO-MECHANICAL TRANSDUCTION

J.-P. Hardelin, J.F. Ashmore and K.H. Iwasa

FIRING UP THE AMPLIFIER: TEMPERATURE, PRESSURE AND VOLTAGE JUMP STUDIES ON OHC MOTOR CAPACITANCE

JOSEPH SANTOS-SACCHI, LEI SONG, XIANTAO LI

Yale University School of Medicine
New Haven, Ct 06510, USA

The outer hair cell (OHC) possesses molecular motors that drive electromotility and cochlear amplification. Here we look at the effects of fast perturbations of biophysical forces that affect the OHC, including voltage, temperature and pressure, on the electrical signature of motor activity, namely nonlinear capacitance (NLC). Under whole cell voltage clamp, we measure changes in NLC at fixed holding voltages following steps in voltage, jumps in temperature induced by ir laser, or jumps in intracellular turgor pressure. In each case we find time dependent changes in NLC resulting from induced shifts of the NLC function across the voltage axis. The time course of these shifts depends on the stimulus, with voltage jumps inducing shifts with fast exponential components less than a millisecond. Those induced by temperature and pressure are within the tens of millisecond ranges and may be limited by cellular and experimental constraints that voltage is able to overcome. The overall observation of time dependent changes to the electromotility function upon stimulation indicates a more complex mechanism than provided by a simple two-state Boltzmann model.

1 Introduction

The outer hair cell (OHC) drives cochlea amplification though a cellular mechanism that works at kilohertz rates [1,2,3,4]. Molecular motors that are housed within the OHC lateral membrane, comprising at least the molecule prestin [5], produce conformational changes in response to voltage which alter the somatic length of the cell[6,7,8]. The deformations of these cells within the organ of Corti feedback energy into the basilar membrane, thus amplifying the auditory stimulus that is delivered to IHC stereocilia [9]. The ultimate effect is a boost in our auditory sensations.

The OHC motor, or prestin transfected into non-auditory cells, produces a signature electrical response analogous to gating currents observed in other voltage dependent membrane bound proteins [2,10,11]. The gating currents, because they are restricted charge movements within the membrane dielectric, can also be measured as a nonlinear capacitance (NLC), bell-shaped as a function of voltage. This charge movement or NLC can be fit most simply by a two-state Boltzmann function, allowing extraction of Boltzmann parameters that provide information on voltage dependence, voltage sensor valence, and maximum charge moved. These steady state parameters do not provide dynamic information about the motor, but the time course of underlying gating charge movements or of shifts in the C-V function can provide hints on the motor's conformational kinetic properties. Here we report on the effects of fast perturbations of biophysical forces that affect the OHC, including voltage, temperature and pressure, on the electrical signature of motor activity, namely NLC. The occurrence of time dependent changes to the C-V (or equivalently, electromotility) function upon such stimulation indicates a more complex mechanism than provided by a simple two-state Boltzmann model.

2 Methods

Guinea pigs were decapitated following anesthetic overdose with halothane. OHCs were freshly isolated from the guinea pig cochlea using Dispase (1 mg/ml) followed by trituration, and were whole-cell voltage clamped at room temperature using an Axon 200B amplifier. Membrane voltages were corrected for the effects of residual series resistance, which ranged from 3-5 MΩ. Ionic blocking solutions were used to remove voltage-dependent ionic conductances so that capacitive currents could be analyzed in isolation [11]. Extracellular solution: 100mM NaCl, 20mM TEA, 20mM CsCl, 2mM $CoCl_2$, 1.48mM $MgCl_2$, 2 mM $CaCl_2$, 10 mM HEPES, and 5mM dextrose, adjusted to pH 7.2 with NaOH, and adjusted to 300mOsm with dextrose. Pipette solution: 140mM CsCl, 10mM EGTA, 2mM $MgCl_2$, and 10mM HEPES, adjusted to pH 7.2 with CsOH, and adjusted to 300mOsm with dextrose. In some cases, TEA, CsCl, $CoCl_2$, $CaCl_2$, and $MgCl_2$ were omitted. Also, in some experiments, the extracellular solution was used in patch electrodes. In a few experiments, 50 μM $GdCl_3$ was included in the extracellular solution to decrease residual currents. At this concentration, gadolinium has an insignificant affect on OHC capacitance [12]. Additional solution modifications are noted in figure legends.

In order to deliver rapid changes in membrane tension, we used a pressure clamp system from ALA instruments. Pipette tip diameters were increased in size for these experiments (~1 MΩ). In order to deliver rapid changes in temperature a Capella IR laser was used to deliver via fiber optic 5-20 ms light pulses (1850 nm).

OHC capacitance was measured with a continuous high-resolution two-sine voltage stimulus protocol (20mV peak at both f1 and f2; f2=2*f1; f1 ranging up to 3906.3 Hz), and subsequent FFT based admittance analysis [13,14]. These small, high frequency sinusoids were superimposed on voltage steps or ramps that spanned up to +/- 200 mV.

OHC C-V plots were fitted with the function $C_m=C_{lin}+C_v$, where C_{lin} is the voltage-independent (linear surface area component) capacitance of the lipid bilayer, and $C_v=dQ_{nonL}/dV$ is the voltage-dependent component of C_m originating from prestin's voltage-sensor activity. The first derivative with respect to V_m of a two-state Boltzmann function $Q=Q_{max}/(1+\exp(-ze(V_m-V_{pkCm})/k_BT))$ was used to fit the C-V data [11]. Q_{max} is total charge moved. The measured apparent valence of prestin voltage-sensor (z) is defined as $z=q_*d$, where d is a normalized perpendicular projection of distance traveled by the voltage sensor charge (q) within the plasma membrane field. V_{pkCm} is the membrane potential (V_m) at which prestin molecules are equally distributed between expanded and contracted states. It corresponds to the peak of the C_v function.

All data collection and most analyses were performed with an in-house developed Window's based whole-cell voltage clamp program, **jClamp** (www.scisoftco.com), utilizing a Digidata 1320 board (Axon, CA). Matlab (Natick, MA) or SigmaPlot was used for fitting the Cm data.

3 Results

3.1 Voltage jumps induce time dependent changes in NLC

When OHC voltage is stepped away from the holding potential, a nearly instantaneous change in C_m is observed because membrane capacitance is a bell shaped function of voltage (Fig. 1). However, unlike memory-less systems, the C-V function is not static but shifts along the stimulus axis to alter C_m over time even though measures are made at a fixed voltage. The change in C_m is multi-exponential. With C_m sampling rates of about 4kHz, we find that the earliest exponential components are below one millisecond, with additional components at order of magnitude intervals.

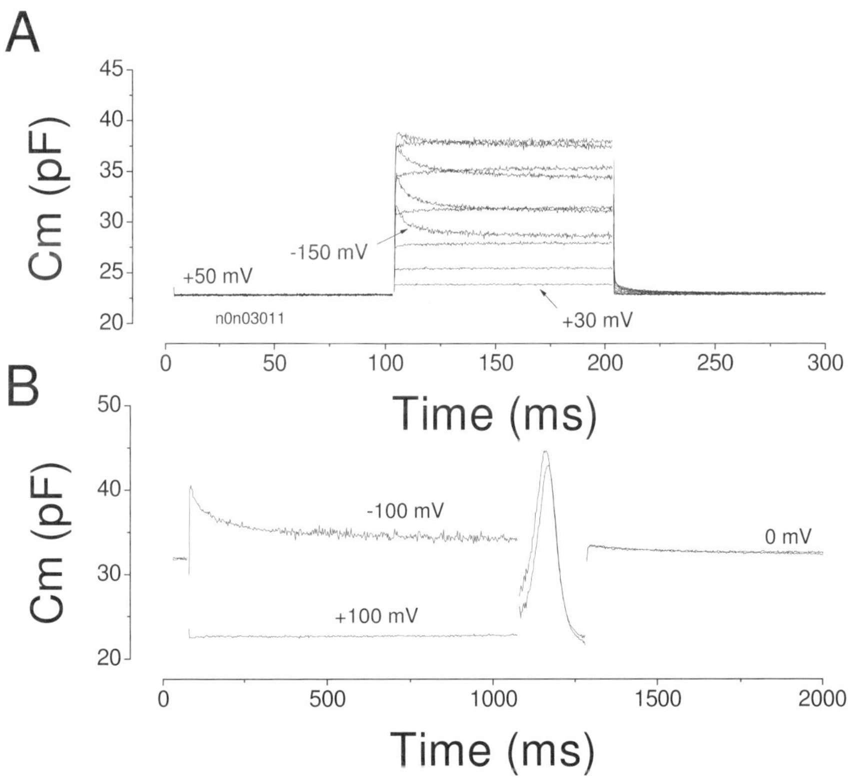

Figure 1. A) An OHC was held under voltage clamp at +50 mV and stepped to 30 (red),10,-10,-30,-50,-70,-90,-110,-130,-150 (blue) for 200 ms. The hyperpolarizing steps induced a time dependent change in Cm, increasing for steps approaching V_{pkcm}, and decreasing for steps beyond V_{pkcm}. B) Another OHC (held at 0 mV) showing that hyperpolarizing steps cause Cm relaxations (red line) due to depolarizing shifts in V_{pkcm}. A ramped voltage from -150 to +150 mV produced the bell shaped Cm response, indicating the rightward shift in V_{pkcm}.

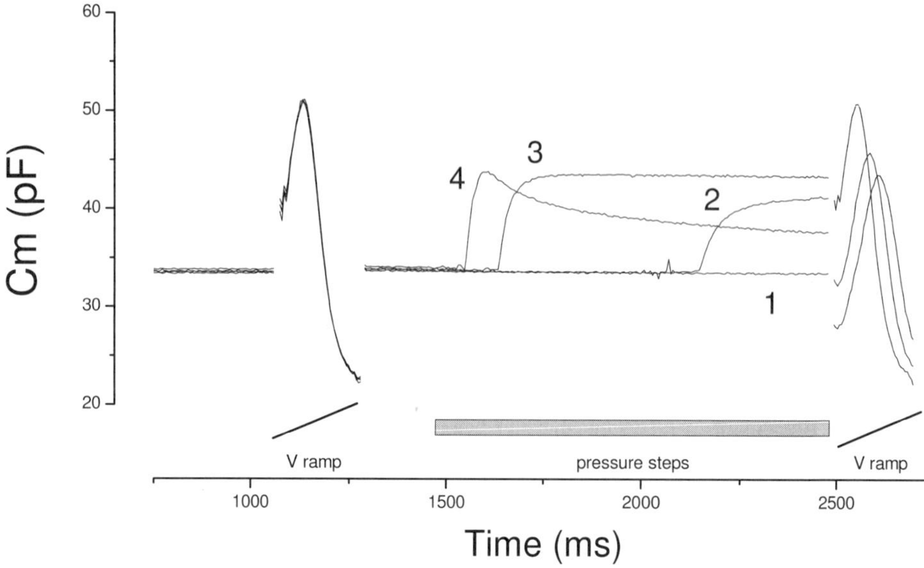

Figure 2. An OHC was held under voltage clamp at 0 mV. A voltage ramp (-100 to +100 mV, 200 ms) was delivered followed by incrementing pressure (1, 2, 3, 4) steps to the patch pipette, and the ramp was repeated (pressure was maintained during the final ramps). 3 sec intervals between traces were allowed for recovery from stimulation. Successful recovery is indicated by overlapping C-V functions. Increments in pressure cause time dependent changes in Cm at the fixed 0 mV holding potential, and occur due to shifts in V_{pkcm}, as the bell shaped C_m functions signify. The final pressure pulse caused the cell to burst; consequently, no C_m ramp data is plotted. In mM: 140 Cl in and out (NaCl 132, $MgCl_2$ 2, $CaCl_2$ 2, Hepes 10. Additional 10 EGTA for Intracellular solution)

3.2 Pressure jumps induce time dependent changes in NLC

Figure 2 shows that rapid changes in membrane tension at a fixed voltage shifts V_{pkcm} in a time dependent manner, mimicking to some extent the C_m changes induced by step voltage changes. The pressure steps produced delayed onset responses because cellular material in the pipette tip obstructed flow. With increasing pressures it can be seen that the pipette unplugged at decreasingly shorter times following onset of the pressure steps (1: 0.8 kPa [fully obstructed], 2: 1.06kPa, 3: 1.33 kPa, and 4: 1.6 kPa). Lower pressures caused an increase in C_m over time, while the largest pressure pulse caused an immediate increase and subsequent relaxation. With this last pressure pulse the cell burst.

3.3 Temperature jumps induce time dependent changes in NLC

We have previously shown that the C-V function of OHCs and prestin transfected HEK cells are very temperature sensitive, shifting about 20 mV to the right with a 10°C increase in temperature [15,16]. Here we find that the effect of temperature jumps induced by IR laser is immediate in onset, followed by relaxation. Unlike steps in voltage and pressure, which could be maintained for unlimited durations, the Capella laser can only deliver up to 20 ms steps. Thus, the time course of C_m changes cannot be attributed to shifts in the C-V function during constant temperature, but could result from return to room temperature. Thus, the shifts in C-V we observed (Fig. 3B) likely correspond to expected changes in temperature as the bath cooled. However, we can glean from these experiments that temperature effects are very rapid, ruling out slow intracellular processes, such as phosphorylation, and likely indicating direct temperature effects on the conformational state of the motor protein prestin.

4 Discussion

By measuring OHC nonlinear membrane capacitance at sub-millisecond resolutions, we have examined the early time course of the OHC motor's dynamic response to jumps in membrane voltage, tension and temperature. We find that the time course of the voltage-induced response is multi-exponential, with the earliest detectable components residing in the kilohertz range, thereby establishing the phenomenon's potential significance in peripheral auditory processing of the mammal. These effects actually extend well into the low frequency range, as well; indeed, altering steady state resting potentials can stably set the operating point of electromotility to different points along the voltage axis.

Our data, though measuring whole cell capacitance, actually provide insight into the initial condition-dependent conformational states of the motor protein, prestin. That is, while capacitance may increase or decrease during incremental unidirectional steps in voltage or tension (Figs. 1 and 2), both correspond to a unidirectional change in motor state probability, as a conversion of capacitance to gating charge would reveal.

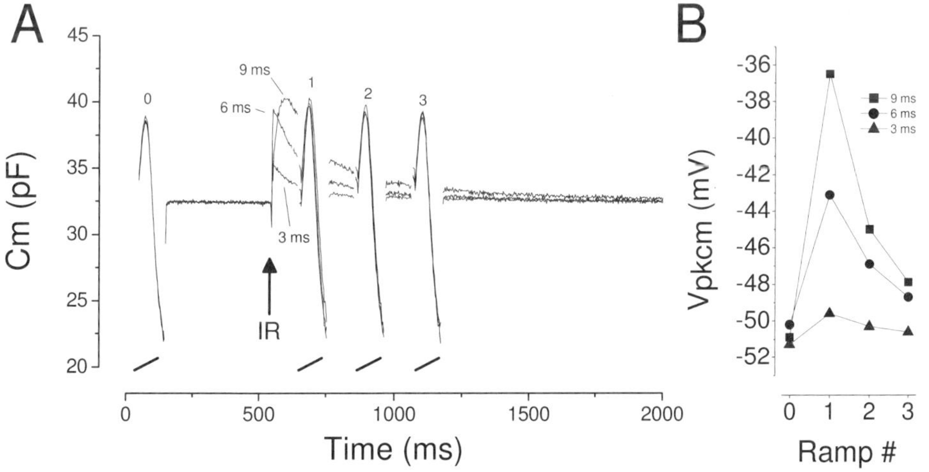

Figure 3. A) An OHC was held under voltage clamp at 0 mV. A voltage ramp (-100 to +100 mV, 200 ms) was delivered (denoted by 0) followed by incrementing temperature step durations to the cell, after which a series of voltage ramps (1, 2, 3) were delivered during the Cm relaxations. 10 sec intervals were used between traces to allow recovery from stimulation. Successful recovery is indicated by overlapping C-V functions. Increments in temperature duration (3, 6, 9 ms) cause time dependent changes in Cm at the fixed 0 mV holding potential, the shifts being plotted in B. Maximal temperature change is estimated to be about 10°C. In this case the slow relaxations may be due to re-cooling to room temperature after laser stimulation.

Consequently, depolarizing voltages, in addition to causing a near instantaneous redistribution of motors to the compact state, will foster a supplemental recruitment of additional motors into that state, leading to an amplification of the initial response. The same amplification effect holds for hyperpolarizing voltages, where recruitment of motors will be into the expanded state. Tension effects appear to follow these rules, as well. However, because of equipment limitations on laser pulse duration we were unable to deliver long steps, and thus cannot confirm that temperature jumps mimic the effects of the other stimuli. We are attempting to modify the design to allow the further comparison.

Anions play a major role in NLC generation, and we were interested whether our results could reflect underlying interactions of chloride with the motor, since we have shown that V_{pkcm} shifts to the right when chloride is replaced with gluconate or other substitutes [17,18,19]. However, we find that modifying the driving force (changing extracellular chloride concentrations) for chloride movement across the OHC lateral membrane through G_{metL} [18] does not modify the extent or time course of the amplificatory shift (data not shown). It should be noted that although driving force for chloride did not significantly affect the motor's amplificatory shift, the interactions of anions with the motor may still play a role. In this regard, we have recently found that the state probability of prestin influences anion binding affinity (Song and Santos-Sacchi, ARO 2007), and that even if concentrations remain the same at the intracellular binding sites of prestin, changes in binding site affinity, as can occur with allosteric modulation, could alter energy profiles, resulting in V_{pkcm} shifts.

Since the initial observations of Iwasa [20], membrane tension effects on the OHC motor have provided important information on this protein's surprisingly efficient piezoelectric activity [21]. Because membrane tension can shift V_{pkcm} and, indeed, in extreme applications, restrict conformational activity [22], we had modeled the voltage-induced amplificatory shift as resulting from motor induced tension [14]. In the present report we attempted to directly test this by applying rapid changes in membrane tension and seeking C_m relaxations that mimic those induced by voltage steps. Though there are indications of C_m relaxations following rapid tension changes, the time courses of voltage-induced and tension-induced C_m relaxations differ, tension effects being slower. It is possible that during our attempts to alter membrane tension through global changes in OHC structure we may have been restricted by the viscoelastic properties of the whole cell. For example, Dong and Iwasa [23] found that mechanical relaxations in the OHC have a time constant on the order of 40 s. However, because the fastest C_m relaxation time constants that we found with pressure steps were about 2 orders of magnitude faster than their measured time course, our pressure clamp driver appears to have been quite successful in overcoming this mechanical impediment. We suggest that tension induced by the molecular motors themselves can better bypass this viscoelastic constraint, thereby providing more rapid tensions to evoke fast amplificatory shifts. Thus, we still view motor-derived membrane tension, possibly through alterations of anion binding affinity, to underlie voltage-induced amplificatory shifts[14]. Nevertheless, as we have shown previously, the membrane environment of prestin can have profound effects on the magnitude and time course of the amplificatory shift [24].

In summary, we show remarkably fast changes in the Boltzmann distribution of prestin motor states induced by fast perturbations of the motor. The shifts along the voltage axis represent amplificatory supplementation to the near instantaneous voltage-induced mechanical response of the cell, and it is expected that this phenomenon will impact on high frequency peripheral auditory processing. Finally, it is expected that distortion in the mechanical response will arise, contributing to one of the hallmarks of mammalian cochlear amplification [25].

Acknowledgments

This work and the Plenary Lecture were supported by NIH grants DC000273, DC008130, DC008115 to JSS.

References

1. Santos-Sacchi, J., 1992. On the frequency limit and phase of outer hair cell motility: effects of the membrane filter. J. Neurosci. 12, 1906-1916.
2. Santos-Sacchi, J., 1990. The Mechanics and Biophysics of Hearing. Dallos, P., Geisler, C.D., Matthews, J.W., Ruggero, M.A. & Steele, C.R. (eds.), Springer-Verlag, Berlin, pp. 69-75.
3. Dallos, P. & Evans, B.N., 1995. High-frequency motility of outer hair cells and the cochlear amplifier. Science 267, 2006-2009.
4. Frank, G., Hemmert, W. & Gummer, A.W., 1999. Limiting dynamics of high-frequency electromechanical transduction of outer hair cells. Proc. Natl. Acad. Sci. U. S. A 96, 4420-4425.
5. Zheng, J. et al., 2000. Prestin is the motor protein of cochlear outer hair cells. Nature 405, 149-155.
6. Brownell, W.E., Bader, C.R., Bertrand, D. & de Ribaupierre, Y., 1985. Evoked mechanical responses of isolated cochlear outer hair cells. Science 227, 194-196.
7. Ashmore, J.F., 1987. A fast motile response in guinea-pig outer hair cells: the cellular basis of the cochlear amplifier. J. Physiol 388, 323-347.
8. Santos-Sacchi, J. & Dilger, J.P., 1988. Whole cell currents and mechanical responses of isolated outer hair cells. Hear. Res. 35, 143-150.
9. Ashmore, J.F., 2008. Cochlear outer hair cell motility. Physiol Rev. 88, 173-210.
10. Ashmore, J.F., 1990. Forward and reverse transduction in the mammalian cochlea. Neurosci. Res. Suppl 12, S39-S50.
11. Santos-Sacchi, J., 1991. Reversible inhibition of voltage-dependent outer hair cell motility and capacitance. J. Neurosci. 11, 3096-3110.
12. Kakehata, S. & Santos-Sacchi, J., 1996. Effects of salicylate and lanthanides on outer hair cell motility and associated gating charge. J. Neurosci. 16, 4881-4889.
13. Santos-Sacchi, J., 2004. Determination of cell capacitance using the exact empirical solution of dY/dCm and its phase angle. Biophys. J. 87, 714-727.
14. Santos-Sacchi, J., Kakehata, S. & Takahashi, S., 1998. Effects of membrane potential on the voltage dependence of motility-related charge in outer hair cells of the guinea-pig. J. Physiol 510 (Pt 1), 225-235.

15. Meltzer, J. & Santos-Sacchi, J., 2001. Temperature dependence of non-linear capacitance in human embryonic kidney cells transfected with prestin, the outer hair cell motor protein. Neuroscience Letters 313, 141-144.
16. Santos-Sacchi, J. & Huang, G., 1998. Temperature dependence of outer hair cell nonlinear capacitance. Hear. Res. 116, 99-106.
17. Song, L., Seeger, A. & Santos-Sacchi, J., 2005. On membrane motor activity and chloride flux in the outer hair cell: lessons learned from the environmental toxin tributyltin. Biophys. J. 88, 2350-2362.
18. Rybalchenko, V. & Santos-Sacchi, J., 2003. Cl- flux through a non-selective, stretch-sensitive conductance influences the outer hair cell motor of the guinea-pig. J. Physiol 547, 873-891.
19. Rybalchenko, V. & Santos-Sacchi, J., 2003. In: Biophysics of the Cochlea: From Molecules to Models. Gummer, A. (ed.), World Scientific Publishing, Singapore, pp. 116-126.
20. Iwasa, K.H., 1993. Effect of stress on the membrane capacitance of the auditory outer hair cell. Biophys. J. 65, 492-498.
21. Dong, X.X., Ospeck, M. & Iwasa, K.H., 2002. Piezoelectric reciprocal relationship of the membrane motor in the cochlear outer hair cell. Biophys. J. 82, 1254-1259.
22. Adachi, M. & Iwasa, K.H., 1999. Electrically driven motor in the outer hair cell: effect of a mechanical constraint. Proc. Natl. Acad. Sci. U. S. A 96, 7244-7249.
23. Ehrenstein, D. & Iwasa, K.H., 1996. Viscoelastic relaxation in the membrane of the auditory outer hair cell. Biophys. J. 71, 1087-1094.
24. Santos-Sacchi, J. & Wu, M., 2004. Protein- and lipid-reactive agents alter outer hair cell lateral membrane motor charge movement. J. Membr. Biol. 200, 83-92.
25. Takahashi, S. & Santos-Sacchi, J., 1999. Distortion component analysis of outer hair cell motility-related gating charge. Journal of Membrane Biology 169, 199-207.

Comments and Discussion

Gummer: The approximately 5-pF adaptation that you demonstrated required a 200-mV change of transmembrane potential. How large is adaptation of nonlinear capacitance for a functionally relevant change – say 10 or 20 mV, corresponding to the maximum saturating receptor potential?

Santos-Sacchi: It is preferable to talk about how large a shift in V_{pkcm} occurs rather than change in C_m, since the magnitude of C_m change will depend on the magnitude of NLC, which varies among OHCs and across voltage. We reported previously (Santos-Sacchi et al., J. Physiology, 1998) that the voltage shift's sensitivity to delta V_m is greatest near V_{pkcm}, where its slope factor, b (for a sigmoidal fit, see Fig. 2c of the 1998 paper) was reported to be 22.2 mV at a membrane potential of -54 mV. These data were collected with high intracellular chloride. We have preliminary evidence that when we keep intracellular Cl near normal (about 10 mM - see Santos-Sacchi et al., J. Neuroscience, 2006) the sensitivity is increased.

MODELING THE COCHLEAR MICROPHONIC IN PRESTIN KNOCKOUT MICE

M.A. CHEATHAM, K. NAIK, J. SIEGEL, P. DALLOS

Communication Sciences and Disorders, The Hugh Knowles Center, Northwestern University, 2240 Campus Dr., Evanston IL 60208 USA

Based on the assumption that individual hair-cell current contributions are proportional to local basilar-membrane amplitude and phase, a model simulating the cochlear microphonic (CM) measured at the round window in mice was constructed, similar to what others have done [1, 2] for guinea pig. Hair-cell contributions were vectorially summed along the length of the cochlear partition. Due to rapidly changing phase in the amplified tip region of the traveling wave, this segment contributes negligibly to the summed response, except at very high frequencies. As a result, the CM is usually dominated by hair-cell outputs on the mechanically linear tails of the traveling wave, i.e., the CM reflects passive cochlear responses. The CM recorded at the round window from mice lacking prestin is consistent with model predictions. Hence, the CM is not a good metric to assay integrity of the cochlear amplifier.

1 Introduction

Mechanical stimulation of cochlear hair cells generates transducer currents, which produce changes in extracellular current flow. The latter is detectable as a voltage change measured across any available resistance, i.e., a recording electrode sees an electrical potential that is a remote manifestation of hair-cell activity. Because an acoustic stimulus activates groups of hair cells, the recording electrode integrates currents produced by a large number of individual generators. Phase changes in the amplified tip region of the traveling wave largely cancel responses from hair cells with characteristic frequencies (CF) near the stimulus frequency [3]. As a result, the CM is thought to be dominated by hair cell receptor currents produced on the linear tails of mechanical excitation patterns. In contrast, the compound action potential (CAP) elicited by low-level inputs reflects responses of spatially localized single neurons innervating frequency-specific regions along the cochlear spiral. Because the CM and the CAP are frequently used to characterize cochlear performance in transgenic mouse models of hearing, a model of CM generation in mice was developed. Model simulations of response patterns in wildtype (WT) and prestin knockout (KO) mice assist our understanding of the conflicting behaviors of summed hair-cell receptor currents (CM) and neural responses when the cochlear amplifier is inactivated.

2 Methods

All recordings were made from surgically anesthetized (pentobarbital) mice. Methods were approved by the National Institutes of Health and by Northwestern University's Institutional Review Committee.

A round-window electrode was used to measure CAP thresholds and CAP tuning curves [4]. In addition, an analog of the CAP tuning curve was generated for the CM. In these experiments, a continuous tone rather than a tone burst was used to generate the CM. This modification decreased the noise floor by ~2 orders of magnitude, allowing one to resolve the CM at low levels. Masker frequencies were then introduced and adjusted in level to produce a small reduction in the CM response. Because of the reduced noise floor, CM iso-response functions could also be collected for low-criterion amplitudes.

3 Results and Discussion

Tuning curves for the CAP are compared for young WT (lower panel) and prestin KO mice at 6, 12, 16 and 19 kHz in Fig. 1. Thresholds for the CAP (bold lines) are appended for reference. Wildtype functions demonstrate the classic tip and tail configuration such that the curves are dependent on probe frequency. In contrast, the tuning curves of KOs are flat with a high-frequency cutoff that approximates that of the CAP threshold curve. In other words, the functions are independent of probe frequency, as if large numbers of contributing nerve fibers were responding to the probe even when it was presented at relatively low levels. As a consequence, the simultaneous masking paradigm does not appear to extract frequency-specific information from discrete locations in KO mice. It is safe to say, however, that single-unit responses in KOs are unlikely to display sharp tip segments.

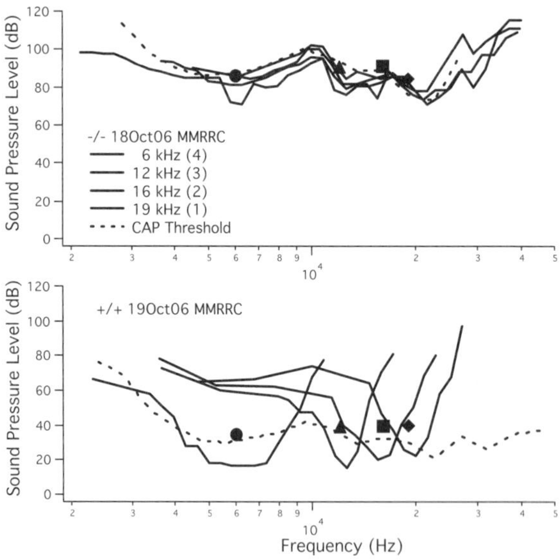

Figure 1. Simultaneous CAP masking curves are provided for a KO (upper panel) and its WT control. The probe alone generated a 25 μV N1/P1, which was then reduced by 3 dB in the presence of the masker.

Iso-response functions for the CM were also recorded for a criterion amplitude of 0.1 μV rms, as shown in Fig. 2A. Average responses from WT controls are provided for comparison. Even though the CM functions are near normal, the CAP thresholds are shifted in sensitivity attesting to the lack of amplification in prestin KO mice. Panel B provides the CM magnitude difference between mice lacking amplification and the mean WT response. Magnitude differences measured relative to WT tend to increase with frequency from ~10 dB to ~25 dB at 20 kHz. If the CM were primarily reflecting quasi-linear contributions from basal hair cells responding on their tails, one would expect that prestin KO and WT mice would produce similar CM responses, assuming they both had comparable numbers of functional OHCs at the base of the cochlea. This prediction has been difficult to demonstrate because KO mice suffer basal OHC loss due to an unknown mechanism [5].

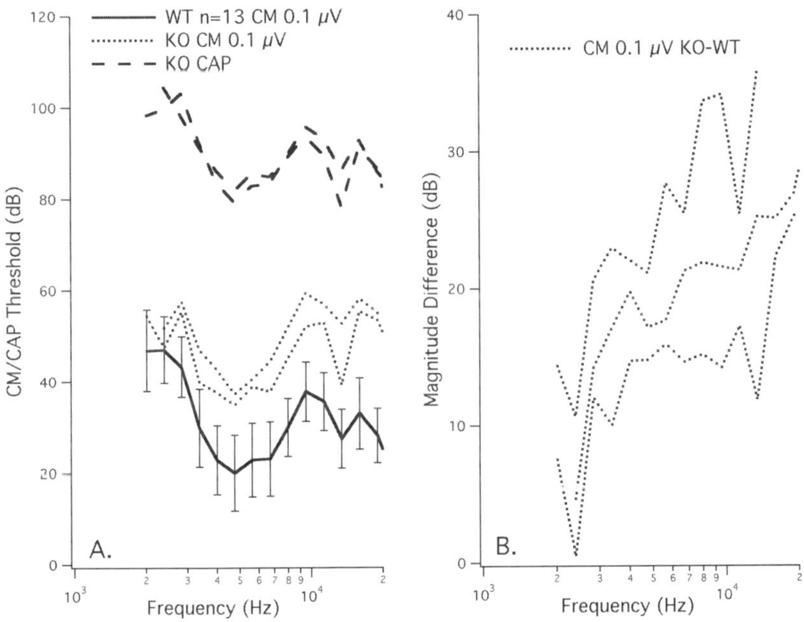

Figure 2. Panel A displays CAP thresholds (dashed lines) and CM iso-response functions (dotted lines) at 0.1 μV rms for two prestin KO mice. Average (+/- SD) WT CM responses (solid lines) are appended for comparison. CM magnitude differences between KO and WT are provided in panel B.

In order to test if near-normal CM responses reflect contributions from basal hair cells responding to stimuli well below CF, we measured simultaneous masking functions in WT mice. In addition to providing the CAP threshold curve, CM masking functions are plotted for probe tones at 6, 12, 19 and 32 kHz in Fig. 3. The low-level probe tones generate an alone response of 0.3 μV rms. Maskers are then introduced to achieve a 0.1 μV criterion decrease. Only for the 32 kHz probe do masker frequencies near the

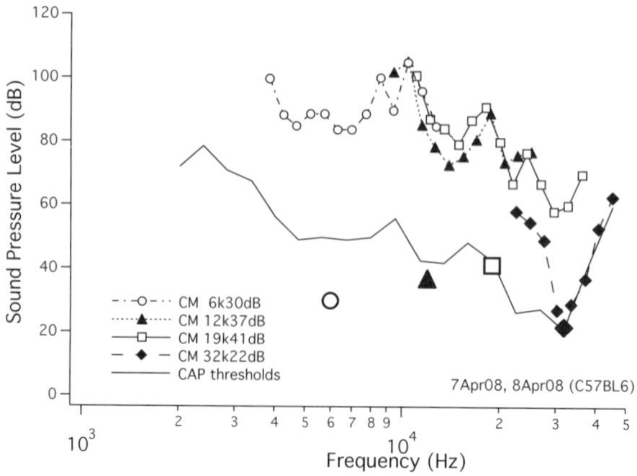

Figure 3. Simultaneous masking functions for the CM in WT mice at different probe frequencies.

probe decrease the CM in a frequency-specific manner when presented at low levels. In all other cases, the near-probe masker levels are high. This result suggests that the CM is dominated by generators at the base of the cochlea independent of probe frequency.

Although CAP responses in WT animals reflect CF-specific activity, the CM is essentially a passive response dominated by OHCs in the base [6, 7]. This result is supported by a model of CM generation specifically designed for comparing WT and prestin KO mice. As shown in Fig. 4, a simple Boltzmann-like nonlinearity compresses amplified tip segments in WT but not in KO mice. Both, however, are modified by an electrical attenuation to account for the observation that apical sources are not only further away from the round-window electrode [8] but they generate smaller receptor currents than do basal OHCs [9, 10]. The attenuation function produces an exponential gradient with maximum signal reduction of 20 dB for section 1 at the helicotrema. In addition, the gain of the cochlear amplifier in WT mice also decreases from section 600 to section 1 by 12 dB, as seen by the reduction in tip segments. Mechanical spatial plots for three frequencies peaking near the apex (section 100 at ~8 kHz), in the middle (section 300 at ~20 kHz) and near the base of the cochlea (section 500 at ~50 kHz) [11] are provided in Fig. 4 for the WT case. KO plots are similar but lack tip segments.

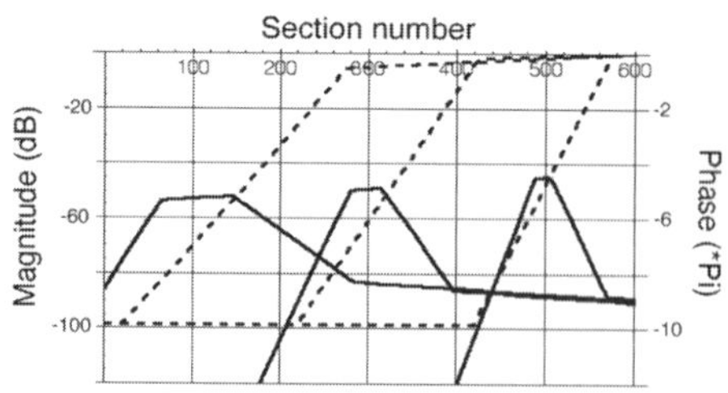

Figure. 4. Mechanical magnitude and phase (dashed lines) patterns are shown for sections 100, 300 and 500.

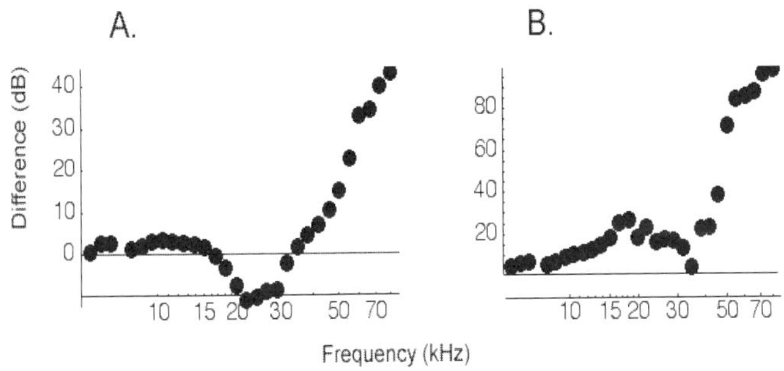

Figure 5. Differences between WT and KO CM as a function of stimulus frequency. The model simulation in panel A represents the low-magnitude case where all OHCs in both WT and KO are considered to be functional. However, in panel B, a prestin-related OHC loss is simulated.

The local CM contribution from any given section is proportional to the local amplitude weighted by the attenuation factors. Individual contributions are then summed vectorially to estimate the CM seen by the round-window electrode positioned at section 600.

Model simulations shown in panel A of Fig. 5 indicate that WT controls produce a larger CM but only at frequencies greater than ~40 kHz. However, when adjustments are made for the loss of OHCs in prestin KO mice, differences in CM are expected, as shown in panel B. In order to simulate the effect of OHC loss in KO mice [7], CM contributions from sections 1-400 were summed versus 1-500 for WT controls. Contributions from sections 500-600 are not included because mice on this 129/C57BL6 mixed background suffer age-related hearing loss, with the result that even WT mice exhibit a CAP threshold shift at very high frequencies. The magnitude differences plotted in panel B indicate that the CM in KO mice is smaller than in WT. This difference increases with frequency such that the difference at 20 kHz is ~25 dB. These predictions are not unlike the results provided for the KO mice plotted in Fig. 2. Taken together, the data suggest that the conventionally measured CM is not a useful tool to quantify the influence of amplification on cochlear function. In fact, for mice with good hair-cell preservation, the CM should be near normal at frequencies less than 40 kHz.

Evidence that the round-window CM, except at very high frequencies, primarily reflects passive basilar membrane mechanics may well explain some paradoxical findings in the literature, since the latter are usually obtained at relatively low frequencies. For example, activation of the crossed olivocochlear bundle results in a sizable reduction in near-CF, low-level neural responses, while the CM response is often seen to increase [reviewed in 12]. This dichotomy is also observed with application of salicylate [13, 14]. Because it is now known that salicylate interferes with prestin function [15], it can be argued that neural responses reflect changes in amplification in the tip regions of tuning curves. In contrast, the passive CM may simply increase due to efferent-induced OHC hyperpolarization [16, 17, 18] or salicylate-induced simulation of a KO-like response, as shown in Fig. 5A between ~17 and 30 kHz.

Acknowledgments

This work was partially supported by grant DC00089 from the National Institutes of Health. C.T. Anderson collected some of the data.

References

1. Patuzzi, R.B., 1987. A model of the generation of the cochlear microphonic with nonlinear hair cell transduction and nonlinear basilar membrane mechanics. Hear. Res. 30, 73-82.
2. Strelioff, D., 1973. A computer simulation of the generation and distribution of cochlear potentials. J. Acoust. Soc. Am. 54, 620-629.
3. Whitfield, I.C., Ross, F.H., 1965. Cochlear microphonic and summating potentials and the outputs of individual hair cell generators. J. Acoust. Soc. Am. 38, 126-131.
4. Dallos, P., Cheatham, M.A., 1976a. Compound action potential (AP) tuning curves. J. Acoust. Soc. Am. 59, 591-597.
5. Wu, X., Gao, J., Guo, Y., Zuo, J., 2004. Hearing threshold elevation precedes hair-cell loss in prestin knockout mice. Brain Res. Mol. Brain Res. 126, 30-37.
6. Dallos, P., Cheatham, M.A., 1976b. Production of cochlear potentials by inner and outer hair cells. J. Acoust. Soc. Am. 60, 510-512.
7. Patuzzi, R.B., Yates, G.K., Johnstone, B.M., 1989. The origin of the low-frequency microphonic in the first cochlea turn of guinea-pig. Hear. Res. 39, 177-188.
8. G. von Békésy, 1960. Experiments in Hearing, McGraw-Hill, New York.
9. Ricci, A.J., Crawford, A.C., Fettiplace, R., 2003. Tonotopic variation in the conductance of the hair cell mechanotransducer channel. Neuron 40, 983-990.
10. He, D.Z.Z., Jia, S., Dallos, P., 2004. Mechanoelectrical transduction of adult outer hair cells studied in a gerbil hemicochlea. Nature 429, 766-760.
11. Müller, M., von Hünerbein, K., Hoidis, S., Smolders, J.W.T. (2005). A physiological place-frequency map of the cochlea in the CBA/J mouse. Hear. Res. 202, 63-73.
12. Guinan, J.J., 1996. Physiology of olivocochlear efferents. In: Dallos, P., Popper, A.N., Fay R.R. (Eds.), The Cochlea, Springer-Verlag, New York, p. 435-502.
13. Stypulkowski, P.H., 1990. Mechanisms of salicylate ototoxicity. Hear. Res. 46, 113-145.
14. Fitzgerald, J.J., Robertson, D., Johnstone, B.M., 1993. Effects of intra-cochlear perfusion of salicylates on cochlear microphonic and other auditory responses in the guinea pig. Hear. Res. 67, 147-156.
15. Oliver, D., He, D.Z.Z., Klöcker, N., Ludwig, J., Schulte, U., Waldegger, S., Ruppersberg, J.P., Dallos, P., Fakler, B., 2001. Intracellular anions as the voltage sensor of prestin, the outer hair cell motor protein. Science 292, 2340-2343.
16. Fex, J., 1967. Efferent stimulation in the cochlea related to hair-cell dc activity: study of postsynaptic activity of the crossed olivo-cochlear fibers in the cat. J. Acoust. Soc. Am. 41, 666-675.
17. Flock, Å., Russell, I.J., 1976. Inhibition by efferent nerve fibres, action on hair cells and afferent synaptic transmission in the lateral line canal organ of the burbot *Lota lota*. J. Physiol. 257, 45-62.
18. Art, J.J., Fettiplace, R., Fuchs, P.A., 1984. Synaptic hyperpolarization and inhibition of turtle cochlear hair cells. J. Physiol. 356, 525-550.

USING A LARGE SCALE COMPUTATIONAL MODEL TO STUDY THE EFFECT OF LONGITUDINAL AND RADIAL ELECTRICAL COUPLING IN THE COCHLEA

PAVEL MISTRÍK

UCL Ear Institute, Gray's Inn Road, London WC1X8EE, UK

JONATHAN ASHMORE

UCL Ear Institute and Division of Neuroscience, Physiology and Pharmacology, UCL, Gower Street, London WC1E 6BT, UK

We describe a large scale computational model of electrical current flow in the cochlea which is constructed by a flexible Modified Nodal Analysis algorithm to incorporate electrical components representing hair cells and the intercellular radial and longitudinal current flow. The model is used as a laboratory to study the effects of changing longitudinal gap junctional coupling, and shows the way in which cochlear microphonic spreads and tuning is affected. The process for incorporating mechanical longitudinal coupling and feedback is described. We find a difference in tuning and attenuation depending on whether longitudinal or radial couplings are altered.

1 Introduction

1.1. *Cochlear amplification in 3-dimensions*

Despite being a recurrent topic throughout workshops in this series, the question remains of how precisely outer hair cells contribute to enhanced cochlear mechanics. Somatic motility of individual cells can be driven at high frequencies and the molecular motor, prestin, seems to be sufficiently simple to allow operation at high frequencies [1]. It is known from specific genetic lesions that there is a major role for OHC electromotility in mechanics. However, as the OHC mechanism is voltage driven, it was recognized early on that the membrane RC time constant, τ, is limiting. Typically $\tau > 1$ ms, restricting the frequency response, and other solutions to the problem need to be considered. The resolution of this conundrum have involved numerous interesting suggestions, and the possibility that mammalian mechanics is driven by hair bundle forces has not been ruled out (reviewed in [2]).

Electromotility is driven by the potential across the OHC basolateral membrane. This is a difficult quantity to measure experimentally *in vivo* [3]. To investigate the transmembrane potential of hair cells in the cochlea *in situ* we have resorted to an *in silico* model which takes into account current flow in the longitudinal direction, while allowing all the cells to be stimulated by a basilar membrane stimulus delivering the correct phase and amplitude stimulus. In this large scale model it then becomes possible to investigate the effects of altering longitudinal and radial coupling, both changes which are potentially critical in understanding the effect of mutations on the sensitivity of the cochlea.

2 Methods

2.1 Computational techniques

The model developed here used conventional electrical circuit elements to generate current flow in the cochlea [4]. The model is a 3-D structure with adjacent sections containing circuit elements representing IHCs and OHCs in the organ of Corti coupled with longitudinal resistances representing the gap junctions between the sections. Specific nodal connections are chosen. With the inclusion of membrane capacitances each section is a standard circuit [5]. Figure 1 shows the way in which these sections are organized to represent the cochlea.

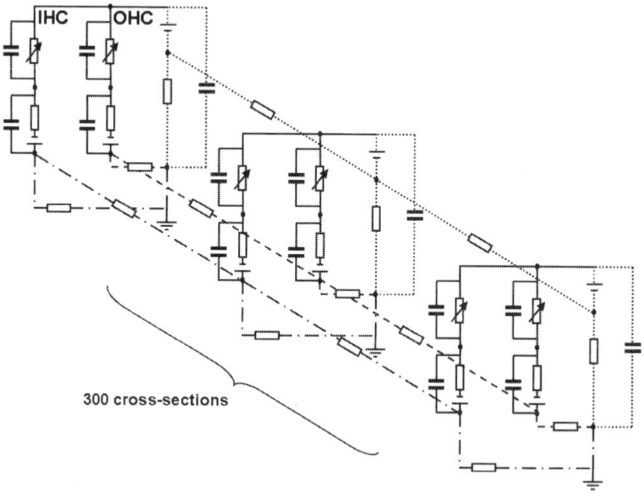

Figure 1. Model of the circuit elements used to build the cochlear model. Two cell types are included (IHC and OHC) . The endocochlear potential is constructed from a voltage source in stria vascularis (···). The variable resistances are the transducer conductances whose instantaneous values are determined by a cochlear mechanical model. The connectivity in the organ of Corti and spiral limbus due to gap junction networks is represented by resistive links (--- and -·-, respectively).

To determine the pattern of current flow within the cochlear equivalent electrical circuit (Figure 1), we developed the solution techniques based on Modified Nodal Analysis (MNA) [6] and modified from SCAM (developed by E. Cheever http://www.swarthmore.edu/NatSci/echeeve1/). MNA allows elements and their connectivities within a circuit of arbitrary complexity to be written down as a table and then to be mapped into a connectivity matrix. Within this approach, the time course of ion flow is described by a set of differential algebraic equations (DAEs) of index 1 as they can be represented as

$$\mathbf{C}dV/dt = -\mathbf{G}V + I \qquad (1)$$

where **C** is the (singular) matrix of capacitances, **G** the matrix of conductances (including the time-independent membrane conductances and the time-dependent contribution from the mechano-transducer channels $\mathbf{g_T}$). I is the the vector of independent current and voltage sources and includes the stria vascularis and the hair cell Nernstian potentials. V is the unknown vector of potentials at each node of the circuit. Both **C** and **G** matrices constructed following the MNA rules are very large and so automatic generation becomes imperative.

The time-dependent input to the set of transducer conductances was determined from a physiologically realistic model of cochlear mechanics [7]. The code allows either complex waveform input in the time domain or single frequency inputs in the frequency domain. All computations were carried out using Matlab v. 7 on a standard desktop PC.

2.2 *Parameter values*

The parameters used for the model are derived from the literature, using mainly data from guinea pig. The circuit values are scaled from isolated cell data to the model section so that 300 sections make up the cochlear length. Thus depending upon the discretization parameters have been rescaled to reflect the true mapping of the physiology onto hair cells, and correspond to 6 IHCs and 18 OHCs per section, Table 1. The circuit elements for each section are self explanatory. R_{SV} , R_{OC} , R_{SL} = resistance of the stria vascularis, the organ of Corti, and spiral limbus (the return path for OHCs and IHCs currents), respectively). C_{SV} (the capacitance of the stria vascularis) was estimated from the surface area of the cells bordering scala media and checked against simulations.

3 Results

The model allows the transmembrane potentials within a cross section any cellular element to be calculated when the input to the array of transducer conductances is known. Both time domain and frequency domain calculations were carried out. In the latter case the computational time is considerably reduced (by over an order of magnitude). Values of the longitudinal resistances were chosen so as to obtain compatible current flow with measurements made of the cochlear microphonic. When just a single section was stimulated, the current spread along the cochlea with an exponential space constant of approximately 1mm, consistent with measured values for CM spread.

The model allows the transmembrane potential $\Delta V = V_m - V_e$ to be computed where V_e is the extracellular potential. V_e contains contributions from the potentials induced by currents generated by neighboring stimulated hair cells. As seen in Table 1, the experimentally observed variation of OHC basolateral conductance with position along the cochlea [9; 11] was included in the calculations below. Then the attenuation of the OHC transmembrane potential ΔV with frequency is significantly reduced [4]. We find that the transmembrane potentials are considerably enhanced over the single isolated cell receptor potentials at frequencies above about 10 kHz.

Table 1. Parameter values used for calculation of current flow in the model for a 300 section discretization. Where two values are given the first value refers to the value at the basal cochlea and the second to the apical values; an exponential interpolation occurs along the duct. Longitudinal value refers to the connectivity of non-hair cell circuit elements.

Component	Node 1	Node 2	Radial value	Longitudinal value R_L	Ref
R_{AI}	1	2	6.7 MΩ	-	Calculated
R_{BI}	2	4	20 MΩ	-	[8]
R_{AO}	1	3	1.4 (70) MΩ	-	Calculated
R_{BO}	3	5	0.6 (30) MΩ	-	[9]
C_{AI}	1	2	6 pF	-	[10]
C_{BI}	2	6	60 pF	-	[8]
C_{AO}	1	3	10.5 pF	-	Estimated
C_{BO}	3	7	176 pF	-	[11]
C_{SV}	1	0	5 nF	-	Estimated
V_{SV}	1	8	80 mV	-	[12]
V_I	4	6	-100 mV	-	[13]
V_O	5	7	-100 mV	-	[14]
R_{SV}	8	0	0.1 MΩ	0.3 MΩ	Variable
R_{SL}	6	0	6.7 MΩ	20 MΩ	Variable
R_{OC}	7	0	0.1 MΩ	0.35 MΩ	Variable

Two changes were made in order to investigate the effects of altered current circulation around the cochlea. In the first case, both the longitudinal and radial conductances corresponding to coupling within the organ of Corti (i.e $1/R_L$ and $1/R_{OC}$ in Table 1) were reduced by up to 10-fold. These changes were chosen so as to mimic the loss of gap junctional connections. It was found that although there was relatively little effect at low frequencies, there was a significant increase in attenuation at high frequencies. In this case the transmembrane receptor potential tends towards a 6 dB/ve roll off as anticipated for isolated cells (Figure 2 A.)

In the second set of changes only the radial coupling (R_{OC}) was scaled to simulate the reduction of transport expression and hence reduction of K^+ transport within the organ of Corti (Figure 2 B). The effect was qualitatively different from the previous case. The attenuation of the receptor potential was increased for intermediate conductance changes and the same progressive reduction as for the gap junction simulation was not found. The shape of the electrical tuning curve was affected for even small differences parameters. The Q_{10dB} of the electrical tuning was also reduced. Sensitive to small changes in radial

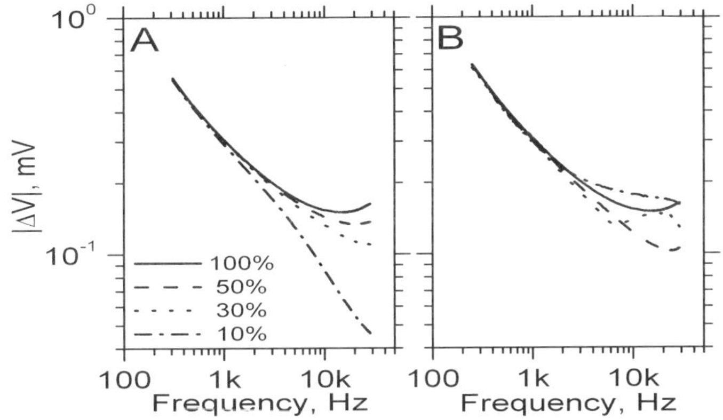

Figure 2. Effect of decreasing both longitudinal and radial coupling (A) and radial coupling alone (B) in the organ of Corti. The curves represent the peak transmembrane potentials, ΔV, of the array of OHCs along the cochlea as a function of their best frequency. 100% refers to values used for the optimal match of the parameters between the model tuning curves and the experimental data. As the sections become progressively decoupled, the attenuated receptor potential approximates the frequency dependence of the OHCs at their best frequencies.

conductances, the Q_{10dB} was much less sensitive to changes in longitudinal conductance changes.

4 Discussion

The model provides a laboratory in which to study the effects of alteration of components on the way in which current flows around the sensory cells. The feedback between the potentials within the circuit elements and the force generation of the OHCs has not been included for reasons of computational expedience. In principle there feedback can be implemented although experience suggests issues of numerical stability are critical in the time domain simulations. Large in silico models allow a study of different scales for current flow. In the case of the cochlea, current flowing around the OHCs generated over a scale comparable to the wavelength of the BM travelling wave near the characteristic frequency point can significantly alter the extracellular current flow around individual cells and thereby modify the transmembrane potential driving the motor element in OHCs. It has often been stated that the filtering property of the OHC membrane precludes many proposals for electromotility as the source of cochlear amplification. The present simulations suggest that other factors need to be considered as well. Nevertheless relatively small perturbations of the coupling which are likely to be involved in K+ recirculation may affect the tuning properties of the cochlear as a whole, and these are the kinds of perturbation which occur for specific gene mutations.

Acknowledgments

This work is supported by European Commission FP6 Integrated Project EUROHEAR, LSHG-CT-20054-512063. We thank Fabio Mammano and Renato Nobili for discussion of cochlear mechanical input and provision of the source code for the basilar membrane mechanics.

References

1. Frank, G., Hemmert, W., Gummer, A.W. 1999. Limiting dynamics of high-frequency electromechanical transduction of outer hair cells. Proc Natl Acad Sci U S A 96, 4420-5.
2. Ashmore, J. 2008. Cochlear outer hair cell motility. Physiol Rev 88, 173-210.
3. Fridberger, A., de Monvel, J.B., Zheng, J., Hu, N., Zou, Y., Ren, T., Nuttall, A. 2004. Organ of Corti potentials and the motion of the basilar membrane. J Neurosci 24, 10057-63.
4. Mistrík, P., Mullaley, C., Mammano, F., Ashmore, J. 2008. 3D current flow in a large scale model of the cochlea and the mechanism of amplification of sound Journal of the Royal Society Interface in press.
5. Dallos, P. 1984. Some electrical circuit properties of the organ of Corti. II. Analysis including reactive elements. Hear Res 14, 281-91.
6. Litovski, V., Zwolinski, M. 1997. VLSI Circuit Simulation and Optimization Kluwer Academic Publishers.
7. Mammano, F., Nobili, R. 1993. Biophysics of the cochlea: linear approximation. J Acoust Soc Am 93, 3320-32.
8. Raybould, N.P., Jagger, D.J., Housley, G.D. 2001. Positional analysis of guinea pig inner hair cell membrane conductances: implications for regulation of the membrane filter. J Assoc Res Otolaryngol 2, 362-76.
9. Mammano, F., Ashmore, J.F. 1996. Differential expression of outer hair cell potassium currents in the isolated cochlea of the guinea-pig. J Physiol 496 (Pt 3), 639-46.
10. Lopez-Poveda, E.A., Eustaquio-Martin, A. 2006. A biophysical model of the inner hair cell: the contribution of potassium currents to peripheral auditory compression. J Assoc Res Otolaryngol 7, 218-35.
11. Housley, G.D., Ashmore, J.F. 1992. Ionic currents of outer hair cells isolated from the guinea-pig cochlea. J Physiol 448, 73-98.
12. Bekesy, G. 1952. Direct observation of the vibrations of the cochlear partition under a microscope. Acta Otolaryngol 42, 197-201.
13. Russell, I.J., Sellick, P.M. 1978. Intracellular studies of hair cells in the mammalian cochlea. J Physiol 284, 261-90.
14. Cody, A.R., Russell, I.J. 1987. The response of hair cells in the basal turn of the guinea-pig cochlea to tones. J Physiol 383, 551-69.

Comments and Discussion

Cheatham: I was confused by some of the nomenclature and the statements that 300 sections make up the cochlear length and that 6 sections are one cell thick. The 2 cannot both be correct. What an I missing?

Ashmore: For reasons of computer power, we could not include the nodes from every cell. We therefore divided the cochlear length into 300 sections, each (model) section containing 6 parallel (anatomical) sections with a total of 6 IHCs and 18 OHCs. The numbers are approximately correct for guinea pig. The rest of our circuit elements follow, we hope, the nomenclature in the literature (e.g. Dallos, 1984; Hear Res 14, 281).

Gardner-Medwin: Two things puzzle me: (1), As I understand it, the model omits extracellular conductance - the major conductance linking some of the nodes. How can this be realistic? And (2),.if the V_m across the basolateral membrane impedance of an OHC falls off less fast than 6dB/octave (or not at all) at high frequencies, then the current through the basolateral membrane (and therefore the apical transducer current) must be increasing with frequency. This is true regardless of the complexity of the electrical network outside the cell. So what accounts for this increase of transducer current with frequency in the model? I also have a related comment: In my poster, I showed that the maximum OHC amplification of a local resonance is achieved (in a simple model), when the membrane time constant is comparable with the time constant for viscous damping of the passive resonance.

Ashmore: The extracellular current paths are included in the model but they are lumped in with the intercellular conductances representing, for example, gap junctions. The grain of this simulation is really quite coarse, but finer details of the circuitry could be added with the MNA technique we use.
Yes, we agree that the apical transducer current in OHCs has to increase in more basal cells. There seems to be supporting experimental evidence for this in recent work in mammalian as well as turtle hearing organs. In this model we scaled up both basolateral and apical OHC conductances towards the base. A similar assumption is also made in the model of Karl Grosh and colleagues in this meeting.

Allen: I have a comment on the model you presented. It is my view that the Dallos and Evans model has a serious flaw that they failed to detect, but that has now been identified by several investigators. The problem is that when current flows out the bottom of the cell, this reduces the voltage drop seen by the membrane, it does not increase it. The reason for this is that the loop voltage sums to zero, so when there is a voltage drop in the fluid return path, this reduces the voltage available to the membrane. If this is not obvious by this brief statement, I would be happy to provide a more formal analysis.

This problem in phase would be addressed only by coupling more basal points to the cell in question, thereby taking advantage of the natural delay in the cochlea. Such a longitudinal coupling would require a complex electrical configuration since only basal current would be allowed, not the local local current, in order to not disrupt the needed 180 degree phase shift. Thus, in my view it, will not be possible to cancel the membrane capacitance in this way. So, some questions:

Have I miss-read the Dallos and Evans paper, or did they appreciate and resolve this essential phase problem?

Why did you not mention this phase problem in your presentation? In fact, you seemed to have verified it in your calculation, if I remember correctly.

Have you included BM phase delay in your equations to solve the problem?

How do you make the local "out of phase" voltage to be smaller than the more basal "in phase" voltage, so as to boost rather than reduce the membrane voltage, thereby canceling the capacitance roll-off with frequency?

Ashmore: We quite agree with this analysis and that the problem of how to generate the phase shift is critical. I do not think that the Dallos and Evans model sorts the problem out properly, as the phase of the basilar membrane displacement is not included consistently. In fact it may be virtually impossible to do this in a lumped model of the sort from which they argue. All we do in our approach is a) to include the effect of the phase delay in the basilar membrane and b) to ensure that sufficient current is propagated from the basal end of the cochlea to influence the voltage drop across the basolateral membrane of the outer hair cell. a) is achieved by using a model which does generate reasonable phase delays in the BM response and hence in the input to the OHC transducer. We chose to use the Mammano-Nobili model for this purpose as an adequate fit to experimental data, rather than say one of the analytical expressions used psychophysically, such as the gamma-chirp. b) is achieved by increasing the current flow through basal OHCs, by increasing the basolateral and transducer conductances. We have assumed that these currents increase exponentially from apex to base, although this growth requirement may not be absolutely necessary. Experimental evidence supports this proposal, both in patch clamp data from isolated mammalian hair cells and, increasingly, from measurements of the transduction currents in mammalian organs of Corti. The final ingredient is to add longitudinal coupling in a way which is consistent with the space constant of the cochlear microphonic and the geometry. This ensures that the both phase and magnitude of the potential at the OHC basolateral membrane is effectively a cancellation of the capacitance roll-off.

So we think that the 'complex electrical configuration' to which you refer is in effect provided by the cochlear gradient of transducer current. It is also worth noting that without the growth in the transepithelial currents, as in the original Dallos and Evans model, the OHC membrane potentials roll-off at 6dB / octave, even with longitudinal coupling. There are more details in our recent paper which has now appeared in J. R. Soc. Interface, doi 10.1098/rsif.2008.0201 (Published online 5/08/08).

VOLTAGE AND FREQUENCY DEPENDENCE OF CHARGE TRANSFER BY PRESTIN: AN ELECTRO-DIFFUSION MODEL

S. SUN[1], B. FARRELL[2], M. CHANA[3], S. FENG[3], G. OSTER[4], W. BROWNELL[2], A. SPECTOR[3]

[1]*Department of Mechanical Engineering, Johns Hopkins University, 3400 N. Charles Street, Baltimore, MD 21218, USA*

[2]*Bobby R. Alford Department of Otolaryngology, Baylor College of Medicine, One Baylor Plaza, Houston, TX 77030, USA*

[3]*Department of Biomedical Engineering, Johns Hopkins University, 720 Rutland Ave, Baltimore, MD 21205, USA*

[4]*Department of Molecular and Cellular Biology, Berkeley, CA 94720, USA*

Prestin is a critical component of the motor complex that generates forces and dimensional changes in cells in response to changes in the cell transmembrane potential. We propose an electro-diffusion model to reveal the frequency and voltage dependence of electric charge transfer by prestin. The movement of the combined charge (including chloride ion and protein charges) across the membrane is described with a Fokker-Planck equation coupled to a kinetic equation that describes the binding of chloride ions to prestin. We find the total transferred charge, its capacitive and resistive components as functions of voltage and frequency.

1 Introduction

Outer hair cells found in the cochlea of the mammalian ear exhibit a unique form of motility, called electromotility, resulting from changes in the transmembrane potential [1]. Prestin, a membrane protein expressed in the plasma membrane of outer hair cells [2], is required for outer hair cell electromotility [3]. This membrane protein is a member of a solute carrier/anion transporter family [2]. Hydrophobic analysis reveals that there are up to 12 transmembrane domains in the protein [4]. Evolutionary trace analysis of the protein shows an important role of residues near the conserved transporter domain [5].

Standard cell lines transfected with prestin exhibit nonlinear capacitance which is measured by electrophysiological techniques [2, 5]. Prestin-associated charge movement requires intracellular chloride ions [6]. Internal protein charges also play a role: replacement of charged residues in prestin causes a shift of charge and nonlinear capacitance functions along the voltage axis [6]. While the composition of the transferred charge and the mechanism by which it moves are actively studied, they have not been finally established.

The displacement charge is often described with a two-state Boltzmann function. Although this discrete model describes the experimental data obtained from outer hair cells and prestin-expressing cells it does not provide a unique solution; it is not possible to determine the number of charged states within the membrane by fitting the experimental data to a two or multiple-state Boltzmann function [7]. Here we introduce a continuous approach and propose an electro-diffusion model of prestin-associated charge transfer. The model includes binding chloride ions to the protein with the subsequent diffusion of charge across the membrane. We interpret this charge as a combination of bound chloride ions together with internal protein charges. Under an AC field, the solution reveals a voltage- and frequency-dependent phase shift between the applied electric field and the transferred charge resulting in capacitive and resistive components of the charge. The total amount of charge and its capacitive component both monotonically decrease with frequency while the resistive component is non-monotonic having a maximum at a frequency of about 40 kHz. These results are consistent with experimental observations [8, 9]. Our findings on the frequency-dependence of the charge movement are important because outer hair cells and prestin provide the active component of hearing at acoustic frequencies. The revealed composition of the transferred charge can also help reconcile the discrepancies between the observed electrical [10] and mechanical [11] frequency responses in outer hair cells.

2 Model

A conceptual sketch of the prestin system is given in Fig. 1. The interaction of the protein is described in terms of the probabilities of binding and detachment and the corresponding rate constants. The total charge movement per molecule probably involves anions and protein charges, and we assume that a collective movement of the protein and chloride results in a net translocation of charge (Fig. 1a). The charge movement across the membrane (z-direction) is described by a function $\rho(z,t)$ defined as the probability density of observing the average position of the charge system at point z and time t (Fig. 1b). While prestin-associated charge may follow a 3-D trajectory, we consider movement in the z-direction reported in experiments. To define the "transferred" charge, we separate total thickness of the membrane (protein), L, into two parts with lengths $(1-l^*)\,L$ and l^*L, respectively, where l^* is a non-dimensional number such that $0 \le l^* \le 1$. The charge is considered "transferred" if it reaches the part l^*L part of the protein (Fig. 1).

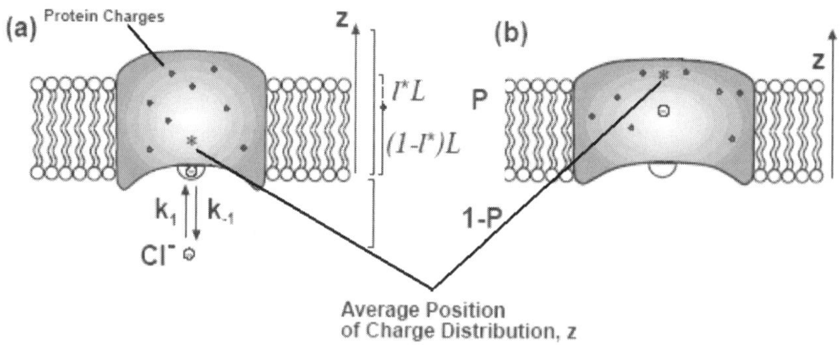

Fig. 1. Conceptual description of prestin-associated electric charge transfer. Chloride ions (minus spheres) bind the protein with association and dissociation constants k_1 and k_{-1} ; P and 1-P are the probabilities that the ion binds to the prestin or is in the cytoplasm. We assume that a combined charge, including chloride and internal protein charges (dark circles), is characterized by the average position of charge distribution z (asterisk), and it is translocated along the vertical axis. We introduce two, upper and lower, parts of the membrane with relative lengths l* and 1-l*. The charge moves to the upper part causing conformational changes in prestin (b).

The binding kinetics of chloride ions is described by the equation:

$$\frac{dP}{dt} = k_1(1-P) - k_{-1}P \quad , \tag{1}$$

and the movement of the charge in the z-direction is described by the Fokker-Planck equation:

$$\frac{\partial \rho}{\partial t} = D\left(-\frac{q\Psi}{kTL}\frac{\partial \rho}{\partial z} + \frac{\partial^2 \rho}{\partial z^2}\right) \tag{2}$$

where D is the diffusion coefficient and q is typical transferred charge. The transmembrane potential, Ψ, is composed of DC and AC components given by the equations

$$\Psi_{DC} = \Psi^i_{DC} + \Psi^e_{DC} \quad \text{and} \quad \Psi_{AC} = \Psi^e_{AC}\sin \omega t \tag{3}$$

The superscripts i and e correspond to the internal protein and external applied electric fields, respectively, and Ψ^i_{DC} is the potential reflecting the protein charges.

Eq. 2 is solved with two boundary conditions:

$$J(t, z=0) = \frac{dP}{dt} \quad \text{and} \quad J(t, z=L) = 0 , \tag{4}$$

where J is the probability flux:

$$J = D\left(\frac{q\Psi}{kTL}\rho - \frac{\partial \rho}{\partial z}\right) \tag{5}$$

The first condition in Eq. 4 describes the continuity of the probability flux at z=0, and the second equation is the condition of no transport out or into the membrane at z=L. The latter assumption can be relaxed when experimental data on transport properties of mammalian prestin become available. After we determine $\rho(z,t)$, the total probability of charge transfer is calculated as:

$$P^* = \int_{L(1-l^*)}^{L} \rho(t,z)dz \ . \tag{6}$$

In the case of application of AC electric fields, the amplitude of probability P^* is used to obtain capacitive, $P_c = P^* \cos\theta$, and resistive, $P_r = -P^* \sin\theta$, components of the transferred charge, where θ is the negative phase shift between the charge and the AC applied field.

3 Results and Discussion

Our model includes three main parameters, the dimensionless length, l^*, typical charge, q, and an effective diffusion coefficient, D, that determine prestin-associated charge

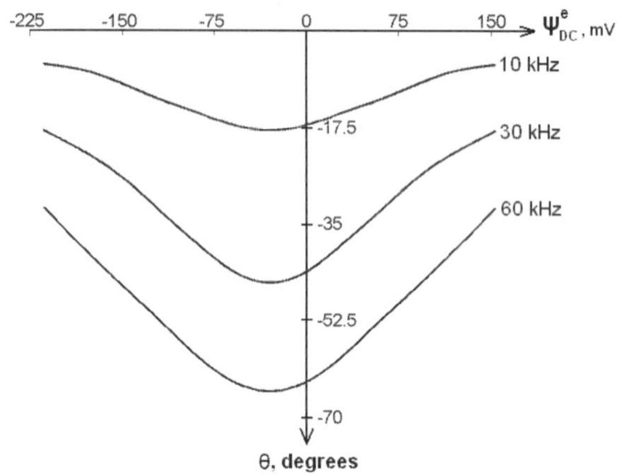

Fig. 2. Phase shift between the applied AC electric field and transferred charge as a function of the DC potential for three different frequencies.

transfer under the action of DC and/or AC electric field. In order to estimate parameters l^* *and* q, we consider the case when the system subjected to a DC field reaches its steady state. We estimate our model parameters fitting a Boltzmann approximation of available experimental data on charge transfer with the sharpness and potential of equal charge distribution close to those in [12]. As a result, we obtain $l^*=0.5$ and $q=1.5e$. To estimate the diffusion coefficient, D, we matched the experimental decrease in nonlinear capacitance (about 50% at 30kHz) in [10] with our model capacitive component of the charge. Assuming the protein spans the thickness of membrane at ~4.5 nm, the binding rates, k_1 and k_{-1}, are on the order of magnitude of 10^5 s^{-1} (for normal, about 10mM, intracellular chloride concentration) , we estimate D of $0.5x10^{-8}$ cm^2/s.

We apply our model and estimate the voltage and frequency effects on prestin-associated charge transfer. We consider prestin-associated charge transfer when an AC electric field of a constant amplitude and variable frequency is applied with a DC membrane potential. Below we present the phase shift between the applied AC field and the transferred charge (Fig. 2), the total charge (Fig. 3a), its capacitive (Fig. 3b) and resistive components (Fig. 3c) as functions of the voltage at three different frequencies. The charge and its both components are normalized by the maximal low-frequency value of P*-amplitude.

The total charge and its capacitive component are both bell-shaped as functions of voltage and monotonically decrease with frequency. As a function of voltage, the phase shift has a single minimum, and for a given voltage, it changes from zero at low frequencies to -90^0 at very high frequencies (e.g., it is about -60^0 and -75^0 at frequencies of 50kHz and 75kHz, respectively). The resistive component has an interesting dependence on frequency. It is small at both low and high frequencies reaching a maximum at frequency of about 40 kHz. Such behavior is explained by the composition

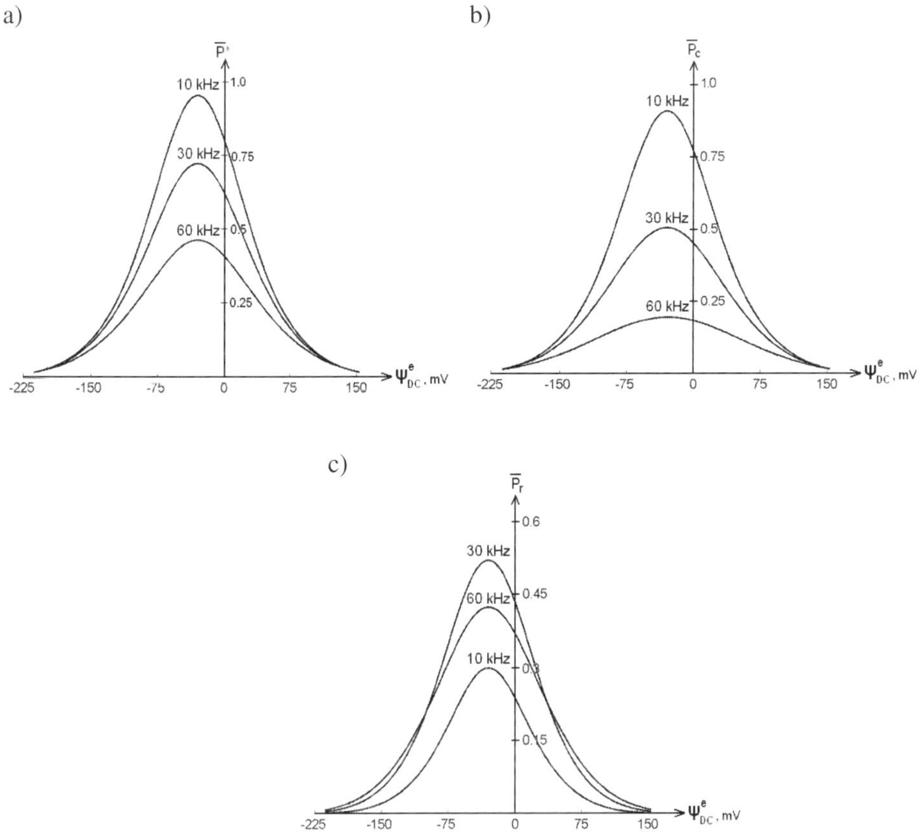

Fig. 3. Total transferred charge (a), its capacitive (b), and resistive (c) components as functions of applied DC potential at three different frequencies.

of the resistive component defined as the product of the total charge (P^*-amplitude) and ($-\sin\theta$). At low frequencies both factors are small, and at high frequencies the total charge tends to zero causing a decrease in the resistive component. As a result, the resistive component exhibits a maximum at an intermediate frequency. Note that the ratio of the resistive to capacitive component (equal to -tan θ) is a monotonic function of frequency. There is a discrepancy in 3dB and 50% frequencies obtained from studies measuring charge transfer [10] and electromotility/active force production [11]. It is possible to reconcile the differences if the resistive component of the charge is included in the interpretation of data of [10]. The total charge, including both components, has a better frequency response (e.g., greater 3dB and 50% frequencies) than the capacitive component only. Thus, the data on charge transfer [10] can be more consistent with the flatter and broader mechanical responses found in [11].

Acknowledgments

This work was supported by research grant R01 DC000354 from NIDCD (NIH).

References

1. Brownell, W.E., Bader, C.R., Bertrand, D., Ribaupierre, Y., 1985. Evoked mechanical responses of isolated cochlear outer hair cell. Science 227, 194-196.
2. Zheng, J., Shen, W., He, D.Z., Long, K.B., Madison, L.D., Dallos, P., 2000. Prestin is the motor protein of cochlear outer hair cell. Nature 405, 149-155.
3. Liberman, M.C., Gao, J., He, D.Z., Wu, X., Jia, S., Zuo, J., 2002. Prestin is required for electromotility of the outer hair cell and for the cochlear amplifier. Nature 419, 300-304.
4. Zheng, J., Long, K.B., Shen, W., Madison, L.D., Dallos, P., 2001. Prestin topology: localization of protein epitopes in relation to the plasma membrane. NeuroReport 12, 1929-1935.
5. Rajagopalan, L., Patel, N., Madabushi, S., Goddard, J.A., Aniat, V., Lin, F., Shope, C., Farrell, B., Lichtarge, O., Davidson, A.L., Brownell, W.E., Pereira, F.A., 2006. Essential helix interaction in the anion transporter domain revealed by evolutionary trace analysis. J. Neurosci 26, 12727-12734.
6. Oliver, D., He, D.Z., Klocker, N., Ludwig, J., Schulte, U., Waldegger, S., Ruppersberg, J.P., Dallos, P., Fakler, B., 2001. Intracellular anions as the voltage sensor of prestin, the outer hair cell motor protein. Science 292, 2340-2343.
7. Scherer, M.P., Gummer, A.W., 2004. How many states can the motor molecule, prestin, assume in an electric field. Biophys. J. 87, L.27-L.29.
8. Lu, C.-C., Kabakov, A., Markin, V.S., Mager, S., Frazer, G.A., Hilgemann, D.W., 1995. Membrane transport mechanisms probed by capacitance measurements with megahertz voltage clamp. Proc. Natl. Acad. Sci. USA 92, 11220-11224.
9. Farrell, B., Ugrinov, R., Brownell, W.E., 2006. Frequency dependence of admittance and conductance of the outer hair cell. In: Auditory Mechanisms. Processes and Models. Nuttall, A.L., Ren, T., Gillespie, P., Grosh, K., de Boer, E., (Eds). World Scientific, New Jersey, pp. 231-332.
10. Gale, J.E., Ashmore, J.F., 1997. An intrinsic frequency limit to the cochlear amplifier. Nature 389, 63-66.
11. Frank, G., Hemmert, W., Gummer, A.W., 1999. Limiting dynamics of high-frequency electromechanical transduction of outer hair cells. Proc. Natl. Acad. Sci. USA 96, 4420-4425.
12. Huang, G., Santos-Sacchi, J., 1994. Motility voltage sensor of the outer hair cell resides within the lateral plasma membrane. Proc. Natl. Acad. Sci. USA 91, 12268-12272.

TOPOLOGICAL CHARACTERIZATION BY ATOMIC FORCE MICROSCOPY OF PRESTIN IN THE PLASMA MEMBRANE OF PRESTIN-TRANSFECTED CHINESE HAMSTER OVARY CELLS USING QUANTUM DOTS

HIROSHI WADA, MICHIO MURAKOSHI, KOJI IIDA, SHUN KUMANO

Department of Bioengineering and Robotics, Tohoku University, 6-6-01 Aoba-yama, Sendai 980-8579, Japan

1 Summary

Somatic electromotility of outer hair cells (OHCs) is thought to be driven by a membrane protein prestin [1]. This protein is believed to cause the OHCs to move reciprocally in their longitudinal direction at microsecond rates. Due to this motility, the basilar membrane is subjected to force, resulting in cochlear amplification and thus leading to the high sensitivity of mammalian hearing.

Morphologically, the lateral wall of the OHCs were shown to be densely covered with particles about 10 nm in diameter by electron microscopy [2-5], these particles being believed to be a motor protein. Imaging by atomic force microscopy (AFM) of prestin-transfected Chinese hamster ovary (CHO) cells revealed particle-like structures 8-12 nm in diameter to possibly be prestin [6]. However, since there are many kinds of intrinsic membrane proteins other than prestin in the plasma membranes of OHCs and CHO cells, it was impossible to clarify which structures observed in such membranes were prestin.

In the present study, an experimental approach combining AFM with quantum dots (Qdots), used as topographic surface markers, was carried out to detect individual prestin molecules. The inside-out plasma membranes were isolated from the prestin-transfected and untransfected CHO cells. Such membranes were then incubated with anti-prestin primary antibodies and Qdot-conjugated secondary antibodies. Fluorescence labeling of the prestin-transfected CHO cells but not of the untransfected CHO cells was confirmed. The plasma membranes of both types of CHO cells were subsequently scanned by AFM. Qdots were not seen in the AFM images of the untransfected CHO cells. But, the Qdots, about 8 nm in height, were clearly seen in the prestin-transfected CHO cells. Squarish ring-like structures, each with four peaks and one valley at its center, were observed in the vicinity of the Qdots, suggesting that prestin forms a tetramer in the plasma membranes of the prestin-transfected CHO cells. Analysis of the size of the observed structures showed two different sizes of prestin, i.e., 9.6 and 13.0 nm, suggesting the existence of the compact- and extended-state prestin.

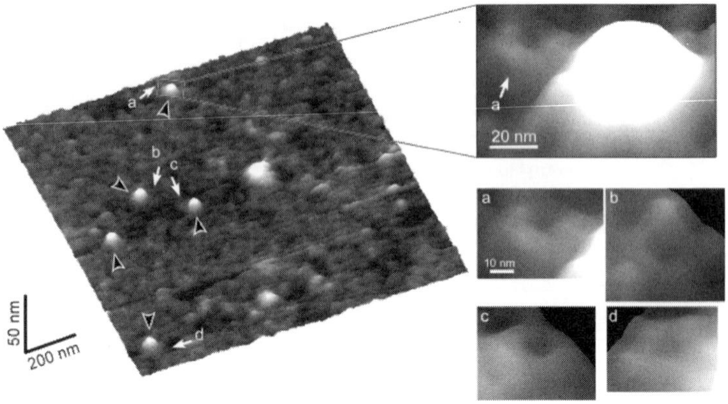

Figure 1. High-magnification 3-D AFM images obtained from an inside-out plasma membrane isolated from a prestin-transfected CHO cell. Qdots 8 nm in height were clearly observed, as indicated by black arrowheads. As shown in the magnification at the top right, a squarish ring-like structure with four peaks and one valley at its center, possibly corresponding to a prestin molecule, was observed in the vicinity of the Qdot (a). Similar structures were observed at the location indicated by arrows (b-d).

Acknowledgments

This work was supported by Grant-in-Aid for Scientific Research on Priority Areas 15086202 from the Ministry of Education, Cultures, Sports, Science and Technology of Japan to H.W., Grant-in-aid for JSPS Fellows 19002194 from the Japan Society for the Promotion of Science and Special Research Grants 11170012 and 11180001 from the Tohoku University 21st Century COE Program of the "Future Medical Engineering Based on Bio-nanotechnology" to M.M.

References

1. Zheng, J., Shen, W., He, D.Z., Long, K.B., Madison, L.D., Dallos, P., 2000. Prestin is the motor protein of cochlear outer hair cells. Nature 405, 149-155.
2. Arima, T., Kuraoka, A., Toriya, R., Shibata, Y., Uemura, T., 1991. Quick-freeze, deep-etch visualization of the 'cytoskeletal spring' of cochlear outer hair cells. Cell Tissue Res. 263, 91-97.
3. Forge, A., 1991. Structural features of the lateral walls in mammalian cochlear outer hair cells. Cell Tissue Res. 265, 473-483.
4. Kalinec, F., Holley, M.C., Iwasa, K.H., Lim, D.J., Kachar, B., 1992. A membrane-based force generation mechanism in auditory sensory cells. Proc. Natl. Acad. Sci. U.S.A. 89, 8671-8675.
5. Souter, M., Nevill, G., Forge, A., 1995. Postnatal development of membrane specialisations of gerbil outer hair cells. Hear. Res. 91, 43-62.
6. Murakoshi, M., Gomi, T., Iida, K., Kumano, S., Tsumoto, K., Kumagai, I., Ikeda, K., Kobayashi, T., Wada, H., 2006. Imaging by atomic force microscopy of the plasma membrane of prestin-transfected Chinese hamster ovary cells. J. Assoc. Res. Otolaryngol. 7, 267-278.

MEMBRANE COMPOSITION TUNES THE OUTER HAIR CELL MOTOR

L. RAJAGOPALAN, J. SFONDOURIS, J.S. OGHALAI, F.A. PEREIRA, W.E. BROWNELL
Baylor College of Medicine, One Baylor Plaza, Houston, Texas 77030, USA

Cholesterol and docosahexaenoic acid (DHA), an ω-3 fatty acid, affect membrane mechanical properties in different ways and modulate the function of membrane proteins. We have probed the functional consequence of altering cholesterol and DHA levels in the membranes of OHCs and prestin expressing HEK cells. Large, dynamic and reversible changes in prestin-associated charge movement and OHC motor activity result from altering the concentration of membrane cholesterol. Increasing membrane cholesterol shifts the q/V function ~ 50 mV in the hyperpolarizing direction, possibly a response related to increases in membrane stiffness. The voltage shift is linearly related to total membrane cholesterol. Increasing cholesterol also decreases the total charge moved in a linear fashion. Decreasing membrane cholesterol shifts the q/V function ~ 50 mV in the depolarizing direction with little or no effect on the amount of charge moved. *In vivo* increases in membrane cholesterol transiently increase but ultimately lead to decreases in DPOAE. Docosahexaenoic acid shifts the q/V function in the hyperpolarizing direction < 15 mV and increases total charge moved. Tuning of cochlear function by membrane cholesterol contributes to the exquisite temporal and frequency processing of mammalian hearing by optimizing the cochlear amplifier.

1 The outer hair cell lateral wall and its membrane-based motor

The OHC lateral wall is specialized for electro-mechanical force transduction. It is an elegant, nanoscale (~100 nm thick), trilaminate structure with the motor mechanism located in the plasma membrane or outermost layer [2]. Previous studies have demonstrated that prestin function and/or electromotility are affected by agents that partition into and alter membrane material properties [3-5]. Labeling studies suggest that lateral wall membranes contain less cholesterol than the OHC apical and synaptic plasma membranes [6]. The relatively low cholesterol level of the OHC lateral wall plasma membrane is unusual among animal cells, and may serve to modulate the function of membrane proteins. Prestin, a polytopic integral membrane protein, is an essential component of the OHC lateral wall motor [7]. Prestin greatly increases charge movement into and out of, as opposed to through, the membrane and results in a voltage-dependent nonlinear capacitance (NLC) [8-10]. There is greater than a 3 order of magnitude increase in NLC in prestin-expressing cells over that seen in prestin knockout OHCs. The increase in charge movement is much larger than the 3-5 times increase in electromechanical force production over the native membrane [8, 11]. Biological membranes are composite structures composed of lipids, proteins, glycolipids, etc. and the generation of non-linear capacitance and electromotility represents an ensemble response. The bidirectional membrane response is both mechanoelectrical in response to membrane tension and electromechanical in response to changes in the transmembrane electric field. We have examined the role of cholesterol [1, 12] and docosahexaenoic acid to characterize the relationships between prestin and its membrane environment and observed the effects of membrane cholesterol at the molecular, cellular and organ level.

2 Results

2.1 Effect of membrane cholesterol on outer hair cell function

We observed the characteristic bell-shaped voltage-dependent capacitance (NLC) with a peak, V_{pkc}, at about -0.05 V in untreated OHCs (Fig. 1A), similar to that reported earlier [1, 13]. Depletion of cholesterol shifted V_{pkc} in the depolarizing direction to about + 0.08 V, while loading excess cholesterol shifted V_{pkc} towards more hyperpolarizing voltages (< -0.13 V, Fig. 1A). Kinetic studies of these phenomena indicated that the shift in V_{pkc} occurred within minutes of adding water-soluble cholesterol (loading) or MβCD (depletion). Further, the effects of depletion and loading were reversible, and the reversal of the V_{pkc} shift occurred over a similar time course of about 5 minutes. [1] These data implicate a direct and dynamic correlation between OHC membrane cholesterol content and V_{pkc}.

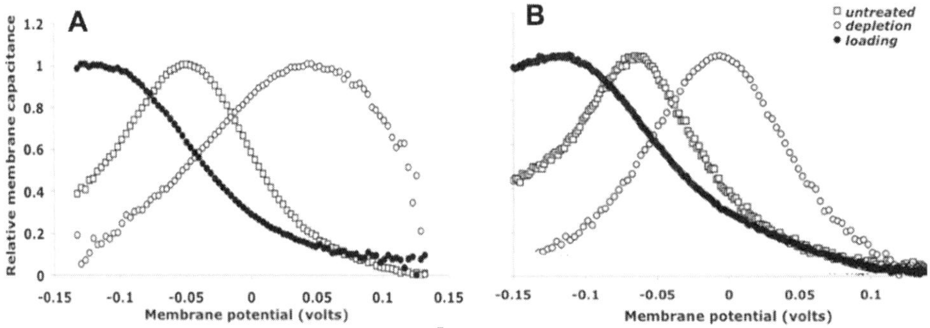

Figure 1. Effect of membrane cholesterol on non-linear capacitance (NLC) in OHCs (**A**) and prestin expressing HEK cells (**B**). Vpkc shifts to depolarized voltages upon cholesterol depletion (MβCD; open circles) and to hyperpolarized voltages upon cholesterol loading, from the control value for untreated cells -0.05 V for OHCs and -0.07 V for HEK cells (open squares). Treatments for OHC/HEK cells were: depletion: 100 μm/10 mM MβCD open circles; loading 1 mM/10 mM water-soluble cholesterol (filled circles). Traces have been normalized relative to peak capacitance. Refer to *Rajagopalan et al., J. Biol. Chem (2007)* for methods [1].

2.2 Membrane cholesterol effects on prestin-expressing HEK 293 cells are similar to those in OHCs

To study the effect of cholesterol on prestin function in particular, we investigated prestin-specific effects by cholesterol in HEK 293 cells. Cholesterol manipulations in prestin-expressing HEK 293 cells produced qualitatively similar results as in OHCs (Fig. 1B). The V_{pkc} of prestin-expressing HEK 293 cells was approximately – 0.07 V. Upon cholesterol depletion, the V_{pkc} shifted towards depolarized voltages, with an average peak at +0.0 V, while upon cholesterol loading, the V_{pkc} shifted towards hyperpolarized voltages, with an average value of about – 0.12 V. Importantly, these effects were reversible. Cholesterol depletion followed by loading, as well as cholesterol enrichment followed by depletion, both shifted the peak voltage towards the control untreated average [1]. The kinetics of these processes were similar to those measured for OHCs, with changes in the V_{pkc} occurring within minutes of addition of cholesterol or MβCD and the process was rapidly reversible [1].

2.3 Effect of cholesterol concentration on peak and magnitude of charge movement

We have measured NLC peak shifts at different membrane cholesterol concentrations, and find that the V_{pkc} shift is linearly correlated with cholesterol concentrations over a range of 5-30 pmol/µg of membrane proteins (Fig. 2A). Our results demonstrate that changes in cholesterol alter prestin-associated charge movement in HEK 293 cells, as in the native environment of the OHC, indicating direct molecular level effects of membrane cholesterol on prestin function. In addition to the effects on V_{pkc}, we have also evaluated the effect of cholesterol on the magnitude of prestin-associated charge movement, as measured by charge density. We find that while cholesterol depletion has no effect on charge density, increasing membrane cholesterol significantly decreases it. The mean charge density decreased linearly with increasing cholesterol, showing a strong inverse correlation to the membrane cholesterol concentration (Fig. 2B).

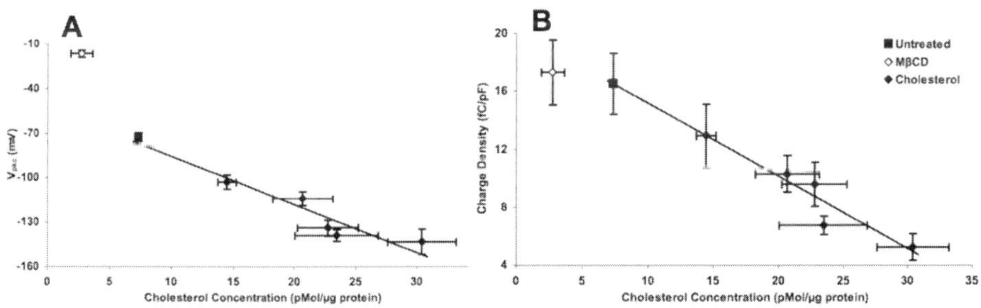

Figure 2. V_{pkc} (**A**) and charge density (**B**) are linearly dependent on the concentration of membrane cholesterol. **A**) Plot of the mean V_{pkc} (mV) versus the membrane cholesterol concentration (pMol/µg protein). In untreated cells (■; membrane cholesterol 7.4 pMol/µg protein), the mean V_{pkc} was approximately -72 mV. In MβCD-treated cells (◇; membrane cholesterol 2.8 pMol/µg protein), the mean V_{pkc} was -17 mV. Significant progressively hyperpolarizing shifts (-103 to -143 mV) in V_{pkc} were seen for each group of cholesterol-loaded cells (◆; 14.4 to 30.4 pMol/µg protein) relative to the control ($p < 0.0001$). **B**) The mean charge density (fC/pF) decreased with increasing membrane cholesterol. The mean charge densities for untreated cells (■) and MβCD-treated cells (◇) were 16.1 and 17.3 fC/pF, respectively. The series of cholesterol-treated cell groups (◆) had mean charge densities ranging from 5.2 to 12.9 fC/pF. Each data point represents data from 7-19 prestin expressing HEK cells. Horizontal and vertical error bars represent 2 standard errors relative to the mean.

2.4 *Effect of docosahexaenoic acid on peak and magnitude of charge movement*

DHA loading (using human serum albumin, HSA, as a vehicle) shifted the V_{pkc} to approximately -85 mV. The shift was statistically significant compared to both untreated ($p < 1x10^{-6}$) and control HSA-treated cells ($p < 0.0002$). We also investigated the effect of DHA on the magnitude of prestin-associated charge movement (Fig. 3). The mean charge density for DHA-loaded cells was significantly higher than both untreated ($p < 0.007$) and HSA-treated cells ($p < 0.05$) at 23.0 fC/pF. DHA loading thus induced a marked increase in the magnitude of prestin-associated charge movement.

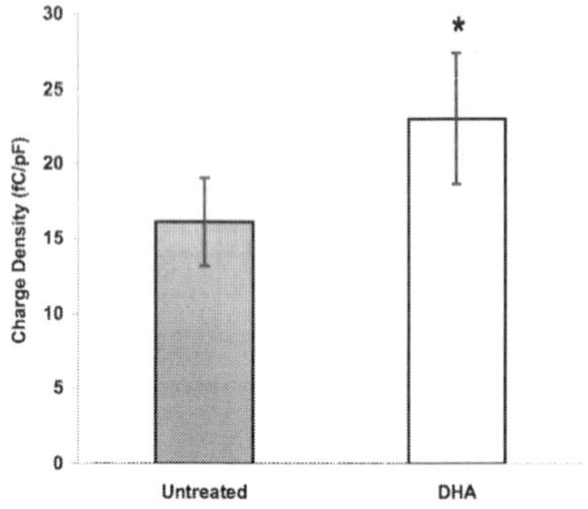

Figure 3. DHA increases charge density. The mean charge density in untreated cells (striped; n = 39) was 16.1 fC/pF. In cells loaded with DHA (black; n = 41), the mean charge density showed a statistically significant increase to 23.0 fC/pF. ($p < 0.007$). Error bars represent 2 standard errors in either direction relative to the mean.

2.5 *Modulation of cochlear function by cholesterol*

We have evaluated cochlear function *in vivo* by measuring DPOAE amplitudes following injection though the round window membrane of compounds that either increase or decrease membrane cholesterol levels (see [1] for details). Cholesterol depletion resulted in a 20 dB decrease in DPOAE amplitudes after 10 minutes from the intracochlear application of the drug. This was slightly longer than the single cell response time (data not shown). Cholesterol loading, on the other hand, resulted in an initial 2-3 dB increase in DPOAE amplitude, followed by a decrease of up to 20 dB (Fig. 4).

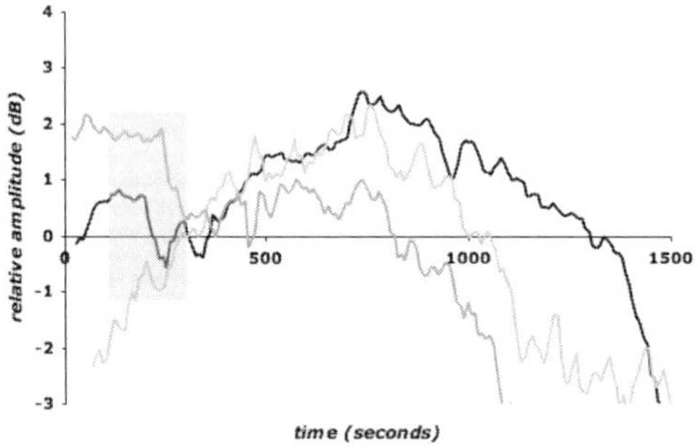

Figure 4. Effect of cochlear cholesterol loading on DPOAE amplitude. DPOAEs were measured continuously during the delivery of MβCD-cholesterol to the perilymph. Each trace represents DPOAE readings from a single animal over the timecourse of the experiment. The gray box indicates time that the micropipette was inserted through the round window membrane and any middle ear fluid aspirated. During this time, the data demonstrate artifactual changes and changes in middle ear mechanics. After this time, the DPOAEs amplitudes are referenced to 0 dB and changes in the data represent changes in cochlear otoacoustic emissions. Cholesterol loading caused an initial 2-3 dB increase in the DPOAE amplitude, followed by a progressive decrease (not shown).

3 Discussion

3.1 OHC membrane cholesterol decreases during development

V_{pkc} and charge density have been shown to change in OHCs during postnatal development [14, 15]. V_{pkc} undergoes a nearly 50 mV depolarizing shift from birth to day 14 and a large increase in charge density which displayed saturation at day 12 in rats. A portion of these shifts may result from decreases in membrane cholesterol that occurs during development [1]. The results presented in Fig. 2 show that a membrane cholesterol decrease would result in both the depolarizing shift and charge density increases previously reported. Other factors such as prestin expression, efferent innervation, changes in membrane tension, and the development of the cortical lattice may contribute to the change in prestin-associated charge movement during maturation but it is likely that cholesterol plays a key role.

3.2 Implication of membrane cholesterol for prestin function

The combination of a depolarizing shift of V_{pkc} and increasing charge density towards the values seen in adult OHCs is consistent with the low cholesterol concentrations and the uniform distribution of prestin observed in adult OHC lateral wall membranes. Prestin is present in microdomains in HEK 293 cells, and its presence in these localized domains may facilitate its interaction with itself, or with other proteins [1, 12]. It is likely that prestin exists in a dynamic equilibrium between monomeric, dimeric and perhaps higher-order oligomeric forms. The effect of cholesterol might be to 'cluster' prestin molecules, shifting the equilibrium towards dimeric or oligomeric species. We have previously reported increased self-association in the presence of increased cholesterol in HEK 293 cells. However, the incredible mechano-sensitivity of the auditory system requires that the sensory epithelium have a small mass (compare the $< 2 \times 10^5$ sensory cells of the organ of Corti to the $>> 10^6$ sensory cells in the retina). The structural benefits of small mass can scale down to the molecular level where a large mass would decrease the ease and speed of conformational changes. The demands of the sensory system may require minimal self-association, and therefore a low cholesterol level as in the mature OHC lateral wall.

3.3 Membrane cholesterol and the cochlear amplifier

The DPOAE results (Fig. 3) demonstrates how prestin function is tuned by membrane cholesterol levels. V_{pkc} is near the maximal gain for the OHC electromotility voltage-displacement function. In normal untreated OHCs, the NLC peak is at approximately -0.050 V which we postulate is slightly depolarized from the OHC resting potential (nominally between -0.060 and -0.080 V). Shifting the NLC peak in the depolarizing direction (as upon cholesterol depletion) would push the prestin-associated charge movement further out of the range of the cell's receptor potential, and reduce DPOAE amplitudes. On the other hand, shifting the NLC peak in the hyperpolarizing direction would transiently place the NLC peak over the cell's resting potential prior to decreases as the NLC peak becomes progressively more hyperpolarized. Electromotility and DPOAE amplitudes would increase as the membrane cholesterol levels increase until the receptor potentials occur during the peak of the NLC and then fall off the "sweet spot" moving towards even more hyperpolarizing values. Our results may provide a partial explanation for the known gender difference in hearing sensitivity and DPOAE magnitudes which are about 3 dB better in women. An estrogen related increase in membrane cholesterol could enhance cochlear amplifier function in the steady state prior to its regulation by other factors such as efferent stimulation, OHC turgor pressure, and stiffness of the lateral wall.

Acknowledgments

The authors thank B. Farrell for valuable suggestions and H. Liu and D. Boatright for technical assistance. Research was supported by NIDCD R01 DC00354 (WEB and FAP), DC006671 (JSO), NSF BES-0522862 (FAP) and the Deafness Research Foundation (LR). LR was supported by a Keck Center for Interdisciplinary Bioscience Training grant. JS was supported by NIDCD T32 DC007367.

References

1. Rajagopalan, L., Greeson, J.N., Xia, A., Liu, H., Sturm, A., Raphael, R.M., Davidson, A.L., Oghalai, J.S., Pereira, F.A., Brownell, W.E. 2007. Tuning of the outer hair cell motor by membrane cholesterol. J Biol Chem 282, 36659-70.
2. Brownell, W.E. 2006. The Piezoelectric Outer Hair Cell. In: Fay, R.R., Popper, A.N., (Eds.), Vertebrate hair cells. Springer, New York. pp. 313-347.
3. Kakehata, S., Santos-Sacchi, J. 1995. Membrane tension directly shifts voltage dependence of outer hair cell motility and associated gating charge. Biophys. J. 68, 2190-2197.
4. Lue, A.J.-C., Zhao, H.-B., Brownell, W.E. 2001. Chlorpromazine alters outer hair cell electromotility. Otolaryngol. Head Neck Surg. 125, 71-76.
5. Murdock, D.R., Ermilov, S.A., Spector, A.A., Popel, A.S., Brownell, W.E., Anvari, B. 2005. Effects of chlorpromazine on mechanical properties of the outer hair cell plasma membrane. Biophys J 89, 4090-5.
6. Nguyen, T.V., Brownell, W.E. 1998. Contribution of membrane cholesterol to outer hair cell lateral wall stiffness. Otolaryngol Head Neck Surg 119, 14-20.
7. Rajagopalan, L., Patel, N., Madabushi, S., Goddard, J.A., Anjan, V., Lin, F., Shope, C., Farrell, B., Lichtarge, O., Davidson, A.L., Brownell, W.E., Pereira, F.A. 2006. Essential helix interactions in the anion transporter domain of prestin revealed by evolutionary trace analysis. J Neurosci 26, 12727-34.
8. Ludwig, J., Oliver, D., Frank, G., Klocker, N., Gummer, A.W., Fakler, B. 2001. Reciprocal electromechanical properties of rat prestin: the motor molecule from rat outer hair cells. . Proc. Natl. Acad. Sci. USA 98 4178-4183.
9. Oliver, D., He, D.Z., Klocker, N., Ludwig, J., Schulte, U., Waldegger, S., Ruppersberg, J.P., Dallos, P., Fakler, B. 2001. Intracellular anions as the voltage sensor of prestin, the outer hair cell motor protein. . Science 292 2340-2343.

10. Zheng, J., Shen, W., He, D.Z., Long, K.B., Madison, L.D., Dallos, P. 2000. Prestin is the motor protein of cochlear outer hair cells. Nature 405, 149-55.
11. Zhang, R., Qian, F., Rajagopalan, L., Pereira, F.A., Brownell, W.E., Anvari, B. 2007. Prestin modulates mechanics and electromechanical force of the plasma membrane. Biophys J 93, L07-9.
12. Sturm, A.K., Rajagopalan, L., Yoo, D., Brownell, W.E., Pereira, F.A. 2007. Functional expression and microdomain localization of prestin in cultured cells. Otolaryngol Head Neck Surg 136, 434-9.
13. Santos-Sacchi, J. 1991. Reversible inhibition of voltage-dependent outer hair cell motility and capacitance. J Neurosci 11, 3096-110.
14. Belyantseva, I.A., Adler, H.J., Curi, R., Frolenkov, G.I., Kachar, B. 2000. Expression and localization of prestin and the sugar transporter GLUT-5 during development of electromotility in cochlear outer hair cells. J Neurosci 20, RC116.
15. Oliver, D., Fakler, B. 1999. Expression density and functional characteristics of the outer hair cell motor protein are regulated during postnatal development in rat. J Physiol 519 Pt 3, 791-800.

MEASUREMENT OF OUTER HAIR CELL ELECTROMOTILITY USING A FAST VOLTAGE CLAMP

MICHAEL G. EVANS

Institute for Science & Technology in Medicine, Keele University, Staffs ST5 5BG, UK

ROBERT FETTIPLACE

Department of Physiology, University of Wisconsin Medical School, Madison, Wisconsin 53706, USA

Isolated outer hair cells from rat pups (P9-P16) were voltage clamped using the whole-cell recording technique, and changes in cell length in response to step voltage changes were measured using a photodiode pair. Cell length changes were rapid and could be reasonably well fitted by single exponential functions with time constants around 0.1-0.2 ms, although particularly with larger steps double exponentials gave a better fit. Replacement of intracellular chloride by sulphate shifted the voltage-dependence of the motility to more depolarised potentials but did not alter the time course of the length changes. Exposure to low calcium (0.1 mM) solution, with or without 0.2 mM dihydrostreptomycin, also had no obvious effect on the time course of motility.

1 Summary

Outer hair cell (OHC) electromotility is thought to provide an active amplification of the travelling wave on the basilar membrane. The motility, due to a motor protein (prestin) packed into the cells' lateral membrane, has been shown to be voltage-dependent, with positive and negative transmembrane voltage changes leading to cell shortening and lengthening respectively [1,2]. In order to amplify basilar membrane vibrations on a cycle-by-cycle basis, the electromotility must be rapid and robust. There have been a number of attempts to measure the speed of electromotility, and estimates using suction electrodes indicate that electromotility can occur up to at least 79 kHz [3]. The whole-cell recording technique, which has the advantage of allowing direct control of transmembrane voltage and ionic gradients under the best conditions, suffers from the disadvantage that the speed of an imposed voltage change is limited by the speed of the voltage-clamp, which depends on the product of the series resistance (typically at least 4-5 MΩ) and the cell capacitance (typically 30 pF) for adult OHCs. These values indicate a voltage-clamp time constant of about 0.15 ms, which is too slow to follow OHC electromotility [2]. In this study we have attempted to improve the speed of the voltage clamp by minimising the series resistance (by using relatively large pipettes) and by using OHCs from the neonatal rat (P9-P16) which have a relatively small capacitance (8-15 pF), but have close to fully developed electromotility [4]. The aim was to investigate electromotility under the fastest clamp conditions, although clamp speed probably remains a limiting factor.

OHCs were isolated using gentle mechanical trituration following 10 min in collagenase (0.5 mg/ml) as previously described [5]. The present study uses data from the same isolated OHCs but focuses on the time course of electromotility. The extracellular solution contained (in mM): 150 NaCl, 6 KCl, 1.5 $CaCl_2$, 1 $MgCl_2$, 2 Na pyruvate, 8 glucose and 10 Na-HEPES, pH 7.4. The low calcium solution was the same, but with 0.1 mM $CaCl_2$. The normal intracellular solution contained (in mM): 142 CsCl, 3.5 $MgCl_2$, 1 EGTA, 5 Na_2ATP, 0.5 Na_2GTP and 10 Cs-HEPES, pH 7.2. A low chloride (sulphate) intracellular solution was also used (110 mM Cs_2SO_4 and 3.5 mM $MgSO_4$ replacing CsCl and $MgCl_2$). In whole-cell recording mode, OHCs were clamped at -80 mV and 10ms voltage steps from -150 mV to at least +45 mV were applied (with the low chloride intracellular solution the holding potential was usually -50 mV). Series resistance compensation was typically 50%. The clamp time constant was 57 ± 13 μs (mean ± SEM, n=10). The OHC image was projected onto a photodiode pair, positioned perpendicular to the cell body, at a total magnification of 330x. The photocurrent was calibrated by manually moving the photodiodes through a known distance in the image plane. Digitized (50 kHz) records were averaged (n = 10) and analysed using the exponential fitting routine in Signal software (version 2, CED), and goodness of fit was assessed visually.

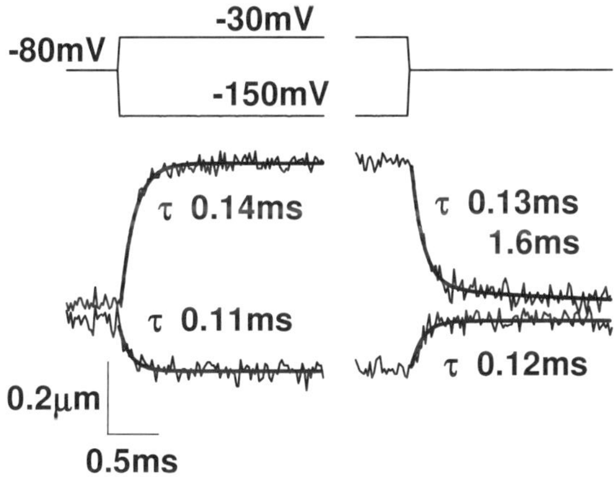

Figure 1. OHC somatic movements in response to voltage steps to -30 mV and -150 mV from a -80 mV holding potential. The onsets (left) and offsets (right) in the photocurrent are shown below the corresponding voltage steps (top). Cell shortening in response to depolarisation is plotted upwards. Single or double exponential fits to the movements are shown superimposed on the records, with the time constant(s) indicated to the right. The voltage clamp time constant was 50 μs and the voltage clamp error was 4 mV (at -150 mV) or less (other voltages). Solutions: CsCl internal, 0.1 mM Ca external.

As previously reported [5], OHCs displayed rapid length changes in response to voltage-steps. For all cells recorded over the 200 mV voltage range, the maximum extent of the change on OHC length was from 0.3 μm to 3 μm in different cells (n = 12). Positive and negative voltage steps produced OHC shortening and lengthening respectively. Most cells showed saturation of motility in both directions, and the manipulations described below did not noticeably affect the amplitude of electromotility, although changes in the voltage-dependence were seen, particularly with low internal chloride [5]. Length changes (relaxations) were fitted with a single exponential function or, particularly for the larger steps a double exponential, with each voltage step producing an *on* and *off* relaxation (see Figure 1). Here we focus on the *on* relaxations (the *off* relaxations were generally similar). Under control conditions mean relaxation time constants measured at -150 mV and -30 mV were 0.20 ± 0.07 ms and 0.25 ± 0.07 ms respectively (mean ± SEM, n = 5). In 0.1 mM external calcium, with or without the addition of 0.2 mM dihydrostreptomycin (DHS), relaxations measured at the same voltages were slightly faster (0.12 ± 0.03 ms; 0.13 ± 0.02 ms respectively (0.1 mM Ca, n = 4) and 0.10 ± 0.02 ms; 0.17 ± 0.03 ms respectively (0.1 mM Ca + DHS, n = 5)). With the Cs_2SO_4 internal solution the corresponding values, measured at more depolarised voltages (-120 mV and +60 mV) to allow for the shift in the voltage-dependence of electromotility were 0.07 ms (n = 2) and 0.15 ± 0.06 ms (n = 3). There was no significant difference between the time constants of the 4 groups at each corresponding voltage (analysis of variance).

Our results indicate that contractions and elongations, in the micrometer range, of isolated OHCs occur rapidly and that the time course can be adequately fitted with a single or double exponential function. The time constants are generally slightly faster and more symmetrical than those reported previously using a similar approach (around 0.3 ms for onsets and 1.0 ms for offsets [2]) and they also show that, at least under the conditions tested, OHC electromotility is intrinsically fast and robust.

Acknowledgments

Supported by a grant from NIDCD (R01 DC 01362).

References

1. Ashmore, J.F., 1987. A fast motile response in guinea-pig outer hair cells: the cellular basis of the cochlear amplifier. J. Physiol. 388, 323-347.
2. Santos-Sacchi, J., 1992. On the frequency limit and phase of the outer hair cell motility: effects of the membrane filter. J. Neurosci. 12, 1906-1916.
3. Frank, G., Hemmert, W., Gummer, A.W. 1999. Limiting dynamics of high-frequency electromechanical transduction of outer hair cells. Proc. Natl. Acad. Sci. USA. 96, 4420-4425.
4. Oliver, D., Fakler, B., 1999. Expression density and functional characteristics of the outer hair cell motor protein are regulated during postnatal development in rat. J. Physiol. 519, 791-800.
5. Kennedy H.J., Evans M.G., Crawford A.C., Fettiplace R., 2006. Depolarisation of cochlear outer hair cells evokes active bundle motion by two mechanisms. J. Neurosci. 26, 2757-2766.

ASSESSMENT OF THE ACTIVITY OF PURIFIED PRESTIN AND THE EFFECT OF SALICYLATE ON PRESTIN-CHLORIDE BINDING STUDIED BY ISOTHERMAL TITRATION CALORIMETRY

KOJI IIDA, MICHIO MURAKOSHI, SHUN KUMANO, HIROSHI WADA

Department of Bioengineering and Robotics, Tohoku University,
6-6-01 Aoba-yama, Sendai 980-8579, Japan

KOUHEI TSUMOTO

Department of Medical Genome Sciences, The University of Tokyo,
5-1-5 Kashiwanoha, Kashiwa 277-8652, Japan

KATSUHISA IKEDA

Department of Otorhinolaryngology, Juntendo University School of Medicine,
2-1-1 Hongo, Bunkyo-ku, Tokyo 113-8421, Japan

TOSHIMITSU KOBAYASHI

Department of Otorhinolaryngology, Tohoku University, Graduate School of Medicine,
1-1 Seiryo-machi, Sendai 980-8575, Japan

IZUMI KUMAGAI

Department of Biomolecular Engineering, Tohoku University,
6-6-07 Aoba-yama, Sendai 980-8579, Japan

1 Summary

Prestin is regarded as the motor protein of cochlear outer hair cells, the somatic motility of which is believed to realize the high sensitivity of the mammalian auditory system. In our previous study, prestin was purified from Chinese hamster ovary cells which had been modified to stably express FLAG-tagged prestin. However, it has remained unclear whether purified prestin is active. Therefore, in the present study, first, to examine the activity of purified prestin, prestin-chloride binding was investigated by isothermal titration calorimetry (ITC) since it has been clarified that chloride binds to prestin, triggering its conformational change [1]. Second, although repression of the conformational change of prestin by salicylate has been reported, the action mechanism of such repression has not been clarified. One possible mechanism is that high-affinity binding of salicylate to prestin interferes with the prestin-chloride binding, and another is that salicylate binds to molecules other than prestin, e.g., phospholipids of the plasma membrane, changing the properties of such molecules and resulting in repression of the conformational change of prestin. In the present study, ITC was employed to analyze the effect of salicylate on prestin-chloride binding. The results are shown in Fig. 1. Figure 1A shows a thermogram obtained when NaCl solution was injected into the chloride-free buffer without prestin. Negative deflections indicating heat generation, resulting from the frictional heat caused by solution movement and the heat of dilution, were observed. Integrated heat calculated from the raw data by integrating the heat flow from the start of the injection to the start of the next injection was plotted versus injection number as shown by the open squares in Fig. 1C, integrated heat caused by the 1st injection being plotted as the datum of injection number 1, and so on. When purified prestin prepared in chloride-free buffer was titrated with NaCl solution (the 0 mM salicylate panel of Figs. 1B and C), heat flow was more positive than that measured in buffer only, indicating that heat was absorbed due to the existence of purified prestin. This result means that chloride binds to purified prestin, thus confirming that purified prestin is active. When the buffer contained sodium salicylate (0.1 mM and 1 mM salicylate panels in Fig 1), the endothermic response caused by the prestin-chloride binding was repressed in a concentration-dependent manner. In the present study, since there were no components of the plasma membrane, our results indicate that salicylate directly interacts with prestin and that the binding affinity of salicylate to prestin is higher than that of chloride to prestin, resulting in interference with the conformational change of prestin.

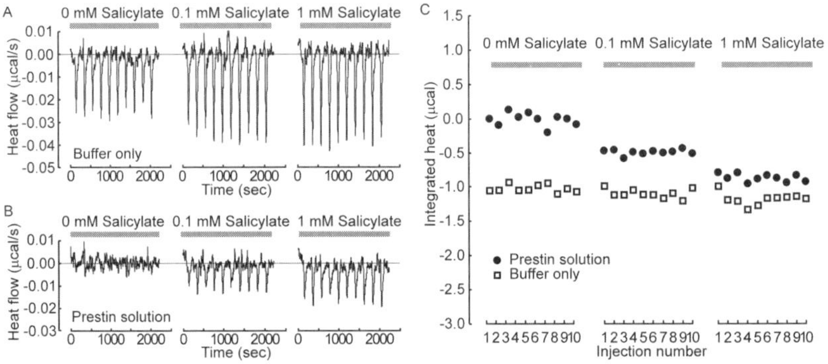

Figure 1. Representative data of ITC. A. Raw data of ITC with NaCl of the buffer only. B. Raw data of ITC with NaCl of prestin solution. Ten microliters of 20 mM NaCl solution was injected into a 1.4 ml-calorimeter cell 10 times, 20 s each time, at intervals of 210 s. Salicylate concentrations are represented at the top of the figures. Positive values indicate heat absorption and negative values indicate heat generation. C. Integrated heat versus the injection number. Closed circles show prestin solution. Open squares show buffer only.

Acknowledgments

This work was supported by Grant-in-Aid for Scientific Research on Priority Areas 15086202 from the Ministry of Education, Culture, Sports, Science and Technology of Japan, by Grant-in-Aid for Scientific Research (B) 18390455 from JSPS, by a grant from the Human Frontier Science Program and by a Grant-in-Aid for JSPS Fellows from JSPS.

References

1. Oliver, D., He, D.Z.Z., Klöcker, N., Ludwig, J., Schulte, U., Waldegger, S., Ruppersberg, J.P., Dallos, P., Fakler, B., 2001. Intracellular anions as the voltage sensor of prestin, the outer hair cell motor protein. Science, 292, 2340-2343.

INCREASE IN THE ACTIVITY BY MUTATIONS OF THE MOTOR PROTEIN PRESTIN

SHUN KUMANO, KOJI IIDA, MICHIO MURAKOSHI, HIROSHI WADA

Department of Bioengineering and Robotics, Tohoku University, 6-6-01 Aoba-yama, Sendai 980-8579, Japan

KOUHEI TSUMOTO

Department of Medical Genome Sciences, The University of Tokyo, 5-1-5 Kashiwanoha, Kashiwa 277-8561, Japan

KATSUHISA IKEDA

Department of Otorhinolaryngology, Juntendo University School of Medicine, 2-1-1 Hongo, Bunkyo-ku, Tokyo 113-8431, Japan

IZUMI KUMAGAI

Department of Biomolecular Engineering, Tohoku University, 6-6-07 Aoba-yama, Sendai 980-8579, Japan

TOSHIMITSU KOBAYASHI

Department of Otolaryngology, Head and Neck Surgery, Tohoku University Graduate School of Medicine, 1-1 Seiryo-machi, Sendai 980-8575, Japan

Prestin is the motor protein of cochlear outer hair cells, which exhibit elongation and contraction in response to acoustic stimuli. In this study, the effects of mutations in unique amino acids, which among members of the SLC26 family were only present in prestin, on the characteristics of prestin were investigated. As a result, a mutation in Met-225 of prestin was found to increase nonlinear capacitance, which is a signature of prestin activity

1 Summary

A conformational change of the motor protein prestin, which is a member of the SLC26 family, is considered to induce the elongation and contraction of outer hair cells (OHCs) [1]. The mechanism of such change has remained unknown. To clarify this mechanism, identification of the amino acids involved in the conformational change of prestin is required. Alignment analysis of amino acid sequences of the SLC26 family showed that Met-122, Cys-192, Met-225, Cys-415 and Thr-428 were only present in prestin among that family. As the conformational change of prestin involved in the motility of the OHCs is a unique function, prestin-specific amino acids may realize such function. In this study, to clarify the roles of those amino acids in prestin, effects of mutations in those amino acids on characteristics of prestin were investigated. Five prestin mutants, namely, M122I, M225Q, C192A, C415A and T428L, were engineered to be expressed in HEK293 cells. The electrophysiological properties and expression level in the plasma membrane of wild-type prestin (WT) and its mutants were then evaluated by the patch-clamp recording and Western blotting, respectively. Contrary to our expectations, results showed that the nonlinear capacitance (NLC) of M225Q, which reflects the amount of functional prestin molecules in the plasma membrane, was significantly larger than that of WT, although the NLC of the other prestin mutants was similar to that of WT as shown in Fig. 1. Meanwhile, the expression level in the plasma membrane of M225Q was similar to that of WT, meaning that the total amount of prestin molecules in the plasma membrane remained unchanged by the mutation in Met-225. Hence, possible explanation for the increase of NLC may be that the mutation in Met-225 may result in an increase in the ratio of functional prestin molecules to the total amount of prestin molecules in the plasma membrane.

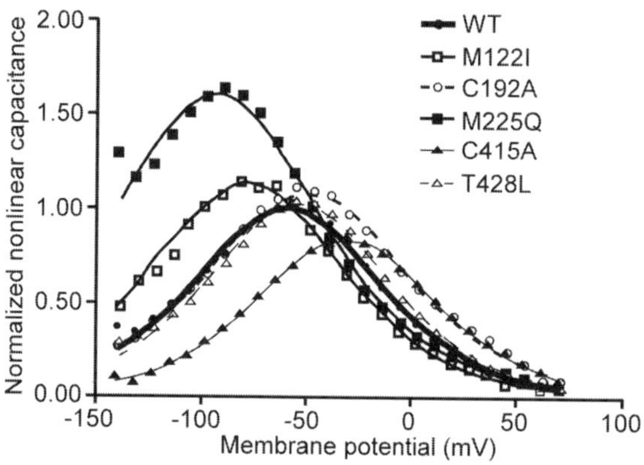

Figure 1. Representative data of the patch-clamp recording for HEK293 cells transfected with either the WT gene or the prestin mutant genes. NLC is normalized by the area of the cell membrane and WT data. NLC of M225Q is significantly larger than that of WT. On the other hand, the other mutants showed NLC similar to that of WT.

Acknowledgments

This work was supported by Grant-in-Aid for Scientific Research on Priority Areas 15086202 from the Ministry of Education, Culture, Sports, Science and Technology of Japan, by Grant-in-Aid for Scientific Research (B) 18390455 from the Japan Society for the Promotion of Science, by Grant-in-Aid for Exploratory Research 18659495 from the Japan Society for the Promotion of Science, by a Health and Labour Science Research Grant from the Ministry of Health, Labour and Welfare of Japan to H.W., and by a Grant-in-Aid for JSPS Fellows from the Japan Society for the Promotion of Science to S.K.

References

1. Zheng, J., Shen, W., He, D.Z., Long, K.B., Madison L.D., Dallos, P., 2000. Prestin is the motor protein of cochlear outer hair cells. Nature 405, 149-155.

PRESTIN DISTRIBUTION IN RAT OUTER CELLS – AN ULTRASTRUCTURAL STUDY

SHANTHINI MAHENDRASINGAM

Institute for Science and Technology in Medicine, School of Life Sciences, Keele University, Keele, Staffordshire, ST5 5BG, UK.

DAVID N FURNESS

Institute for Science and Technology in Medicine, School of Life Sciences, Keele University, Keele, Staffordshire, ST5 5BG, UK

ROBERT FETTIPLACE

Department of Physiology, University of Wisconsin Medical School, Madison, Wisconsin 53706, USA

CAROLE M HACKNEY

Department of Physiology, Development and Neuroscience, University of Cambridge, The Physiological Laboratory, Downing Street, Cambridge, CB2 3EG, UK

The density and distribution of prestin in pre-hearing (postnatal day 7) and hearing (postnatal day 16) rats has been investigated in apical and basal cochlear locations using post-embedding immunogold labelling. The results suggest that there is a gradient in the labelling before the onset of hearing with apical cells having a higher density of prestin in their basolateral membranes than basal cells. However, in hearing animals there is no obvious difference in density although total levels of prestin are higher in each region than in pre-hearing animals. Also, apical hair cells are longer than basal ones and therefore have a higher total amount of prestin.

1 Introduction

The mammalian cochlea contains two types of hair cell, inner hair cells (IHCs) and outer hair cells (OHCs). These are embedded in a sensory epithelium, the organ of Corti that sits on the basilar membrane which vibrates in response to sound. The vibrations are, however, enhanced by active mechanisms; depolarisation of the IHCs results in them sending neural signals via the cochlear nerve to the brain but the OHCs boost the stimulus by electromechanical feedback. This mechanism has been called 'the cochlear amplifier' because it increases both the amplitude and frequency selectivity of basilar membrane vibrations for low level sounds (see reviews by Dallos, 1992; Fettiplace and Hackney, 2006)). The OHCs change the length via a somatic motor powered by changes in membrane potential. Dissociated OHCS *in vitro* shorten in response to depolarization and extend in response to hyperpolarization (Brownell et al., 1985) at speeds that follow sound stimuli to frequencies spanning the auditory range (Ashmore, 1987). This somatic motility involves a membrane protein, prestin (Zheng et al., 2000)) which belongs to a solute carrier (SLC26) family of anion transporters and which is densely packed in the basolateral membrane of OHCs (Belyantseva et al., 2000). The importance of prestin has been demonstrated by construction of mutations that reduce its expression. Targeted deletion of prestin in mice causes loss of OHC electromotility and at least a 100-fold reduction in cochlear sensitivity (Liberman et al., 2003) without affecting forward mechanical transduction (Cheatham et al., 2004). It has been suggested that higher frequency hair cells may have a greater density of prestin in their lateral walls than those from lower frequency regions of the cochlea on the basis of the density of motility related charges in guinea-pig OHCs from different regions of the cochlea (Santos-Sacchi et al., 1998). It has also been suggested that the amount (Jensen-Smith and Hallworth, 2007) and distribution of prestin changes with the maturation of hearing (Winter et al., 2006). In this study, these two suggestions have been investigated in rat cochleas using post-embedding immunogold labelling of the C-terminus of prestin at the ultrastructural level. This technique permits comparisons of the density of the labelling in OHCs from apical and basal regions of the cochlea and comparisons between animals of different ages.

2 Materials and Methods

Cochlear segments were obtained from Sprague Dawley rats using procedures approved by the Animal Care Committee at the University of Wisconsin-Madison and in accordance with the UK Animals (Scientific Procedures) Act 1986. Cochleae were obtained from animals of both sexes on postnatal day 7 (P7, pre-hearing) and 16 (P16, hearing). Preyer reflex tests indicated that the P16 animals had an acoustic startle response but that the P7 animals did not.

For cochlear tissue fixation, animals were deeply anaesthetised with ketamine (50-100 mg/kg), xylazine (4-8 mg/kg) or with sodium pentobarbitone (Pentoject: 100 mg/kg) injected intraperitoneally and then were perfused transcardially for 1 min with a vascular flush consisting of buffered saline containing 5% dextran or heparin (7.2 U/ml) and 0.15% procaine HCl and then for 10 min with 4% paraformaldehyde and 0.1% glutaraldehyde in 0.1 M sodium phosphate buffer (PB), pH 7.4. Each bulla was removed and the cochleas fixed further by immersion in the same fixative for 2h at room temperature.

Cochlear segments from both apical and basal locations were dissected out and washed in PB and then dehydrated in a graded series of ethanols. The pieces of tissue were infiltrated with LR-White resin for 24 h and polymerized in resin at 50°C for 24 h.

For immunogold labelling, nickel grids bearing ultrathin sections were washed in 0.05M Tris-buffered saline (TBS), pH 7.4, and non-specific protein binding was blocked using 20% goat serum and 0.2% Tween 20 in TBS. They were then incubated overnight at 4°C in an affinity purified rabbit polyclonal anti-prestin antibody (custom prepared by BETHYL Labs to the C-terminus of rat prestin; peptide sequence CTV LPP QED MEP NAT PTT PEA) diluted 1:5,000-1:50,000 in 0.05M TBS containing 1% bovine serum albumin and 0.02% Tween 20 (BSA-T20-TBS), washed in BSA-T20-TBS and incubated in goat anti-rabbit IgG conjugated to 15 nm gold particles diluted 1:20 in BSA-T20-TBS for 2 h at room temperature. The grids were then washed again in TBS and distilled water, stained in aqueous uranyl acetate and examined using a JEOL JEM 1230 transmission electron microscope operated at 100kV.

The density of gold labelling was determined in radial sections of the cochleas taken from an apical coil (low frequency region), and a basal coil (high frequency region) of P7and P16 rats. Grids containing sections from apical and basal locations of the cochlea were placed in the same drop of solution and sections from P7 and P16 rats were incubated in the same batch of solution. Immunogold labelling was also performed simultaneously. To count the number of gold particles along the basolateral wall of OHCs, micrographs of OHCs were taken at the same magnification (15,000 X or 50,000 X). Gold particles on either side of the basolateral wall (within the spatial resolution distance of 24 nm) were counted along every micrometer of membrane starting from the base of the tight junction on the *stria vascularis* side and ending at the base of the tight junction on the modiolar side of the hair cell. Graphs were plotted using moving averages on the Y-axis, the average number of gold particles at distances of 1-5 μm, 2-6 μm, 3-7μm, 4-8 μm and so on, from the tight junctions were calculated.

3 Results

In pre-hearing (P7) rats, immunogold labelling for prestin was localized on the basolateral wall and in the cytoplasm of OHCs and not in any other subcellular compartments (Fig. 1A-D) and in hearing (P16) rats it was localized only on the basolateral wall of OHCs (Fig. 1E, F). IHCs were not labelled (not shown).

In P7 rats, qualitatively, the basolateral wall of apical OHCs was more heavily labelled for prestin than the basal one (compare Fig. 1A, C with Fig.1B, D) and semiquantitative analysis indicated the same trend. The particle densities for basal OHCs were less than those in apical ones. However, apical OHCs were, as expected longer than basal ones and thus have more prestin in total (Fig. 2A). The density of labelling for prestin in the different OHC rows in the apex did not appear to be different and the same was true for the rows in the base (not illustrated). The cytoplasm below the cuticular plate of both apical and basal OHCs and the cytoplasm below the nucleus were also labelled for prestin.

In P16 rats, qualitatively the basolateral walls above the nucleus of OHCs from both apical and basal locations showed similar intensities of labelling for prestin (Fig. 1E, F) and semiquantitative analysis indicated the same. The basolateral walls of OHCs from both locations showed less labelling in the subnuclear region compared to that of the supranuclear region and very few or no gold particles were localized in the region where the nerve terminals made contact (not shown).

To calculate how the density changes along the membrane, a moving average was determined. At P7, although the length of the apical OHCs is greater than that of the

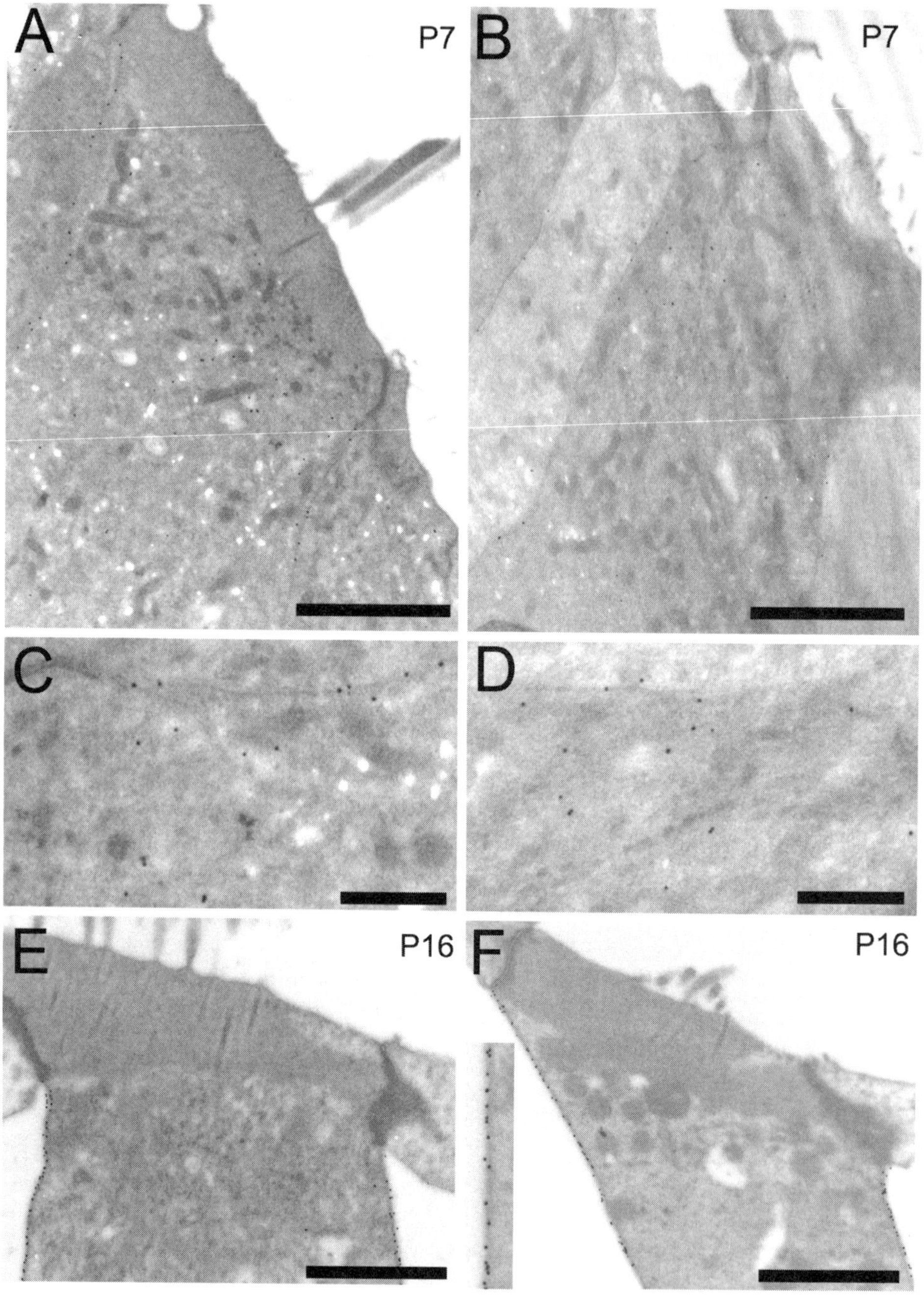

Fig. 1. Immunogold labelling for prestin in rat OHCs. A, C - P7 apical; B, D - P7 basal. Labelling is observed in cytoplasm and lateral membrane, and is stronger in apical OHCs. E, F. P16 apical (E) and basal (F) OHCS. There is no cytoplasmic labelling, and the lateral wall labelling is denser and similar in both locations. The inset in F is an enlargement of the lateral wall to show the particles more clearly. Scale bars, A, B, E, F = 2 μm; C, D = 0.5 μm.

basal OHC, the semiquantitative analysis of particle densities for apical and basal OHCs showed some differences in the pattern of labelling (Fig. 2A). However, basolateral wall labelling density in apical P7 hair cells was substantially greater than basal ones. Labelling towards the base of the cell became less dense in both cases, but was still present to some degree in the sub-nuclear region.

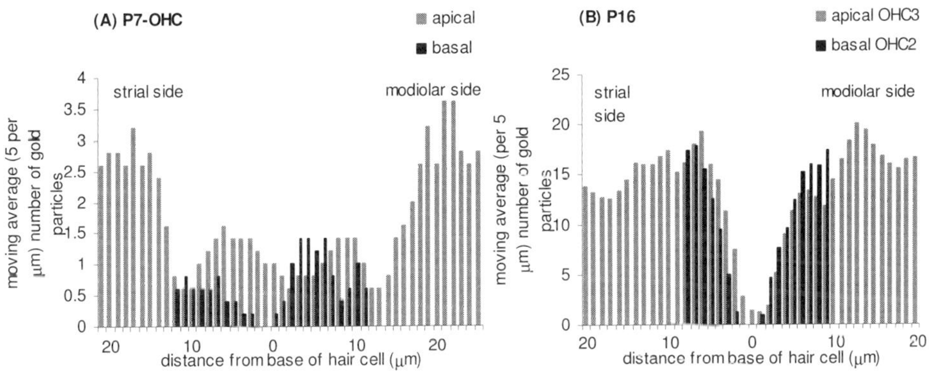

Fig. 2. Histograms comparing the density of labelling in apical and basal P7 (A) and P16 (B) hair cells. Bars represent a moving average calculated in 5 µm bins along the side of the hair cell. The distances are calculated from the approximate base of the hair cell; apical hair cells which are longer consequently have more bars.

At P16, there was relatively little difference between apical and basal hair cells in labelling densities in equivalent parts of the membrane, in contrast to the P7. Labelling density was also substantially greater in P16 than P7 in both locations (Fig. 2).

4 Discussion

These results suggest that in hearing animals there is little, if any, difference in the density of prestin in the basolateral walls of apical and basal hair cells despite the well documented difference in the total length of the cells in these regions. At first sight this appears to be different from the findings of Santos-Saachi et al (1998). There may be limitations however, in the sensitivity of the immunogold technique, so further investigation is needed. There is, however a substantial difference between apical and basal levels of prestin during development with apical hair cells being more heavily labelled than basal ones. In pre-hearing rats, prestin labelling is seen in the cytoplasm presumably because it is being manufactured for insertion into the basolateral wall whereas in hearing rats, little labelling is seen suggesting a lower turn-over. Labelling for prestin is seen all along the basolateral wall but although it is reduced below the nucleus, the only membrane region where it is not present is where there are synaptic contacts with the cell and above the base of the tight junction at the apex of the cell.

Further ultrastructural studies using double- or triple-labelling and smaller gold particles could be used to ascertain the relationship of prestin with the cytoskeletal network that lies just below the plasma membrane.

Acknowledgments

Supported by a grant from DRUK to CMH and DNF, the Steenbock Professorship (RF, SM) and NIDCD Grant RO1 DC01362 to RF.

References

1. Ashmore, J.F., 1987. A fast motile response in guinea-pig outer hair cells: the cellular basis of the cochlear amplifier. J. Physiol. (Lond) 388, 323-47.
2. Belyantseva, I.A., Adler, H.J., Curi, R., Frolenkov, G.I. and Kachar, B., 2000. Expression and localization of prestin and the sugar transporter GLUT-5 during development of electromotility in cochlear outer hair cells. J. Neurosci. 20, RC116.
3. Brownell WE, Bader CR, Bertrand D, de Ribaupierre Y, 1985. Evoked mechanical responses of isolated cochlear outer hair cells. Science 227, 194-196.
4. Cheatham MA, Huynh KH, Gao J, Zuo J and Dallos P, 2004. Cochlear function in Prestin knockout mice. J. Physiol. (Lond) 560, 821-830.
5. Dallos P, 1992. The active cochlea. J Neurosci 12, 4575-4585.
6. Dallos P, Zheng J, Cheatham MA, 2006. Prestin and the cochlear amplifier. J Physiol (Lond) 576, 37-42.
7. Fettiplace R, Hackney CM, 2006. The sensory and motor roles of auditory hair cells. Nature Rev Neurosci 7, 19-29.
8. Jensen-Smith H, Hallworth R, 2007. Lateral wall protein content mediates alterations in cochlear outer hair cell mechanics before and after hearing onset. Cell Motil Cytoskeleton 64, 705-717.
9. Liberman MC, Gao J, He DZ, Wu X, Jia S, Zuo J, 2002. Prestin is required for electromotility of the outer hair cell and for the cochlear amplifier. Nature 419, 300-304.
10. Oliver D, He DZ, Klöcker N, Ludwig J, Schulte U, Waldegger S, Ruppersberg JP, Dallos P, Fakler B., 2001. Intracellular anions as the voltage sensor of prestin, the outer hair cell motor protein. Science 292, 2340-2343.
11. Rybalchenko V and Santos-Sacchi J, 2003. Cl^- flux through a non-selective, stretch-sensitive conductance influences the outer hair cell motor of the guinea-pig. Physiol (Lond) 547, 873-891.
12. Santos-Sacchi J, Kakehata S, Kikuchi T, Katori Y and Takasaka T, 1998. Density of motility-related charge in the outer hair cell of the guinea pig is inversely related to best frequency. Neurosci Lett 256, 155-158.
13. von Békésy G, 1960. Experiments in Hearing. McGraw-Hill, New York.
14. Winter H, Braig C, Zimmermann U, Geisler HS, Fränzer JT, Weber T, Ley M, Engel J, Knirsch M, Bauer K, Christ S, Walsh EJ, McGee J, Köpschall I, Rohbock K, Knipper M., 2006. Thyroid hormone receptors TRalpha1 and TRbeta differentially regulate gene expression of Kcnq4 and prestin during final differentiation of outer hair cells. J Cell Sci. 119, 2975-84.
15. Zheng J, Shen W, He DZ, Long KB, Madison LD, Dallos P, 2000. Prestin is the motor protein of cochlear outer hair cells. Nature 405, 149-155.

SECTION VII

HAIR BUNDLES AND MECHANO-ELECTRICAL TRANSDUCTION

K. Dierkes, D. Ó Maoiléidigh and M. Fleischer

ACTIVE HAIR-BUNDLE MOTILITY BY THE VERTEBRATE HAIR CELL

J-Y. TINEVEZ[†], P. MARTIN
Laboratoire Physico-Chimie Curie, CNRS, Institut Curie, UPMC
26, rue d'Ulm, F- 75248 Paris cedex 05, France

F. JÜLICHER
Max-Planck-Institute for the Physics of Complex Systems
Nöthnitzer Str. 38, 01187 Dresden, Germany

The hair bundle is both a mechano-sensory antenna and a force generator that might help the vertebrate hair cell from the inner ear to amplify its responsiveness to small stimuli. To study active hair bundle motility, we combined calcium iontophoresis with mechanical stimulation of single hair bundles from the bullfrog's sacculus. A hair bundle could oscillate spontaneously, or be quiescent but display non-monotonic movements in response to abrupt force steps. Extracellular calcium changes or static biases to the bundle's position at rest could affect the kinetics of bundle motion and evoke transitions between the different classes of motility. The calcium-dependent location of a bundle's operating point within its nonlinear force-displacement relation controlled the type of movements observed. A unified theoretical description, in which mechanical activity stems from myosin-based adaptation and electro-mechanical feedback by Ca^{2+}, could account for the fast and slow manifestations of active hair-bundle motility.

1 Introduction

Sensory hair cells in the inner ear make use of mechanical amplification to enhance their sensitivity and sharpen their frequency selectivity to faint stimuli [1]. The mechanism that would explain this active process remains a central issue of auditory biophysics. In the mammalian ear, somatic electromotility of outer hair cells [2] appears to be necessary for cochlear amplification [3; 4] but its exact contribution to the cochlear amplifier is still under debate [5]. Non-mammalian hair cells lack electromotility but nevertheless display sensitive and frequency selective hearing [6]. There, the motor of the amplificatory process most probably resides in the hair bundle, the tuft of extended actin-rich microvilli that adorns the apical surface of each hair cell.

Spontaneous hair-bundle oscillations demonstrate that a hair cell can power movements of its mechano-sensory organelle, even in the absence of stimulation (reviewed in [7]). Moreover, active oscillations provide a characteristic frequency near which a hair bundle displays increased sensitivity to small periodic mechanical stimuli [8; 9]. Other forms of activity have been reported under conditions for which hair bundles are quiescent. When stimulated by an abrupt force step with a flexible glass fiber, a hair bundle can indeed produce a non-monotonic "twitch" in which motion in the direction of

[†] Jean-Yves Tinevez's present address is Max-Plank-Institute of Molecular Cell Biology and Genetics, Pfötenhauerstrasse 108, 01307 Dresden Germany.

the applied force is interrupted by a fast recoil [10-12]. In other cases, the hair bundle monotonically relaxes towards a new steady state position that can sometimes exceed that of the stimulus fiber's base [13]. These various regimes of hair-bundle motility are intimately related to adaptation, the active process that provides negative feedback on the open probability of the transduction channels. We review here recent evidence [14] that clarify the role of gating compliance, the mechanical correlate of channel gating, and of the Ca^{2+} component of the transduction current in shaping the type of active movements that a hair bundle can produce. By so doing, we provide a minimal and unified physical description of active hair-bundle motility.

2 Methods

2.1 Mechanical stimulation of single hair bundles and Ca^{2+} iontophoresis

Each experiment was performed at a room temperature of 21-24 °C with hair cells from the saccule of an adult bullfrog (*Rana catesbeiana*) by published methods [14]. In short, the preparation was mounted on a two-compartment chamber. While the basolateral surface was bathed by standard saline consisting of 110 mM Na^+, 2 mM K^+, 4 mM Ca^{2+}, 122 mM Cl^-, 3 mM **D**-glucose, 2 mM creatine phosphate, 2 mM sodium pyruvate and 5 mM HEPES, hair bundles projected in NMDG artificial endolymph containing 2 mM Na^+, 3 mM K^+, 0.075-4 mM Ca^{2+}, 110 mM *N*-methyl-**D**-glucamine, 111 mM Cl^-, 3 mM **D**-glucose, and 5 mM HEPES. After digestion with the protease subtilisin (Sigma, type XXIV) at a concentration of 50-67 $\mu g{\cdot}ml^{-1}$ in endolymph for 25 mn, the otolithic membrane was peeled off the macula to get access to individual hair bundles. Each solution had a pH of ~7.3 and an osmotic strength of ~230 $mmol{\cdot}kg^{-1}$.

The experimental chamber was secured to the stage of an upright microscope (BX51WI, Olympus). The preparation was viewed using bright-field illumination through a ×60 water-immersion objective lens and a ×1.25 relay lens. The tip of a hair bundle, or of a stimulus fiber attached to it, was imaged at a magnification of ×1,000 on a displacement monitor which included a dual photodiode (PIN-SPOT2D, UDT Sensor Inc.). This photometric system was characterized by a bandwidth of 6.5 kHz at half the maximal power and yielded an output linearly proportional to the displacement of the stimulus fiber in a range of ±500 nm with a resolution of ~1 μm, corresponding to a ~1 nm resolution at the specimen plane.

Hair-bundles were stimulated with flexible glass fibers that were fabricated from borosilicate capillaries and coated with a ~100 nm layer of gold-palladium to enhance optical contrast [15]. The stiffness K_F and drag coefficient λ_F of the fibers were respectively, 150-500 $\mu N{\cdot}m^{-1}$ and 40-200 $nN{\cdot}s{\cdot}m^{-1}$. The fiber was secured by its base to a stack-type piezoelectric actuator (PA 8/14, Piezosystem Jena) driven by a low-noise power supply (ENV 150 and ENT 150/20, Piezosystem Jena). When powered by an unfiltered, abrupt voltage step, the actuator with an attached rigid fiber displayed a mechanical resonance at ~4 kHz. By using displacement-clamp circuitry to apply a series of step displacements to a hair-bundle, we measured the forces that had to be exerted to hold the hair bundle at these positions, thereby describing the intrinsic force-displacement relation of the hair bundle [15; 16]. For each displacement step, we estimated the elastic response of the hair bundle by recording the force ~3 ms after the onset of stimulation to allow just enough time for the viscous response to vanish and minimize the mechanical relaxation provided by adaptation [16].

We used iontophoresis to rapidly change the local Ca^{2+} concentration near a hair bundle. Coarse microelectrodes were fabricated from borosilicate capillaries, bent through an angle of ~90° in their tapered region and then filled either with 2.5 M $CaCl_2$ or 350 mM disodium ATP (here used as a Ca^{2+} chelator). At a distance $r = 3$ µm from the hair bundle, this technique allowed an increase of ~20 µM of Ca^{2+} concentration per nanoampere of iontophoretic current.

2.2 *Theoretical description of active hair-bundle mechanics*

This description (detailed in [7; 14]) is based on the gating-spring theory of mechano-electrical transduction [17] and a myosin-based model of adaptation [18; 19].

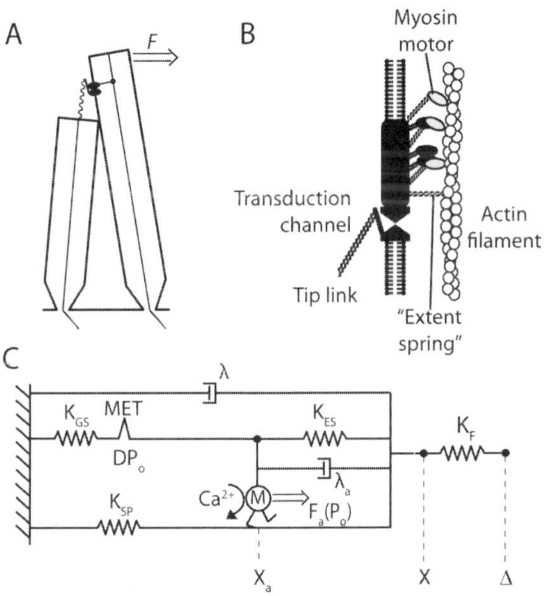

Figure 1. A: Schematic representation of a hair bundle. We assumed that N transduction elements, here lumped into a single element, operate in parallel within a hair bundle. An external force F is applied to the tip of the hair bundle and defined positive when oriented as depicted here. B: Functional view of a transduction element (adapted from [18]). At steady state, the active force exerted towards the tip of the stereocilia by a group of ~60 myosin molecules is balanced by an elastic restoring tension in the tip link, thereby defining the resting open probability of the mechano-sensitive transduction channel. C: Mechanical arrangement. Transduction channels (MET) of open probability P_o are connected to gating springs of stiffness K_{GS} and anchored to the actin cytoskeleton of the stereocilia, dynamically by adaptation motors (M) that can change their position X_a and statically by extent springs of stiffness K_{ES} that limit the extent of adaptive movements of the motors. The external force $F = K_F(\Delta - X)$, exerted by means of a flexible fiber of stiffness K_F, affects tension in both gating and pivot springs, the later being of stiffness K_{SP} and operating in parallel to the former. The speed of hair-bundle motion is inversely proportional to the friction coefficient λ. Opening of a transduction channel evokes a decrease of gating-spring extension that amounts to a motion of size D of the bundle's top [17]. When the combined tension in gating and extent springs differs from the active force F_a that the adaptation motors produce at stall, the motors are moving. Motor speed is inversely proportional to λ_a, which has units of a friction coefficient and represents the slope of the force-velocity relation of the motors near stall condition. The motor force F_a is down-regulated by the Ca^{2+} concentration at the motor site and thus depends on P_o. All variables are expressed at the top of a hair bundle along the stimulation axis.

The dynamic interplay between the positions X of the hair bundle and X_a of the adaptation motor (Fig. 1) is described by two coupled equations:

$$\lambda \frac{dX}{dt} = -\mathrm{K}_{GS}\left(X - X_a - D\,P_o\right) - \mathrm{K}_{SP} X + F \tag{1}$$

$$\lambda_a \frac{dX_a}{dt} = \mathrm{K}_{GS}\left(X - X_a - D\,P_o\right) - \mathrm{K}_{ES}\left(X_a - X_{ES}\right) - F_a\,, \tag{2}$$

in which the open probability of the transduction channels is given by:

$$P_o = \frac{1}{1 + A\exp\left(-\dfrac{Z(X - X_a)}{k_B T}\right)}. \tag{3}$$

Here, X_{ES} is the value of X_a for which the extent springs bear no tension, $P_o = 1/(1+A)$ when the gating springs are severed and $Z = \mathrm{K}_{GS}\,D/N$ is the gating force of a single transduction element.

To account for the regulation of adaptation by Ca^{2+} [20; 21], we assumed here that the only effect of Ca^{2+} was to down-regulate the active force F_a produced by the motors at stall. By demonstrating that the ATPase cycle of myosin-1c, the putative adaption motor [23], is regulated by Ca^{2+}, recent biochemical evidence in vitro [22] provided support for this assumption. For simplicity, we further supposed that F_a depends instantaneously and linearly on the Ca^{2+} concentration at the motor site and that this concentration equilibrates within microseconds of channel gating. These assumptions yield the relation:

$$F_a \cong F_{\max}\left(1 - S\,P_o\right), \tag{4}$$

in which $F_{\max}$ is the maximal force that the motors can generate at stall and the dimensionless parameter S defines the strength of the Ca^{2+} feedback on the motor force. Note that S is proportional to the extracellular Ca^{2+} concentration, which could be modified experimentally.

3 Results

3.1 Three classes of active hair-bundle movements

A hair bundle could be quiescent but display non-monotonic movements in response to step stimuli that were either negatively or positively directed, but not both (Fig 2A and C, respectively; see also Fig 3). These "twitches" were associated with force-displacement relations in which linear branches for large positive or negative displacements were separated by a region of reduced slope near the position that the hair bundle assumed at rest. Interestingly, this operating point lay on one side of the compliant region of the force-displacement relation. Correspondingly, twitches were observed only when the external stimulus had the appropriate directionality to force the hair bundle to explore the nonlinear region of its force-displacement relation. A hair bundle could also oscillate spontaneously (Fig 2B). As recognized before [16], this active behavior was associated with an intrinsic force-displacement relation containing a region of negative slope and an operating point located within this unstable region. It was thus possible to classify active hair-bundle movements in three categories according to the properties of the intrinsic force-displacement relation of the corresponding hair bundles.

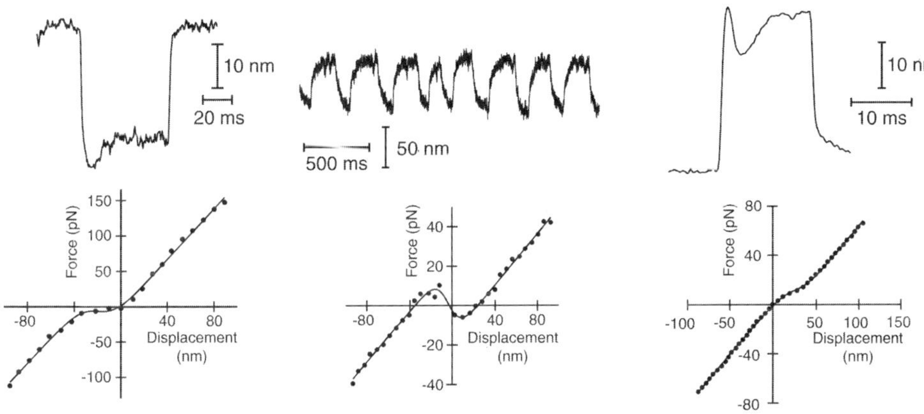

Figure 2. Active hair-bundle movements. A: Negative twitch elicited by a -60 nm step displacement of the stimulus fiber's base. The fiber had a stiffness of $K_F = 739\ \mu m/N$. B: Spontaneous oscillations of hair-bundle position at ~12 Hz with a peak-to-peak magnitude of 63 nm. C: Positive twitch in response to a +137 nm displacement of the fiber's base ($K_F = 270\ \mu m/N$). Note the difference in kinetics between positive (fast) and negative (slow) twitches. Each hair bundle was characterized by a force-displacement relation that is shown below the corresponding active behavior. The Ca^{2+} concentration in the fluid bathing the hair bundle was 250 μM in (A) and (B), and 4 mM in (C).

3.2 Ca^{2+} involvement in active hair-bundle motility

Spontaneous oscillations were most often observed under two-compartment ionic conditions that mimicked native physiological conditions with 250 μM Ca^{2+} in endolymph, whereas positive and negative twitches were favored by high and low Ca^{2+} concentrations, respectively. Extracellular Ca^{2+} changes are known to affect the kinetics of active hair-bundle movements [11; 14; 15]. Higher Ca^{2+} concentrations indeed result in sharper twitches or faster spontaneous oscillations, whereas reduced Ca^{2+} concentration result in slower active movements. For large enough Ca^{2+} changes, we could evoke, with the same hair bundle, transitions between the three classes of active hair-bundle motility (Fig. 3A-D). Our classification of active hair-bundle movements (Fig. 2) suggests that these transitions should be associated with a shift of the corresponding force-displacement relation. By using Ca^{2+} iontophoresis, we could indeed demonstrate that increased Ca^{2+} concentrations evoked a positively-directed shift of the force-displacement relation of a given hair bundle (Fig. 3E). The converse behavior was also observed (not shown).

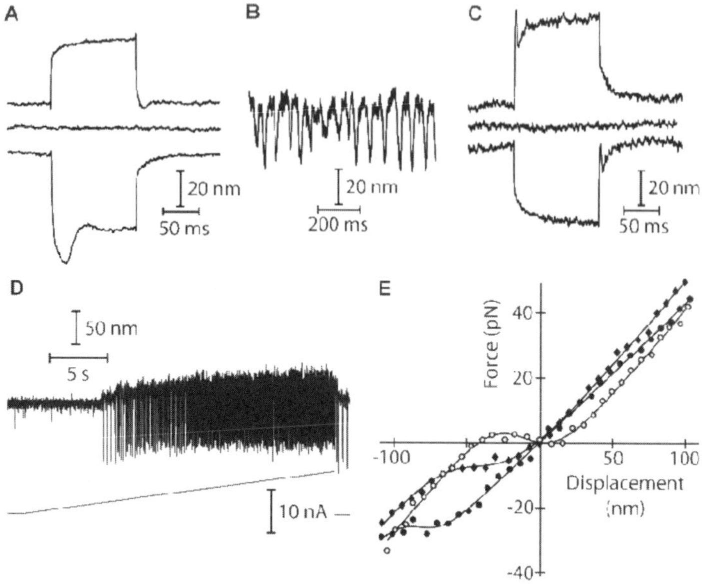

Figure 3. Effects of Ca^{2+} on hair-bundle motitily. A-C: A quiescent hair bundle that displayed a negative twitch when exposed to 75 µM Ca^{2+} (A) could also show spontaneous oscillations under the same conditions (B) but produced a positive twitch (C) when subjected to 4 mM Ca^{2+}. D: Another hair bundle was quiescent at rest but became oscillatory when iontophoresis was used to increase the Ca^{2+} concentration in its vicinity (ramp of iontophoretic current shown at the bottom). The oscillation frequency almost tripled from ~4 Hz to ~11 Hz, whereas the amplitude of the negatively-directed spikes remained roughly constant at ~100 nm. This hair bundle was characterized by a force-displacement relation that resembled that shown in Fig. 2C. E: Measurement of the force-displacement relation for increasing iontophoretic currents (●: -5 nA, holding current; ♦: +3 nA; ○: +6 nA) revealed a positively-directed shift. Although the hair bundle was stable under control conditions (●), the bundle's operating point belonged to an unstable region of negative stiffness of the force-displacement relation for an iontophoretic current of +6 nA (○). Under such circumstances, the hair bundle oscillated spontaneously (not shown).

Calcium changes not only affected the operating point of a hair bundle within its intrinsic force-displacement relation (Fig. 3E) but also evoked net movements. In the case of non-oscillatory hair bundles, we studied the time course of Ca^{2+}evoked movements by using iontophoresis of Ca^{2+} or of a Ca^{2+} chelator [14]. Notably, the directionality of a Ca^{2+} evoked movement varied from cell to cell within the same epithelium and could even be non-monotonic. Moreover, the directionality of bundle movement could be reversed by imposing an offset to the resting position of the hair bundle before applying the iontophoretic pulse (Fig. 4). Polarity reversal of Ca^{2+}-evoked movements was associated with a change in kinetics: for a given Ca^{2+} change, movements in one direction were always slower than when they occurred in the opposite direction.

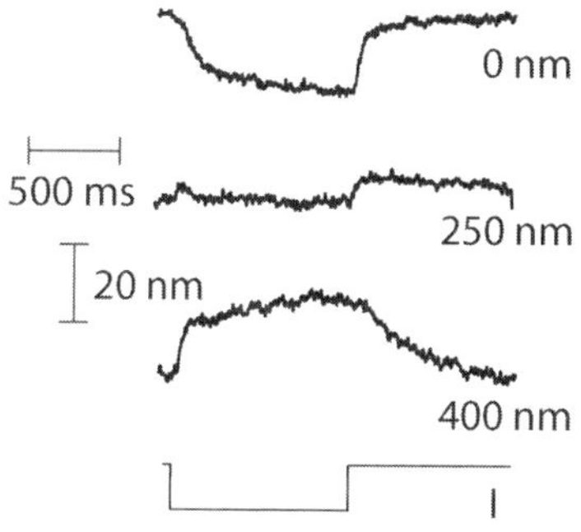

Figure 4. Ca^{2+} evoked movements at various resting positions of a quiescent hair bundle. In response to an iontophoretic pulse of a Ca^{2+} chelator (I = -50 nA; shown at the bottom), here ATP, this quiescent hair bundle displayed a slow movement in the negative direction. With a +250 nm bias, the same hair bundle produced a small biphasic motion with no net deflection at steady state. In the presence of larger biases, here +400 nm, the ATP-evoked movement displayed faster kinetics and opposite directionality than those measured with no offset. The hair bundle was immersed in artificial endolymph containing 250μM Ca^{2+}.

3.3 *Simulations*

Numerically solving the dynamic Eqs. 1 and 2 for the position X of the hair-bundle and X_a of the adaptation motor, together with Eq. 4 that describes Ca^{2+} feedback on the motor's activity, we could mimic the three classes of active hair-bundle movements that we have observed in the bullfrog (Fig. 5) as well as Ca^{2+}-evoked movements (not shown) [14].

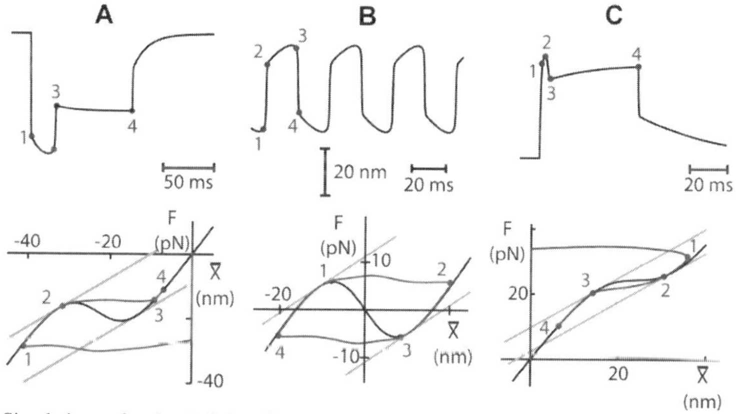

Figure 5. Simulations of active hair-bundle movements. A: Positive twitch in response to a step force of +35 pN. B: Spontaneous oscillations. C: Negative twitch in response to a force step of -30 pN. In each case, we enforced a force-displacement relation, shown at below each type of movement, and an operating point within this relation that resembled those we measured (Fig. 2), thereby constraining most parameters (see reference [14] for parameter values). The trajectory superimposed on each force-displacement relation represents the force $F - K_{SP}\overline{X}_a$ as a function of $\overline{X} - \overline{X}_a$, in which $\overline{X}$ and $\overline{X}_a$ are the hair-bundle and motor positions, respectively, with respect to their steady-state values. The grey oblique lines have a slope of K_{SP}, the stiffness of the stereociliary pivots.

4 Discussion

Linear-stability analysis of the dynamical system defined by Eqs. 1 and 2 indicates that the hair bundle can become oscillatory if two necessary conditions are met. First, gating compliance must be strong enough that the force-displacement relation displays a region bundle positions for which the slope stiffness $K_{HB} = \overline{K}_{GS} + K_{SP}$ of the system is negative: the apparent gating-spring stiffness $\overline{K}_{GS} = K_{GS}\left(1-\left(ZDP_o\left(1-P_o\right)\right)/\left(k_B T\right)\right)$ must be negative enough to dominate the pivot stiffness K_{SP}. Second, the Ca^{2+}-feedback strength S must be neither too large nor too small in order to position the system's operating point within the unstable region of negative stiffness. In such instance, the myosin-based adaptation motor can never reach stall condition and thus produces an oscillation (Fig. 5B). Note that without Ca^{2+} feedback, the system is always stable, even in the presence of negative stiffness. When operating on the verge of an unstable region of negative stiffness, however, the system is excitable: pushed across the unstable region of negative stiffness by an external force, the system relaxes back towards its operating point by displaying a non-monotonic twitch-like movement whose time-course looks like one half-cycle of a spontaneous oscillation (Fig. 5A).

When $K_{HB}>0$, the system is stable but can nevertheless display twitches in response to abrupt step forces (Fig. 5C). In the limit where hair-bundle motion is fast compared to motor movements, Eq. 1 indeed indicates that an adaptive movement of the motor elicits a movement of the hair bundle that can be approximated by:

$$\delta X = \frac{\overline{K}_{GS}}{K_{HB}}\,\delta X_a\,. \tag{5}$$

A positive adaptive movement of the motor towards the base of the stereocilia, as would be expected in response to a positive force step or to a sudden increase of the extracellular Ca^{2+} concentration, can thus evoke a negative deflection of the hair bundle (a recoil), provided that $\overline{K}_{GS}$ or equivalently that the hair-bundle stiffness is locally smaller than that of the stereociliary pivots. This condition can be met only if gating-compliance is strong enough that $ZD > 4k_B T$ and, if the case, is more easily met near $P_o = 0.5$ where the mechanical effects of channel gating are the greatest. Note that in contrast to movements of the adaptation motors, the internal forces that result from the conformational changes associated to channel gating during adaptation provide positive feedback on gating-spring tension. Non-monotonic movements are the result of a non-linear and dynamic tradeoff between opposing forces that are produced by adaptation motors on one end and those exerted by adaptive channel gating on the other end.

The kinetics of active hair-bundle movements and of the associated transduction currents (Eq. 3) are controlled in part by how fast the adaptation motors can react to an external perturbation. Adaptation kinetics can easily be discussed in the simple case where the bundle's position is clamped and a small step displacement δX is applied. Then, Eq. 2 indicates that the time-course of the adaptive shift δX_a can be described by a single exponential with the characteristic timescale

$$\tau_a = \lambda_a \big/ \left(\tilde{K}_{GS} + K_{ES}\right)\,. \tag{6}$$

in which the effective gating-spring stiffness $\tilde{K}_{GS}$ is given by:

$$\tilde{K}_{GS} = K_{GS}\left(1 - \frac{ZD}{k_B T}\left(1 - S\frac{F_{\max}}{K_{GS} D}\right) P_o\left(1-P_o\right)\right)\,. \tag{7}$$

In the case where Ca^{2+} feedback is strong enough that $S > K_{GS} D / F_{max}$, intracellular Ca^{2+} changes that result from channel gating effectively stiffen the gating springs ($\tilde{K}_{GS} > K_{GS}$), thereby providing an effect opposite that of gating compliance. Adaptation is thus expected to be significantly faster when the system operates near $P_o = 0.5$ than near $P_o = 0$ or 1, where the stimulus evokes little channel rearrangements. With parameters appropriate for transduction in the bullfrog sacculus [14], fast and slow adaptations are characterized by timescale of 4 and 24 ms, respectively. Because S is proportional to the extracellular Ca^{2+} concentration, fast adaptation will occur preferentially at high Ca^{2+}. At low Ca^{2+} concentrations, Ca^{2+} feedback on the motor force is too weak to fully overcome the effects of gating compliance ($\tilde{K}_{GS} < K_{GS}$), and adaptation is thus always slow, reaching timescales of several tens of milliseconds or more near $P_o = 0.5$.

Because adaptation is incomplete, a static deflection of the hair bundle can displace the bundle's operating point within its force-displacement relation and, correspondingly, change the open probability of the transduction channels at steady state. By so doing, the directionality of active hair-bundle movements evoked by a given stimulus can be inverted and its kinetics altered, as we have observed experimentally (Fig. 4). These effects can be described by a change in sign of $\overline{K}_{GS}$ and in magnitude of $\tilde{K}_{GS}$, respectively.

In conclusion, only three ingredients suffice to account for the different manifestations of active hair-bundle motility: strong gating-compliance, myosin-based adaptation and Ca^{2+} feedback on the motor's activity.

Acknowledgments

J.-Y.T. and P.M. supported by HFSP grant RGP0051/2003 and by European Commission FP6 Integrated Project EUROHEAR, LSHG-CT-2004-512063.

References

1. Hudspeth, A.J. 1997. Mechanical amplification of stimuli by hair cells. Curr. Opin. Neurobiol. 7, 480-486.
2. Santos-Sacchi, J. 2003. New tunes from Corti's organ: the outer hair cell boogie rules. Curr Opin Neurobiol 13, 459-68.
3. Dallos, P., Wu, X., Cheatham, M.A., Gao, J., Zheng, J., Anderson, C.T., Jia, S., Wang, X., Cheng, W.H., Sengupta, S., He, D.Z., Zuo, J. 2008. Prestin-based outer hair cell motility is necessary for mammalian cochlear amplification. Neuron 58, 333-9.
4. Mellado Lagarde, M.M., Drexl, M., Lukashkina, V.A., Lukashkin, A.N., Russell, I.J. 2008. Outer hair cell somatic, not hair bundle, motility is the basis of the cochlear amplifier. Nat Neurosci 11, 746-8.
5. Chan, D.K., Hudspeth, A.J. 2005. Ca2+ current-driven nonlinear amplification by the mammalian cochlea in vitro. Nat. Neurosci. 8, 149-55.
6. Manley, G.A. 2001. Evidence for an active process and a cochlear amplifier in nonmammals. J. Neurophysiol. 86, 541-549.

7. Martin, P. 2008. Active hair-bundle motility of the hair cells of vestibular and auditory organs. In: Manley, G.A., Popper, A.N., Fay, R.R., (Eds.), Active processes and otoacoustic emissions. Springer, New York. pp. 93-144.
8. Martin, P., Hudspeth, A.J. 2001. Compressive nonlinearity in the hair bundle's active response to mechanical stimulation. Proc. Natl. Acad. Sci. USA 98, 14386-14391.
9. Martin, P., Hudspeth, A.J. 1999. Active hair-bundle movements can amplify a hair cell's response to oscillatory mechanical stimuli. Proc. Natl. Acad. Sci. USA 96, 14306-14311.
10. Howard, J., Hudspeth, A.J. 1987. Mechanical relaxation of the hair bundle mediates adaptation in mechanoelectrical transduction by the bullfrog's saccular hair cell. Proc. Natl. Acad. Sci. USA 84, 3064-3068.
11. Ricci, A.J., Crawford, A.C., Fettiplace, R. 2000. Active hair bundle motion linked to fast transducer adaptation in auditory hair cells. J. Neurosci. 20, 7131-7142.
12. Benser, M.E., Marquis, R.E., Hudspeth, A.J. 1996. Rapid, active hair bundle movements in hair cells from the bullfrog's sacculus. J. Neurosci. 16, 5629-5643.
13. Kennedy, H.J., Crawford, A.C., Fettiplace, R. 2005. Force generation by mammalian hair bundles supports a role in cochlear amplification. Nature 433, 880-3.
14. Tinevez, J.Y., Julicher, F., Martin, P. 2007. Unifying the various incarnations of active hair-bundle motility by the vertebrate hair cell. Biophys J 93, 4053-67.
15. Martin, P., Bozovic, D., Choe, Y., Hudspeth, A.J. 2003. Spontaneous oscillation by hair bundles of the bullfrog's sacculus. J. Neurosci. 23, 4533-48.
16. Martin, P., Mehta, A.D., Hudspeth, A.J. 2000. Negative hair-bundle stiffness betrays a mechanism for mechanical amplification by the hair cell. Proc. Natl. Acad. Sci. USA 97, 12026-12031.
17. Howard, J., Hudspeth, A.J. 1988. Compliance of the hair bundle associated with gating of mechanoelectrical transduction channels in the bullfrog's saccular hair cell. Neuron 1, 189-199.
18. Hudspeth, A.J., Gillespie, P.G. 1994. Pulling springs to tune transduction: adaptation by hair cells. Neuron 12, 1-9.
19. Assad, J.A., Corey, D.P. 1992. An active motor model for adaptation by vertebrate hair cells. J. Neurosci. 12, 3291-309.
20. Hacohen, N., Assad, J.A., Smith, W.J., Corey, D.P. 1989. Regulation of tension on hair-cell transduction channels: displacement and calcium dependence. J. Neurosci. 9, 3988-3997.
21. Crawford, A.C., Evans, M.G., Fettiplace, R. 1991. The actions of calcium on the mechano-electrical transducer current of turtle hair cells. J. Physiol. 434, 369-398.
22. Adamek, N., Coluccio, L.M., Geeves, M.A. 2008. Calcium sensitivity of the cross-bridge cycle of Myo1c, the adaptation motor in the inner ear. Proc Natl Acad Sci U S A 105, 5710-5.
23. Stauffer, E.A., Scarborough, J.D., Hirono, M., Miller, E.D., Shah, K., Mercer, J.A., Holt, J.R., Gillespie, P.G. 2005. Fast adaptation in vestibular hair cells requires Myosin-1c activity. Neuron 47, 541-53.

IN VIVO DISSECTION OF FLY AUDITORY MECHANOTRANSDUCTION

J.T. ALBERT

UCL Ear Institute, University College London, 332 Gray's Inn Road London, WC1X 8EE, United Kingdom

B. NADROWSKI

VW Foundation Research Group, Institute of Zoology, University of Cologne, Weyertal 119 50923 Cologne, Germany

T. EFFERTZ

VW Foundation Research Group, Institute of Zoology, University of Cologne, Weyertal 119 50923 Cologne, Germany

M.C. GÖPFERT

VW Foundation Research Group, Institute of Zoology, University of Cologne, Weyertal 119 50923 Cologne, Germany

Hearing relies on specialized, mechanosensitive ion channels that transduce sound-induced vibrations into electrical signals. Linking this transduction to identified proteins has proven difficult because of the scarcity of native auditory transducers and their tight functional integration into ears. We have devised an *in vivo* paradigm for the non-invasive study of auditory transduction. By investigating displacement responses of the *Drosophila* sound receiver, we find that the gating of mechanotransducer channels impacts on the mechanics of the ear. This impact includes a nonlinear compliance that correlates with electrical nerve responses, shifts with adaptation, and conforms to the gating spring model of vertebrate auditory transduction. The discovery of spring-operated, mechanically adapting transducer channels in the *Drosophila* ear– the function of which can be probed in intact flies- now offers an attractive model for the targeted molecular dissection of an auditory transducer channel complex.

1 Introduction

Based on sub-millisecond response latencies [1, 2], which seem to short to be compatible with an indirect, second messenger-mediated mode of activation, sensory mechanotransduction (MET) is generally assumed to rely on transducer channels that are *directly* gated by stimulus forces[3-5]. In *Drosophila*, sound receiver deflections are followed by compound action potential (CAP) responses in the antennal nerve within less than 0.6 ms (see Fig. 1)[6]. As the of CAPs delays include the preceding generation of action potentials and their propagation to the site of recording, the latency between mechanical stimulus and first electrical response (i.e. the transducer current) can be expected to be considerably shorter. This fast response time speaks in favour of a direct mode of mechanotransducer activation in the *Drosophila* ear. The most compelling evidence for directly gated ion channels, however, is the demonstration of a *gating compliance*[3, 7]: as all components which contribute to funneling forces from receiver to transducer must be mechanically coupled, the receiver will inevitably be more compliant over that range of forces (and displacements) at which the channels gate. We tested whether gating compliances are detectable in the mechanics of the antennal receiver of a fly.

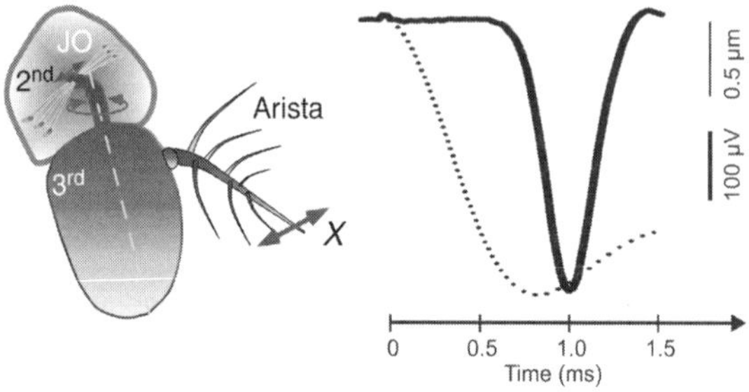

Figure 1. (left) In response to sound the *Drosophila* antennal sound receiver (3rd antennal segment) starts rotating about its longitudinal axis (dashed line). This movement is picked up by two opposing populations of sensory neurons in Johnston's Organ (JO, 2nd segment) (right) Receiver movements (dotted) are followed by CAP responses (solid) in the antennal nerve with submillisecond delays. Data from Ref. [6]

2 Methods

Wild-type flies (OregonR) were used for all experiments. The flies were mounted on top of a Teflon rod, with their halteres, wings, legs, and mouthparts stabilized by wax and dental glue. The second antennal segment, which harbours the fly's auditory neurons, was immobilized by dental glue to prevent muscle-based antennal movements. All experiments were carried out in a vibration-isolated environment at temperatures between 20-25°C. Stimulus forces were applied electrostatically via an external electrode positioned ca. 300 μm behind the antenna's arista. Displacements of the receiver were measured at the tip of the arista using a PSV-400 Laser Doppler vibrometer CAP responses were recorded as the difference potential between a tungsten electrode at the base of the antenna and a second tungsten electrode inserted into the animal's thorax[6].

3 Results

3.1 The fly's auditory transducers are mechanically gated, spring-operated ion channels

When actuated by a force-step, the fly's antennal receiver (Fig. 1) displays a characteristic pattern of deflection (see Fig. 2a): An initial overshoot leads to a displacement maximum, X_{peak}, which is followed by a fast recoil and a subsequent slower excursion to a steady-state displacement, X_{steady}. In contrast to X_{steady} which is *linearly* related with the external force F_{ext}, X_{peak} scales *nonlinearly* with F_{ext} (Fig. 2b). The peaking displacement response at the onset of the force-step coincides with a transitory CAP response in the antennal nerve and corresponds to a transitory increase of the receiver's compliance. The nonlinear force dependence of the receiver's

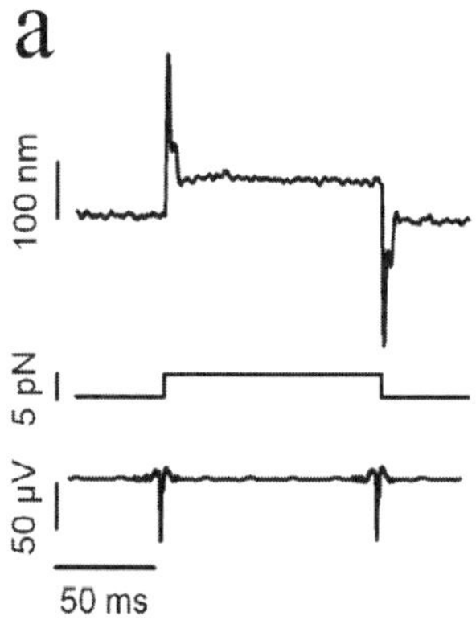

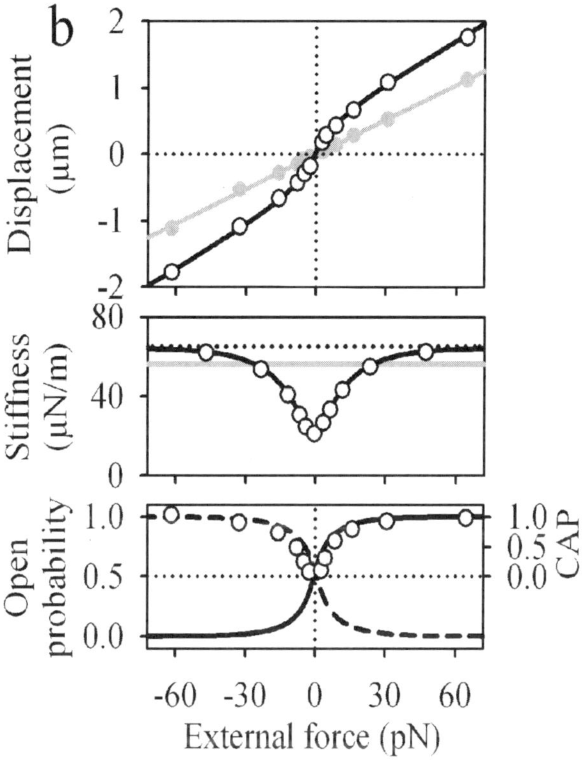

Figure 2. (a) Receiver displacement response (top) to force step (middle). Both the onset and the termination of the stimulus evoke a CAP response (bottom). (b) Steady-state displacements (top, grey) scale linearly with force, initial displacement peaks (top, white symbols) scale nonlinearly with force. The initial nonlinearity is well described with a two-sate gating spring model (black lines in all) and (middle) corresponds to a stiffness drop for small forcing amplitudes. (bottom) Measured CAP amplitudes (symbols) and open-probabilities as predicted from receiver mechanics [solid line: $P_O(X_{peak})$ dashed line: $P_O(-X_{peak})$ Data from Ref. [6].

compliance was found to accord with a two-state gating-spring model [3, 7], that expresses F_{ext} as a function of X_{peak} and takes the receiver's inertia into account:

$$F_{ext} = K_\infty X_{peak} - P_o(X_{peak})Nz + F_0 - m\ddot{X}_{peak} \qquad (1)$$

Here, K_∞ is the receiver's dynamic stiffness if all channels are either open or closed (Fig. 3C), N is the number of channels, z is the change in force in a single gating spring when the channel opens as seen in the receiver's mechanics, F_0 is a constant offset term, m, the receiver's apparent mass and $\ddot{X}_{peak}$ the receiver's acceleration at X_{peak}. The open probability is given as $P_o(X_{peak}) = 1/1 + e^{-z(X_{peak} - X_0)/k_B T}$, whereby X_0 is the displacement at which $P_o(X_{peak}) = 0.5$. Using Eq. (1) to fit the nonlinear force-displacement curve of the *Drosophila* sound receiver (Fig. 2b) predicts a resting open probability of approximately 0.5 (corresponding to $X_0 \approx 0$). If the nonlinearity in the receiver's mechanics is introduced by the gating of transducer channels the open probabilities deduced from the receiver's mechanics should be reflected by the electrical activity of the fly's auditory neurons, ideally the compound currents flowing through the transducer channels. In substitute for such a compound current, which can not be recorded yet, we measured the CAPs from the antennal nerve. In accordance with the existence of two opposing neuronal populations- which attach to opposite sides of the receiver's hook-like extension into the second segment (Fig. 1)- both anteriad and posteriad receiver deflections evoked CAP responses. We found that the peak amplitudes of these CAPs closely followed the open probabilities inferred from the receivers' mechanics for $P_o > 0.5$ (Fig. 2b, bottom), which is consistent with the idea that stimulus-correlated CAPs are evoked only when the open probability exceeds its resting value. Taken together these results suggest that the transitory compliance seen in the receiver's displacement response to force-steps is indeed introduced by the mechanical gating of transducer channels and thus represents a *gating compliance*.

3.2 The fly's auditory transducers are mechanically adapting ion channels

One of the most peculiar features of the spring-gated ion channels in vertebrate hair cells is their association with motor proteins that operate to release tension from the gating-springs[8, 9], thereby (i) allowing for the channels' reclosure in the presence of an external force and (ii) fully or partly restoring the channels' open-probabilities to the pre-stimulus levels. To test for mechanical transducer adaptation in *Drosophila* we exposed the fly's receivers to test stimuli (force steps) before and 50 ms after the onset of an adapting stimulus of constant offset force (Fig. 3a). In the presence of an adapting

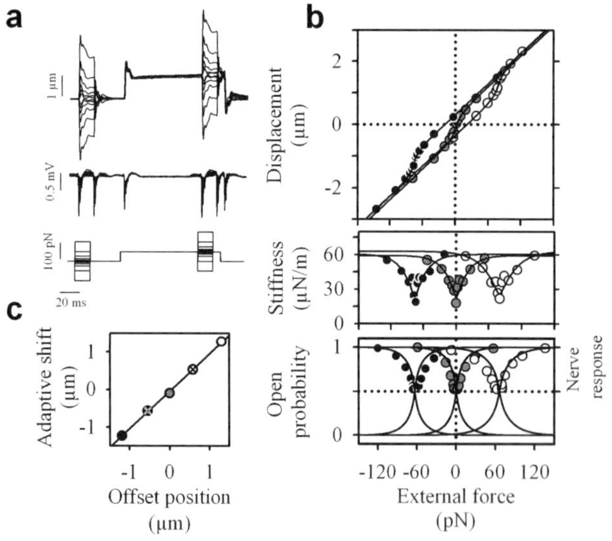

Figure 3. (a) Receiver displacement (top) and CAP (middle) responses to test stimuli before and 50 ms after the onset of a maintained receiver deflection (bottom). (b) X_{peak} (top), and resulting dynamic stiffnesses (middle), and CAP responses (bottom) as a function of the external force for three different offset forces. (c) The position of maximum gating compliance X_0 shifts with the offset position along a line of unity.

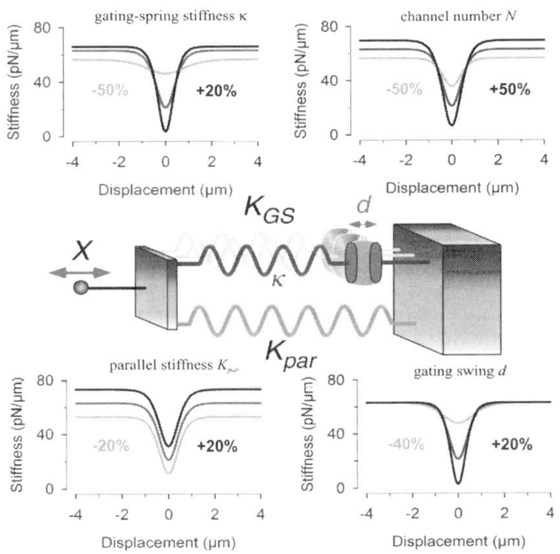

Figure 4. Probing transducer function in vivo. (sketch at centre) Gating spring operated transducer module with a gating spring of stiffness κ. The combined stiffness of all the gating springs, K_{GS}, and a parallel stiffness, K_{par}, determine the asymptotic stiffness of the system, K_{∞}. Because each gating spring shortens by a distance d as the associated channel opens (the gating swing), the stiffness of the system will drop at those displacement (force) amplitudes at which the channels gate ('gating compliance'). (graphs) Predicted gating compliances, as betrayed by the fly's antennal receiver, for changes in κ, N, K_{par} and d. Black curves illustrate increases, light-gray curves decreases in the respective parameter. Changes are given in percentage of wildtype condition (grey curve). For more details see Ref. [10].

stimulus, both the working range of the CAP response and the region of the receiver's nonlinear compliance jointly moved in direction of the offset force (Fig. 3b). This adaptive shift was complete, with the position of maximum compliance (X_0 in Eq. (1)) moving exactly to the imposed offset position (Fig. 3c)[6].

4 Discussion

Neither for vertebrates nor invertebrates has an auditory transducer channel been identified molecularly. Given the vast amount of information that has accumulated concerning the biophysical properties of these channels this specific lack of knowledge seems surprising. It is partly explained by the direct mode of gating: As a consequence of a spring-operated activation, auditory transducer channels crucially depend on the integrity of accessory components involved in transmitting forces to the channels, such as intra- and extracellular tethers. Given this interdependence, functional transducer complexes will be difficult to restore in conventional heterologous expression systems, for example. The discovery of spring-operated, mechanically adapting transducer channels in the *Drosophila* ear– the function of which can be probed in intact flies- now provides an attractive additional approach to the molecular dissection of an auditory transducer channel complex. The gating-spring model, on the one hand, makes clear and testable predictions for the phenotypes that will associate with distinct transducer defects[10] (see Fig. 4 for an illustration), *Drosophila*, on the other hand, provides the genetic and molecular tractability that is required to specifically manipulate identified candidate proteins. The fruitful and mutual interplay between model and mutants is expected to help shedding light on the molecules of auditory transduction.

Acknowledgments

This work is supported by grants from the Volkswagen Foundation (to B.N. and to M.C.G.), the German Bernstein Network for Computational neuroscience (to M.C.G.) and a research fellowship from Deafness Research UK (to J.T.A). We thank Michael Dübbert and Oliver Hendrich for technical assistance.

References

1. Corey, D.P., and Hudspeth, A.J. 1979. Response latency of vertebrate hair-cells. Biophysical Journal 26, 499-506.
2. Thurm, U. 1963. Die beziehungen zwischen mechanischen reizgrossen und stationaren erregungszustanden bei borstenfeld-sensillen von bienen. Zeitschrift fur Vergleichende Physiologie 46, 351-382.
3. Hudspeth, A.J., Choe, Y., Mehta, A.D., and Martin, P. 2000. Putting ion channels to work: Mechanoelectrical transduction, adaptation, and amplification by hair cells. P Natl Acad Sci USA 97, 11765-11772.
4. Gillespie, P.G., and Walker, R.G. 2001. Molecular basis of mechanosensory transduction. Nature 413, 194-202.
5. Christensen, A.P., and Corey, D.P. 2007. TRP channels in mechanosensation: direct or indirect activation? Nat Rev Neurosci 8, 510-521.
6. Albert, J.T., Nadrowski, B., and Gopfert, M.C. 2007. Mechanical signatures of transducer gating in the Drosophila ear. Curr Biol 17, 1000-1006.
7. Howard, J., and Hudspeth, A.J. 1988. Compliance of the hair bundle associated with gating of mechanoelectrical transduction channels in the bullfrogs saccular hair cell. Neuron 1, 189-199.
8. Howard, J., and Hudspeth, A.J. 1987. Mechanical relaxation of the hair bundle mediates adaptation in mechanoelectrical transduction by the bullfrogs saccular hair cell. P Natl Acad Sci USA 84, 3064-3068.
9. Eatock, R.A. 2000. Adaptation in hair cells. Annual Review of Neuroscience 23, 285-314.
10. Albert, J.T., Nadrowski, B., and Gopfert, M.C. 2007. Drosophila mechano-transduction - Linking proteins and functions. Fly 1, 283-241.

TRANSDUCER-BASED ACTIVE AMPLIFICATION IN THE HEARING ORGAN OF DROSOPHILA MELANOGASTER

BJÖRN NADROWSKI

Zoological Institute, University of Cologne, Weyertal 119
50923 Cologne, Germany

JÖRG T. ALBERT

UCL Ear Institute, 332 Gray's Inn Road
London WC1X 8EE, UK

MARTIN C. GÖPFERT

Zoological Institute, University of Cologne, Weyertal 119
50923 Cologne, Germany

In *Drosophila*, hearing is mediated by the antenna. Stimulus forces acting on the antennal receiver are coupled to dedicated neurons that comprise the molecular machinery for mechanosensory transduction, adaptation and amplification. Because the action of this machinery is reflected in the receiver's mechanics, the latter can be used to probe the molecular mechanisms that bring about hearing in an intact ear. Here, we show that these mechanisms closely resemble those that are at work in hair cells in vertebrate ears. Based on the gating-spring model of transduction in vertebrate hair cells, we have developed an extended, symmetric gating-spring model reflecting the anatomy of the antennal receiver, that describes the physical processes that make flies hear. This model explains the ear's performance, including the receiver\s mechanics and the electrical response of the afferent nerve. These findings suggest that while the auditory anatomy is vastly different, the molecular components and mechanisms that promote fly and vertebrate hearing may be evolutionarily conserved.

1 Introduction

Ears rely on an active amplification process in order to achieve their exquisite sensitivity and frequency selectivity [1-7]. Inside the ear, hair cells transduce sound-induced vibrations into electrical signals while generating forces which augment the vibrations they transduce [2-8]. This mechanical feedback, known as the cochlear amplifier [9, 10], accounts for vital characteristics in auditory system performance, including compressive nonlinearity, frequency–selectivity, amplification of weak stimuli, and self-sustained oscillations [2-11].

In vertebrates, these four key characteristics of the cochlear amplifier have been observed in the hair bundle mechanics of inner–ear hair cells [12-15]. A feedback-regulated [16,17] gating-spring model of mechanotransduction [18-22] can explain these properties while tracing the origin of active hair bundle motility to the transduction modules themselves. This gating-spring model consists of a serially arranged force-gated ion-channel, a gating-spring and an ensemble of molecular adaptation motors characterized by a linear force-velocity relation, whereby the motor's stall force is linearly coupled to the open probability of the channels [16, 17]. Here, we show that the active performance of the entire *Drosophila* hearing organ can be explained by the assumption that its transduction process is performed by the same hair cell gating-spring module which explains active hair bundle mechanics in a variety of vertebrate species [16, 17].

2 Methods

Oregon R wild-type flies, two to three days after eclosion, were used for the experiments. The fly's were mounted ventrum-down on top of a Teflon rod, with their halteres, wings, legs, and mouthparts being stabilized by wax [23]. The second antennal segment was immobilized by dental glue to prevent muscle-based antennal movements. All experiments were carried out on a vibration isolation table at room temperature (20-25°C). Stimulus forces were applied electrostatically via an external electrode positioned ca. 300 μm behind the antenna's arista [24]. Displacements of the receiver were measured at the tip of the arista using a PSV-400 Laser Doppler vibrometer. Compound action potentials were recorded from the antennal nerve via an electrolytically tapered tungsten electrode inserted between the antenna and head.

3 Results

3.1 Transducer-based model of the Drosophila hearing organ

Consistent with previous observations [24, 25], we describe the *Drosophila* auditory system [26-28] by two mirror-symmetric opposed gating-spring transducer populations as proposed to be at work in vertebrate hair bundles [16, 17] that couple to a simple harmonic oscillator that represents the sound receiver (**Fig. 1A, B**). The dynamic properties of the model are described by four coupled differential equations with nine free parameters, three of which refer to the harmonic oscillator representing the fly's

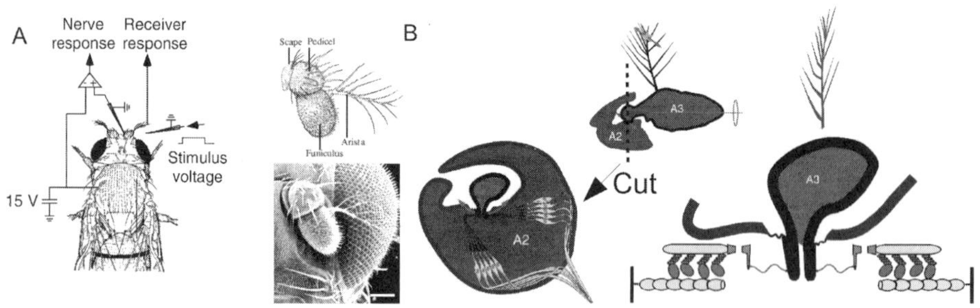

Figure 1. A, the experimental setup. The sound-receiver (arista) is stimulated using a non-contact electrostatic method. B, Details of the ear's anatomy and sketch of the model's constituents. The arista and the funiculus (A3) form a unit which rotates around the long axis of the funiculus. The sound-induced rotational movement stimulates two distinct sets of sensory neurons situated at both sides of the anchoring point inside the pedicelus (A2). The model reflects this symmetry by coupling a mass-spring system (arista) to two identical populations of transducer-modules.

antennal sound receiver. Fluctuating forces were included by adding white Gaussian noise with an intensity corresponding to the fluctuation-dissipation theorem [16]. We fitted the model simultaneously to an ensemble of 10 steps, the systems' linear response function, and its spectral density of free fluctuations. The nine resulting fit parameter thereby define the model's representation of the fly's ear. In all figures displayed in this article, the dashed lines correspond to simulations using the same parameter set obtained by a fit to one typical animal.

3.2 *Response to force steps*

When deflecting the receiver by means of a force step, the mechanical response resembled that observed in vertebrate hair bundles [22, 24, 29, 30]: an initial overshoot in the forcing direction was followed by a rebound and a damped oscillation, before reaching a constant steady–state position (**Fig. 2A**). The fly's receiver displayed a transient nonlinearity at the onset of the stimulus, the so-called gating-compliance [22, 24] (**Fig. 2B**), the system being most compliant for zero forcing (**Fig. 2C**). As judged from our model, and as expected for a gating-spring system, this transient compliance results from the opening of channels and the ensuing relaxation of tension in the associated gating-springs.

Stimulus-evoked displacements of the fly's antennal receiver associate with compound action potentials (CAPs) in the antennal nerve. Because stimulus-correlated action potentials will only be generated when the transducer channels' open probability exceeds their resting value, we defined an excess channel open probability: this is the sum of the difference between actual and steady-state open probability for each transducer population, whereby only positive values are being considered. This excess open probability was found to be approximately proportional to the CAP amplitude (**Fig. 2C,3C** bottom panel); in addition, CAP and excess open probability displayed similar temporal patterns (**Fig. 2A**, bottom panel).

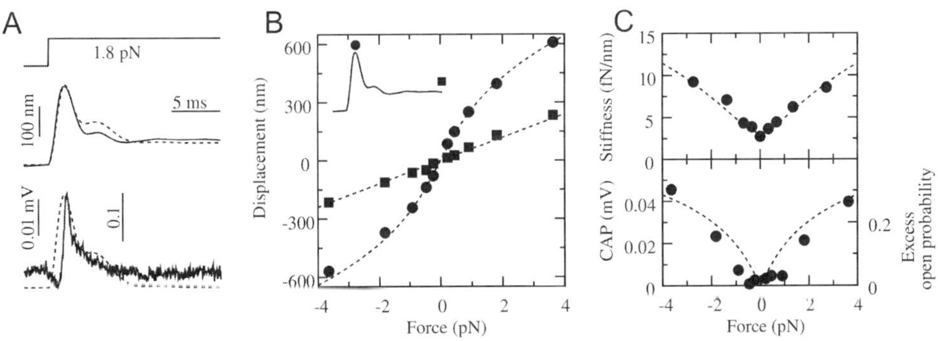

Figure 2. Response to force steps. Experiments are designated by symbols and solid lines, simulations by dashed lines. A, single force step. Top, applied force; middle, receiver response; bottom, CAP compared to the excess open probability (dashed lines and rightmost scalebar). B, gating compliance. Squares correspond to the displacement at the end of a force step, circles to the peak displacement. C, corresponding stiffness and nerve response. Top, stiffness for the peak displacements in B. Bottom, CAP and excess open probability amplitudes.

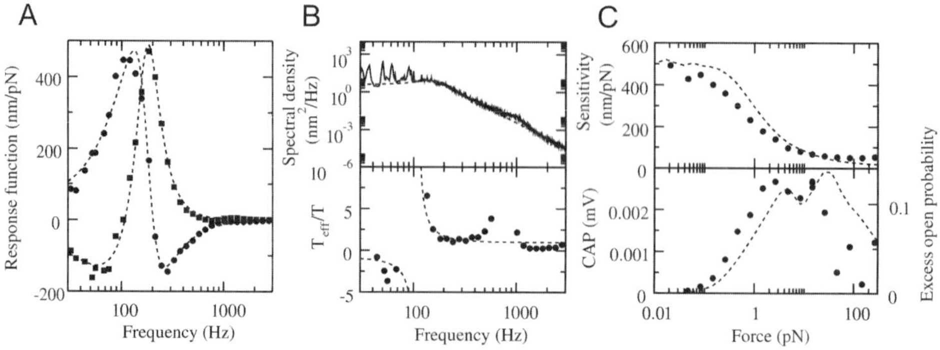

Figure 3. Response to sinusoidal stimuli. Experiments are designated by symbols, simulations by dashed lines. A, linear response function. Circles, real part. Squares, imaginary part. B, free fluctuations and effective temperature. C, nonlinear compression. Stimulus frequency was 156 Hz. Top, sensitivity at different stimulus amplitudes. Bottom, corresponding CAP and excess open probability amplitudes

3.3 Linear response function, free fluctuations, deviation from thermal equilibrium

The linear response function of the receiver resembled that of a simple harmonic oscillator for frequencies ≥ ca. 100 Hz (**Fig. 3a**). At lower frequencies, however, the imaginary part became negative, indicating that energy was withdrawn from the receiver at these frequencies [14, 16]. The independent measurements of both linear response function and the spectral density of free fluctuations allows to quantify the system's deviation from thermal equilibrium by using the fluctuation-dissipation theorem [14, 16] in order to define an effective temperature, T_{eff}, which for a system in thermal equilibrium equals the ambient temperature *T*. The fly's receiver deviates from thermal equilibrium at all frequencies, not only at the frequencies characterized by a negative imaginary part of the linear response function (**Fig. 3B**), and must therefore be characterized as active.

Although our model only contains one nonlinear relation (the channels' open probability as a function of the gating-spring tension [16, 18]), key features of the receiver's nonlinear behaviour, including nerve activity, are reproduced by the model (**Fig. 3C**).

4 Discussion

Profiting from the experimental accessibility of the Drosophila hearing organ and linking measurements and theory, this study shows that transduction modules as proposed for vertebrate hair cells explains the active performance of an entire ear. Our experimental data documents hair bundle-like active properties for the fly's antennal sound receiver, including deviations from thermal equilibrium [14, 16], transient nonlinear compliance in response to force steps [17, 18, 22], and a compressive nonlinearity in response to sinusoidal stimuli [15] (**Fig. 3C**). Our theoretical analysis, in turn, shows that, with some minor additions, a transduction model as used to describe active hair bundle mechanics [16, 17] suffices to explain all these active properties of the fly's sound receiver, whereby activity arises from the feedback between open probability and adaptation motor dynamics: when this feedback is eliminated from the model, the system obeys the fluctuation-dissipation theorem, becoming entirely passive. The good quantitative agreement between the model and the experimental data supports the idea that vertebrate and fly hearing relies on functionally equivalent, possibly evolutionarily conserved transduction modules. In addition, the finding that properties of the electrical activity are similar to the excess open probability argue for relatively simple coding characteristics connecting mechanical input and action potential generation.

Acknowledgments

This work is supported by grant 1/82 331 (to B.N.) and 1/79 147 of the Volkswagen Foundation (to M.C.G). We thank Thomas Effertz for help with the measurements, Frank Jülicher for fruitful discussions, and Michael Duebbert and Oliver Hendrich for technical assistance.

References

1. Gold, T. 1948. Hearing II. The physical basis of the action of the cochlea. Proc. R. Soc. Lond. B Biol. Sci. 135, 492-498.
2. Hudspeth, A.J. 1989. How the ear's works work. Nature 341, 397-404.
3. Dallos, P. 1992. The Active Cochlea. J. Neurosci. 12, 4575-4585.
4. Robles, L., and Ruggero, M.A. 2001. Mechanics of the mammalian cochlea. Physiol. Rev. 81, 1305-1352.
5. Eguiluz, V.M., Ospeck, M., Choe, Y., Hudspeth, A.J., and Magnasco, M.O. 2000. Essential nonlinearities in hearing. Phys. Rev. Lett. 84, 5232-5235.
6. Camalet, S., Duke, T., Jülicher, F., and Prost, J. 2000. Auditory sensitivity provided by self-tuned critical oscillations of hair cells. Proc. Natl. Acad. Sci. USA 97, 3183-3188.
7. Lukashkin, A.N., Walling, M.N., and Russell, I.J. 2007. Power amplification in the mammalian cochlea. Curr. Biol. 17, 1340-1344.
8. Fettiplace, R., and Hackney, C.M. 2006. The sensory and motor roles of auditory hair cells. Nat. Rev. Neurosci. 7, 19-29.
9. Davis, H. 1983. An Active Process in Cochlear Mechanics. Hear. Res. 9, 79-90.
10. Ashmore, J., and Gale, J. 2004. The cochlear amplifier. Curr. Biol. 14, R403-R404.
11. Chan, D.K., and Hudspeth, A.J. 2005. Ca2+ current-driven nonlinear amplification by the mammalian cochlea in vitro. Nat. Neurosci. 8, 149-155.
12. He, D.Z.Z., Jia, S., and Dallos, P. 2004. Mechanoelectrical transduction of adult outer hair cells studied in a gerbil hemicochlea. Nature 429, 766-770.
13. Martin, P., and Hudspeth, A.J. 1999. Active hair-bundle movements can amplify a hair cell's response to oscillatory mechanical stimuli. Proc. Natl. Acad. Sci. USA 96, 14306-14311.

14. Martin, P., Hudspeth, A.J., and Jülicher, F. 2001. Comparison of a hair bundle's spontaneous oscillations with its response to mechanical stimulation reveals the underlying active process. Proc. Natl. Acad. Sci. USA 98, 14380-14385.
15. Martin, P., and Hudspeth, A.J. 2001. Compressive nonlinearity in the hair bundle's active response to mechanical stimulation. Proc. Natl. Acad. Sci. USA 98, 14386-14391.
16. Nadrowski, B., Martin, P., and Jülicher, F. 2004. Active hair-bundle motility harnesses noise to operate near an optimum of mechanosensitivity. Proc. Natl. Acad. Sci. USA 101, 12195-12200.
17. Tinevez, J.-Y., Jülicher, F., and Martin, P. 2007. Unifying the various incarnations of active hair-bundle motility by the vertebrate hair cell. Biophys. J. 93, 4053-4067.
18. Hudspeth, A.J., Choe, Y., Mehta, A.D., and Martin, P. 2000. Putting ion channels to work: mechanoelectrical transduction, adaptation, and amplification by hair cells. Proc. Natl. Acad. Sci. USA 97, 11765-11772.
19. Martin, P., Bozovic, D., Choe, Y., and Hudspeth, A.J. 2003. Spontaneous oscillation by hair bundles of the bullfrog's sacculus. J. Neurosci. 23, 4533-4548.
20. Corey, D.P., and Hudspeth, A.J. 1983. Kinetics of the Receptor Current in Bullfrog Saccular Hair-Cells. J. Neurosci. 3, 962-976.
21. Howard, J., and Hudspeth, A.J. 1987. Mechanical Relaxation of the Hair Bundle Mediates Adaptation in Mechanoelectrical Transduction by the Bullfrogs Saccular Hair Cell. Proc. Natl. Acad. Sci. USA 84, 3064-3068.
22. Howard, J., and Hudspeth, A.J. 1988. Compliance of the hair bundle associated with gating of mechanoelectrical transduction channels in the bullfrog's saccular hair cell. Neuron 1, 189-199.
23. Albert, J.T., Nadrowski, B., and Göpfert, M.C. 2006. Mechanical tracing of protein function in the Drosophila ear. Nat. Protoc.
24. Albert, J.T., Nadrowski, B., and Göpfert, M.C. 2007. Mechanical signatures of transducer gating in the Drosophila ear. Curr. Biol. 17, 1000-1006.
25. Göpfert, M.C., Humphris, A.D.L., Albert, J.T., Robert, D., and Hendrich, O. 2005. Power gain exhibited by motile mechanosensory neurons in Drosophila ears. Proc. Natl. Acad. Sci. USA 102, 325-330.
26. Göpfert, M.C., and Robert, D. 2002. The mechanical basis of Drosophila audition. J. Exp. Biol. 205, 1199-1208.
27. Caldwell, J.C., and Eberl, D.F. 2002. Towards a molecular understanding of Drosophila hearing. J. Neurobiol. 53, 172-189.
28. Kamikouchi, A., Shimada, T., and Ito, K. 2006. Comprehensive classification of the auditory sensory projections in the brain of the fruit fly Drosophila melanogaster. J. Comp. Neurol. 499, 317-356.
29. Crawford, A.C., and Fettiplace, R. 1985. The Mechanical-Properties of Ciliary Bundles of Turtle Cochlear Hair-Cells. Journal of Physiology-London 364, 359.
30. Russell, I.J., Kössl, M., and Richardson, G.P. 1992. Nonlinear mechanical responses of mouse cochlear hair bundles. Proc Biol Sci 250, 217-227.

THE DYNEIN MOTOR IS THE BASIS OF ACTIVE OSCILLATIONS OF MOSQUITO ANTENNAE

B. WARREN, A. N. LUKASHKIN, I. J. RUSSELL

School of Life Sciences, University of Sussex, Falmer, Brighton, East Sussex, BN1 9QG, UK

The driver responsible for spontaneous oscillations of the mosquito (*Culex quinquefasciatus*) antennae was investigated. The activation energy derived from the temperature dependence of the spontaneous oscillation frequency is 30 kJ/mol suggesting a dynein ATPase is responsible. Colchicine application abolished spontaneous oscillations but left transduction intact. Hence, the transduction apparatus is thought not to be responsible for the spontaneous oscillations of the antennae.

1 Summary

The sensory cells at the base of the antennae, which transduce mechanical displacement into electrical impulses, are thought to underlie spontaneous antennal oscillations [1]; evidence supporting the hypothesis that motility and mechanosensitivity of cilia in sensory cells are a complementary feature of mechanosensory transduction [2]. Cilia motility is suggested to increase the sensitivity of mechanotransduction by reducing the inertia needed to push the system past threshold. The oscillations themselves are hypothesized to be a feature of the mechanotransduction channel [3]. An alternative hypothesis is that dynein, which resides along the length of the cilia, is responsible for the oscillations. Dynein has been shown to be an oscillating force generator [4]. By sliding the microtubule doublets relative to one another dynein is able to induce ciliary bending; the same ancestral mechanism that powers flagella in single-celled organisms.

The mechanical oscillations of the flagellum and the electrical oscillations of the Johnston's organ (JO), which share the same frequency, have a temperature dependence of the oscillation frequency that could be fitted by a straight line in the Arrhenius coordinates (Fig. 1) above 15°C. The driver of these oscillations is therefore composed of a single rate limited chemical reaction for these temperatures with an activation energy of 30 kJ/mol. This value is similar to other dynein driven biological systems, such as the flagellum of protozoa [5] but lies below that found for myosin ATPases [6].

The amplitude of the mechanical and electrical oscillations recorded in the Johnston's organ (Fig. 2) show similar temperature dependence, revealing that they are strongly coupled. The amplitude is maximal over a narrow temperature range with steep temperature dependences on either side which become reduced at low temperatures. The relationship could be well fitted by a single harmonic oscillator model with temperature dependant changes in the frequency of the driving force (Fig. 2).

Dynein function was disrupted by injection of colchicine into the pedicel. Colchicine, which disrupts the polymerization of microtubules, also blocked active oscillations (Fig. 3). Sound evoked electrical responses were recorded from the

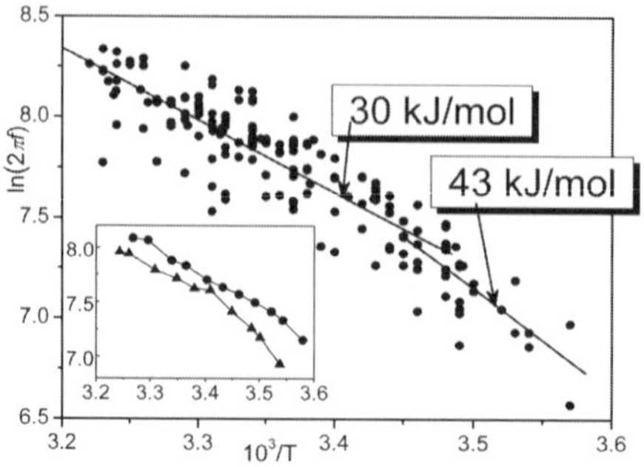

Figure 1. Temperature dependence of the frequency of mechanical oscillations of the flagellum of the antennae. Average data for 12 mosquitoes. Inset is the Arrhenius plots for two individual mosquitoes.

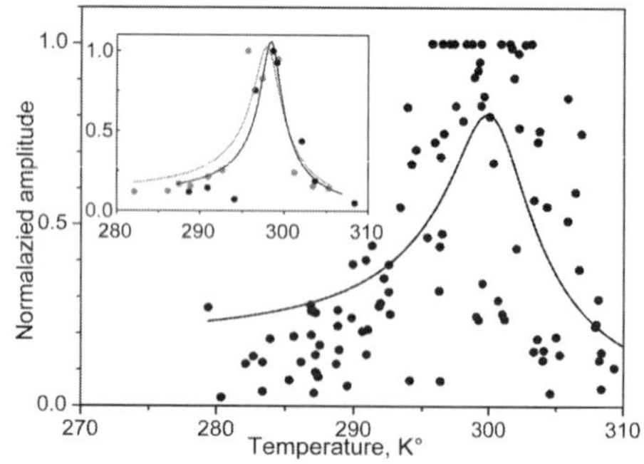

Figure 2. Temperature dependence of the electrical oscillations of JO fitted with a single harmonic oscillator model. Average data for 10 mosquitoes. Inset: single harmonic oscillator model fitted to two individual recordings.

JO after injection of colchicine and loss of spontaneous oscillations. Hence, the transduction apparatus is not affected by the colchicine. The transduction apparatus is therefore not responsible for the active oscillations of the mosquito antennae.

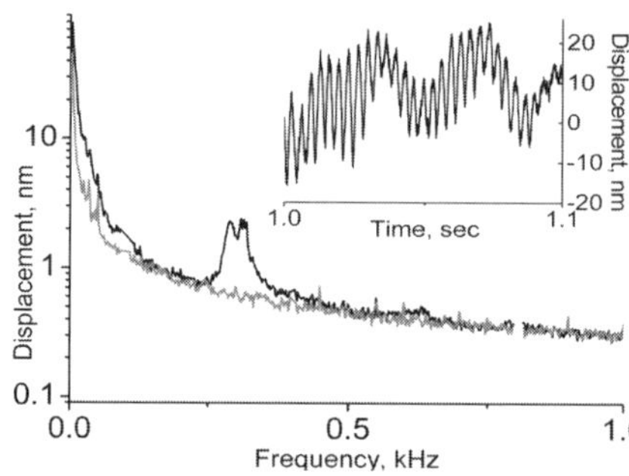

Figure 3. Amplitude spectrum of flagellum displacement before (black) and after (grey) injection of colchicine solution. Inset: oscillogram trace of displacement of flagellum before injection.

Acknowledgments

This work was supported by the BBSRC and MRC. We would like to thank James Hartley for his technical assistance.

References

1. Gopfert, M., 2001. Active auditory mechanics in mosquitoes. Proc. R. Soc. Lond. 268, 333-339.
2. Wiederhold, M.L., 1976. Mechanosensory transduction in sensory and motile cilia. Ann. Rev. of Biophy. Bioeng. 5, 39-62.
3. Gopfert, M.C., Robert, D., 2003. Motion generation by *Drosophila* mechanosensory neurons. PNAS. 100, 5514-5519.
4. Shingyoji, C., Higuchi, H., Yoshimura, M., Katayama, E., Yanagida, T., 1998. Dynein arms are oscillating force generators. Nature. 393, 711-714.
5. Holwill, M.E.J., Masi J., 1979. Thermodynamic and hydrodynamic studies relating to the mechanochemical cycle in the flagellum of *Crithidia-oncopelti*. J. Exp. Bio. 82, 177-195.
6. Anson M., 1992. Temperature-dependence and arrhenius activation-energy of f-actin velocity generated *in vitro* by skeletal myosin. J. Mol. Bio. 224, 1029-1038.

TRAFFICKING OF AMINOGLYCOSIDES INTO ENDOLYMPH *IN VIVO*

QI WANG, PETER S STEYGER

Oregon Hearing Research Center, Oregon Health & Science University Portland, Oregon, 97239, USA

In vitro, aminoglycosides increase the stiffness of cochlear hair cell stereocilia, altering bundle motion and transduction kinetics. Aminoglycosides also permeate the mechanosensitive transduction channel and rapidly initiate cytotoxicity in hair cells. If these effects occur *in vivo*, aminoglycosides would need to enter endolymph. The most direct route for systemically-administered aminoglycosides to enter endolymph is by trafficking from strial capillaries across the stria vascularis. An as-yet-unidentified active transporter is required to translocate aminoglycosides from the intra-strial space into the cytoplasm of marginal cells. Once in marginal cells, aminoglycosides would passively flow down the electrochemical gradient into endolymph. We present data that support a trans-strial trafficking route of aminoglycosides into endolymph, where they can then interfere with the mechanosensitive hair bundles.

1 Introduction

In vitro, aminoglycosides increase the stiffness of stereocilia of cochlear hair cells, altering bundle motion and micromechanics [1]. Aminoglycosides also block the mechano-electrical transduction channel, disrupting auditory sensitivity [2]. Aminoglycosides also permeate the mechano-electrical transduction (MET) channel and rapidly initiate cytotoxicity in hair cells [3, 4]. If these phenomena are to occur *in vivo*, it would require aminoglycosides to enter endolymph. In this study, we determined the distribution of fluorescently-conjugated gentamicin in the cochlea over time following systemic administration using confocal microscopy.

2 Materials and Methods

Mice were injected intra-peritoneally with 2 mg/kg purified gentamicin-Texas Red (GTTR) in PBS, pH 7.4, [5, 6]. At specific time-points after injection (10, 20, 30 minutes, 1, 3, and 9 hours, mice were anesthetized, and serum collected prior to cardiac-perfusion with 4% paraformaldehyde. Serum GTTR concentrations were determined using particle-enhanced turbidimetric inhibition immunoassay [7]. After fixation, cochleae were either embedded in OCT and cut as 8 μm cryostat sections, or dissected into cochlear surface preparations and mounted for confocal microscopy [6]. Prior to coverslipping, all specimens were phalloidin-labeled. Control mice were injected with hydrolyzed Texas Red [8]. The cytoplasmic intensity of Texas Red fluorescence in cells was statistically compared between cytoplasmic pixel data sets from the basal coils of at least 3 experimental and control cochleae [6]. Each time point ($n \geq 3$) was conducted simultaneously with the standard time point (30 minutes) group and ratiometric intensity

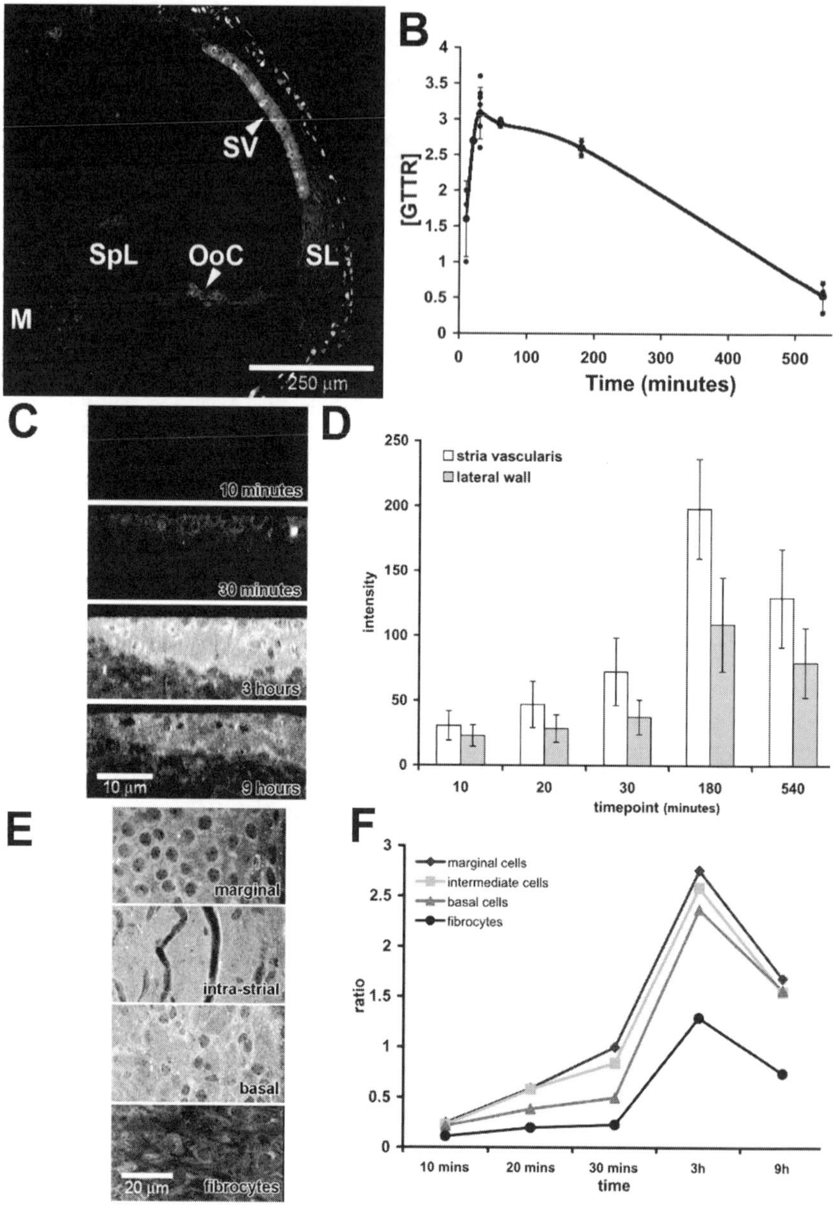

Fig. 1: (A) Distribution of GTTR fluorescence (white) in a cryostat section of the murine cochlea, 30 minutes after i.p. injection. M, modiolus; OoC, organ of Corti; SL, spiral ligament; SpL, spiral limbus; SV, stria vascularis. (B) Murine serum kinetics of GTTR (in IIIg/ml) over time. (C) Cross-sections show GTTR fluorescence in the stria vascularis compared to the lateral wall over time after GTTR administration. (D) Pixel intensity analysis of GTTR in the stria vascularis and lateral wall over time. (E) A focal series through a wholemounted stria vascularis and lateral wall, 9 hours after i.p. injection. Individual focal planes show marginal cells, the intra-strial tissues (composed of both intermediate cells and intra-strial space), basal cells, and fibrocytes. Individual cell types were identified by phalloidin labeling (not shown). (F) Comparative pixel intensities of GTTR fluorescence in different focal planes over time. All values are ratioed against marginal cell intensity at 30 minutes.

analysis conducted. Optical sections from each experimental set were identified, and regions of interest (marginal cells [minus nuclei], intra-strial tissues [putatively intermediate cells], basal cells and fibrocytes) were manually segmented for pixel intensity determination (ImageJ, NIH). To normalize data between experimental sets, the mean intensity was ratioed against the standard (30 minutes GTTR alone) and plotted. Student's t-test was used to determine any significant difference between time points, dosing groups or for regions of interest in the same specimens.

3 Results

In cryostat sections, GTTR fluorescence was preferentially localized in the stria vascularis within 30 minutes, compared to adjacent lateral wall tissues, or other intra-cochlear structures (Fig. 1A). Cryostats from mice treated with hydrolyzed Texas Red had negligible fluorescence (data not shown). GTTR levels in serum following intra-peritoneal injection reached approximately 1.5 μg/ml within 10 minutes, and peaked at 3 μg/ml in 30 minutes. GTTR levels remained elevated above 2.5 μg/ml for 3 h before falling to barely detectable levels at 9 hours (Fig. 1B).

In sections of the lateral wall, GTTR labeling was always more intense in the stria vascularis (location verified by phalloidin labeling) compared to the lateral wall at any time point (Fig. 1C). Comparative intensity analysis of GTTR fluorescence revealed that the stria vascularis had significantly more GTTR fluorescence than adjacent lateral wall tissues at any individual time point (Fig. 1D; $p< 0.001$; $n \geq 3$ at any time point). In addition, GTTR fluorescence in the stria vascularis (or lateral wall) at any single time point was significantly different than at any adjacent time point ($p < 0.001$).

In high resolution focal series through wholemounted surface preparations of the stria vascularis, individual cells and distinct regions could be identified (Fig. 1E). Comparative intensity analysis for each major cell type in the stria vascularis and lateral wall revealed that strial cells consistently displayed more GTTR fluorescence compared to fibrocytes. Marginal cells were consistently brighter than other strial cell types, except for endothelial cells. GTTR fluorescence peaked at 3 hours, and declined thereafter.

4 Discussion

GTTR levels in serum peaked at 30 minutes, and remained elevated above 2.5 μg/ml for at least two hours before falling, with a half life of 130 minutes. In contrast, unconjugated gentamicin has a serosal half-life of 53 minutes [9]. Thus, cochlear vascular structures are exposed to micromolar concentrations of systemic GTTR for longer periods than for gentamicin. However, these serum kinetics for GTTR mimic the intra-venous infusion kinetics of aminoglycosides in neonates and infants that are thought to be at non-ototoxic levels [10, 11]. Our microscopic data consistently show that the stria vascularis preferentially takes up GTTR compared to the lateral wall. Within the stria vascularis, GTTR fluorescence was typically more intense in marginal cells compared to other strial cells, except for endothelial cells of strial capillaries.

The most direct route for systemically-administered aminoglycosides to enter endolymph is by trafficking across the stria vascularis from strial capillaries. An unidentified transporter likely translocates GTTR into marginal cells [6]. Once in marginal cells, GTTR could passively flow down the electrochemical gradient into endolymph via aminoglycoside-permissive cation channels, e.g., TRPV4, located on the luminal surface of marginal cells [12]. The cationic aminoglycosides would then enter hair cells through both apical endocytosis and permeation of the mechanosensitive transduction channels at the tips of their stereocilia [3, 13]. This postulated trafficking route is being tested using cochlear perfusion of perilymphatic scalae during systemic administration of fluorescently-tagged aminoglycosides.

Acknowledgments

Funded by National Institute of Deafness and other Communication Disorders (grants DC004555 and DC005983).

References

1. Brummett, R.E., Fox, K.E., Bendrick, T.W., Himes, D.L. 1978. Ototoxicity of tobramycin, gentamicin, amikacin and sisomicin in the guinea pig. J Antimicrob Chemother 4 Suppl A, 73-83.
2. Dai, C.F., Steyger, P.S. 2008. A systemic gentamicin pathway across the stria vascularis. Hear Res 235, 114-24.
3. Dai, C.F., Mangiardi, D., Cotanche, D.A., Steyger, P.S. 2006. Uptake of fluorescent gentamicin by vertebrate hair cells in vivo. Hearing Research 213, 64-78.
4. Gale, J.E., Marcotti, W., Kennedy, H.J., Kros, C.J., Richardson, G.P. 2001. FM1-43 dye behaves as a permeant blocker of the hair-cell mechanotransducer channel. J Neurosci 21, 7013-25.
5. Hansen, A., Forbes, P., Arnold, A., O'Rourke, E. 2003. Once-daily gentamicin dosing for the preterm and term newborn: proposal for a simple regimen that achieves target levels. J Perinatol 23, 635-9.
6. Hashino, E., Shero, M. 1995. Endocytosis of aminoglycoside antibiotics in sensory hair cells. Brain Res 704, 135-40.
7. Karasawa, T., Wang, Q., Fu, Y., Cohen, D.M., Steyger, P.S. 2008. TRPV4 enhances cellular uptake of aminoglycoside antibiotics. J Cell Sci (in press).
8. Kroese, A.B., Das, A., Hudspeth, A.J. 1989. Blockage of the transduction channels of hair cells in the bullfrog's sacculus by aminoglycoside antibiotics. Hear Res 37, 203-17.
9. Marcotti, W., van Netten, S.M., Kros, C.J. 2005. The aminoglycoside antibiotic dihydrostreptomycin rapidly enters mouse outer hair cells through the mechano-electrical transducer channels. J Physiol 567, 505-21.
10. Murphy, J.E. 2005. Prediction of gentamicin peak and trough concentrations from six extended-interval dosing protocols for neonates. Am J Health Syst Pharm 62, 823-7.
11. Myrdal, S.E., Johnson, K.C., Steyger, P.S. 2005. Cytoplasmic and intra-nuclear binding of gentamicin does not require endocytosis. Hear Res 204, 156-69.

12. Newman, D.J., Henneberry, H., Price, C.P. 1992. Particle enhanced light scattering immunoassay. Ann Clin Biochem 29 (Pt 1), 22-42.
13. Richardson, G.P., Russell, I.J., Wasserkort, R., Hans, M. 1989. Aminoglycoside antibiotics and lectins cause irreversible increases in the stiffness of cochlear hair-cell stereocilia. In: Wilson, J.P., Kemp, D.T., (Eds.), Cochlear Mechanisms - Structure, Function and Models. Plenum Press, New York. pp. 578-66.

BIG AND POWERFUL: A MODEL OF THE CONTRIBUTION OF BUNDLE MOTILITY TO MECHANICAL AMPLIFICATION IN HAIR CELLS OF THE BIRD BASILAR PAPILLA

CHRISTINE KÖPPL

School of Medical Sciences (Physiology) and Bosch Institute, University of Sydney, NSW 2006, Australia

KUNI H. IWASA, BORA SUL

Biophysics Section, NIDCD, NIH, 5 Research Ct, Rm 1B03 Rockville, Maryland 20850, USA

Stimulus amplification by hair cells contributes to sensitivity and frequency selectivity in all vertebrate hearing organs. Motility of the hair-cell bundle is likely to be an important contributor to this amplification; in non-mammalian hair cells, bundle motility is thought to be the sole amplification mechanism. The largest hair-cell bundles - in terms of stereovillar numbers - are found in the avian basilar papilla (cochlea). We examined these morphological data to evaluate the energy produced by the hair bundle, specifically by comparing this energy with the viscous loss in the subtectorial space. Hair-bundle forces are predicted to be effective into the kHz range and are thus a realistic candidate mechanism for the cochlear amplifier in birds. In contrast to the smaller bundles of mammalian hair cells, the effect of bundle motility should be considerably larger in birds.

1 Introduction

Stimulus amplification by hair cells is now a widely accepted concept thought to contribute to sensitivity and frequency selectivity in all vertebrate hearing organs. Two distinct sites of motility have been identified as possible mediators of amplification. 1) Membrane-based electromotility found only in outer hair cells of the mammalian cochlea and 2) Motility of the hair-cell bundle which is inherent to the molecular transduction and adaptation machinery associated with each stereovillus and is thus present in all hair cells. For mammals, the relative contribution of each site of motility to cochlear amplification is a controversial issue (e.g., [1; 2]). For non-mammalian hearing, hair-bundle motility appears to be the only option, however, the precise mechanism (or mechanisms) still remains to be clarified. There are several components to bundle motility, with different timecourses (review in [3]).

We explored the possible contribution of hair-bundle motility known as fast adaptation, which includes twitch and release mechanisms (e.g., [4]). The force generated by hair bundles is expected to increase with increasing numbers of mechanoelectric transduction (MET) channels and with decreasing bundle height. As an extreme example for both parameters, we chose hair cells of the bird basilar papilla as the basis of the model. Their hair bundles are the largest of any hair cells - in terms of stereovillar numbers – and are also among the shortest. Furthermore, hair cells in a highly specialised bird, the barn owl, are known to operate at the highest frequencies that non-mammalian cochleae are capable of, about 10 kHz [5].

2 Methods

We made three basic assumptions: 1. Hair bundles produce force based on channel re-closure (or twitch) due to the binding of Ca^{++} introduced during channel opening. 2. The energy generated by the twitch must be sufficient to counteract viscous loss in the subtectorial space. 3. The mechanical stimulus is a continuous sinusoid with infinitesimal amplitude. With these assumptions, the limiting frequency is expressed by

$$f_{\lim} = \frac{\Phi N(\gamma f_g)^2 h}{2\pi^2 \eta A k_B T} \quad (1)$$

where k_B is Boltzmann's constant, T the temperature in Kelvin, Φ a phase factor, determined by channel kinetics (~ 0.07), η the viscosity of the fluid, N the number of tip links (which we assume is the same as the number of MET channels), h the height of the tallest stereovilli, A the surface area of the hair cell exposed to the subtectorial space, and f_g the gating force of the MET channel. The factor γ is the geometric factor, which is approximately s/h [17], where s is the separation of stereovillar rootlets. Thus, the dependence of the limiting frequency on the morphological factor F_m is given by

$$F_m = \frac{Ns^2}{hA} \quad (2)$$

We made an effort to use, wherever possible, specifically avian values for the various experimental parameters entering the model, and to also take into account differences between chicken and owl and between different locations within the basilar papilla. We supplemented published data with new scanning-electron microscopy data, obtained with routine methods. Hair-cell samples were taken at defined, normalized positions both along and across the papilla. Stereovillar counts were carried out in 4 barn owl and 2 chicken basilar papillae, from images taken at a minimal magnification of x8500. Hair-cell surface areas were determined in one barn owl papilla (aged P41), from images taken at x2500. Custom-made modifications to the SEM stage ensured a perpendicular surface view onto the strongly curved owl papilla at all locations evaluated.

3 Results

Two birds for which a host of morphological and physiological data are available, were contrasted: The chicken as a bird with unspecialized hearing and the barn owl as a high-frequency specialist.

3.1 Values for morphological parameters

We first summarize the sources of data used, as well as uncertainties associated with them that have important consequences for the model.

Hair-cell surface area (A): In the chicken, presumed tall hair cells (at neural positions) are smallest, while presumed short hair cells (at abneural positions) are larger at all longitudinal positions and reach a peak midway between apex and base [6; 7]. In the barn owl, the relative sizes of tall and short hair cells uniquely reverse in the basal, high-frequency regions [8]. While there is agreement on these species-specific patterns, data sets for the same species differ by a factor of about 1.5 in their absolute values. We believe this is due to the difficulties of obtaining a perfectly perpendicular view onto the curved surface of the basilar papilla. This would tend to underestimate surface areas. For model calculations, we have adopted Manley et al.'s [7] larger values for the chicken and our own for the owl, all corrected for 30% linear shrinkage due to SEM processing.

Height of tallest stereovilli (h): Bundle heights reported for both chicken and owl are in very good agreement between different studies. In the chicken, bundle height decreases nearly linearly from apex to base. We adopted the most comprehensive dataset by Tilney et al. [9] (SEM data, corrected for 30% linear shrinkage). In the owl, stereovillar height decreases linearly over the apical half of the papilla, but, uniquely, stays nearly constant within the basal half – a correlate of the foveal frequency representation [5]. We used the TEM measurements of Fischer [10], without correction.

Number of MET channels per hair bundle (N): The number of tip links was obtained by subtracting the number of stereovilli in one row from the total count of stereovilli. In birds, stereovillar numbers typically increase both from apex to base and, at a given longitudinal position, from abneural to neural; absolute numbers, however, differ substantially between different studies. For the chicken, our own counts are intermediate between the reported extremes, which differ by a factor of about 1.5 [6; 11]. For the owl, Fischer et al. [8] first showed the pronounced non-linear pattern along the basilar papilla which correlates with the foveal frequency representation [5]. Our own counts indicated, in addition, an extraordinarily steep rise in stereovillar numbers towards the neural edge of the papilla which had previously been missed. Numbers up to 350 stereovilli were typical for hair cells at the neural edge. For the model, we adopted our own data, supplemented by Tilney and Tilney [12] for the most apical positions in the chicken.

Stereovillar rootlet separation (s): Very few data are available for bird hair cells. Pickles [13] gave a value of 0.67 μm for a chicken tall hair cell. Our own estimates from published TEM micrographs indicate 0.27-0.39 μm for chicken basilar-papilla hair cells [14], his Figs. 2, 4) and 0.36-0.53 μm for the barn owl [10], his Fig.3). We therefore assumed a constant stereovillar separation of 0.45 μm for both chicken and owl.

3.2 The morphological factor for upper frequency limit

We plotted the morphological factor F_m ($=Ns^2/(hA)$ in Eq. 2) as a function of longitudinal (tonotopic) position and compared it with the known frequency maps for the chicken ("Greenwood" map from review [15]) and for the owl [5] (Fig. 1). The morphological factor reproduced the characteristic features of the tonotopic map for both birds. While, in the barn owl, tonotopic frequency and F_m rise sharply in the apical part and then flatten

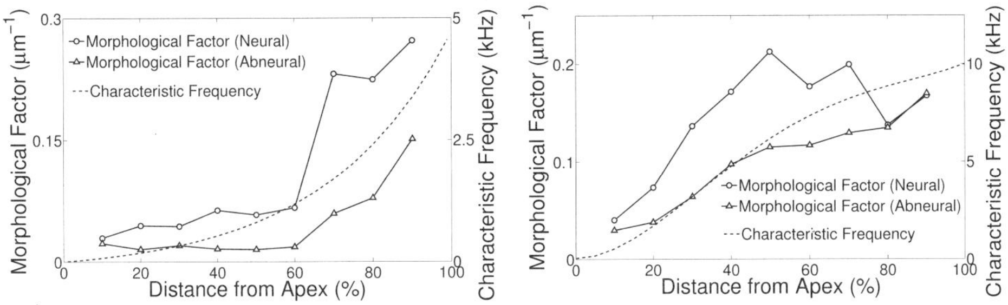

Figure 1. Morphological factor as a function of longitudinal (tonotopic) position, separately for neural and abneural positions across the papilla, and compared to the known tonotopic frequency gradient. Left graph shows data for the chicken, right graph for the owl.

off, the reverse pattern is seen in the chicken. The values for F_m, however, did not reflect the factor-of-2 difference in the upper limits of the tontopic maps of these birds.

3.3 Predicted upper frequency limit of effective hair-bundle forces

The morphological factor F_m was converted into frequency by assuming a gating force f_g of 4.3 pN at the MET channel and a phase factor Φ of 0.07 (see Eq. 1). The value for f_g was based on Le Goff et al. [16] who measured 470 fN at the tip of the tallest stereovilli in bullfrog saccular hair cells. This converts to 4.3 pN at the tip-link due to the geometrical factor [17]. Using that value, we obtained 4.3 kHz and 2.6 kHz for the frequency limits of the most basal-neural hair cells in chicken and owl, respectively.

4 Discussion

4.1 Are hair-bundle forces likely to be effective in the avian frequency range?

The estimated frequency limits, using parameter values that we judged to be the most realistic, were close to the known upper tonotopic limits of birds with average hearing, such as the chicken. We conclude that hair-bundle forces remain a realistic candidate mechanism for the cochlear amplifier in birds. Preliminary data indicate that chicken hair bundles display pronounced mechanical oscillations in response to step displacements, i.e. show evidence of such an active process [18].

We observe that the frequency range highly depends on the value of the gating force. Gating force F_g measured at the top of the hair bundle has been determined in a number of hair-cell preparations. Those values are mostly ~400 fN. For preparations whose geometrical factor γ can be estimated, the gating force f_g at the MET channel is about 4 pN [17]. Oddly, estimates for F_g of chicken basilar-papilla hair cells are unusually low, at about 40 fN [20; 21]. We presume that those low values could be due to incoherence in gating because the temporal resolution for observing synchronous gating is very high (Bozovic, personal communication). We assume here that the gating mechanism is shared between all types of hair cells and that a realistic value for the gating force of bird hair cells is nearer the high end of the measured scale (review in [22]). An even higher value of the order of 10 pN has been suggested as plausible [19], which would be more than adequate for explaining the hearing range of the barn owl.

4.2 Can the model account for species-specific differences?

Our predictions for chicken and owl reproduced the species-specific shapes of the known tonotopic maps very well. Importantly, the *combination* of a number of salient parameters produced the match, which goes beyond the individual correlations with bundle height and stereovillar number that have previously been shown. In absolute terms, however, the estimated frequency limits approximated only the auditory range of up to ~5 kHz in the chicken, while the estimate for the barn owl was significantly lower and contradicted the fact that the auditory range for the barn owl is twice as wide, reaching to 10 kHz. As pointed out above, some of the data that form the basis of the model calculations show significant variation across studies. We resolved several uncertainties by obtaining additional data, however, two parameters remain with large ranges of error: stereovillar rootlet separation and MET channel gating force. If chicken and owl hair cells should differ in one or both respects, it could reverse the predicted upper frequency limits and thus match the real situation. We believe the most likely possibility for obtaining a realistic frequency limit near 10 kHz in the owl is a higher MET channel gating force.

Acknowledgments

This work was supported by the Deutsche Forschungsgemeinschaft, grant KO 1143/11 and the Intramural Research Program of NIDCD, NIIH.

References

1. Chan, D.K., Hudspeth, A.J. 2005. Ca2+ current-driven nonlinear amplification by the mammalian cochlea in vitro. Nature Neuroscience 8, 149-155.
2. Liberman, M.C., Gao, J.G., He, D.Z.Z., Wu, X.D., Jia, S.P., Zuo, J. 2002. Prestin is required for electromotility of the outer hair cell and for the cochlear amplifier. Nature 419, 300-304.
3. Martin, P. 2008. Active Hair-Bundle Motility of the Hair Cells of Vestibular and Auditory Organs. In: Manley, G.A., Fay, R.R., Popper, A.N., (Eds.), Active Processes and Otoacoustic Emissions in Hearing. Springer Science+Business Media, LLC, New York. pp. 93-144.
4. Cheung, E.L.M., Corey, D.P. 2006. Ca2+ changes the force sensitivity of the hair-cell transduction channel. Biophys. J. 90, 124-139.
5. Köppl, C., Gleich, O., Manley, G.A. 1993. An auditory fovea in the barn owl cochlea. J. Comp. Physiol. A 171, 695-704.
6. Tilney, L.G., Saunders, J.C. 1983. Actin filaments, stereocilia and hair cells of the bird cochlea I. Length, number, width and distribution of stereocilia of each hair cell are related to the position of the hair cell on the cochlea. J. Cell Biol. 96, 807-821.
7. Manley, G.A., Meyer, B., Fischer, F.P., Schwabedissen, G., Gleich, O. 1996. Surface morphology of basilar papilla of the tufted duck *Aythya fuligula*, and domestic chicken *Gallus gallus domesticus*. J. Morphol. 227, 197-212.

8. Fischer, F.P., Köppl, C., Manley, G.A. 1988. The basilar papilla of the barn owl *Tyto alba*: A quantitative morphological SEM analysis. Hear. Res. 34, 87-102.
9. Tilney, M.S., Tilney, L.G., DeRosier, D.J. 1987. The distribution of hair cell bundle lengths and orientations suggests an unexpected pattern of hair cell stimulation in the chick cochlea. Hear. Res. 25, 141-151.
10. Fischer, F.P. 1994. Quantitative TEM analysis of the barn owl basilar papilla. Hear. Res. 73, 1-15.
11. Tilney, L.G., Tilney, M.S., Saunders, J.S., DeRosier, D.J. 1986. Actin filaments, stereocilia, and hair cells of the bird cochlea III. The development and differentiation of hair cells and stereocilia. Dev. Biol. 116, 100-118.
12. Tilney, L.G., Tilney, M.S. 1988. The actin filament content of hair cells of the bird cochlea is nearly constant even though the length, width and number of stereocilia vary depending on the hair cell location. J. Cell Biol. 107, 2563-2574.
13. Pickles, J.O. 1993. A model for the mechanics of the stereociliar bundle on acousticolateral hair cells. Hear. Res. 68, 159-172.
14. Fischer, F.P. 1992. Quantitative analysis of the innervation of the chicken basilar papilla. Hear. Res. 61, 167-178.
15. Gleich, O., Manley, G.A. 2000. The hearing organ of birds and crocodilia. In: Dooling, R.J., Fay, R.R., Popper, A.N., (Eds.), Comparative Hearing: Birds and Reptiles, Vol. Springer Handbook of Auditory Research, Vol. 13. Springer Verlag, New York. pp. 70-138.
16. Le Goff, L., Bozovic, D., Hudspeth, A.J. 2005. Adaptive shift in the domain of negative stiffness during spontaneous oscillation by hair bundles from the internal ear. Proc. Natl. Acad. Sci. USA 102, 16996-17001.
17. Jacobs, R.A., Hudspeth, A.J. 1990. Ultrastructural correlates of mechanoelectrical transduction in hair cells of the bullfrog's internal ear. Cold Spring Harbor Symp.Quant.Biol. 55, 547-561.
18. Hudspeth, A.J., Choe, Y., Mehta, A.D., Martin, P. 2000. Putting ion channels to work: Mechanoelectrical transduction, adaptation, and amplification by hair cells. Proc. Natl. Acad. Sci. USA 97, 11765-11772.
19. Ricci, A.J., Kachar, B., Gale, J., Van Netten, S.M. 2006. Mechano-electrical transduction: New insights into old ideas. Journal of Membrane Biology 209, 71-88.
20. Zhao, Y.D., Yamoah, E.N., Gillespie, P.G. 1996. Regeneration of broken tip links and restoration of mechanical transduction in hair cells. Proc. Natl. Acad. Sci. USA 93, 15469-15474.
21. Si, F., Brodie, H., Gillespie, P.G., Vazquez, A.E., Yamoah, E.N. 2003. Developmental assembly of transduction apparatus in chick basilar papilla. J. Neurosci. 23, 10815-10826.
22. Fettiplace, R. 2006. Active hair bundle movements in auditory hair cells. J. Physiol. (London) 576, 29–36.

Comments and Discussion

de Boer: Your suggestions on how to explain the unexpected result for the barn owl are too vague. Please be more specific.

Iwasa: Hair cells may have some physiological constraints that affects the morphological factor. For example, a large surface area would be needed for glucose uptake to support energy consumption of highly stimulated hair cells. Although I am not quite certain about the difference between the barn owl and the chicken in those physiological constraints, I imagine the barn owl may have stronger constraints because it depends more on hearing for its survival than does the chicken. Such constraints may require optimizing other parameters, such as molecular gating force, a quantity that is hard to determine. Among more accessible parameters, the rootlet separation is a good candidate for further examination because it affects the morphological factor on the second power.

THE INTERPLAY BETWEEN ACTIVE HAIR BUNDLE MECHANICS AND ELECTROMOTILITY IN THE COCHLEA

DÁIBHID Ó MAOILÉIDIGH, FRANK JÜLICHER

Max Planck Institute für Physik komplexer Systeme,
Nöthnitzerstr. 38, 01187 Dresden, Germany

We present a physical description of cochlear mechanics by examining the interaction between active hair bundle motion and electromotility of the outer hair cell. We use a model for hair bundle mechanics which has been shown to account quantitatively for spontaneous oscillations and non-linear amplification observed in bullfrog hair cells. In the mammalian cochlea outer hair cell electromotility provides an additional mechanical feedback to the hair bundle which couples stereocilia and cochlear mechanics. We show that this combined system can exhibit spontaneous oscillations and can provide active amplification and cochlear nonlinearity.

1. Introduction

At present the mechanism by which the cochlea amplifies sound is unknown. However, several features of the "cochlear amplifier" have been identified. The discovery of otoacoustic emissions and in particular spontaneous otoacoustic emissions provides the most direct support for the proposition that the "cochlear amplifier" is active [1, 2], a hypothesis which is based on the observation that an active process is necessary in order to overcome the effects of viscous damping in the inner ear [3]. Moreover, amplification is frequency selective and nonlinear, the latter allows the ear to function over 120 dB of sound pressure levels [4, 5, 6].

The hair bundle has been proposed to be the key component for active amplification in non-mammalian vertebrates and recent observations indicate that active hair bundle motility may play a role as part of the "cochlear amplifier" in mammals [7, 8]. Indeed, individual hair bundles exhibit many of the features of the cochlea as a whole. Spontaneous oscillations have been observed and have been shown to be active [9, 10, 11, 12, 13, 14, 15]. The hair bundle can amplify mechanical stimulation in a frequency dependent manner[13, 16]. Finally, the response of the hair bundle to external stimulus is nonlinear and compressive [17]. However, it is possible that the hair bundle alone is not sufficient for amplification once it is embedded within the cochlea.

Since the first observations of outer hair cell electromotility, [18, 19], there has been much investigation into how the inner ear can utilise such motility. While the electromotile response is in phase and undiminished for frequencies up to at least 79 kHz with the electric potential difference across the basolateral membrane of the outer hair cell, there is very significant attenuation of the receptor potential created by deflecting the hair bundle [20]. In other words, the outer hair cell membrane capacitance acts to low pass filter the voltage changes across the membrane due to the current through the hair bundle, with a corner frequency which is several octaves below the place frequency of each outer hair cell [21]. This problem would

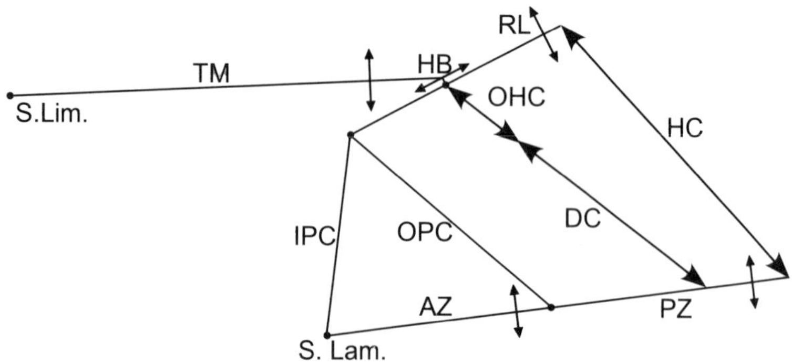

Fig. 1. Organ of Corti slice. Filled circles – pivot points, arrows – displacements or length changes. S. Lam. – basilar membrane meets spiral lamina, S. Lig. – tectorial membrane meets spiral limbus, IPC – inner pillar cell, OPC – outer pillar cell. Beams: TM – tectorial membrane, HB – hair bundle, RL – reticular lamina, AZ – basilar membrane arcuate zone, PZ – basilar membrane pectinate zone. Springs: OHC – outer hair cell, DC – Deiter cell, HC – Hensen cells

appear to prohibit the electromotile response from amplifying the motion of the organ of Corti on a cycle by cycle basis as is required[22, 23]. However, a targeted deletion in prestin, the membrane protein responsible for electromotility, produces a 40-60 dB loss in cochlear sensitivity *in vivo*[24, 25]. Thus, electromotility would appear to be an essential element of the "cochlear amplifier".

2. Methods

We study the interaction between outer hair cell electromotility and the active hair bundle by considering the interaction of a single effective electromotile outer hair cell with a cross section of the cochlea. The mechanics of the organ of the Corti is described as a system of springs attached to rigid beams (Fig. 1), which pivot about one end. This system is linear for small deflections of the beams about their pivot points. This is expected to be the range over which the organ of Corti operates *in vivo* as, for example, the deflection of a 4 μm outer hair cell hair bundle by only three degrees constitutes its operating range [26].

The hair bundle is described by a model for active hair bundle mechanics which has been shown to quantitatively describe isolated hair bundles from the sacculus of the bullfrog and to be capable of producing the same behaviours which are observed for the hair bundles of the turtle and the rat[27, 28]. The model consists of two coupled differential equations for the position of the hair bundle and the average position of motors, which apply forces on the transduction channels of the hair bundle as they walk along the actin filaments inside the stereocilia of the bundle. Here we employ a version of this model without noise, although a consideration of noise is required for a quantitative understanding of isolated hair bundles[27].

Electromotility is nonlinear over a range of about 200 mV [29]. However, the receptor potential changes by at most 5 mV around the resting potential which allows us to linearise about this point [21]. The electromotile outer hair cell is described as

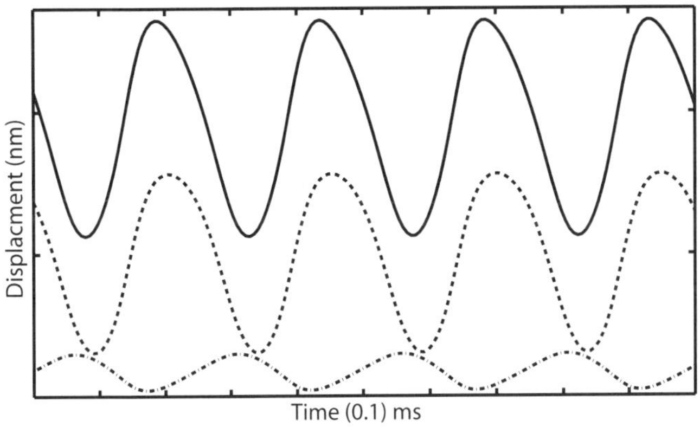

Fig. 2. Oscillations. From top: hair bundle displacment, pectinate zone displacment, arcuate zone displacement.

a linear piezoelectric material. Indeed, the outer hair cell satisfies the same reciprocal relations which characterise a piezoelectric material within the linear response regime [30].

Finally, we introduce a single effective ion species as an idealisation for the many ion species which are found within the cochlea. We assume that the stria vascularis maintains a constant ion concentration within the endolymph and perilymph which bathe the apical and basal portions of the outer hair cell, respectively. We describe the rate of change in the displacement of charge in the outer hair cell from a reference state as being a linear function of both the open probability of the mechanoelectrical channels of the hair bundle and the electric potential relative to the constant potentials of the scala media and scale tympani.

We find that the dynamics of the entire system may be described by three coupled equations for the deflection of the hair bundle the average motor displacement and the change in the change in the effective charge inside the outer hair cell, in the limit where the other system variables' (Fig. 1) intrinsic relaxation times are small in comparison to these three slower variables. We defer a detailed description of the model to a future publication and simply note that the elastic coupling of the organ of Corti (Fig. 1) introduces a feedback from the electromotile basolateral wall of the outer hair cell to its hair bundle. This feedback may be described, to linear order, as an additional effective stiffness and damping for the hair bundle. The signs of these effective terms depends on the elastic properties of the surrounding organ of Corti.

3. Results

We consider the case where the electromotility of the outer hair cell provides effective negative damping to the hair bundle. Using parameters corresponding to the 4 kHz region of the cochlea we find that this system may oscillate spontaneously at about 4 kHz (Fig. 2). The spontaneous oscillation frequency is not limited by

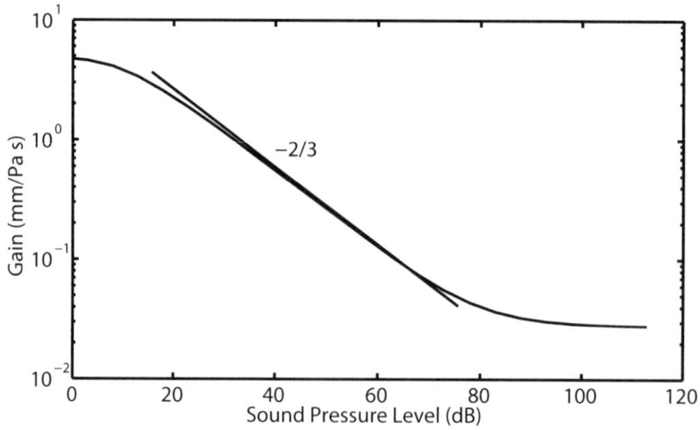

Fig. 3. Basilar membrane gain.

the membrane capacitance of the basolateral wall of the outer hair cell due to the presence of the feedback.

Small changes in the parameters result in a quiescent system which resides near a critical instability known as a Hopf bifurcation [31, 32]. The response magnitude of the system to a periodic pressure difference across the basilar membrane at a frequency of about 4 kHz displays a region with a power law dependence on stimulus magnitude characteristic of a system near such a critical point (Fig. 3) [31, 32]. This response is very similar to experimental observations of basilar membrane motion [5].

4. Discussion

The gating spring hypothesis is that the open probability of the hair bundle's mechanoelectrical channel is a non-linear function of the displacement of the hair bundle [33, 34, 35]. This is the single essential nonlinearity in our description of the system which in concert with negative feedback from the Ca^{2+} ions, which enter the channel, allows for a region of parameter space for which the system oscillates [27]. The region of nonlinear compression we report (Fig. 3) with a slope of -2/3 is generic to any system which resides near the transition between the region of oscillations and the quiescent region[31, 32]. Such a transition from a quiescent system to an oscillatory one, as a control parameter is varied, is known as a Hopf bifurcation[31, 32].

Our work suggests that amplification in the cochlea may result from a combination of active hair bundle mechanics with electromotility.

References

1. Kemp, D. T., 1978. Stimulated acoustic emissions from within the human auditory system. J. Acoust. Soc. Am. 64, 1386–1319.
2. Wilson, J. P., 1980. Evidence for a cochlear origin for acoustic re-emissions, threshold fine structure and tinnitus. Hear. Res. 2, 233–252.

3. Gold, T., 1948. Hearing. II. The physical basis of the action of the cochlea. Proc. R. Soc. B 135, 492–498.
4. Rhode, W. S., 1971. Observations of the vibrations of the basilar membrane using the Mössbauer technique. J. Acoust. Soc. Am. 49, 1218–1231.
5. Ruggero, M. A., Rich, N. C., Recio, A., Narayan, S. S., Robles, L., 1997. Basilar-membrane responses to tones at the base of the chinchilla cochlea. J. Acoust. Soc. Am. 101, 2151–2163.
6. Robles, L., Ruggero M. A., 2001. Mechanics of the mammalian cochlea. Physiol. Rev. 81, 1305–1352.
7. Kennedy, H. J., Crawford, A. C., Fettiplace, R., 2006. Force generation by the mammalian hair bundle supports a role in cochlear amplification. Nature 433, 880–883.
8. Chan, D. K., Hudspeth, A. J., 2005. Ca^{2+} current-driven nonlinear amplification by the mammalian cochlea in vitro. Nat. Neurosci. 8, 149–155.
9. Crawford, A. C., Fettiplace, R., 1985. The mechanical properties of ciliary bundles of the turtle cochlear hair cells. J. Physiol. 364, 359–379.
10. Rusch A., Thurm, U., 1990. Spontaneous and electrically induced movements of ampullary kinocilia and sterovilli. Hear. Res. 48, 247–263.
11. Howard, J., Hudspeth, A. J., 1987. Mechanical relaxation of the hair bundle mediates adaptation in mechanoelectrical transduction by the bullfrog's saccular hair cell. Proc. Natl. Acad. Sci. U.S.A. 84, 3064–3068.
12. Denk, W. Webb, W. W., 1992. Forward and reverse transduction at the limit of sensitivity studied by correlating electrical and mechanical fluctuations in frog saccular hair cells. Hear. Res., 60, 89–102.
13. Martin, P., Hudspeth, A. J., 1999. Active hair-bundle movements can amplify a hair cell's response to oscillatory mechanical stimuli. Proc. Natl. Acad. Sci. U.S.A. 96, 14306–14311.
14. Martin P., Hudspeth, A. J., Jülicher, F., 2001. Comparison of a hair bundle's spontaneous oscillations with its response to mechanical stimulation reveals the underlying active process. Proc. Natl. Acad. Sci. U.S.A. 98, 14380–14385.
15. Spontaneous oscillations by the hair bundles of the bullfrog's sacculus. J. Neurosci. 23, 4533–4548.
16. Martin, P., Mehta, A. D. Hudspeth, A. J., 2000. Negative hair bundle stiffness betrays a mechanism for mechanical amplification by the hair cell. Proc. Natl. Acad. Sci. U.S.A. 97, 12026–31.
17. Compressive nonlinearity in the hair bundle's active response to mechanical stimulation. Proc. Natl. Acad. Sci. U.S.A. 98, 14386–14391.
18. Brownell, W. E., 1984. Microscopic observation of cochlear hair cell motility. Scan. Electron Microsc., 1401–1406.
19. Brownell, W. E., Bader, C. R., Bertrand, D., De Ribaupierre, Y., 1985. Evoked mechanical responses of isolated cochlear outer hair cells. Science 227, 194–196.
20. Frank, G., Hemmert, W., Gummer, A. W., 1999. Limiting dynamics of high-frequency electromechanical transduction of outer hair cells. Proc. Natl. Acad. Sci. U.S.A. 96, 4420–4425.
21. Preyer, S., Renz, R., Hemmert, W., Zenner, H, Gummer, A. W., 1996. Receptor potential of outer hair cells isolated from base to apex of the adult guinea-pig cochlea: Implications of cochlear tuning mechanisms. Aud. Neurosci. 2, 145–157.
22. Neely, S. T., Kim, D. O., 1983. An active cochlear model showing sharp tuning and high sensitivity. Hear. Res. 9, 123–130.
23. DeBoer, E., 1986. Mechanics of the cochlea: Modeling efforts. In: Dallos, P., Popper, A. N., Fay, R. R. (Eds.), The Cochlea. Springer, New York, pp. 258–317.

24. Liberman, M. C., Gao, J., He, D. Z. Z., Wu, X., Jia, S., Zuo, J., 2002. Prestin is required for electromotility of the outer hair cell and for the cochlear amplifier. Nature 419, 300–304.
25. Dallos, P., Wu, X., Cheatham, M., Gao, J., Zheng, J., Anderson, C., Jia, S., Wang, X., Cheng W., Sengupta, S., 2008. Prestin-based outer hair cell motility is necessary for mammalian cochlear amplification. Neuron 3, 333–339.
26. Kros, C. J., Lennan, G. W. T., Richardson, G. P., 1995. Transducer currents and bundle movements in outer hair cells of neonatal mice. In: Flock, Å, Ottoson, D., Ulfendahl, M. (Eds.), Active Hearing. Elsevier Science, Oxford, pp. 113–125.
27. Nadrowski, B., Martin, P., Jülicher, F., 2004. Active hair-bundle motility harnesses noise to operate near an optimum of mechanosensitivity. Proc. Natl. Acad. Sci. U.S.A. 101, 12195–12200.
28. Tinevez, J., Jülicher, F., Martin, P., 2007. Unifying the various incarnations of active hair-bundle motility by the vertebrate hair cell. Biophys. J. 93, 4053–4067.
29. Santos-Sacchi, J., 1992. On the frequency limit and phase of outer hair cell motility: effects of the membrane filter. J. Neurosci. 12, 1906–1916.
30. Dong, X., Ospeck, M., Iwasa, K. H., 2002. Piezoelectric reciprocal relation of the membrane motor in the cochlear outer hair cell. Biophys. J. 82, 1254–1259.
31. Camalet, S., Duke, T., Jülicher, F., Prost, J., 2000. Auditory sensitivity provided by self-tuned critical oscillations of hair cells. Proc. Natl. Acad. Sci. U.S.A. 97, 3183–3188.
32. Eguílez, V. M., Ospeck, M., Choe, Y., Hudspeth, A. J., Magnasco, M. O., 2000. Essential nonlinearities in hearing. Phys. Rev. Lett. 84, 5232–5235.
33. Corey, D. P., Hudspeth, A. J., 1983. Kinetics of the receptor current in bullfrog saccular hair cells. J. Neuosci. 3, 962–976
34. Howard, J., Hudspeth, A. J., 1988. Compliance of the hair bundle associated with the gating of mechanoelectrical transduction channels in the bullfrog's saccular hair cell. Neuron 1, 189–199.
35. Howard, J., Roberts, W. M., Hudspeth, A. J., 1988. Mechanoelectrical transduction by hair cells. Ann. Rev. Biophys. Chem. 17, 99–124.

MECHANICAL PROPERTIES OF COUPLED HAIR BUNDLES

KAI DIERKES, BENJAMIN LINDNER, FRANK JÜLICHER

Max-Planck-Institut für Physik komplexer Systeme,
Nöthnitzer Str. 38, 01187 Dresden, Germany

We present numerical results concerning the stochastic dynamics of arrays of coupled hair bundles. Our findings indicate that elastic coupling of hair bundles could render spontaneous oscillations more coherent and improve signal detection properties.

1. Summary

Hair bundles from the sacculus of the bullfrog have been shown to amplify weak periodic stimuli in a frequency selective manner [1]. For stronger stimuli their response is marked by a region of non-linear compression [2]. They can also exhibit spontaneous oscillations [3].

These features are also recognized as signatures of the cochlear amplifier. But while e.g. the gain of the cochlear amplifier is about 1000 [4], the gain of a single hair bundle is about 10. Single hair bundle dynamics thus cannot account quantitatively for the performance of the cochlear amplifier.

However, in many inner ear organs hair bundles are attached to overlaying structures such as tectorial or otolithic membranes. And thus hair bundles are often elastically coupled to each other. Here we investigate what effect such a coupling could have on hair bundle dynamics. In a first approach we describe coupling by linear springs (see Fig. 1). We describe single hair bundle dynamics by means of a biophysical model that earlier has been shown to capture quantitatively the main features of stochastic hair bundle motility [5].

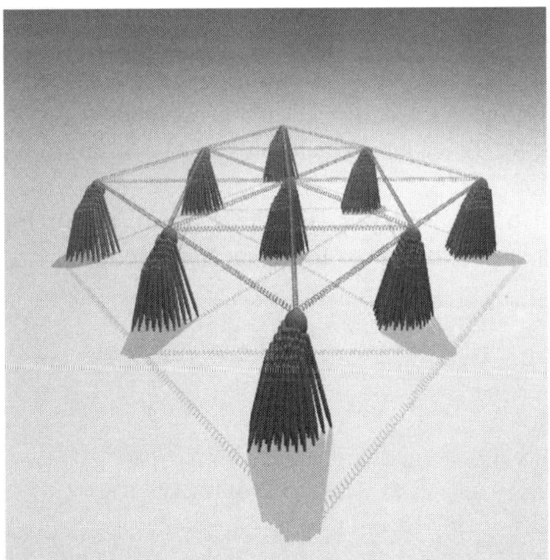

Fig. 1. Schematic view of a system of hair bundles coupled by springs.

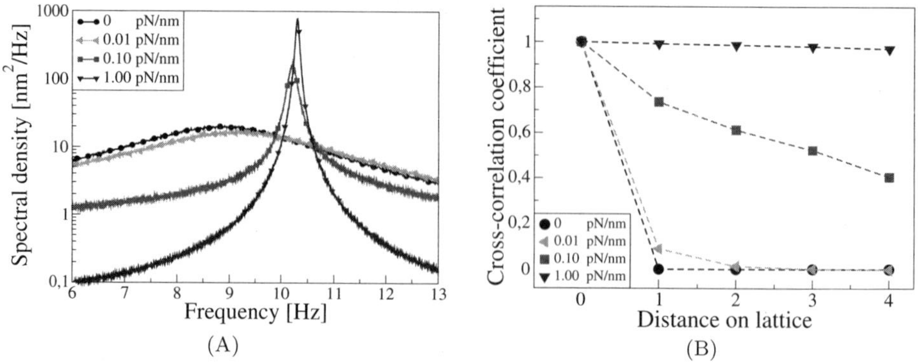

Fig. 2. (A) Various power spectra for the most central hair bundle in a 9x9 lattice for various coupling strengths. For increasing coupling strength the spectral peak sharpens. Also, there is a shift of the characteristic frequency of the oscillation. (B) Cross-correlation coefficient between the most central hair bundle in a 9x9 lattice and neighbouring hair bundles at various distances. Note that for a coupling of about 1 pN/nm hair bundle movements are synchronized.

Our simulations show that coupling could, indeed, alter the spectral characteristics of spontaneous movements, rendering stochastic oscillations more coherent (see Fig. 2A). Also, coupled hair bundles can synchronize (see Fig. 2B). As far as signal detection properties are concerned, we find that sensitivity to weak periodic driving is increased considerably when the strength of coupling is chosen in an appropriate way.

References

1. Martin, P., Hudspeth, A. J., 1999. Active hair-bundle movements can amplify a hair cell's response to oscillatory mechanical stimuli. Proc. Natl. Acad. Sci. U.S.A. 96, 14306–14311.
2. Martin, P., Hudspeth, A.J., 2001. Compressive nonlinearity in the hair bundle's active response to mechanical stimulation. Proc. Natl. Acad. Sci. U.S.A. 98, 14386–14391.
3. Martin, P., Bozovic, D., Choe, Y., Hudspeth, A.J., 2003. Spontaneous oscillation by hair bundles of the bullfrog's sacculus. J. Neuroscience 23(11), 4533–4548.
4. Robles, L., Ruggero, M. A., 2001. Mechanics of the mammalian cochlea. Physical Reviews 81(3), 1305-1352.
5. Nadrowski, B., Martin, P., Jülicher, F., 2004. Active hair-bundle motility harnesses noise to operate near an optimum of mechanosensitivity. Proc. Natl. Acad. Sci. U.S.A. 101, 12195–12200.

CONNECTIONS BETWEEN STEREOCILIARY ROOTLETS AND LATERAL WALL: A POSSIBLE ROUTE FOR INTERACTIONS BETWEEN BUNDLE AND PRESTIN BASED COCHLEAR AMPLIFICATION?

DAVID N FURNESS, SHANTHINI MAHENDRASINGAM

Keele University, Institute of Science an Technology in Medicine, Keele, Staffs., ST5 5BG, UK

CAROLE M HACKNEY

University of Cambridge, Department of Physiology, Development and Neuroscience, Downing Street, Cambridge, CB2 3EG, UK

The stereocilia in a hair bundle have dense rootlets that extend down into the cuticular plate. Using scanning and transmission electron microscopy the position of stereocilia and their rootlets was examined. This revealed attachment of some of the rootlets of stereocilia at the ends of the hair bundle to the lateral wall in outer hair cells in mice. This attachment suggests firm anchoring of the rootlet, and thus probably the stereocilia, to lateral wall membranes. It is possible that this attachment represents a means of physically coupling the lateral wall to the stereociliary bundle providing a potential route for coupling bundle amplification with prestin-based motility.

1 Introduction

Hair-cell stereocilia are stiff actin-containing rods that appear to pivot around their narrowed ankle region [1] and have dense rootlets extending into the apical cuticular plate. During acoustic stimulation, the stereociliary bundles are displaced at high frequencies, placing the pivot point under considerable stress. It is probable that the dense rootlet acts as an anchoring structure and may thus also strengthen the ankle to provide durability given that such stimulation and therefore bending occurs throughout the lifetime of the hair cell.

The cochlea also delivers enhanced frequency selectivity and sensitivity via the cochlear amplifier. The general consensus is that this is based on the motor protein prestin in the lateral wall of the outer hair cells (OHCs); however there has been recent evidence suggesting a contribution to amplification by hair-bundle force production (see review by Hackney and Fettiplace[2]). In the present study, the structure of the apex of OHCs was examined to determine whether there was a possible link between the structures involved in lateral wall-based motility and bundle-based amplification.

2 Materials and Methods

Adult CD1 mice outer hair cells were examined by scanning and transmission electron microscopy. The mice were killed with an overdose of sodium pentobarbital, decapitated and the cochleae exposed. A hole was made in the apex and then fixative (2.5% glutaraldehyde in 0.1 M sodium cacodylate buffer contaiing 2 mM calcium chloride) was gently perfused through the round window. Cochleae were immersed in fixative for 2 h, washed in buffer and then perfused with 1% osmium tetroxide in the same buffer, and immersed for 1 h.

For scanning electron microscopy, the bony wall of the cochleae was removed and they were then prepared using the OTOTO technique [3]. After dehydration and critical-point drying, they were mounted on specimen stubs and examined using a Hitachi S4500 field emission scanning electron microscope.

For transmission electron microscopy, fixed cochleae were decalcified for 3 days in 5 mM EDTA, dehydrated through an ethanol series and embedded in Spurr resin. After bisecting the cochlea in a mid-modiolar plane, ultrathin sections were cut and mounted onto copper grids, stained with uranyl acetate and lead citrate and examined in a JEOL 100CX transmission electron microscope.

3 Results

In the scanning electron microscope, the OHC stereociliary insertions could be seen in preparations where the stereocilia had sheared off during processing (Fig. 1A). This revealed not only the three rows of stereocilia typical of rodents but also that the bundle occupied the full width of the hair cell, the arms of the W-shape approaching to the front and lateral edges of the cell.

Examination of radial sections of OHCs by transmission electron microscopy also revealed the pattern of stereocilia in three rows, the shafts converging together and with a tilt of the tallest stereocilia in the inhibitory direction (Fig. 1B). Each stereocilium possessed a rootlet that was approximately proportional to its height. The rootlets consisted of dense material that extended through the cuticular plate, and, in the case of the tallest stereocilia (Fig. 1C), often projected below the latter's lower surface in the central parts of the bundles. In sections through the periphery of the cell, rootlets of all three rows showed evidence of connection to the lateral wall of the OHC in the junctional region (Fig. 1D), made possible by the close approach of the stereocilia to the cell edge. The dense material of the rootlet appeared to be contiguous with the dense material of the junctional complex (Fig. 1E).

4 Discussion

The present results show that the rootlets have a connection to the lateral wall in the junctional complex. The anatomical information does not indicate directly what role this connection plays, but there are at least two possibilities.

Firstly, it seems likely that the major function of the rootlets is to anchor the stereocilia within the cuticular plate; they appear to be matched in length to the stereociliary height, at least in any given hair cell. The connection to the lateral wall is consistent with that view, providing a potentially strong anchor for the stereocilia.

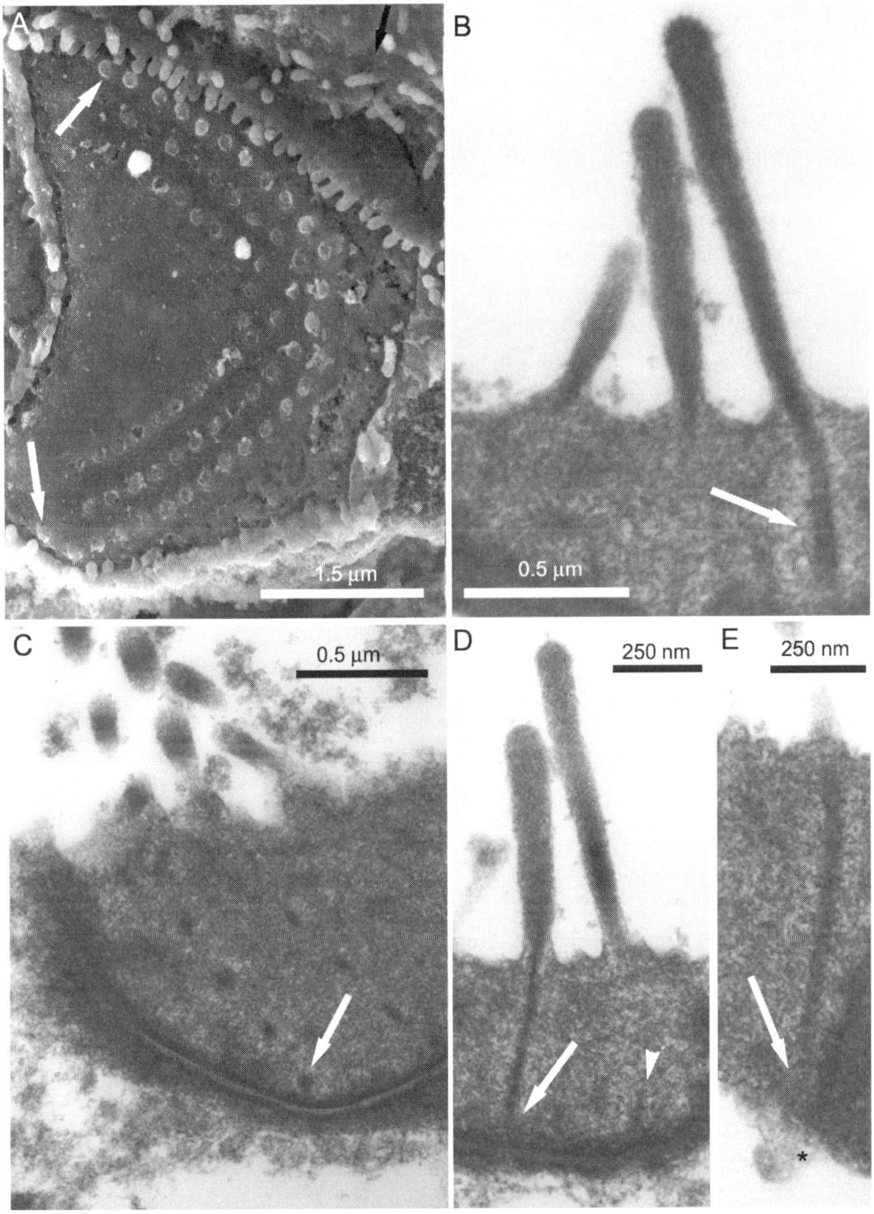

Figure 1. A. Scanning electron micrograph of a mouse OHC apex where the stereocilia have been sheared off to reveal their apical insertion points. The stereocilia at the ends of the arms of the 'W' closely approach the cell periphery (white arrows). B – E. Transmission electron micrographs of mouse OHC apices. B. The three rows of stereocilia are seen with an example of a rootlet extending from the tallest one (arrow). C. The close approach of the row of rootlets of the tall stereocilia to the cell periphery (arrow) can be seen, similar to the view of the stereociliary insertions in A. D. The rootlets of both intermediate (arrow) and tall (arrowhead) appear to end on the dense material of the junctional complex in the lateral wall. E. A tall root penetrating to the base of the cuticular plate where it approaches the lateral wall and the start of the prestin-bearing membrane (*).

The second possibility is that the connection to the lateral wall allows for an interaction between the hair bundle, via the rootlets, and motility generated by prestin-associated movements of the lateral wall, thought to be the origin of the cochlear amplifier (see Dallos et al [4]). In many of our images (for example Fig. 1E here), the rootlet more or less traverses the depth of the cuticular plate and attaches low down in the junctional complex. The nature of the junctions may be to tighten connections between cells of the organ of Corti and it may not be possible for substantial movement of the structures to occur in their vicinity. However, it is not known how forces may be transmitted through them so the possibility needs to be investigated. Thus, it may be that if the membrane pulls and pushes on the junction during prestin-evoked movements, it could result in pulling or pushing on the rootlets and the stereocilia (see Fig. 2 for a schematic illustration). As all the stereocilia are linked by extracellular cross-links [3], pulling on a subset of the rootlets may result in movement of several or all of the stereocilia.

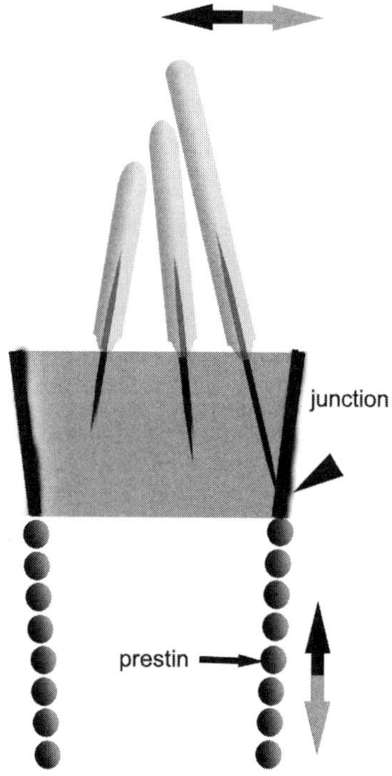

Figure 2. Schematic diagram of the rootlet linkage to the lateral wall and a speculative model of how the prestin-based amplifier may interact with the bundle. Contraction of the cell (downward grey arrow) could pull on the junction just below the tall rootlet anchor point (arrowhead). Depending on the precise geometry and properties of the connection and the rootlet within the cuticular plate this could conceivably pull the stereocilia in an excitatory direction (horizontal grey arrow). On this scheme, the opposite would occur with elongation of the cell (black arrows).

The geometry of the stereocilia is crucial in terms of this model. In most of our micrographs, the stereocilia seem at rest to converge together and lean in the negative direction. This geometry will help to determine in which direction the relevant forces might operate.

In addition, the hair bundle appears to possess its own amplification system [5]. Thus a potential connection between the bundle and the lateral wall via the rootlets may enable coupling between the two forms of force production. To confirm or refute this hypothesis, more data are needed on the properties of the rootlets, their connection to the lateral wall, and the properties of the junction.

Acknowledgments

Funded by Deafness Research UK.

References

1. Å. Flock, B. Flock and E. Murray, 1977. Studies on the sensory hairs of receptor cells in the inner ear. Acta Otolaryngol. 83, 85-91.
2. R. Fettiplace and C.M. Hackney, 2006. The sensory and motor roles of auditory hair cells. Nat. Rev. Neurosci. 7, 19-29.
3. D.N. Furness and C.M. Hackney, 1985. Cross-links between stereocilia in the guinea pig cochlea. Hear. Res. 18, 177-188.
4. P. Dallos, J. Zheng and M.A.Cheatham, 2006. Prestin and the cochlear amplifier. J. Physiol. 576, 37-42.
5. H.J. Kennedy, M.G. Evans, A.C. Crawford and R. Fettiplace, 2006. Depolarization of cochlear outer hair cells evokes active hair bundle motion by two mechanisms. J. Neurosci 26, 2757-66.

SECTION VIII

DISCUSSION

L. Bian, M.A. Ruggero, S.J. Elliott, M. van der Heijden and Y. Liu

EDITED TRANSCRIPTS OF THE OPEN DISCUSSION SESSION HELD AT KEELE UNIVERSITY ON JULY 31, 2008

Session Chairs: C.A. SHERA, D.C. MOUNTAIN
Transcription: J. NORTON, N.P. COOPER, D.T. KEMP

1 Introduction

Shera: Welcome to "Mechanics of Hearing 102", a follow-on discussion to "Mechanics of Hearing 101". The purpose of this session is to give us a snapshot of the field, to give people an impression of how people are thinking about important issues. I think this is really valuable for senior investigators, but also especially for young people. I know that when I first went to my first Mechanics of Hearing meeting in Boston, I was really stunned to discover that there were people who disagreed with things that I thought were completely obvious, and so it was really valuable for me to learn that, and to understand that there were disagreements not only about details, but also about fundamental issues. And I hope that some more of that will come out here.

The way that Dave Mountain and I have organized this, we have sort of split it roughly into two parts. The first part is the traditional question and answer session about the burning issues of the day ..., and then the second part will be more forward looking, talking about what we as a group might want to be the goals of the field, what we might like to have learned by the time that we reconvene in Williamstown in 2011.

There are some ground rules. Our session will be recorded and transcribed in the book and so it will be important for everyone, before they say anything, to identify themselves to help with the transcription. I will be primarily moderating the first part of it and then later on I will pass that responsibility onto David.

So I thought I would just begin by getting something out of the way that always comes up and occupies an inordinate amount of time in these discussion sessions, and that is, "what is the definition of the cochlear amplifier?"... So, rather than just debate that, I thought I would just tell you what it is... then hopefully you will all agree with me and we can get on to more interesting things.

So I actually wrote down a little definition so that I wouldn't get anything wrong. I define it here as *a process*, so in other words it's not a protein, it's not prestin, it's not the transduction channel, its nothing like that. It is *a process* that provides cycle by cycle power amplification of the vibrational response in the normal hearing organ. This vibrational response, in the mammalian pathway at least, might include, or definitely does include, but is certainly not limited to, the vibrational response of the basilar membrane. So hopefully everyone agrees with that... Alright excellent, so at least for the purposes of this discussion, I would like to use that as the working definition of the cochlear amplifier. One virtue of this is that it is a definition that is consistent with what has been demonstrated to be true in non-mammals - mainly as we have seen, through the work of Martin and Hudspeth, for example, and also in papers at this meeting, that the stereocilia, the hair bundles of non-mammalian species can do mechanical work, they can provide cycle by cycle, power amplification of the vibrational response.

2 Experimental evidence of power amplification

Shera: So with that definition, I would like to begin by asking what is the experimental evidence that the mammalian cochlea actually does have a cochlear amplifier in this sense, in other words that it provides power amplification? And before I open it up, I would just like to say two things that I think are *not* evidence for this, and one of those is non-linearity. Just because a mechanical response is non-linear, it does not imply that the mechanical response has been boosted due to power amplification. So non-linearity in itself is no evidence for power amplification. Another thing that I think is definitely *not* evidence for power amplification is physiological vulnerability. Just because something is physiologically vulnerable also doesn't mean that the thing arose through processes involving actual power amplification.

So with those two things off the table, what is the evidence that the mammalian cochlea actually involves, or makes use of power amplification?

Santos-Sacchi: May I simply begin this with the question – why did the cochlear amplifier arise? In other words, according to Gold it was to counter viscous damping, and yet when I have talked to some people here, they disagree with that and I would like to get that settled. Is that why, or why not? So, Diek, Jont…

Duifhuis: I think the arguments that Gold put forward in his paper to present the cochlear amplifier are not completely valid, or are completely invalid. He put it forward as the means to screw up the Q, from the value that we think it is to a value that's between a couple of hundred to a thousand, because he thought that was necessary to account for the psychophysical $\Delta f/f$ accuracy. That is not what we believe any more, so the argument that he needed, that he used to introduce the amplifier doesn't exist any more.

Allen: So […] I am always the one that says there is no cochlear amplifier, and that does not mean that I do not believe in non-linearity in the cochlea. I very much believe in non-linearity in the cochlea. So I spent a lot of time talking to people about this… but, I think the real problem here is a definitional one. There are at least three different ways that people use this term. [interrupted]

Shera: I think we have already settled on the definition, it means power amplification.

Allen: Not everybody is going to buy into that definition.

Shera: I think for the purposes of this discussion, let's just buy into that.

Allen: But that's the problem, definitions.

Shera: The question is – what is the evidence that the mammalian cochlea actually involves power amplification?

Duifhuis: What is the evidence that there is power amplification?

Shera: Yes. What is the evidence? We have seen it in non-mammalian species where people have looked at the hair bundles and actually shown that they can do mechanical work.

Chadwick: So besides Gold, then it seems to me that the one that actually coined the phrase, but may be I am not right here, but I thought that Halliwell Davis is the one who found that we also have to talk about a cochlear amplifier, but what I don't know - maybe someone can answer, is why did he think we needed something called the cochlear amplifier? There was some experimental measurement that he was worried about and I don't remember, exactly, the details of it.

Shera: Well Steve Neely was down the hall, right?...

Neely: Yes, I was at Washington University and just across the street from Halliwell Davis at the time he wrote that paper and coined the phrase. What he was trying to do is summarize an emerging new viewpoint of the way the cochlea works. He presented this first in terms of what he called the unicorn model. These excitation patterns seemed to have a sharp horn on them, that was his unicorn. And there were experiments going on - modelling experiments that Duck On Kim and I were doing that we were discussing with him at the time, and Eric LePage was there making capacitive probe measurements. And this was his term to summarize what he felt was an emerging point of view - that there was something *active* going on in the processing of sounds, and the cochlear amplifier was the term he coined to summarize what he felt was an emerging point of view.

Long: I was at CID [Washington University] at the time too and there was also the spontaneous emission data from Pat Zurek, that would also fit into this.

Duifhuis: I think the spontaneous emission data were the data that convinced a couple of people at the time that there really was some extra power coming out of the cochlea – that it was a source.

Shera: But... spontaneous emissions sort of indicate that the ear is capable of amplification, but do they really indicate that the ear makes use of that in its normal operation?

Allen: Can you tell the difference in a lossless ear and a cochlear amplifier ear? A lossless ear would also have spontaneous emissions.

Guinan: Not of the kind we have.

Shera: Some people have argued that spontaneous emissions arise through some sort of pathology, that some sort of control system has gone awry, and maybe that is the way they arise in non-mammals. So if cochlear amplifiers exist in normal hearing, I am not so sure that spontaneous emissions are really good evidence for it in mammals.

de Boer: I don't think that the question – "what is the evidence for a cochlear amplifier?" - is a good question. As far as I know there is no evidence for the mammalian cochlea yet. In my view, the cochlear amplifier is a hypothesis [...interrupted... "but hypotheses can have evidence to support them"] ...and I use in my brain a very simple model of the cochlea: I can compute the basilar membrane's impedance, and from the curves I get I can deduce that there is a limited region wherein the real part of that impedance is negative, and that is where the basilar membrane or the organ of Corti gives off energy to the fluid wave. So it is a hypothesis. And the other thing is, I want to stress that it is always limited in length along the length of the basilar membrane.

Shera: So you raise solutions to the inverse problem. You take a mechanical response and you try to determine what is required in order to reproduce that response within the framework of some sort of model assumptions, and you find a negative...

de Boer: A model without amplification cannot simulate the measured response curves. But that is subject to the restriction of the type of model. I have not found in my long life any possibility of a model that is not active in this sense, and can simulate a measured response curve from the basilar membrane.

Shera: So I don't see why you say that's not evidence for it. It is just evidence that comes with certain assumptions like all evidence.

de Boer: Exactly, that is what I want to point out.

LePage: Chris, I would like to address the question of "cycle by cycle" (amplification), because you included that in your definition and I just wanted people to consider what is the evidence for cycle by cycle. Most of the direct mechanical evidence involves considerable averaging of the signal, and I wonder whether it can be said that it is cycle by cycle if you can go down to a very low level, average for 200 seconds or whatever it requires, and see a response. That is not necessarily evidence for cycle by cycle.

Hackney: I was at a conference where a modeller called Paul Kolston was present and Robert Fettiplace, who has been mentioned a few times here, and I said "does it have to be cycle by cycle?". When I push my son, who was then much smaller, on a swing, I don't push him every time... So I have often wondered whether it is necessary, and the modellers must be able to tell me... whether it is necessary to input power cycle by cycle or just at the appropriate points, periodically in the cycle?

Shera: Can anyone address that?

Neely: I think it is implicit in the definition of the cochlear amplifier that Chris proposed that the power amplification happen at frequencies that are relevant to the auditory system, at audible frequencies and I think that is what we mean by cycle by cycle. It happens at a frequency that's an audible frequency. So whether it's every cycle or every other cycle, it's less important than the frequency at which power amplification occurs.

Allen: The thing is, it's reasonable what you are saying, but there isn't going to be a proposal that is in that category. I mean that, if you were operating neurally and you only needed to get a spike on the cycle, basically once every time you got ten neurones that's fine, you can apply this kind of idea; but we are talking about processes that are going to be continuous to the hair cell motility, and it will be applied every cycle if it can be and won't if it can't, so it really won't end up being relevant.

Grosh: I just want to address Carole's question because I agree with what Jont was saying. It's on average over a cycle we should input power if it's an amplifier. So I would also like to address the point of whether there is direct evidence, or evidence, for a cochlear amplifier. I am not expert in the reptilian or other hair bundles, but there is evidence for outer hair cells that they can do work. They can convert electrical work to mechanical work and they can transform energy from one domain to another. So there is direct evidence that a hair cell can do it. There is also direct evidence [...] that a hair bundle can do work too, and there is also ample evidence that electromotile processes are critically important for the healthy operation of a mammalian cochlea. That said, Egbert's comments about evidence, is that there is *indirect* evidence for a cochlear amplifier. I don't know that there has been direct evidence for a cochlear amplifier in the mammalian cochlea...

de Boer: in a dish...

LePage: Well just following up on Jont's comment. With no other candidate for it, well we are all the time borrowing from electronics and electronic theory and all we need is a blocking oscillator to account for an analogy [...] once every 10 cycles.

Ashmore: Can I just interject a little point here as a physiologist. I was interested that Chris excluded physiological vulnerability as one of the criteria. If you have an amplifier you have to get the energy from somewhere. It may be that the word "vulnerability" is a terrible phrase but you need some source of energy and those of us who are interested in physiological mechanisms actually worry about where the energy source actually is.

Santos-Sacchi: Here, here…

Ashmore: One other point on that is that, again, I am a relative newcomer to the field, so I wasn't around when Halliwell Davis was developing these ideas. I was interested in what Steve had to say, but I do remember Davis coming to give a talk at UCSF in the late 1970's and what he was really interested in then was why it was that the cochlea microphonic went away when you killed the animal, and that was part of his evidence, as far as I remember, for the idea that there was some sort of amplification process.

Shera: If I could just explain the reason why I included physiological vulnerability. It seemed to me, at least in principle, one could have the response of the basilar membrane depend on the d.c. component of outer hair cell receptor potentials, or something like that, something that would be physiologically vulnerable, but which would not necessarily imply that there was actual power amplification. The stiffness of the basilar membrane, the d.c. stiffness, could depend on some physiologically vulnerable process, as people like Jont Allen have proposed, but without any power amplification. So I am not saying that there is necessarily no power source, just that [physiological vulnerability] is not direct evidence for power amplification.

Santos-Sacchi: It is required, but not necessary?

Steyger: I don't think that we can say that the cochlear amplifier is a discrete unit, it is the sum of all the component parts of the cochlea, and you take one part away and it has a domino effect on every other single aspect. You know if you just take the endolymphatic potential (E.P.) away, no matter how sensitive the hair cell is, it's not going to have sensitive hearing because there is no E.P. there; and that can come from many, many different disruptions of the generation of E.P.

Chadwick: Well I want to ask another question, related of course... People here who measure tuning curves of anything, can you *ever* answer the question from the tuning curve, amplitude and phase, of whether there is a cochlear amplifier?

Shera: I can answer that question if no one else will. I know that you can analyze tuning curves within the framework of a model like Bert de Boer has done, and if you do that you find that within the framework of those assumptions, standard cochlear model assumptions, you need to have a negative real part to the impedance. Now one of the big assumptions that has come to be more and more questioned, especially at this meeting, is that those models all have [...] point impedances and they ignore longitudinal coupling along the cochlea.

Gardner-Medwin: If the energy flux along the cochlea exceeds the incident energy, that is good evidence. Is there not good evidence that this is actually true?

Shera: Can anyone address that. Is David Kemp here? David, I think you wrote a paper with Brass...

Kemp: Well, I don't know what people think of it, but that was an attempt – and that was no different than what de Boer was talking about. Basically you can convince yourself it is necessary - but have you got the assumptions right?

Shera: Exactly, so there are assumptions there too.

Duifhuis: I want to go back a little bit to the theory, that there's some sort of an oscillator that you keep running by pushing it every now and then. Spontaneous acoustic emissions also looks like an oscillator, and one of the important properties of an oscillator is that you are unable to disentangle, to get apart the active behaviour and the non-linearity. If you take away the non-linearity you get an unstable system, so I feel a little bit uncomfortable with leaving out non-linearity.

de Boer: The type of models that I have in mind, and I think many other people have in mind also, they are stable, despite that locally they are active. That's why I stressed the word "locally". The region for a particular frequency where the real part of this impedance is negative is limited. Every bit of power that is internally generated is dissipated somewhere else, and the whole system is stable. It may be, if you design such a system in the wrong way, that it becomes unstable and perhaps this is the case in a cochlea that emits sounds, but we do not know exactly.

Allen: There is some indirect evidence, which Egbert refers to and Shera has also done, which they have made using several key assumptions such as scale invariance, and gotten negative resistance out of the calculated basilar membrane. And I would acknowledge that as *indirect* evidence - I think it is indirect anyway, may be you think it is direct evidence. But one of the problems with this is that it is a very, very small amount of negative resistance. Now I have asked each of the two people in question how much, I don't know what the number is, but how much power gain is there in these negative resistances that they have calculated? And I think the numbers are surprisingly small. We need to account for a very large amount of non-linearity at the tip, like 40dB, if you go to higher frequencies. And so I always thought that that meant you had to have 40 dB of power gain. But Steve Neely tells me he has got a model where he only has 6 dB of power gain and yet he gets 40 or 50 dB of non-linearity. He has not successfully explained to me how it works yet.

de Boer: May I add, this is a very good point. We have addressed that question somewhere around 2001 I believe, in a conference at Miro in Holland, and we found to our surprise that the maximum power gain in our model was limited to 17 dB, whereas the actual difference in level between the dead and the living animal was much more.

van der Heijden: I would just like to react to what Egbert de Boer said [about local activity]. In a system like the fruit fly that was addressed at this meeting, that would be very problematical because the whole system is local – there's just one antenna, and I believe these people claim that there is amplification there. Also the question of stability there becomes really problematical. Maybe that's a simpler system there to study the claim of amplification.

Shera: So it seems to be the general opinion that the evidence, such that it is, for power amplification in the mammalian ear is basically limited to the solutions to the inverse problem. In other words, mechanical regiments interpreted within the framework of a model.

Fridberger: I would like to ask a question to those that have spoken already. How should we decide on an experiment to prove or disprove the power gain?

Duifhuis: Well, measure the power, measure the pressure and the velocity simultaneously.

Allen: Of course there is this paper by Allen & Fahey, for measuring the power gain on the basilar membrane, and I had a lot of fun with George Zweig in the audience saying there's good news and bad news - the good news is there is a cochlear amplifier, the bad news is that its gain is 6 dB, which he didn't think was very funny. Now Chris has criticized this and I think there is some validity to it. The method that we developed is accepted apparently as being a reasonable thing to do, several people have repeated the experiment and, they got basically the same answer we got, but several people have criticized it saying that the local extension - we assumed there is a point source on the basilar membrane – that the local extension creates contrary phase excitations which will cancel each other out. I think that this needs to be further explored and if there is a method that theoretically can work, we need to push on it as hard as we can to get at the question, and if we can solve the problem of the spread by either having a point source excitation or by some kind of signal processing means or more details on what the spread is, and whether it's a really significant one or not. We did try to address the question of suppression and, I am sure that suppression is not a problem.

Santos-Sacchi: Maybe there's an experimental approach, probably the creation of a mosaic mouse that has outer hair cells alternating, working in opposite fashion to their neighbour, so that you are not removing any energy source, you are just using it in reverse and cancelling out? So you might be able to engineer a prestin molecule that could be put in every other mouse [...laughter] that will cancel out the neighbours.

Wilson: Is it not possible to agree, as it were, between the 6 dB gain and the 40 dB gain, by taking into consideration that the tuning curve is getting narrower? I am just thinking of the power amplification system here (in the lecture theatre) - if you turn it up until it nearly oscillates you get a ringing and sharp peaks and just a change of half a dB will make a tremendous amount of difference in the amplitude of the peak, if you use it as a centre detector.

Aranyosi: I just want to expand on something that Egbert de Boer alluded to a little while earlier about solutions to the inverse problem, which is that one of the assumptions regarding power gain is that you are looking at a purely local impedance. That is, if you see a negative real part, that means you are adding energy locally. But one of the things that we have seen at this conference is that there is a lot of longitudinal coupling in the cochlea. So what looks like the addition of energy locally could really just be energy transfer from other parts of the cochlea. And that may or may not be active.

Shera: Or it may not involve power amplification.

Santos-Sacchi: Can you ask the opposite question, what is the evidence against the cochlear amplifier?

Shera: Well that would just bring up the Allen & Fahey experiment again and I was trying to avoid that.

Elliott: To amplify this question of energy drops, then there is an experiment in structural acoustics that if you have a structure which has got a wavelength short compared to the pressure, you see power coming out at some points and just going straight back in at other points. So just because you see power coming out of the structure then that doesn't mean to say that that is the source if there is longitudinal coupling in the cochlea.

Martin: I was just wondering if it was possible to measure the fluctuations of the basilar membrane, are they large enough that they can be measured? [...interrupted]

Santos-Sacchi: Noise you mean?

Martin: Displacement fluctuations of the basilar membrane; is it beyond the noise of the instrument that has been used to measure it?

Shera: Are you thinking of applying fluctuation dissipation?

Martin: Exactly, because that is the only way to determine whether the system is active without any assumptions.

Chadwick: Compute the effective temperature. Measure the effective temperature.

Olson: Didn't Fred Nuttall and Tianying Ren measure noise in the basilar membrane?

Martin: Then what you have to do is record that, and then stimulate the basilar membrane at a level that is low enough that you are sure that you are in a linear regime, and then compare the two. And that will give you a definite answer.

3 Overcoming the outer hair cell's time constant

Shera: It sounds like there is some agreement - that maybe the evidence for actual power amplification in the mammalian cochlea is perhaps not absolute, but there does seem to be agreement that individual hair cells can do it, outer hair cells. But if they are going to do it in the context of the cochlea they have to solve the RC time constant problem. So what do people think about that? Have there been any suggestions for that, that have been disproved.

Ashmore: It's solved, it's solved, what's the problem?

Shera: Its solved? So how is it solved?

Ashmore: At least three papers at this meeting have suggested how it could be solved.

Shera: But there's "could be solved" and there's solved.

Santos-Sacchi: They were conflicting papers though.

Shera: Well has it at least been shown, to peoples' satisfaction, that the proposal of Dallos & Evans has been ruled out?

Ashmore: The Dallos & Evans proposal needs to be modified, but the basic idea is there.

Santos-Sacchi: So, I think we should look at one of the bases of this problem, and that is that we don't know what the mid-point… of the Boltzmann electromotility function is, where it sits on the voltage axis for an outer hair cell. The outer hair cell has a membrane potential of -70, and according to *in vitro* measures the electromotility function has a low sensitivity to a change in voltage there. So if the membrane potential and the V_H of the non-linear capacitance or electromotility function were in line, the gain would be 30 nm per mV, much better than the 20 dB loss that you would predict with this offset. So the movement of this function and the determination of where this function sits *vis-a-vis* the resting membrane potential will define how much of a problem we have. So it's predictable that the problem could be reduced 20 dB. So think of it in that way, and anything that might help promote the receptor potential, may… you may not need to put so much in. We just don't know the answer as far as the *in vivo* alignment of those two important functions, resting membrane potential and the maximum voltage sensitivity of electromotility. So that's a critical experiment to do *in vivo*.

Allen: So I am confused by something. People, Pascal just said that it's proved that it is active. We know the cochlea is active, it has a battery, that's not what we are asking. We are asking if there is a cochlear amplifier that's not active. I mean a cochlear amplifier has to be active, but just because the system is active, it doesn't mean that you have a cochlear amplifier.

Shera: Do violations of fluctuation dissipation imply power amplification, Pascal?

Martin: That's a tough one. To be rigorous it just shows you that the mode you are probing, the particular movement that you are probing at a specific frequency, is not at thermal equilibrium.

Shera: One can imagine many ways of going away from thermal equilibrium that perhaps don't involve power amplification.

van der Heijden: And being bi-stable is one of them, so any type of nonlinearity, even a diode, violates the fluctuation dissipation theorem, which is only rigorously validated for linear systems. Also I am surprised by Pascal's earlier remark about the linear regime of the basilar membrane. I thought that in this Hopf bifurcation there were none. Isn't that the hallmark of that Hopf bifurcation, that there is no linear regime at low levels?

Martin: The Hopf bifurcation is a mathematical concept, it exists only in the absence of noise. In any biological system you will have noise, and if you reach lower levels that are low enough you always get back to a linear regime.

van der Heijden: But that's not the linear regime of the system – it's linearized by the noise. But I am not sure that you can properly call that the linear regime of the system, which is what you want to probe, not the noise.

Mountain: I just want to go back to Chris's original question. I think that most of us, if not all of us, would agree that if we have an outer hair cell in isolation, and it's in pristine condition, that the amount of work that we have to do on the hair bundle is going to be a lot less than the work the somatic motility can produce. So at the individual hair cell level we have power amplification and the battery is the electrochemical gradient across the cell membrane. But you know the question was "what happens in the real cochlea because of membrane capacitance?" and so forth. And Jonathan and other people have proposed various models that have some of the same problems that Bert pointed out - they are very dependent on the assumptions that go in the models. And as an electrophysiologist I find that most of these models are making assumptions that I am pretty uncomfortable with. But at the same time we have other forms of longitudinal coupling, that some people have claimed have helped out as well. So my question both as a modeller and experimentalist, to all of you, is… that we have heard proposals from Al Hubbard about coupling inside the organ of Corti; we have heard proposals about longitudinal coupling through the tectorial membrane; we may have longitudinal coupling through the reticular lamina; we have models of electrical longitudinal coupling… But what kind of coupling do we need to make this thing work? So now let's try to abstract a little bit. What do we need? I know a couple of you have worked on this problem.

Chadwick: Well how about coupling of the rows, that would help. Just getting the three rows of hair cells doing the same thing and synchronised.

Santos-Sacchi: You think they don't?

Chadwick: I think they do, but to be sure that they do, you know it's a kind of assumption that people make. But we think that the tectorial membrane damper is helping.

Duifhuis: Isn't there a layer underneath the tectorial membrane that hits the longest row of stereocilia on the three outer hair cell rows, and that would be coupling.

Chadwick: But suppose one hair bundle is spontaneously oscillating. So then this mechanism would couple.

Duifhuis: But if you look at the anatomy they are coupled.

Chadwick: There were posters about coupling of hair bundles through the tectorial membrane that shows a boost in gain from spontaneous oscillation.

Grosh: In a 2007 paper we showed that longitudinal coupling in the electrical domain gives something like a 5 dB boost in electromotility. In the results today, which aren't published, we showed that tectorial membrane coupling and the viscosity associated with that gives you broader tuning curves but it also reduces the gain so we have to have either more MET functionality or a little bit more sensitivity of the electromotile process. And so what Joe said is very important, you may make certain assumptions in that model about the electromotility, we basically assume that both the hair bundles and the soma are at their best place, and that's the low level response and we don't have that validated yet experimentally, because it is such a difficult measurement. So we find that those forms of longitudinal coupling help. We found that the basilar membrane longitudinal coupling didn't help as much, in terms of broadening the response and we didn't see much boost and gain there, I don't think we saw any.

Decraemer: We saw a lot of very nice models at this conference. Have they ever been used to calculate the entire energy balance, what is going in, what is used up. They have low frequency, mid frequency, high frequency tuning curves, but some models still allow to calculate what is the entire energy balance. Has it ever been done, or is it possible?

Duifhuis: Sure.

Shera: You say sure and what was the answer?

Duifhuis: Some have, that's the answer.

Shera: Do you know what they found?

Duifhuis: Well if you set it up properly you, you… [interrupted]

van der Heijden: The numbers - How many Watts?

Duifhuis: Well it depends what you put in… you have to control what you want mathematically. If you put up a physical model with the proper boundary – there's no way around it…[interrupted]

Allen: So what's the power amplification, just the same question, but just for an outer hair cell. On a good working outer hair cell it's a lot simpler calculation. You made the claim that there is power gain in an outer hair cell, and it certainly looks like a transistor to me, so I suspect that there is, but of course it depends on the frequency. Once you get above 3 kHz where you loose the phase locking then it's going to drop like a stone. But at least somebody should do the calculation. Has somebody done the calculation?

Shera: Stephen, have you done the calculations?

Neely: On single hair cells? - No.

Shera: Or even on a model?

Neely: I was going to try to answer Wim's question. In whole cochlear model solutions, we usually try to avoid spontaneous oscillations, if we don't have spontaneous oscillations then the energy balance is zero. The cochlear amplifier can put in a lot of extra energy and the cochlea is still very good at absorbing all of it, so all of the energy that the cochlear amplifier puts in is absorbed somewhere. This is the point that Egbert was making about *locally* active. There is power being put in some places, but all of that power is absorbed elsewhere, and so you could define the power gain as the total amount of power that gets absorbed in the cochlea divided by the power that enters from the stapes. And just to repeat a point that was made earlier. It may be 6-12 dB in a typical cochlear model solution and not the 40-50 dB of displacement gain that some people associate with a cochlear amplifier.

Furness: I just want to come back to the coupling issue for a minute or two. We found when we damaged the cochleas with aminoglycoside antibiotics that cause hair cell degeneration, that the stereocilia could remain attached to the tectorial membrane even when the rest of the hair cell had gone. That seemed to us to imply quite a strong coupling for outer hair cell stereocilia into the tectorial membrane, so that's both longitudinal and radial, and to throw in a slight wobbly for some people perhaps, there is a study back in 1970 or 71, not by me - I was a bit young then - where inner hair cell stereocilia seem to have imprints into the tectorial membrane, in kittens, so that might modify some peoples perception.

Hubbard: What I have to say is much like what Steve Neely was saying. Given a pretty extensive model, it's very difficult to answer your question, even though you can measure all the losses within the model. Think of the model: you know exactly what's coming in the front door, it's the stapes motion, then the velocity, you note I call it the front door. Now you got to remember the power doesn't just get dumped down the sewer, or the hall, power only goes up as heat (I^2R) and it is delivered by batteries (VxI). Still after you can measure all of these variables, you cannot directly answer this question as you posed it. You can't tell, you really don't know if that proves there is a cochlear amplifier, but I'll say certainly that more power gets dissipated than comes in the front door. Certainly there can be a velocity gain, stapes versus displacement at a point, that doesn't necessarily, as Steve was saying - and I think he is absolutely right - the power gain or the power dissipated may be 12 dB in terms of power, but you can get huge amplitude increases. We have tried that, we have all the variables, and we still cannot answer that question. But remember only resistors dissipate power.

Ruggero: I have just heard the numbers 50-60 dB of amplitude gain, dead versus alive, or low level versus high level. It's not that much. That number, quite a bit larger than those numbers actually, are correct numbers when you measure at the original characteristic frequency. But if you measure the peak velocity or displacement, peak alive versus peak dead, the gain may only be 30 dB. Even in the most beautiful preparations. So it is a much smaller problem than 60 or 80 dB make it sound.

Shera: Al mentioned power dissipation in resistors. Where do people think that the major viscous losses are in the cochlea? Are they as Jont Allen calculated in the sub-tectorial space or are they in the TM? Where do people think the major viscous losses are?

Iwasa: Depends on the mode of vibration, right?

Shera: Depends on the mode of vibration?

Allen: Where the shear is...

Aranyosi: I'll comment on that… With regard to our model of TM waves, we saw a lot more dissipation within the TM itself because of shearing, than in the sub-tectorial space, and it turns out that there is a fairly simple argument for why that should be so. There's more shear in the sub-tectorial space, but more viscosity in the TM, and those two sorts of things tend to cancel, but you end up with a bigger cross-sectional area shearing on the TM for typical model parameters, and that ends up making the TM be the biggest source of loss. But of course this was a model of an isolated TM in a dish and we don't know, for example, whether the TM is really moving *in vivo* or not. So I would say if the TM is moving there's probably more loss in the TM than in the sub-tectorial space.

Duifhuis: I would hope that not all of the power that is going into the ear is lost. I mean the ear after all is a sensory system, and the benefit of the sensory system should be that it is converting some of the energy to something that is going on in the inner hair cell and to the auditory nerve, and all of the rest is there to help it work as efficiently as possible, I would say. So it can be important, but it is not the only thing.

Ashmore: No biological processes are 100% efficient, even when you get down to the single cell, it's perfectly clear that an outer hair cell is actually not using the energy that is being put into the basolateral membrane to turn that into a force. There have been calculations made about what the efficiencies of this particular process are, and they are low. So one expects dissipation at every level of this whole system biological project, as it were, everything from molecules through to the entire cochlea itself.

Iwasa: Simple calculation shows that killing the viscous drag in the sub-tecorial space is not very simple. It requires a lot of energy to compensate for, and then there is the additional drag in the tectorial membrane, and how that can be cancelled is not obvious to me.

Chadwick: Being a mechanical person, I am asking a question about something that I don't know anything about, and that's the auditory nerve. Is there a way to measure the power flowing in the auditory nerve?

Shera: Associated with the action potential for example?

Chadwick: Yeah, some spike train. Is that a way to get at an answer to the question if there's an amplifier?

Lampacher: My question is "is it really required that all the power is being dissipated in the cochlea?", because I know as fact that you can record sound pressure from the round window. That would suggest that, I don't know how much, a large amount of power is passing through the cochlea and getting out again through the round window.

Shera: We also know that there are such things as otoacoustic emissions, indicating power coming back out of the cochlea.

van der Heijden: So there is some confusion that all the dissipation has to be overcome and I think it's quite the opposite to me. I think it is crucial that the energy that gets into the cochlea gets dissipated there because you don't want to bounce it around once it has been detected. So dissipation is may be a bigger problem for the cochlea than the amplification.

4 Stimulating inner hair cells

Shera: So we have had several mentions of the auditory nerve, maybe we should talk a little bit about how the inner hair cell is stimulated. What is the stimulus to the inner hair cells? Does anyone have anything to say about that?

Hubbard: [joking] It's the basilar membrane.

Allen: It's the shear between the tectorial membrane and the reticular lamina.

Ruggero: OK, no! I think that for many years now the proposal that Peter Dallos made at least as far back as 1970, or thereabouts, that the stimulation of the inner hair cell must be via fluid has held sway correctly. My own measurements at the apex of the cochlea in chinchilla, and of inner hair cells by Peter and Mary-Ann in the guinea pig apex, agree fundamentally that excitation seems to be occurring in phase with velocity of scala vestibuli, which fits basically Peter's idea very well. Now, there is conflicting evidence for the base on this issue, and my suspicion is that may be due to some kind of micromechanical interactions that we are not aware of. [interrupted…] There are data that suggest that there is a clear separation of components in the way that the inner hair cell is being stimulated, some of it was presented by John Guinan yesterday, which is very much in line with things I have been getting blue in the face talking about for the last 25 years, namely that about 85 dB something drastic is happening and whatever was normal at low levels becomes 180 degrees abnormal at high levels. Clearly, there is not just one answer to what is stimulating the inner hair cell. I think it is probably getting stimulated by two different things at different levels.

Furness: So that comes back again to the point I made earlier, that in some places the inner hair cell stereocilia may actually be attached or coupled to the tectorial membrane and that's that evidence I said before. But the other thing that may be of interest is that we noticed that adjacent hair bundles are linked together physically by cross-links. So individual hair bundles - the inner hair cells don't necessarily act as having individual hair bundles. I don't know if that makes any difference, but you might expect a whole row of them for several hair cells to be stimulated simultaneously, to move together.

Gummer: Surely the fluid force is stimulating the inner hair cell stereocilia and the question is what is producing the fluid force. The fluid force is presumably being produced by these two mechanisms; one a shearing motion and one counter-phasic motion depending upon the frequency between reticular lamina and tectorial membrane.

Hallworth: The spiral sulcus is a structure carved out in the mammalian cochlea, and it is only in the mammalian cochlea. What influence is the spiral sulcus going to have on the fluid influence on the inner hair cell stereocilia? Is it going to reduce coupling between neighbouring regions? Is it going to act like a reservoir? Like a helicotrema?

Chadwick: I don't think one needs to come up with two separate mechanisms because something different happens above 85 dB. I think with fluid forcing, just the way the hydrodynamics of the bundle is affected by the forcing, you don't have all the tip links opening at the same time, or you may get peak splitting and things like that. It can all be explained by the hydrodynamics. I don't think you need fundamentally different mechanisms.

Allen: Let me address Rick's question. If you think about the spiral sulcus, it's pretty big in area relative to the sub-tectorial space, so the acoustic impedance, longitudinally, is going to be very, very low, relative to the acoustic impedance for flow in the sub-tectorial space. So you have contra opposed phasic operations, as the basilar membrane is going up and down, and so it's got a half a wavelength to go, which isn't very far away, to equalize these pressure variations in fluid flow. So I think given the relative impedance of the sub-tectorial space to the longitudinal spiral sulcus, this is going to be an exceedingly small effect and one we don't need to worry about too much, that's my opinion.

Shera: So Mario said that strange things start happening at 85 dB and above, but if I recall John Guinan's data it looks like strange things were happening at relative low levels.

Ruggero: No, not above, *at* 85 dB.

Shera: At 85 dB. But it seems that there were still strange things happening at other levels.

Guinan: Well, there is a difference in the first peak response versus later responses. The first peak is inhibited by efferents, and the outer hair cells are involved as shown by the bias tones affecting it twice a cycle, and so something else I think is happening in addition to just say a passive first part of the travelling wave coming in, and although it has been said that fluid moves the inner hair cell stereocilia, the movement of structures moves the fluid. And so my phrasing of the question would be "what structural movements are moving that fluid, and are they right *at* the same radial section as that particular inner hair cell, or are they partly separated in distance from that inner hair cell?"

Mountain: I just want to add my 2 cents to this. As some of you know Alan Cody and I, a few years back, tried to develop a model for peak splitting. We treated it largely as a curve fitting exercise, except that we had one piece of information, and that was that we recorded from inner hair cells and we also had the organ of Corti potential right next to the outer hair cells and we had the cochlear microphonics in scala media, which at lower frequencies are generally accepted to be a good indication of outer hair cell receptor current, and we found at low and moderate levels that the only way that we could explain the actual wave shape of the inner hair cell receptor potential, and this is before you get to levels where peak splitting occurs, was to assume that the [inner] hair cell was seeing a high-pass filtered version of the outer hair cell receptor potential because normally with a Boltzmann type transmitter function you would expect as you raised level, depending on the offset, you might end up with a half wave rectified sign wave and eventually a square wave. But instead if you raised level in inner hair cells at low frequencies you get a very peaky response, and we can predict that response quite accurately by just assuming that at low and moderate levels, the outer hair cells through mysterious means were driving the inner hair cells directly. And then as we got to higher levels we had to bring in something we call the basilar membrane component as well as a third component that actually had an expansive non-linearity, and that non-linearity seemed to kick in at estimated basilar membrane displacements of 200 nm irrespective of frequency. It seemed like the basilar membrane displacement and not the sound pressure level that was the important factor. So we ended up with these 3 different terms, if you will, in order to replicate these receptor potentials, and so I have been watching with great interest as John Guinan's work progresses because although it is a very different kind of experiment, also seeing multiple components, and so when we ask this question of what stimulates the inner hair cells, well ultimately we have got to get some tilt to those hair bundles, but it could be the reticular lamina that is tilting, when the outer hair cells contract pressures build up in the organ of Corti, you saw examples from the finite element modelling, things can bulge out, we've got perhaps the whole reticular lamina moving back and forth, the tectorial membrane is moving back and forth. We have got a lot going on inside that organ. I find it really exciting that we have several groups now who are bringing different kinds of imaging techniques to bear on this problem and at least giving us little windows into what different parts of that organ is doing.

van der Heijden: I can add to this, that from the neural data that looks at level effects like notches and stuff, are very different in the apex than the base, and I think that also matches some of Nigel's observations in the apex.

Ashmore: Can somebody just remind us how the brain uses the phasing information in an action potential train. It strikes me that the nature of the coupling to the inner hair cell is going to determine the phasing of an action potential relative to a stimulus... onset, and it is just worth thinking about what that is used for.

van der Heijden: Directional hearing is one important thing.

Ashmore: Below 3 kHz. But what about above 3kHz though?

van der Heijden: It dwindles so there is no phase locking. So the envelope information, plus the place information, becomes increasingly important. So you would think that once the system gives up to do phase locking, other priorities kick in and also that will have its repercussions on the cochlear design, if you will, because it seems that a lot of effort has gone into pushing upward the limits of phase locking but obviously the hard limit's there, so in the base you have envelope coding and place coding, sharpness of tuning is important, and a wide dynamic range, that is compression, is very important. So that it [phase-locking] is less important than in the apex.

Ruggero: Just to be fair, because I think that Mary-Ann Cheatham is not around anymore, I mean she is in Paris... [interrupted]

Mountain: She's up there.

Ruggero: Ah, Mary-Ann. speak, speak about your experiments with Peter on notches and things.

Cheatham: We did find the inner hair cells that had best frequencies of 1 kHz, this behaviour, the 180 degree phase reversals. These were at a slightly more basal position than was Nigel's mechanical data.

5 Differences between the base and apex of the cochlea

Shera: So we seem to have transitioned very appropriately into a discussion about possible differences between the base and the apex, and perhaps their implications and their origin and constraints of neural coding and whatnot. I guess I am interested in hearing people's opinions about how these differences that we see, for example, in the auditory nerve, might originate in the mechanics and what the implications for cochlear models are. Almost all the models that were presented are really models based on data collected from the base.

Duifhuis: One of the things that definitely tells you that the cochlea does not behave as scale invariant.

Shera: That's absolutely true. We know that from the shapes of neural tuning curves, for example, that they change more or less in line with each other, at least in the cat, until 3 kHz, and then they start changing shape quite dramatically.

Duifhuis: There is also an historical, human data, also in the frequency accuracy - the Δf/f data that Gold was referring to – [that] had the same relation.

van der Heijden: This morning there was a talk by Al Hubbard where he remarked that we don't have any 2 kHz data, and I was really offended, because there are 4 decades of neural data there, and they are surely a lot better fit to basilar membrane models than scaling down 14 kHz mechanical data. So I think that people should be aware that there is a heavy bias of mechanical data to the base because that happens to be measurable, and don't scale it down to the apex because it's plainly wrong.

Hubbard: I will comment a little bit about that. I have to say that both David and I grew up in the laboratory of Joe Hind and Rose and Anderson, and certainly we are familiar with a lot of low frequency nerve work. I find it very difficult to compare with auditory nerve results, I just can't quite get it, but I will try because it is a good suggestion.

One thing that I was going to say, that with respect to the apex and base, what we find in the models, which may mean nothing, is that uniformly the apex cannot take nearly the gain, and when I talk about gain, I am talking about the mV receptor potential across the transformer piezo-electric device, if you want to think of it that way, into nanoNewtons in all the models. This morning I mentioned that I wasn't sure about the model I was talking about today, but my student sent me some email back, he said absolutely the gain is cut down from base to apex by about a factor of 10 and we have always, I'll say always found that the apex cannot take the same kind of gain functions that the base can without going unstable. So it would seem to me that the cochlear amplifier, if there is such a thing, doesn't need so much power down in the apex.

Grosh: We found in our model... so we mostly concentrated on the base because that is where a lot of the mechanical data is and we would love to have some more mechanical data at the apex but I realize how difficult that is. What we found in our model is, right now, if we turn off activity the gain goes down so it's like Zinn & Gummer, but then slight changes in parameters and you get slight changes in that, so it's like in the base lower Q, lower gain and so we are looking for more data there and if we can look at neural data, we haven't tried that yet either as Allyn and Marcel mentioned.

Allen: In case you don't know speech information is from 300 Hz or just below 300 Hz up to about 7.5 kHz, that's where the information bandwidth is for speech perception. Of course we want to understand how the cochlea works so we aren't going to limit ourselves to that frequency range, but do keep in mind that speech communication is over that band and obviously music is over a wider band. I think that is relevant here.

Shera: So the bit of the cochlea that we know the least about, for us is the most important in terms of communication.

6 The role of compression waves in forwards transduction

Shera: We have heard a lot about compressional waves, fast waves in the cochlear fluid and their possible role in otoacoustic emissions, but what we haven't heard a lot about is what role if any, they might play in the normal operation of the cochlea, in other words, sort of forward waves stimulating the inner hair cells. Do they have any role to play there?

Ruggero: No.

Shera: No? Why do you say that?

Ruggero: I think that otoacoustic emissions are epiphenomena.

Shera: I'm not talking about otoacoustic emissions.

Ruggero: I'm sorry what did you say.

Shera: My question was: It has been hypothesized that compressional waves play a role in otoacoustic emissions, but setting aside otoacoustic emissions, do compressional waves, if they exist at all, and obviously they do exist, do they have any functional role to play in normal hearing? Stimulating the hair cells for example.

Ruggero: First of all, obviously they have a role, they are responsible for the basilar membrane moving, that's the first in mammals.

Mountain: Compressional waves?

Ruggero: Yes, at least I call sound in the cochlea compressional waves. And that is the source of energy for the passive motion of the basilar membrane in mammals.

Shera: But you are not claiming, are you, that the basilar membrane is directly moved by compressional waves or the hair cells are directly stimulated by compressional waves?

Santos-Sacchi: They are, by the way [...] the piezo-electric properties of the basolateral membrane of the outer hair cells, they are.

Ruggero: No, I see the stimulus to the cochlea at the beginning is a compressional wave and then somehow, like Chuck Steele was pointing out, some parts you get a very small pressure difference and in other parts large pressure differences. The large pressure differences we call the traveling wave. The small pressure differences we call the plateau. Okay that's for mammals. For frogs, it is not clear, but there is some evidence that there may be specialised mechanisms to create some sort of fluid motion. I am plagiarising directly from Weber.

Guinan: I certainly agree that the stapes pushes in and there is a pressure wave and that's part of what creates the movement of the basilar membrane at the base. When you go a little further up, say the apical end of the cochlea in auditory nerve fibres, the shortest latency you get when you put in a click up to 120 dB SPL is a considerable latency [...correct...] and so it is very hard to imagine that even at that very high click level, you are getting an excitation by the direct pressure wave locally that produces an effect locally that you don't see then for many half cycles of the response.

Cooper: I am not entirely sure what you are talking about, but I think I know what you are talking about. There's a paper about exactly what John was taking about, the latency in the auditory nerve fibres in the apex, by Ted Evans in 1972 believe it or not. It's in his classical paper about guinea-pig auditory nerve fibres: when they were doing their experiments back then they occasionally made surgical mistakes, and those surgical mistakes could crack the wall of the cochlea, and Ted has got about 3 sentences in his paper saying they got [900] Hz nerve fibres with 2 ms delays, but when he cracked the wall of the cochlea he got one [900] Hz nerve fibre with an abnormally short delay [<1 ms]. And I think that abnormally short delay is likely to be the response to the compression wave getting to the apex, because in our mechanical recordings we get artifacts caused by opening the cochlea like he would get by cracking the cochlea. So I think, however many years ago it was, that was consistent evidence to say that it isn't the compression wave that causes normal auditory nerve fibre responses.

Chadwick: Well I just wanted to bring up - I think Charles Steele said something close to this today. Suppose the round window were calcified or not able to move. Clearly the slow wave would probably be suppressed and the compression wave will... will move the basilar membrane.

Allen: I think Susan Voss showed otherwise.

Shera: Just to clarify what Jont is referring to. Susan Voss did an experiment where she was able to stimulate the cochlea by controlling the pressure independently outside the two windows and she found, by recording from auditory nerve fibres, that she did not get auditory nerve responses when the pressures outside the two windows were the same. And unfortunately, because of the signal to noise in her experiment, she couldn't really lower those pressures lower than 60 dB SPL, but at 60 dB SPL and above the cochlea only responded when there was a measurable pressure difference load.

Puria: So, talking about the compressional wave being very small, but on the other hand that's one of the putative mechanisms for bone conduction hearing. Normally, you know you vibrate the skull and compress both sides, but because of the unevenness in the area of the scalae or some mechanical properties...

Ruggero: First of all I fully agree with what Nigel said with the data being what they are. It's clear if you have studied auditory nerve fibres to clicks like Kiang did many years ago, we know that there is a real delay wall, which I like to call the signal front delay. If you take a 1 ms asymptote at the high CFs, subtract that from the entire response, you get probably what von Békésy meant by the delay of the travelling wave, which he illustrated for human cochleas in one figure in one of his publications. That would be the true equivalent of what we mean by signal front, by which we mean something that starts out of nowhere, seemingly, and then there is something that you can call filter delay, if you are so disposed, which may approximate something like a minimum phase response. And by the way, Nigel's apical data in the chinchilla, and my apical data in the auditory nerve match perfectly, once you make these assumptions.

Allen: [in response to Puria's comment] We don't know how bone conduction works. It's not a demonstration at all. You even use the word putative. We do not know that the compressive, you know when you drive bone conduction, that this is an example of cochlear fast wave excitation. So that is an open question that needs to be resolved, and I am pretty sure that by the time we do get around to figuring that out we will find that it's not relevant.

Anon: Just like the cochlear amplifier then…

Puria: I agree that's one of the proposed things, there are a couple of other ones…

Guinan: So you could have compression which puts a compression wave in, and fills the cochlea with that, and the pressure difference then seen at the round window produces the excitation that you see.

van der Heijden: So really to the point of the compression wave, I think that is a question for the middle ear people, because in text books you can read that the middle ear is a transformer between the characteristic impedance of air to the characteristic impedance of fluid. Is that true?

Duifhuis (and others): No, it's the input impedance of the cochlea.

Ruggero: So just to put this whole thing together. As far as I can tell from the time of Békésy there is a very good reason to believe that bone conduction, however it occurs, up to the cochlea, once it's at the cochlea is exactly the same as stapes, with the proviso that Jont Allen's put forth, that somehow you have to get the pressure difference generated and that is where there is dispute in the literature. Then, the next point is that once you have generated an acoustic wave in the cochlea then the basilar membrane and the organ of Corti take over, and you do get a real delay of 2-3 ms in an animal like the chinchilla, and then you get filter delays of various sorts.

Olson: I just wanted to point out for anybody who doesn't know that the pressure next to the basilar membrane also shows the same delays as the basilar membrane velocity, so the real pressure there at the BM also has the same few cycles of delay, so the thing that's driving the BM also has the delay - that it isn't the compression wave that's driving the basilar membrane.

7 Robustness of measurement techniques

Shera: Any other comments about compression waves? Alright, so here is a question we got obviously from a sceptic: "How much confidence should one place in experimental measurements when one is not sure what or where one is measuring, e.g. basilar membrane motion measured without beads?". Can anyone comment on that?

Ruggero: Basically somebody alluded to the advantages and disadvantages of the methods. When you have a bead you have a target, you know where you are, there is no problem in recognising the depth of field. The problem is that typically you would be in the hands of randomness where you are going to be making your measurements, and if you want to be making some of the measurements like Tianying Ren has been making, longitudinal ones with a scanner which is difficult to orient in a proper angle and so forth, then clearly you don't want beads. They can be made but they are not easy. So without beads it would be ideal, and this is what Ian Russell and his group do, but I think there are prices to be paid there too, because the method depends on the energy coming back, not on a modulation of the energy, and that means that you do get the possibility of having artifacts that have to do with the quality of the messenger rather than with the quality of the message.

Shera: Does anyone else with experience of measuring these things want to comment.

Ashmore: Briefly, it still strikes me that the bead is not the panacea. There is still a problem of whether the bead is sticking to the thing that you are interested in.

Ruggero: Oh, well let me say the last thing. Nigel Cooper's experiments. He did one experiment, or a series of them, in which he measured and compared the responses to a bead and the responses with this very fancy homemade equipment, and they were the same.

Guinan: But that doesn't mean that they are the same all the time in measurements. I mean one big question is, when you are shining a light on something and you see the basilar membrane, but underneath the basilar membrane, say the bottom of a hair cell, is much more reflective and is moving more, that's what ends up controlling your laser site, and so without a bead you can be measuring anything along the depth of the organ of Corti and you don't know what it is.

Nuttall: That of course depends on whether you can localise that measurement along the axis.

Guinan: Yes. So if you use your new method that would get around that, but with a typical continuous laser method you don't have that depth ability, and particularly if you have made a small hole.

Olson: I have a comment on this too, because I sort of inherited this interferometer… so we had already started looking at this problem with Shyam Khanna, because he has an interferometer, or invented one, that was meant to have very good optical sectioning. But it is still a question though if you have competing surfaces like people have said, you don't really know what will be giving you your response, and one thing I just wanted to say is that you can get very funny responses if you have a competing surface, even if they are both individually moving very simply, it depends whether they are moving close to the wavelength of light or much less, but as soon as they are getting towards a wavelength, over 2 pi, you can end up getting very distorted signals if you have competition.

Shera: So that pretty much exhausts the list of questions that were submitted in advance. So I think we'll open it up and see if anyone has any questions that have arisen out of this discussion, or things that they didn't get to submit in advance, or may be things that just got forgotten about.

8 Tectorial membrane motion

Decraemer: I have a question for the experimentalists now. The question of the tectorial membrane came up. Whether there is also a vibration set up there that could help the functioning of the cochlea. So I think that if you don't look for certain things in experiments you overlook, easily, certain findings. So maybe if you are looking at this membrane at a certain angle and if you vary the angle, and you apply a cosine correction, and it doesn't turn out that you have exactly the same response, that there might be already some hidden in-plane motion of the tectorial membrane. So I would like to ask the experimentalists if in their experiments they have come across some of the possibilities that there is something happening that they could explain by this kind of motion.

Richter: I showed I think two meetings ago, we looked in the hemicochlea and we particularly looked at the vibration patterns of the tectorial membrane in a cross-section of the organ of Corti. Keep in mind that hemicochlea is a very specialised preparation. But at least in this preparation we can also distinguish between three different modes of vibration of the tectorial membrane. Namely in the gerbil: on the marginal band, which is a specific structure not seen in the guinea pig or mouse for example; and the body of the tectorial membrane; and then the region around the Hensen's stripe. Otherwise, the rest, the big part of body the Hensen's stripe is pretty much doing the same thing as the reticular lamina would do, or the rest of the organ of Corti, so we didn't see in the vibration a real phase reversal or phase shift, phase difference between the movements of the underlying structure and the above structure, basically.

Mountain: In our isolated preparation, which was an isolated coil as opposed to a hemicochlea, when we electrically stimulated the outer hair cells what we have noticed is that the underside of the tectorial membrane moves in response to the hair cell contractions but the upper surface does not. The tectorial membrane is not a homogeneous structure, it is really quite complex and at least in our preparation we are really seeing a different kind of motion, and that undersurface is of course the important surface for the hair bundles as opposed to the upper surface which is the access for the beads that one might want to use for conventional measurement.

Cooper: So, can I just ask a question about that mode? Do you mean that the tectorial membrane…

Mountain: …this is radial motion, not vertical motion.

Chadwick: You're saying it's getting sheared

Mountain: Yes.

Richter: But you are working in the middle section of the cochlea - at that part, the marginal band stretches very far towards the mid-section of the tectorial membrane. Did you distinguish between the more lateral parts, towards the Hensen's cells, or, versus the more central parts.

Mountain: It's both, I can't remember off the top of my head, whether there was much difference between the marginal band and the region over the centre of the reticular lamina. All were moving, whereas the upper portions were not, and I cannot remember off the top of my head whether there was much differential motion between the marginal band and […]

Gummer: To answer your question directly. A few years ago we measured the 3 dimensional motion of the tectorial membrane and found different vibration modes.

Chadwick: Following up on that we computed in a complex model, complex vibrational modes that agreed with Gummer's measurements.

Cooper: Wei Dong did some PhD studies with me doing much the same as Tony, but *in vivo,* and we didn't get the same result The data aren't extremely high quality or robust, so we don't want to say that Tony is wrong, but there are contradictory data out there in her thesis.

Shera: I have a related question. We have seen evidence for travelling waves on the tectorial membrane, it seems to me that people who have realistic models of the cochlea should have seen those in their models. I was just wondering if anyone has, or whether they have looked for them? Are there waves on the tectorial membrane in your models?

Grosh: Now there are. [Laughter]

Iwasa: Can I ask a silly question?

Shera: Is that it? [Laughter]

Iwasa: What is the goal of auditory mechanics?

Mountain: That's a good transition, because what we wanted to do for the second half of this session was to start thinking about what we would like to accomplish, or have our friends accomplish for the next meeting in 3 years time. So I thought, you know, the modellers are always asking the experimentalists to do this experiment and that experiment and so on. So I would like to start with those of you who consider yourselves as primarily experimentalists, anatomists, people who don't play with computers all day long. What would you like the modellers to tell you in 3 years time? And then we will reverse this a couple of times for the next 40 minutes or so.

9 Directions for future study

Guinan: Actually first before I say anything, I think you are mistaken to think that people who do experiments don't also sit before computers... [Laughter]

One thing I already alluded to at the end of my talk, I presented some click data on measurements from auditory nerve fibres that suggest that the first peak reacts very differently at high levels versus low levels, and I would like to know the origin of that, which of course has both an experimental part in people doing micromechanics but certainly also a modelling part. What are the structural things that are either giving way at high levels that change the mode of motion, or is it just that you are getting so high that you overwhelm the motion that the outer hair cells produce? I think a lot of those questions are amenable to a modelling approach.

Mountain: Does anyone want to follow up on that?

Ashmore: I would like a library of models that us experimentalists can play with. With accessible front ends, which don't really require getting too much into the fine detail, but you gradually do as you play with them. Its sort of the hands-on experience of modelling.

Mountain: We've done that but more for the central nervous system and I understand what you are talking about. You would like something, perhaps even web based, that you can just twiddle a couple of knobs and move a couple sliders and not have to worry about paying for MatLab.

Ashmore: Or they could be MatLab models, but there needs to be a whole different layers of complexity, from the very simple to the more complicated.

Chadwick: I would like to turn that around. Can you give me some little neat experiments that I could play with? [Laughter]

Hallworth: I would like to see the modellers come up with a standard notation for all the different quantities. Everybody use the same letter for the same variable, preferably not

Greek letters and things with not too many curls in them. So when we look at the slides at meetings we have some fighting chance of understanding what you are saying.

Mountain: You're asking a lot.

Nuttall: I would like to see the modellers come up with a definitive experiment to tell whether we have a stereocilia and/or basolateral force.

Mountain: That was going to be my next question.

Decraemer: May be in terms of modellers, can they be accessible, can we ask them questions, like this morning it was very nice that a modeller had modelled an empirical situation, an experimental situation by putting in a sensor probe into his model and showing what the result was to the sound field or pressure field in scala vestibuli. The thing is that just showing one position and one sensor is not enough. If you want to do an experiment then you open a big hole in the side of the cochlea, what will that be given as an effect on, what you can expect as a change in the pressure field. So not just having access to this model, but being able to do some kind of experimental modification let's say, mimic experimental situations, is very important.

Liu: I am very new here and I'm more of a theorist, I don't do any experiments. So for people like me who would like to interact with experimentalists one thing I have found very hard, because I haven't seen how an experiment is done, so I am wondering may be for future meetings, can we organise like a lab tour for modellers, just to see how things are being done in the lab.

Shera: May be we would like to reconsider the location of our next meeting? [Laughter]

LePage: We have spent a lot of time figuring out how the cochlea codes very low level amounts of energy, and so there are a lot of people out there who are very concerned about putting huge amounts of energy into the cochlea. This particularly occurs with hearing aids. How does the cochlea handle very large amounts of energy?

Santos-Sacchi: By dying! [laughter]

Mountain : So basically, I think what you're asking for is also... failures and types of stimuli that are applied to hearing aids and similar devices?

LePage: For example, somebody mentioned that energy can come out of the round window. Does the energy path get shunted out the round window if you put in a lot of energy?

Guinan: I am not specifically going to answer that but add more to the question. One of the points about reversal that one sees for tones at 80-90 dB in the cat at 1 kHz, if that kind of thing happening in an impaired ear where you have impaired the cochlear amplifier and it does happen in the cat if you degrade the cochlear amplifier, someone with a hearing aid is amplifying the sound right into the region where that is happening. And I just wonder what that does to the hearing, how one normally deals with it? That's not a mechanics of hearing question, that's a central nervous system question, but understanding how pervasive that is and what happens, I think is quite an important problem.

LePage: Its an amazing business that you've got a multibillion dollar industry producing hearing aids, and they are all do it on the basis of behavioural data, there are no physiological studies behind them.

Mountain: Do we have any more questions from the experimentalists before we turn to more computational types.

Gummer: Perhaps we can't answer it without knowing how the cycle by cycle cochlear amplifier works but it would be nice to have a definitive answer on how emissions are produced and how they get back to the stapes. Do you think you can do that in 3 years?

Shera: I think it has already been done, but I'll try to repeat myself in 3 years time.

Brownell: One question that Richard touched on briefly before was why are there three outer hair cells and I think that some of the models that we have seen at this meeting might allow an exploration as to what happens when you remove one or two or a bunch of them from a given area and end up with a single row or two rows or something like that.

Mountain: In humans and primates you often see 4 right?

Brownell: But the models that we have been seeing are getting to a stage where they could address this and it might be good to see what the results are.

Long: Mine is a bit lateral to what we are talking about. Somebody asked at the back about the possibility of going to see experiments and somebody else asked about us having cochlear models available on the web, and they are both addressing the issue that we need more interaction between modellers and experimentalists, and this is probably one of the very few environments where we can encourage something like that. So I would like to say to the person that asked the question, just send an email to people who are doing stuff nearby and just see if you can come by, its not the type of thing that's portable, we can't bring it here or Williamstown or something like that, but similarly the same way I am neither a modeller or an experimentalist, but I would like to have access to more models too, and I think that somehow this organisation probably is in a position better than anyone else to try and set up a collaboration. Not formally but informally, we meet each other here, so get together.

Mountain: May be we can set up a satellite symposium after the Williamstown meeting where everyone comes to Boston and visits the multiple labs that we have there.

van der Heijden: I would like the modellers to enlarge their test set up. I have seen so many times this single tone frequency sweep at 2 or 3 levels, and if you look at the experimental literature there is a rich variety of stimuli including noise and what have you, but I think that 90% of the test stimuli in models are one or two tones and I think that is a little backwards.

Wolter: I have a question that is more oriented towards the general audience. I am fairly new to the field, my background is from computer graphics, mathematics and computer science. What I would like to know is what should be appropriate features that future models should have such that they would be helpful to be used in cochlear implant controlling. This appears to be a very important question and there is now enough research effort going into the improvements of cochlear implants and the question would be "how could we make contributions to that community that should in the end profit from the theoretical understanding of cochlear models to be offered"?

Mountain: So are you asking what features from more of a signal processing point of view that would be important to incorporate into the processor for the implant, because often cochlear models get involved in modelling the electrical properties of the cochlea but they often don't talk about what are the key properties that might be incorporated in the processor. Is that what you are thinking about?

Wolter: OK, maybe I should expand a bit more in detail. Are there already activities that in constructing cochlear implant processors, signal processors, that make use of say computational models that have been presented at this conference or on some better models that really use that. Do they use the mechanics of hearing models that have been reported here or have been done research in the past? OK that's one question. The other question would be "what kind of features should people who create models look into that might be helpful to be used in processor used to control cochlear implants?"

Mountain: Good questions... Shall we flip it now and get a couple of questions from the modelers?

Bergevin: This might be a mute point, but for modellers one thing that I'd like to mention is for experimentalists to do in this day and age with information being easy to store, if there was some way they could put experimental results on-line whether it's a journal on-line supplementary material or on their own personal web site, but such that modellers could access that data in a relatively easy and straightforward way, including things like clear delineation, quantification of uncertainty, so it would be easy to pull off all these auditory nerve fibres responses…

Wolter: […] Was that a comment on my question?

Mountain: No. The question sort of was, could experimentalists and analysts make their data available perhaps on the web in some easily available format or perhaps through supplementary information in a journal. I think what Chris is getting at is many modellers, you know in the old days we got out our rulers and measured where the data points were on a graph, nowadays some of us have more sophisticated computer-based methods for doing that. We tried to do that on the EarLab project, more focussed on the central nervous system, one of the problems we had was getting our funding renewed because it is very hard to document who uses something like that, but we are going to try again and I hope you will all write testimonial letters when I send in the next grant application.

LePage: I just wanted to come back to the rows of outer hair cells, the number of outer hair cells. We have spent a lot of time thinking that may be the outer hair cells are increasing the gain, at maximal function, to provide the required data to achieve a particular sensitivity. The other notion is of redundancy. That maybe the cochlea has the capability to achieve that sensitivity without all the outer hair cells required and the extra hair cells are there to provide redundancy. So you can afford to lose a lot of them. That's never really been explored in any kind of model.

Mountain: Unless there are any more questions from the experimentalists, I want to start transitioning now so that the modellers get a chance to ask the experimentalists.

Chadwick: I want to talk about hair cells. You said something about the more hair cells the better, but actually mutant mice that have too many hair cells are usually deaf. They have other problems, because it reflects a lack of organisation and structure, and usually when there is too many there are none of the proper rows, and so I think it's carefully there to be a certain way. I know that it's not exactly three rows in all species but it's almost like a crystal, the way it's set up. I mean Bell was talking about that, and at NIH in Matt Kelly's lab he is getting incredible you know studies of this structure of the epithelium. So I think there is something very special about the way it is, and we need to understand why.

Mountain: The human ear stands out as being more irregular than most of the animals I am familiar with but that's so we can generate stimulus frequency emissions, right?

Grosh: I would love it if the experimentalists could measure, *in vivo*, tectorial membrane motion and reticular lamina motion. Anywhere, even at 18 kHz.

Ruggero: I would like to just point out that it has been done. A number of the recordings that are referred to, apical recordings from Nigel and Bill Rhode were from the tectorial membrane. And basically the story, if I remember correctly, is that they were identical to the ones that you got from Hensen cells above the basilar membrane.

Cooper: Yes, but I don't think that quite answers the question, I think he wants to know what the tectorial membrane is doing and the reticular lamina.

Fulton: As was mentioned earlier the tectorial membrane is not a solid mass, you really need to record on both the Kimura's membrane and the webbing on the back and differentiate between those two.

Hallworth: As an experimenter, I would like to pose six questions to other experimenters. First of all are the particles seen in freeze fracture in the outer hair cell basolateral membrane really prestin? In other words when are we going to see the freeze fracture on the prestin knock-out mouse? Second question, is the voltage to length change curve in outer hair cells really the same as the voltage to non-linear capacitance curve? There is only one published figure that even addresses that question that I am aware of. Third question, what is the point of endolymphatic potential when it gives you only a 6 dB gain in the receptor potential? Fourth question, why does efferent stimulation reduce cochlear amplification? What is the mechanism for that? Fifth question, what is the point of all that energy cost of putting potassium in endolymph when sodium goes through the transduction channel just as well as potassium? Finally, why do outer hair cells die before inner hair cells and why do basal turn outer hair cells die before apical outer hair cells?

Baumgart: As a modeller, I would appreciate, as the models are getting more and more complex, that the experimental analysis we are not just getting simplified Young's modulus, but we are also getting the raw data on displacement field or what you really measured, what were your distances if you do a bending probe, or things like this...

Elliott: Can I come back to the question about making models available on the web, which I think is a really good idea. I can't really regard myself as a modeller since I stole my model from Steve Neely. But I think the fact that I was able to [steal it] demonstrates that all of the parameters were in the paper for me to be able to do that. And I think that as models become more elaborate both the computational methods and the parameters that are used within the method become less accessible to other people to do it, and I kind of question whether that is real science if you can't repeat the simulations that someone else has done, whether people should be making such a big deal about it. I don't really know what the solution is but I think if those parameter sets and computational methods were made available then other people could check them and compare across models.

Ruggero: I wanted to answer one of the questions that has been posed, because I think it was answered back in 1991. When you inject furosemide into the ear directly, it has a miserable insensitive effect because it can't get to the stria vascularis. On the other hand if you inject systemically like I did, and I must confess that I did not measure the endocochlear potential, but in a couple of cases I did have recovery which probably means that there was not much more than 6 dB of reduction upon the furosemide, the basilar membrane basically became dead.

Hallworth: But that's how! I know how, I wanted to know why?

Mountain: We had another couple of hands up and I want to get a good collection of questions so that we can all go home whether we are experimentalists or computational types, because also I have a couple of high level questions that I want to ask the group before we leave.

Decraemer: Its never nice to ask questions and have no answers. So there was this question of how the cochlear models - could they help to develop cochlear implants - but I think mainly the cochlear implant is there when the cochlea is so much failing that you have direct electrical stimulation and it's the electrical field around the electrode that is important and there are a lot of people doing those kind of calculations so in fact there is nothing to blame on the cochlear people here.

Duifhuis: I don't completely agree - I think the pre-processing that is done to get the best feature extraction is something for instance along the line that Jont Allen is proposing that definitely makes sense. I think something like that should be pursued, I am not saying that you should buy his patent but I think that is a very good line.

Ashmore: Just to return to the tectorial membrane theme. What we need is a good model of the tectorial membrane which links its molecular structure to its macroscopic structure. And that may make contact with a whole variety of things that have been going on at this meeting.

de Boer: But then we have to know more about its actual movements so we come back to the question that there is a requirement for experimentalists as well as for theoreticians.

Ashmore: It may be an example where theory can actually constrain the way in which the membrane is structured and that will make sense of the molecular biology of it.

Mountain: I want to ask a high level question to everybody, and that is we have heard a lot of different questions throughout the meeting. In your mind, if this collective group were to solve one major problem in cochlear mechanics over the next 3 years, what would be the one problem that you would like a large group of this audience to focus on? So I want something that might be hard to do in 3 years but that you could see we were actually getting the answer to the question. So what would be your favourite question for this group to answer in 3 years time, that we don't have resolved right now.

Chadwick: I would like to come back to the question of Fred Nuttall which is can someone think of a way to decide the contribution of the bundle and the contribution of somatic motility to the way we hear. Is there a way to think about an experiment, or we heard some models today that were in the right direction, but is there a definitive experiment to do? I know Fred has been really thinking about it a long time.

Mountain: The question is what are the relative contributions of hair bundle motility and somatic motility to the cochlear amplifier, cochlear processes?

Evans: I would like to comment on that. It's very difficult because how one works depends on the other. So when you have got two processes that are interacting, it's difficult.

Mountain: They are probably in series and not parallel, is that what you are saying? [Yes…] But still, if you think of it as a cascade of amplifiers, so you can say that this guy's got 5 dB of gain and this guy's got 15dB of gain or something like that, so may be we can think of it in that broader context.

Allen: I'll admit that this is already very self-serving but I would like somebody to take some speech that I give them that sounds like /ta/ or sounds like /ka/ and I modify it so that it subtly becomes the other thing and then we put that into the auditory nerve at various signal to noise ratios and tell me what is going on at the auditory nerve level, because I can't trust a model to tell me this. I want neurophysiological correletes…

van der Heijden: To Jont, well, contact me.

The question I would like to ask is how special is the mammalian cochlea - because it seems to be very special - what can it do that the non-mammalian ear doesn't do?

Mountain: Can you distill that a little more, when you say what's special about the mammalian cochlea?...

van der Heijden: Outer hair cells, basilar membranes, and so forth. Modelers attribute all kind of roles to those structures that are quite unique to mammals but functionally the other vertebrates may have the same or comparable features, so it's very important to tease that out. So what is so special, if anything, about mammalian cochleae?

Neely: I would just like to extend that question at little bit and ask what is special about human cochleas? Are human cochleas different from other species?

Ruggero: Yes.

Köppl: Regarding the developing roles of bundle motility and somatic motility, I am delighted that we are finally debating those on equal terms now. I would like to encourage more modellers to broaden their minds to non-mammalian systems, in part to solve that question. We may all learn a lot along the way, and it may help us dissect the relative roles by looking at systems that do not have electromotility.

Steyger: I would like to have an answer to "where is the major source of energy coming from in the cochlea, that is driving the cochlear amplifier, that is driving the sensitivity of the cochlea?". Is it coming from hair bundle motion, is it coming from the somatic motility, or is it coming from elsewhere in the cochlea?

Long: I just wanted to go on from what Christine said and remind people that there is not *a* mammalian cochlea, and that there are variations within the mammalian system as well as beyond the mammalian system that could give us some information as well.

Mountain: So we should include the platypus as well?

Ruggero: Yes!

Long: And the kiwi? No, really I was thinking of some of the bats in particular.

LePage: There are tremendous insights to be obtained from looking at all the maturation literature. When you are look at all the people who study whether the cochlea grows in utero, whether it grows out as a whole structure or whether it elongates from the apical end, and it seems to me that this sort of data should be used in modelling this process.

Bergevin: This might be a bit of a stretch given that we are mechanics people and not molecular biology, but with respect to Glenis's comment, it would be really interesting to know why non-mammalian ears in general are capable of regeneration, and mammalian ears are not, and is there some way to take that one step further and learn how to make mammalian ears regenerate in a clinically beneficial way?

LePage: And conversely if the mammalian ear doesn't reproduce, what has it given up, what has it traded in order that it does not reproduce the hair cells?

Mountain: Well we wanted to make it big and may be one or two of those questions will be answered in 3 years time. Do we have any other questions, I am surprised that some of you prestin-types haven't raised the issue of how does prestin really work?

Ashmore: So you want molecular questions as well do you?

Mountain: When they tie in the mechanics, absolutely.

Ashmore: Like for example, you want the structure for prestin, and we want to know what the mechano-electrical transducer channel is as well, or the complex.

Santos-Sacchi: The holy grail.... The crystal structure of prestin, and by the way any other molecules that are working with prestin is another big issue.

Mountain: And likewise for the mechano-electric transduction channel, we have heard a lot about tip-links and things like that and we are getting to the point where maybe we will know exactly "is that a braided cord?" and that kind of questions.

Long: I have just noticed that we have kind of neglected part of the mechanics of hearing, the role of the middle ear and how that interacts with the cochlear amplifier.

Mountain: We used to have middle ear stuff in this meeting... but it kind of got cut out.

Long: Well I want to bring it back in again.

Santos-Sacchi: Didn't it split off as another meeting?

Mountain: To some extent, yes.

Long: But they are not independent and we need to have them together.

Nuttall: I think another question of mechanics that we might think about is what the role of all of these different kinds of supporting cells are. There are many, many of them, and they are structurally similar from one mammal to another. We know they have something to do perhaps with potassium circulation pathways and so forth, but there's a lot of tremendous variability in the structure of the individual cells, some of them have a lot of structural protein and others look clear inside. I think there is an interesting story there in the supporting cells.

Brownell: I think that brings together a previous question about regeneration, because it is more than likely the specialty cells that we have in the organ of Corti that make it so beautiful, aesthetically beautiful, are the same things that keep the stem cells from going in those various directions. As in the other epithelial hair cell organs you don't see anything like spaces of Nuel, Deiters' cells.

Mountain: It turns out that that is not quite true because... I have been looking into that. But I think it brings us back full circle to these other species, and other vertebrates, because there is so much diversity across the animal, vertebral kingdom and it's really amazing that birds hear pretty well and they don't have all this fancy structure, and we should be paying more attention to that.

LePage: I would like to address the Dieters' cell processes. I think we have seen one model where they are modelled as struts and I wonder whether they are struts or strings and what would happen if you modelled them as strings? I want to refer back to one of the earlier meetings probably the one in Delft where Voldrich turned up with his 3D structure model, and he poked on the underside of the basilar membrane and basically produced a hump, and what appeared at the reticular membrane was that spatially differentiated, and I wonder whether that could possibly tie up with the Dieters' cell processes and whether this has any advantage in terms of space constant?

Chadwick: I think this group should encourage more understanding and co-operation of cell biology. I think we need to understand how cell biology interacts with mechanics much better than we do. The cell biologists are going like mad and they don't have much intuition about mechanics, they make up crazy things, and we are totally ignorant about what they are doing and we need to come together with that.

Nuttall: Another mechanics question. The fibres of the basilar membrane are related to the fibrocytes of the lateral wall and we have talked in this meeting about homeostatic mechanisms and changes in hearing with various kinds of hormones and so forth. The fibrocytes have something to do with tension in the basilar membrane, and so some studies about the mechanics of the basilar membrane in relation to the lateral wall would be interesting.

Mountain: It seems like people are kind of winding down - I want to thank everybody for these wonderful suggestions, questions and comments, and we will have more time this evening over drinks and dinner to continue the conversation.

de Boer: Thank you both for chairing [Applause]

Cooper: OK, to formally close this meeting I'd just like thank Chris Shera and David Mountain for doing this; because if there was one part of this workshop that we got anywhere near right, I'm sure that this was it. As I'm sure that you will all agree, they have done an absolutely wonderful job. [Applause]

AUTHOR INDEX